Foundations of Statistical Algorithms

With References to R Packages

Chapman & Hall/CRC
Computer Science and Data Analysis Series

The interface between the computer and statistical sciences is increasing, as each discipline seeks to harness the power and resources of the other. This series aims to foster the integration between the computer sciences and statistical, numerical, and probabilistic methods by publishing a broad range of reference works, textbooks, and handbooks.

SERIES EDITORS
David Blei, Princeton University
David Madigan, Rutgers University
Marina Meila, University of Washington
Fionn Murtagh, Royal Holloway, University of London

Proposals for the series should be sent directly to one of the series editors above, or submitted to:

Chapman & Hall/CRC
4th Floor, Albert House
1-4 Singer Street
London EC2A 4BQ
UK

Published Titles

Semisupervised Learning for Computational Linguistics
Steven Abney

Design and Modeling for Computer Experiments
Kai-Tai Fang, Runze Li, and Agus Sudjianto

Microarray Image Analysis: An Algorithmic Approach
Karl Fraser, Zidong Wang, and Xiaohui Liu

R Programming for Bioinformatics
Robert Gentleman

Exploratory Multivariate Analysis byExample Using R
François Husson, Sébastien Lê, andJérôme Pagès

Bayesian Artificial Intelligence, Second Edition
Kevin B. Korb and Ann E. Nicholson

Computational Statistics Handbook with MATLAB®, Second Edition
Wendy L. Martinez and Angel R. Martinez

Published Titles cont.

Exploratory Data Analysis with MATLAB®, Second Edition
Wendy L. Martinez, Angel R. Martinez, and Jeffrey L. Solka

Clustering for Data Mining: A Data Recovery Approach, Second Edition
Boris Mirkin

Introduction to Machine Learning and Bioinformatics
Sushmita Mitra, Sujay Datta, Theodore Perkins, and George Michailidis

Introduction to Data Technologies
Paul Murrell

R Graphics
Paul Murrell

Correspondence Analysis and Data Coding with Java and R
Fionn Murtagh

Pattern Recognition Algorithms for Data Mining
Sankar K. Pal and Pabitra Mitra

Statistical Computing with R
Maria L. Rizzo

Statistical Learning and Data Science
Mireille Gettler Summa, Léon Bottou, Bernard Goldfarb, Fionn Murtagh, Catherine Pardoux, and Myriam Touati

Foundations of Statistical Algorithms: With References to R Packages
Claus Weihs, Olaf Mersmann, and Uwe Ligges

Computer Science and Data Analysis Series

Foundations of Statistical Algorithms

With References to R Packages

Claus Weihs
Olaf Mersmann
Uwe Ligges

TU Dortmund University
Germany

CRC Press
Taylor & Francis Group
Boca Raton London New York

CRC Press is an imprint of the
Taylor & Francis Group, an **informa** business
A CHAPMAN & HALL BOOK

CRC Press
Taylor & Francis Group
6000 Broken Sound Parkway NW, Suite 300
Boca Raton, FL 33487-2742

First issued in paperback 2019

ISBN-13: 978-1-4398-7885-9 (hbk)
ISBN-13: 978-0-367-37909-4 (pbk)

Visit the Taylor & Francis Web site at
http://www.taylorandfrancis.com

and the CRC Press Web site at
http://www.crcpress.com

To
Heidrun and Max,
Sabine,
and Sandra

Contents

Preface

This book is largely based on a yearly lecture, "Computer Supported Statistics" (Computergestützte Statistik), for statistics students regularly held and improved by the authors since winter 1999/2000 at the TU University Dortmund (Germany). The exercises are based on those prepared for this lecture and other related lectures. For the book, the material of this lecture was thoroughly revised, extended, and modernized. This is particularly true for Chapters 4 and 8.

This book is not "yet another treatise on computational statistics". In fact, there is, as of this writing, no other book on the market that has a similar emphasis, for at least three reasons.

1. All the textbooks on computational statistics we know of present concise introductions to a multitude of state-of-the-art statistical algorithms without covering the historical aspect of their development, which we think is instructive in understanding the evolution of ever more powerful statistical algorithms. Many of the older algorithms are still building blocks or inspiration for current techniques. It is therefore instructive to cover these as well and present the material from a historical perspective before explaining the current best-of-breed algorithms, which naturally makes up the main body of the book.
2. With the chosen chapter titles, we try to emphasize certain recurring themes in all statistical algorithms: Computation, assessment and verification, iteration, deduction of theoretical properties, randomization, repetition and parallelization and scalability. Students should not only understand current algorithms after reading this book, but also gain a deeper understanding of how algorithms are constructed, how to evaluate new algorithms, which recurring principles are used to tackle some of the tough problems statistical programmers face, and how to take an idea for a new method and turn it into something practically useful.
3. The book contains two chapters on topics neglected in other books. One chapter is dedicated to systematic verification, a topic that is not covered in any other statistical computing book we know of. Instead of focusing on

contrived test examples, we show how to derive general classes of worst case inputs and why it is important to systematically test an algorithm over a large number of different inputs. And another chapter covers the upcoming challenge of scaling many of the established techniques to very large data sets and how the availability of many CPU cores will change the way we think about statistical computing.

To summarize, this book is based on a new and refreshingly different approach to presenting the foundations of statistical algorithms. Therefore, this book provides a great resource for both students and lecturers teaching a course in computational statistics.

Acknowledgments

We thank Daniel Horn, Sarah Schnackenberg, and Sebastian Szugat for their tireless critical proof reading, Pascal Kerschke for investigating historical literature, John Kimmel for his powerful realization of the review process, the unknown reviewers and the copy-editor for their valuable comments on draft copies of the manuscript, and last but not least Marcus Fontaine for LaTeX support in preparing the final manuscript.

The Authors

Prof. Dr. Claus Weihs studied mathematics in Bonn (Germany). After his studies, he developed a large software system at the economics department in Bonn (Germany). He received his PhD (Dr. rer. nat.) in numerical mathematics from the University of Trier (Germany) in 1986. He then practiced statistics and mathematics for 9 years as a consultant for Ciba-Geigy (Basel, Switzerland). He took on his current position as the chair of Computational Statistics in Dortmund (Germany) in 1995. So far, among more than 150 publications, Prof. Weihs has published two monographs and edited four other books. Moreover, he has served as president of the German Classification Society (GfKl) from 2004 to 2013, and he is one of the editors of the journal *Advances in Data Analysis and Classification (ADAC)*.

Olaf Mersmann has studied physics and statistics at the TU Ilmenau, University of Marburg, and the TU Dortmund University. He finished his MA in data analysis in 2011 at the department of statistics of the TU Dortmund University. Starting with his BA, Mr. Mersmann has been researching new and innovative ways to objectively test benchmark computer algorithms. He has contributed eight packages to CRAN, the R software package repository, and worked on several more.

Dr. Uwe Ligges is junior-professor for Data Analysis and Statistical Algorithms at the department of statistics, TU Dortmund University. He is author of the (German) textbook *Programmieren mit R (Programming in R)* (Springer Verlag, Heidelberg), which was first published in 2004 and is currently available in its third edition. A Japanese translation of this book was published in 2006. Uwe Ligges is also known as a member of the R Core Team and the CRAN maintainer for Windows binaries of contributed packages. Additionally, he acts as one of the editors for the *Journal of Statistical Software* as well as a column editor for *The R Journal*.

List of Algorithms

Notation

Unless otherwise noted, we use the following symbols and notation throughout the book.

Abbreviations

Abbreviation	Meaning
API	Application Programming Interface
BCV	Bootstrap Cross-Validation
BFGS	Broyden-Fletcher-Goldfarb-Shanno
BLAS	Basic Linear Algebra Subroutines
$\mathrm{cov}(\cdot,\cdot)$	covariance of two random variables
$\mathrm{Cov}(\cdot)$	covariance matrix of a vector of random variables
cp.	compare
CV	Cross-Validation
DFP	Davidon-Fletcher-Powell
$\mathrm{E}(\cdot)$	expected value
EM	Expectation Maximization
gcd	greatest common divisor
GFLOPS	Giga (billions of) Floating-Point Operations Per Second
GPGPU	General Purpose Graphics Processing Unit
GPU	Graphics Processing Unit
GS	Gram-Schmidt
HPC	High-Performance Computing
ICMCMC	Independence Chain Markov Chain Monte Carlo
iff	if and only if
iid	independent identically distributed
LCS	Linear Congruential Sequence
LHD	Latin Hypercube Design
LAPACK	Linear Algebra PACKage
LINPACK	LINear algebra PACKage

LLS	Linear Least Squares
LM	Levenberg-Marquardt
LOOCV	Leave-One-Out Cross-Validation
MCMC	Markov Chain Monte Carlo
MFLOPS	Mega (millions of) Floating-Point Operations Per Second
MGS	Modified Gram-Schmidt
MPI	Message Passing Interface
MWC	Multiply-With-Carry
NLS	Nonlinear Least-Squares
NN	Nearest Neighbor
PLS	Partial Least Squares
R	Software R
RCV	Repeated Cross-Validation
RNG	Random Number Generator
RS	Random Sequence
SCV	Stratified Cross-Validation
SIMD	Single Instruction Multiple Data
SS	SubSampling
TM	Turing Machine
URS	Uniformly distributed Random Sequence
$\text{var}(\cdot)$	variance

Basic Symbols

Symbol	Meaning
$:=$	equal by definition, defined by
$\forall$	universal quantifier: for all
$\exists$	existential quantifier: there exists
$\oslash$	empty set
$\sim$	distributed as
x, y, z	Lower case letters represent scalar variables
X, Y, Z	Upper case letters denote random variables
$\boldsymbol{x}, \boldsymbol{y}, \boldsymbol{z}$	Vectors are represented using bold lower case letters
$\boldsymbol{X}, \boldsymbol{Y}, \boldsymbol{Z}$	Upper case bold letters are used for matrices
$\boldsymbol{X} = [x_{ij}]$	matrix $\boldsymbol{X}$ with elements x_{ij}
$\boldsymbol{x} = [x_i]^T$	(column) vector $\boldsymbol{x}$ with elements x_i

Mathematical Functions

Symbol	Meaning
$\hat{+}$	finite precision addition with appropriate round-off
$\hat{-}$	finite precision subtraction with appropriate round-off
$\hat{*}$	finite precision multiplication with appropriate round-off
$\hat{\div}$	finite precision subtraction with appropriate round-off
L(m,n)	set of all real-valued $m \times n$ matrices
$\nabla f(\boldsymbol{\beta})$	gradient (vector) of function $f(\boldsymbol{\beta})\colon \mathbb{R}^n \to \mathbb{R}$ in $\boldsymbol{\beta}$
$\nabla^2 f(\boldsymbol{\beta})$	Hessian (matrix) of function $f(\boldsymbol{\beta})\colon \mathbb{R}^n \to \mathbb{R}$ in $\boldsymbol{\beta}$
$\boldsymbol{J}_f(\boldsymbol{\beta})$	Jacobian (matrix) of function $f(\boldsymbol{\beta})\colon \mathbb{R}^n \to \mathbb{R}^m$ in $\boldsymbol{\beta}$
$\|\|\boldsymbol{\beta}\|\|_1$	L1 norm of $\boldsymbol{\beta} \in \mathbb{R}^n$
$\|\|\boldsymbol{\beta}\|\|_2$	L2 norm of $\boldsymbol{\beta} \in \mathbb{R}^n$.
$\|\|\boldsymbol{\beta}\|\|$	L2 norm of $\boldsymbol{\beta} \in \mathbb{R}^n$.
$\|\|\boldsymbol{\beta}\|\|_\infty$	L∞ norm of $\boldsymbol{\beta} \in \mathbb{R}^n$
$\|\|\boldsymbol{X}\|\|_2$	spectral norm of matrix $\boldsymbol{X}$
$\|\|\boldsymbol{X}\|\|$	spectral norm of matrix $\boldsymbol{X}$
$\|\|\boldsymbol{X}\|\|_F$	Frobenius norm of $\boldsymbol{X} \in L(m,n)$
$\boldsymbol{X}^T, \boldsymbol{x}^T$	transpose of $\boldsymbol{X}$, $\boldsymbol{x}$
$\text{im}(\boldsymbol{X})$	image of the matrix $\boldsymbol{X}$
$\ker(\boldsymbol{X})$	kernel of the matrix $\boldsymbol{X}$
$\det(\boldsymbol{X})$	determinant of the matrix $\boldsymbol{X}$
$\text{rank}(\boldsymbol{X})$	rank of the matrix $\boldsymbol{X}$
$Bin(n,p)$	binomial distribution with n replications and success probability p
$Exp(\lambda)$	exponential distribution with parameter λ
$\mathcal{N}(\mu,\sigma^2)$	normal distribution with expected value μ and variance σ^2
$R[a,b]$	continuous uniform distribution on the interval $[a,b]$

Chapter 1

Introduction

This book gives an overview of the most important foundations of statistical computing with particular emphasis on the historical development. The book will not provide a description of all numerical algorithms in current use for statistical computations. Instead, the book focuses on the most important construction principles for these algorithms. Our aim is to enable the reader, after working through the material covered in the book, to quickly understand the main ideas of modern numerical algorithms because he or she has seen and understood the underlying principles of these algorithms. We feel this capacity is much more valuable to both practitioners and theorists than having memorized the current, and soon to be outdated, set of popular algorithms from computational statistics.

In Chapter 2 we lay the basis for statistical computation. When we use a computer for solving a statistical problem, most of time we trust the computer that the solution will be (at least approximately) correct. Today, nearly no statistician thinks about such basic questions like "What can a computer compute?", "How does a computer compute?", or "How exact does a computer compute?". We just "trust". Knowledge about such basic facts threatens to disappear since we nearly always act as if the computer always produces the correct answer. To understand that this is not true is one of the aims of this chapter. In particular, we will discuss the above questions and what the answers mean for statisticians.

Systematic verification of the results of numerical algorithms is one of the most important and, at the same time, one of the most neglected tasks in the development of such algorithms. On the one hand, there is the well-established field of software engineering that studies how to design and verify large software systems so that they adhere to formal specifications. On the other hand, the verification of the exactness of numerical results is obviously restricted to problems for which the correct solution is well-known a priori. Moreover, in order to be able to verify the results in the general case, there is a need for such correct solutions for all degrees of (numerical) difficulty. For this, one has to fully understand the numerical problem to be solved, and there has to be a general theory for the generation of test problems with exactly known solutions. For this purpose, a systematic approach is necessary, i.e. some sort of an experimental design of testing. Testing sporadic examples will likely show a completely distorted image. In particular, the well-established practice of testing new algorithms on standard problems from literature does not in any way assess the general capabilities of the algorithm under test.

Chapter 3 demonstrates how a general verification procedure can be constructed. In order to be able to rigorously understand the problem to be solved for being able to assess the difficulty of data situations for this problem, we concentrate on probably the most used model in statistics, the linear model $\boldsymbol{y} = \boldsymbol{X}\boldsymbol{\beta} + \boldsymbol{\epsilon}$. For its popularity alone, it should be of particular importance to analyze the numerical problems in the estimation of unknown coefficients for this model.

For the linear least squares (LLS) problem we could derive a closed-form analytic solution. When we used this solution to derive an algorithm to solve the LLS, numerical instabilities surfaced and different algorithms that try to mitigate these problems were introduced. In Chapter 4 we will study solution strategies for problems where no closed-form analytic solution is known - in fact, where even no closed-form solution may exist.

We will therefore resort to methods that improve an initial solution in each iteration of the algorithm. Hence all the methods presented in this chapter are, at their core, methods that, given a solution, return a new, improved solution. We then iterate these until we reach either a fixed-point or some other termination criterion. This idea is a powerful general concept. Instead of trying to solve a problem in one big step, we can develop a, usually simpler, method that only improves a given solution. By iteration, this method will then reach something akin to a locally optimal solution. While we will focus on classical statistical optimization problems in this chapter, this concept can be applied to a much broader set of problems.

Often, intuition comes first when building an algorithm. "Couldn't we compute this in that way?" This leads to many so-called heuristics, which do not stand the practice test if their theoretical properties are unsatisfactory. Only if one can prove favorable theoretical properties for an algorithm, then there is a very good chance that it will be used in practice for a longer time. In Chapter 5, we will study two meanwhile established algorithms regarding their theoretical properties:

- The Partial Least Squares (PLS) algorithm regarding its optimality properties and
- the Expectation Maximization (EM) algorithm regarding its convergence.

On the one hand, randomness is the basis for statistics. If you do not accept the concept of randomness, then you cannot practice statistics. On the other hand, statistical algorithms are, for the most part, deterministic. Even though they might produce so-called random numbers, they are usually designed to generate reproducible numbers. Indeed, reproducibility is even one of the intrinsic requirements for scientific studies. This kind of pseudo-randomness will be discussed in Chapter 6.

First, we will introduce a method for the generation of sequences that one might be willing to call (uniformly distributed) random. We will discuss criteria for randomness of such sequences, and we will demonstrate how to choose certain parameters adequately so that these criteria may be fulfilled. Then, we will introduce methods with which uniformly or otherwise distributed random sequences can be generated.

In Chapter 6 we will introduce two different kinds of methods for the generation of random numbers. By the first kind, only random realizations of the desired distribution are generated. Unfortunately, such methods fail, especially in the multidimensional case. Since the 1980s, however, quite another type of methods has been more and more in use which try to generate tentative points of a desired density by means of simplified approximating densities that are either accepted or rejected. Two of the most flexible and promising types of such methods are the rejection method and the Markov Chain Monte Carlo (MCMC) methods, which will be discussed in Chapter 6 up to implementation details.

If you wish to obtain an impression of the distribution of, say, an estimator without relying on too many assumptions, you should repeat the estimation with different unique samples from the underlying distribution. Unfortunately in practice, most of the time only one sample is available. So we have to look for other solutions. New relevant data can only be generated by means of new experiments, which are often impossible to conduct in due time, or by a distribution assumption (see Chapter 6 for random number generation). If we do not have any indication of what distribution is adequate, we should beware of assuming just any, e.g. the normal, distribution. So what should we do? As a solution to this dilemma, resampling methods have been developed since the late 1960s. The idea is to repeatedly sample from the only original sample we have available. These repetitions are then used to estimate the distribution of the considered estimator. This way, we can at least be sure that the values in the sample can be realized by the process. In Chapter 7 we will study how to optimally select repetitions from the original sample. After discussing various such methods, the ideas are applied to three kinds of applications: Model selection, feature selection, and hyperparameter tuning.

In Chapter 2 we studied what is computable and how much effort it takes to compute certain solutions. For this we used the Turing machine, an abstract model of a computer that sequentially executes a sequence of instructions stored on a tape. Real-world central processing units (CPUs) in computers are conceptually similar. They sequentially read instructions and data from memory, process them, and write them back to memory. Increasing the execution speed of such a machine amounts to increasing the number of instructions or data words read and processed per second.

In Chapter 8 we will study the empirical scalability of statistical algorithms and especially how the availability of parallel computing resources has changed and will continue to change the way we develop and deploy statistical methods.

There are two main reasons why statisticians should know more about parallel computing and scalability. First, we are now living in a world with very large and unstructured data sets from which we derive our information. Second, the development of new statistical methods has shifted away from studying designs that are purely grounded in theory to methods that incorporate and possibly even actively build upon the vast computing resources available today. A good example of this is the new breed of ensemble learning methods being developed which actively exploit the availability of parallel computing resources by training many classifiers or regression models in parallel on, possibly smaller, data sets, and then combining these models into a more powerful predictor.

History

Let us now arrange the different chapters and their contents in history. Table 1.1 gives the, to our knowledge, first appearances of some of the basic ideas introduced in this book. In at least one aspect the book follows history, namely that we start with the historical basics in Chapter 2 ("Computation") and end with the very new developments on scalability and parallelization" in Chapter 8. The oldest ideas, discussed in the book and relevant even today, are the golden section (300 b.c.), the algorithm (825), and the Newton method (1668). The most modern ideas and algorithms discussed are likely the evolutionary strategies (1971) and their current state-of-the-art incarnation the CMA-ES (1996). The main basics of Chapters 3, 4, and 5 ("Verification", "Iteration", and "Deduction of Theoretical Properties") were founded in the

1960s and 1970s. Some ideas from Chapter 6 ("Randomization") are newer, namely from the 1980s and 1990s, and a modern view of evaluation by repetition (Chapter 7) was first mentioned in the early 2000s. By 1970 the idea of parallel computing had emerged as the dominant idea to increase performance (see Chapter 8, "Scalability and Parallelization"). However, only the introduction of powerful graphics processors in the 2000s that could be programmed to perform general arithmetic allowed scientists to perform calculations on their desktop computers that previously required medium-sized computing clusters. Note that Table 1.1 only gives the first appearances of some ideas. Newer developments in the fields are given in the chapters themselves.

The Structure of the Chapters
Each chapter contains examples and exercises. Solutions to the exercises can be found in a separated Solutions volume. Where appropriate, tables and figures are included to reinforce the concepts presented in the text. We provide a website with supplementary material, e.g. program code for selected figures, simulations, and exercises.

Each chapter contains the following elements:

1. Motivation and relevance
2. Preliminaries
3. Selected theory
4. Examples, applications, and simulations
5. Implementation in R
6. Summaries, conclusions, and outlook
7. Exercises

Note that the mathematics mentioned in the preliminaries of the chapters is assumed to be well-known in advance and is thus neither proven nor cited.

The exercises in each chapter are organized by section. So for example Exercise 2.3.1 is related to the material covered from Section 2.3. Most exercises are focused on applying the covered concepts in practice by implementing relevant algorithms and studying their behavior. Typically the exercises are meant to be solved using the programming language R and some even require the use of R.

Covered Statistical Methods
Once more, notice that this book does not aim at full coverage of algorithms for statistical methods. Nevertheless, we cover many different statistical topics, in particular

Table 1.1: Contents and history

Chap.	Title / Problem	Keywords	Author and Year	Cross-ref.
2.	**Computation**			
2.2	What can a computer compute?	algorithm	al Chwarizmi (830)	
		time complexity	Landau (1909)	
		Turing machine	Turing (1936)	
2.3	How do computers compute?	floating-point calculation	Zuse (1936)	6.2
2.4	How exact are computer computations?	condition numbers	Turing (1948)	3.1.2, 4, 6
3.	**Verification**			
3.2	linear least squares	condition number	Turing (1948)	
3.3	test data generation	test matrices	Zielke (1973)	6.
4.	**Iteration**			
4.3	univariate optimization	golden section	300 b.c. (Euclides and Thaer, 1937)	
		convergence rate	Traub (1964)	
		Newton method	1668 (Newton, 1711)	
4.4	multivariate optimization	gradient descent	Cauchy (1847)	
			DFP (Fletcher and Powell, 1963)	
			BFGS (Fletcher, 1970)	
		nonlinear least squares	Gauß (1809)	
		direct search	Hooke and Jeeves (1961)	
		stochastic search	Nevel'son et al. (1976)	2.4, 3, 6, 7.3.1
4.5	artificial neural nets	identifiability	Hwang and Ding (1997)	7.4
4.6	constrained optimization	linear programs	Dantzig (1949)	
		quadratic programs	Frank and Wolfe (1956)	3, 7.3.2
4.7	evolutionary optimization	evolutionary strategy	Rechenberg (1973)	5.2.3, 6.4

5.	**Deduction of Theoretical Properties**			
5.1	partial least squares	predictive power	Wold (1980)	7.4
5.2	EM algorithm	partly unknown data	Dempster et al. (1977)	
6.	**Randomization**			3, 4
6.2	univariate random	linear congruential series	Lehmer (1951)	
		tests on randomness	Yuen (1977)	
		inversive generators	Eichenauer and Lehn (1986)	
		multiply with carry	Marsaglia (1996)	
		inversion method	Devroye and Devroye (1986)	
		rectangle-wedge-tail	Marsaglia et al. (1964)	
6.3	multivariate random	rejection method	von Neumann (1951)	
		Metropolis Hastings	Metropolis (1953), Hastings (1970)	
		Gibbs sampler	Geman and Geman (1984)	
6.4	stochastic modeling	BUGS	Gilks et al. (1994)	
7.	**Repetition**			
7.2	model selection	decision theory	Hothorn et al. (2003)	
		cross-validation	Lachenbruch and Mickey (1968)	
		bootstrap	Efron (1979)	6.2.2
7.3	classification	error estimates	Smith (1946)	4.6.4, 5.2.4, 6.4
7.4	continuous models	predictive power	Cattin (1980)	4.5.3
8.	**Scalability and Parallelization**			
8.3	optimization	vectorization	Iverson (1962)	
		compiled languages	Backus et al. (1956)	
		BLAS/LAPACK	Lawson et al. (1979) / Anderson et al. (1999)	
8.4	parallel computing	Amdahl's law	Amdahl (1967)	
		MPI	Message Passing Interface Forum (2009)	
		OpenMP	OpenMP Architecture Review Board (1997)	

- the determination of univariate statistics like median, quantiles, and variance,
- univariate and multivariate sampling,
- various kinds of regression methods like linear, nonlinear (especially neural nets), and L1,
- various kinds of classification methods like data independent rules, Bayes rules, nearest neighbors, Linear Discriminant Analysis (LDA), Support Vector Machine (SVM), and logistic regression,
- maximum likelihood estimation,
- multivariate modeling like Partial Least Squares (PLS),
- clustering like k-means clustering and
- the parallelization of some of these methods.

Notice, however, that we do not discuss these methods in blocks concerning there purpose, but in connection with the foundation oriented structure of this book. If you are interested in algorithms for special purposes, please refer to the Index of the book.

Webpage

Some data, R code, particularly for examples and exercises, possibly a list of errata we try to avoid, and other additional material you need to conveniently follow the contents of the book can be found under the following URL: `http://www.statistik.tu-dortmund.de/fostal.html`.

Chapter 2

Computation

2.1 Motivation and History

When we use a computer for solving a statistical problem, most of the time we trust the computer that the solution will be (at least approximately) correct. Today, nearly no statistician thinks about such basic questions like "What can a computer compute?", "How does a computer compute?", or "How exact does a computer compute?". We just "trust." Knowledge about such basic facts threatens to disappear since we nearly always act as if the computer always produces the correct answer. To understand that this is not true is one of the aims of this chapter. In particular, we will discuss the above questions and what the answers mean for statisticians.

What Can a Computer Compute? (see Section 2.2)
Computability is the basis of all computational statistics. Naturally, by means of a computer only such problems can be solved whose solutions are computable by a machine. Fortunately, the following two theses (in a way) guarantee that computers can do what we want them to do:

Church-Turing thesis: The class of intuitively computable functions is equivalent to the class of Turing computable functions (Church and Turing, 1930s).

This thesis is not provable since the term **intuitively computable** is not well defined and cannot be exactly formalized. By this we understand all functions that could in principle be calculated by humans. It is standard to assume that this thesis is true. This leads to the possibility to prove that a function is not computable.

Therefore, in this book we will concentrate on Turing computability, also since there is another thesis that relates general machine computability to this term:

Thesis M: Whatever can be calculated by a machine is Turing (machine) computable (Gandy, 1980).

A function is defined to be **Turing computable** if the function's value can be computed with a Turing machine. So, as an abstract example of a mechanical calculating machine, we will introduce the **Turing machine** which is an idealized computer named after **A.M. Turing** (1912 – 1954). Though Turing machines are surely a very old idea and in their pure form they are not utilized in practice, nevertheless, the basic ideas of today's computers can be easily represented by Turing machines. The introduction of Turing machines, both facilitates the discussion of computability and builds the bridge to today's real computers.

The term **Turing computable** is fundamental to the term **algorithm**, summarizing instructions for the solution of a problem. Indeed, computable functions are the formalized analogue of the intuitive notion of an algorithm. In order to assess the quality of an algorithm, at least two different aspects are important, complexity and accuracy. **Complexity** means the calculation effort to solve the problem and is obviously important for the applicability of the algorithm. We will introduce so-called complexity measures, and will exemplify the complexity of different algorithms for the solution of the same problem by the practically very important problem of **sorting** n items. The other important quality aspect, i.e. the accuracy of the computed result, will be discussed when we study the question "How exact does a computer compute?".

How Does a Computer Compute? (see Section 2.3)
Today's computers mainly utilize so-called **floating-point numbers**. Therefore, the realization of floating-point operations is basic for all statistical calculations. In particular, the notion of a **rounding error** is based on the definition of floating-point operations. Rounding errors are fundamental for the error propagation caused by typical algorithms of computer statistics, and the extent of error propagation is important for the trustworthiness in such algorithms.

How Exact Does a Computer Compute? (see Section 2.4)
Floating-point calculations are inexact by nature. One of the main problems of numerical methods is the determination of the **accuracy** of their results. Basic for today's view of rounding errors on computers is the work of Wilkinson in the early 1960s (Wilkinson, 1965). He used a simple but generally useful way to express the error behavior of floating-point operations, namely the concept of significant digits, i.e. relative errors. We are interested in the error propagation caused by algebraic operations, i.e. to what extent relative errors already present in the inputs are amplified (increased) or damped (decreased) by floating-point operations.

One consequence of the inexactness of floating-point operations is that some mathematical laws of arithmetical operations are not valid anymore on computers. We will derive that the associative laws are not valid for floating-point operations with important consequences for the formulation of algorithms. We will see that such laws are only **approximately true** leading to new operations for **floating-point comparison** aiming at the assessment of the relative difference between floating-point numbers.

In order to discuss the second quality aspect, i.e. the accuracy of results, we will introduce so-called **condition numbers** characterizing the amplification of errors caused by the algorithm. We will exemplify this notion by looking at different ways of numerically calculating the **sample variance** of n observations.

Having answered the above three questions, we have built the basis for higher order characterizations of statistical algorithms in the next chapters.

2.1.1 Preliminaries

Definition 2.1: Vector Norm
We call a function $||\cdot||: \mathbb{R}^n \to \mathbb{R}^+$ a **vector norm** if

1. For $\beta \in \mathbb{R}^n$, $||\beta|| = 0$ iff $\beta = 0$ and for $\beta \neq 0$, $||\beta|| > 0$.
2. For any scalar $a \in \mathbb{R}$, $||a\beta|| = |a|||\beta||$.
3. For $\beta_1, \beta_2 \in \mathbb{R}^n$, $||\beta_1 + \beta_2|| \leq ||\beta_1|| + ||\beta_2||$.

In particular we will use the following common vector norms:

– $||\beta||_1 := \sum_{i=1}^n |\beta_i|$, the **L1 norm** of $\beta \in \mathbb{R}^n$.

– $||\beta||_2 := \sqrt{\sum_{i=1}^n \beta_i^2}$, the **L2 norm**[1] of $\beta \in \mathbb{R}^n$.

– $||\beta||_\infty := \max_{1 \leq i \leq n} |\beta_i|$, the **L$\infty$ norm** of $\beta \in \mathbb{R}^n$.

2.2 Models for Computing: What Can a Computer Compute?

2.2.1 Algorithms

2.2.1.1 Motivation and History: Models for Computing[2]

Computability is the basis of all computational statistics. Naturally, by means of a computer, which is not more than a big calculator, only such problems can be solved whose solutions are computable.

The various reflections about the term **computability** lead to very different concepts and definitions. All these definitions, however, proved to be by

[1]The index 2 is usually omitted if there is no risk for confusion.

[2]Partly based on Maurer (1969, pp. 14 – 16) and on Böhling (1971, pp. 50 – 53).

and large equivalent. In this book we choose to use the **Turing computability** as our notation of general computability, since Turing machines are a concrete and intuitive model of computation.

A function is defined to be **Turing computable** if the function's value can be computed with a Turing machine. For more details on Turing machines see Section 2.2.2.

The Church-Turing thesis (following Alonzo Church and Alan Turing) makes a statement about the capabilities of a Turing machine:

Church-Turing thesis: The class of intuitively computable functions is exactly the same as the class of Turing computable functions.

As already mentioned above, this thesis is not provable since the term intuitively computable is not well defined and cannot be exactly formalized. It is standard, however, to assume that this thesis is true. This leads to the possibility to prove that a function is not computable.

Do not confuse the Church-Turing thesis and the different proposition Thesis M with regard to the capabilities of Turing machines (Gandy, 1980):

Thesis M: Whatever can be calculated by a machine is Turing (machine) computable.

It is this thesis, however, which allows us to concentrate on Turing computability in this book, since this thesis relates general machine computability to Turing computability.

The term **algorithm** is fundamental for the definition of what Turing computable means. We will now give a somewhat intuitive definition. Based on this notion we will then define the Turing machine in the next section.

Definition 2.2: Algorithm
An **algorithm** is a procedure with which an answer is given to a certain complex of questions by a prefixed method. Algorithms have to be specified unambiguously in every detail. In particular, the instructions stating the algorithm have to be given by a text of finite length.

Example 2.1: **Euclidean Algorithm** to Determine the Greatest Common Divisor (GCD) of Two Natural Numbers
For two natural numbers a_1 and a_2 with $a_1 \geq a_2 > 0$, natural numbers $q_2, q_3, \ldots$ can be determined by continued division with remainder so that

$$\begin{aligned} a_1 &= a_2 q_2 + a_3 \text{ with } a_3 < a_2 \\ a_2 &= a_3 q_3 + a_4 \text{ with } a_4 < a_3 \\ a_3 &= a_4 q_4 + a_5 \text{ with } a_5 < a_4 \\ &\vdots \end{aligned}$$

The sequence (a_i) reaches the value 0 after a finite number of steps.
If $a_n > 0$ and $a_{n+1} = 0$, i.e. a_n is the last element of the sequence (a_i) different from 0, then $a_n = gcd(a_1, a_2)$.

Sample calculation: Let $a_1 = 810$, $a_2 = 246$, then:

$$
\begin{array}{ccccccc}
 & & & & 810 & :246 & =3 \\
 & & & & \mathbf{738} & & \\
 & & & 246 & :72 & =3 & \\
 & & & \mathbf{216} & & & \\
 & & 72 & :30 & =2 & & \\
 & & \mathbf{60} & & & & \\
 & 30 & :12 & =2 & & & \\
 & \mathbf{24} & & & & & \\
12 & :6 & =2 & & & & \\
12 & & & & & & \\
\mathbf{0} & & & & & &
\end{array}
$$

Verify by reformulating the above scheme as in the proof below:

$$246 = 216 + 30 = 3 \cdot 72 + 30 = 7 \cdot 30 + 3 \cdot 12 = 17 \cdot 12 + 7 \cdot 6 = 41 \cdot 6 \quad \text{and}$$
$$810 = 3 \cdot 246 + 72 = 3 \cdot 246 + 2 \cdot 30 + 12 = 3 \cdot 246 + 5 \cdot 12 + 2 \cdot 6 = 135 \cdot 6.$$

The **Euclidean algorithm** for the calculation of the greatest common divisor of two natural numbers delivers the answer after a finite number of steps and is therefore a terminating algorithm.

Proof. The algorithm delivers the $gcd(a_1, a_2)$ because of the following arguments: If $a_3 = 0$, then a_1 is divisible by a_2 and therefore $a_2 = gcd(a_1, a_2)$. If $a_3 \neq 0$, then a_3 is the greatest candidate for the gcd. If then $a_2 = a_3 q_3$ (i.e. $a_4 = 0$), then $a_1 = (q_2 q_3 + 1) a_3$, and $a_3 = gcd(a_1, a_2)$. ... □

Example 2.2: Calculation of All Integer Roots of a Polynomial
Let $x \in \mathbf{Z}$ with $y(x) = 0$ a root of the polynomial y. We are looking for an algorithm to calculate all integer roots of a polynomial.

Consider, e.g., a polynomial $y(x) = x^3 + ax^2 + bx + c$ (where a, b, and $c \neq 0$ are integers). A possible algorithm would be:

> Among all divisors of c, determine all those divisors $d_1, d_2, \ldots, d_n$ for which $y(d_i) = 0$.

Obviously, this is not an algorithm in the strict sense since the instructions do not describe every detail of the calculation. For example, there are no instructions on how to find the divisor of a number or how to check if $y(d_i) = 0$. However, the given instructions can be amended so that they are complete.

Some examples:

$$y(x) = x^3 - 1 = 0 \quad \Rightarrow \quad x = 1$$
$$y(x) = x^3 - 8 = 0 \quad \Rightarrow \quad x = 2$$

Generally, it has to be true that $x^3 + ax^2 + bx = (x^2 + ax + b)x = -c$. This motivates to search only among the divisors of c.
Notice: This algorithm is obviously superior to the naive approach of checking all $x \in \mathbf{Z}$ until all roots are found.

Algorithms as Transformations
In the theory of algorithms mainly algorithms are analyzed that transform **words** (finite character strings) over an **alphabet** into words over a (possibly) different alphabet.

Example 2.3: Continuation of Example 2.2
The instructions to calculate all the integer roots of a third-degree polynomial can be regarded as follows: The word

$$x \uparrow 3 + 2 \cdot x \uparrow 2 + 2 \cdot x + 1$$

over the alphabet

$$\{x, 0, 1, 2, 3, 4, 5, 6, 7, 8, 9, +, -, \cdot, \uparrow\}$$

is transformed into the word -1 over the alphabet

$$\{+, -, 0, 1, 2, 3, 4, 5, 6, 7, 8, 9\},$$

since -1 is the only integer root of the polynomial $x^3 + 2x^2 + 2x + 1$.

By the above intuitive Definition 2.2 it is expressed that an algorithm is not just a method but also the method's description. This can be realized in a natural or an artificial **(algorithmic) language** (e.g. in a programming language) or by means of a **graph** (e.g. a flowchart). Essential for this is the uniqueness of the description by means of a finite text.

In the theory of algorithms it is assumed that linear character strings (**words** over a given alphabet) are transformed by the analyzed algorithm. Such words may be interpreted as code words for specific items worked upon.

In the context of an (algorithmic) language a sequence of words over the alphabet of the algorithm is generated from an **initial word** (starting word, input word) by means of a given system of so-called rules (**transformation rules**). One gets a so-called **deduction** corresponding to the considered algorithm.

Algorithms as a Sequence of Operations

An algorithm is uniquely determined by the specification of a **set of items** (input items, intermediate items, and output items), a **set of basic operations** (elementary operations), and the **instruction** that specifies in which order (or step of an iteration) the operations are to be performed.

The system of operations has to be complete, i.e. each operation specified in the instruction has to be executable in the desired step. Also, the order of the operations has to be uniquely determined.

Executing operations (for the manipulation of items) are distinguished from **testing operations** (comparing operations, logic operations), which generally result in binary decisions.

Characterization of Algorithms

In summary, algorithms can be characterized as follows:

1. An algorithm operates on specific items (e.g. characters).
2. An algorithm is given by a finite instruction I.
3. The operations prescribed by I are followed stepwise.
4. I specifies the execution of the operations in every detail.
5. The execution of I has to be reproducible.
6. The execution of I does not need any additional information.

Notice: The results of the stochastic algorithms, which are discussed in the following, are also reproducible if all the parameters determining the stochastic behavior are fixed. This includes, e.g., starting values of a random number generator (see Chapter 6).

Classes of Algorithms

Algorithms can be classified according to various aspects. We will discuss only two of them.

Definition 2.3: Classes of Algorithms

An algorithm is called **terminating** if it delivers an answer for the considered complex of questions in finitely many steps. Otherwise, the algorithm is called **non-terminating**.

An algorithm is called **unbranched** if it is executed linearly and **branched** if it contains jumps.

An unbranched algorithm is always terminating. If a branched algorithm leads into a cycle in which it steadily remains, the process is non-terminating. If an algorithm describes an infinite process (according to time measurement) the algorithm is also non-terminating.

Example 2.4: Algorithm for the Calculation of a Square Root
No **algorithm for the calculation of the square root** of a natural number in decimal representation can be terminating if only one digit of the solution can be generated per step (cp. Section 2.2.2), since a square root normally has an infinite number of digits.

Besides the natural termination of an algorithm (e.g. of the Euclidean algorithm for the determination of the greatest common divisor of two natural numbers), **artificial terminations** are in use (e.g. with all mathematical approximation processes).

For algorithmic control, though, e.g. in process control, a termination is undesired, since the process ideally should continue without break.

Universal Algorithm
Algorithms should not only be applicable to a specific task but to a whole **class of tasks**. This is achieved by defining as many as possible operations for variables with deliberately selectable values.

Superposition and linking of special algorithms may lead to more general ones. This leads to the question of whether there is something like a **"universal" algorithm**. However, the question cannot be answered without further specification of the term **algorithm**.

The notion of an algorithm is specified very differently, e.g., by A. Church, A. A. Markov, E. L. Post, and A. M. Turing. These notions mainly differ with regard to different specifications of the transformation rule system and the allowed formation of deductions. Although these different specifications are based on very different ideas, nevertheless, the **fundamental theorem of the classical algorithm theory** proves that all the proposed notions of algorithms are equivalent. Therefore, these notions of algorithms are called **Turing complete**. We thus generally define so-called computable functions as follows.

Definition 2.4: Computable Functions
Let M be the set of permissible input words and N be the set of permissible output words. A function $f : M \to N$ is called **(effectively) computable** iff there is a terminating algorithm delivering the corresponding function value $f(a) \in N$ for each $a \in M$.

This notion is further specified in Section 2.2.2 in terms of Turing.

History
Following the Uzbek mathematician **Ben Musa al-Chwarizmi**, who published the first Arabian book on algebra in the 9th century, the term **algorithm** was originally only relevant for purely numerical methods.

The Euclidean algorithm for the determination of the greatest common divisor of two natural numbers is a classical example.

However, also outside mathematics there are algorithmic processes in which, following prescribed instructions based on a set of given operations, certain input items are stepwise transformed via intermediate items into output items.

The modern theory of algorithms is very much influenced and stimulated by the requirements and results of computer engineering and data processing, by information theory and newly developed research areas like bioinformatics and learning theory. Moreover, algorithmic processes play an important role in control theory.

2.2.1.2 Theory: Complexity of Algorithms[3]

The example of the Euclidean algorithm for the determination of the gcd already shows that the purpose of an algorithm is not always easily visible. In other words, the relationship between algorithm and problem is potentially unclear.

In particular, different algorithms may be developed for the same problem and all of them may significantly differ in the proximity to the problem. Indeed, our first important criterion for the quality of an algorithm is by no means its proximity to the problem but the so-called **complexity** of an algorithm, meaning the calculation effort to solve the problem.

Before illustrating proximity to the problem and complexity by means of sorting algorithms, we will define so-called **complexity measures**. A natural candidate is, e.g., the calculation time needed by the algorithm. Therefore, the term **time complexity** is used.

Time Complexity
Calculation time naturally depends on the used computer. Therefore, the number of (elementary) operations (approximately) needed for solving the problem is often used instead. For our purposes, **operations** are arithmetic operations like addition, subtraction, multiplication, and division, but also comparison- and interchange-operations. If operations are used that need different effort, a common unit for the effort might be useful. Obviously, the **number of operations** depends on the amount of data to be processed. Naturally, the number of operations increase with the amount of data.

The following **complexity measures** mainly indicate how strong the number of operations increase with the size of the data. Thus, complexity

[3]Partly based on Wegener (1992, pp. 16 – 19).

measures are functions mapping the size of the input data to the number of operations needed by the algorithm.

Generally, only the **order** of the number of operations is important. A typical statement is, e.g.: "The number of operations increases on average with the number of data n" instead of "The number of operations is on average $c_1 \cdot n$." There is hope that the relative complexity does not depend on the exact prefactors, since very large and very small prefactors are seldom. However, the function $20n$ is only smaller than the function $\lceil n^3/125 \rceil$ if $n > 50$. Here, the order estimation $n \leq n^3$ is only valid for medium-size data sets because of the prefactors 20 and $1/125$. Even more problematic can be orders that only differ in logarithmic factors. For example, the function $10n$ is only smaller than the function $\lceil (1/2)n \log_2 n \rceil$ if $n > 2^{20}$. In such cases the exact prefactors might be interesting as well.

However, prefactors are often nevertheless ignored since:

- An exact determination of prefactors is often impossible.
- Calculation time is most of time only relevant for large data sets.
- Most relevant is the speed of growth of calculation time dependent on the size of the data set.

Definition 2.5: Characterization of Complexity
Let $f, g : \mathbb{N} \to \mathbb{R}_0^+$ be **complexity measures** for the dependency of the number of operations of two algorithms on the number of data n. Then we call:

1. $f = O(g)$ (in words: f increases not faster than g) if
   $\exists c, n_0 \forall n \geq n_0 : \quad f(n) \leq cg(n)$.
2. $f = \Omega(g)$ (f increases at least as fast as g), if $g = O(f)$.
3. $f = \Theta(g)$ (f and g are of the same order) if $f = O(g)$ and $g = O(f)$.
4. $f = o(g)$ (f increases slower than g), if $f(n)/g(n)$ is a null sequence.
5. $f = \omega(g)$ (f increases faster than g) if $g = o(f)$.

The two symbols capital O and small o are often called Landau symbols after the mathematician Edmund Landau (1877-1938). For better understanding, all these terms will be illustrated by examples:

Lemma 2.1: Complexity Properties

1. $\forall k > 0 : n^k = o(2^n)$.
2. Let p_1 and p_2 be polynomials of degree d_1 and d_2 of n, where the coefficients of n^{d_1} and n^{d_2} are positive. Then,
   (a) $p_1 = \Theta(p_2) \Leftrightarrow d_1 = d_2$.
   (b) $p_1 = o(p_2) \Leftrightarrow d_1 < d_2$.

(c) $p_1 = \omega(p_2) \Leftrightarrow d_1 > d_2$.

3. $\forall k > 0,\ \varepsilon > 0 : \log^k n = o(n^\varepsilon)$.
4. $2^{\frac{n}{2}} = o(2^n)$.

Proof. By graphical illustration. □

Table 2.1 reinforces for typical calculation times that the complexity order is decisive for the assessment of calculation time. Notice that all statements relate to the arbitrarily fixed 1000 operations per second in the case $f(n) = n$.

Table 2.1: Maximum Number of Data for Different Calculation Times

	Calculation Time =		
$f(n) =$	1 sec	1 min	1 h
n	1000	60000	3600000
$n \cdot \log_2 n$	140	4893	204433
n^2	31	244	1897
n^3	10	39	153
2^n	9	15	21

Figures 2.1 and 2.2 illustrate the relationship between time budget (runtime) and possible problem size. Let a machine be able to carry out 1000 elementary operations per second, and let the time budget be t seconds. How large is the largest problem that an algorithm that needs n, $n \cdot \log_2 n$, n^2, n^3, or 2^n operations to solve the problem can solve in t seconds?

That means given a time budget and an exact complexity of an algorithm, the graphics show the largest problem one can solve. Especially interesting is the comparison between the problem sizes n for t, $2t$, or t^2, showing what can be gained by doubling the time budget (or optionally by a jump from 1000 to 2000 operations/sec) or by squaring the budget.

In order to verify the curves in Figure 2.1, the following sample calculations might be helpful:

$$f(n) = 1000 \cdot 200 \text{ min} = 1000 \cdot 12{,}000 \text{ sec} = 1.2 \cdot 10^7 = n$$

$$f(n) = 1000 \cdot 330 \text{ min} \approx 1000 \cdot 20{,}000 \text{ sec} = 2 \cdot 10^7 \approx 10^6 \cdot \log_2(10^6)$$

$$f(n) = 1000 \cdot 1666 \text{ min} \approx 1000 \cdot 100{,}000 \text{ sec} = 10^8 = (10^4)^2$$

Figure 2.2 shows the same relations on log scale and Table 2.2 shows how technological progress influences the magnitude of calculation time. Note that Figure 2.2 can be derived directly from Table 2.2. How much can the number

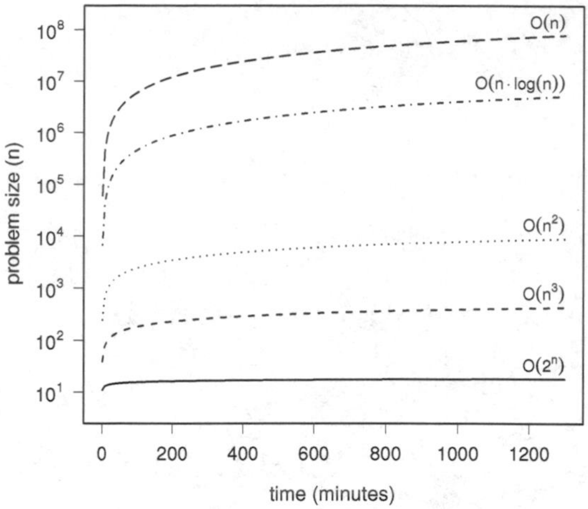

Figure 2.1: Problem size dependence on runtime.

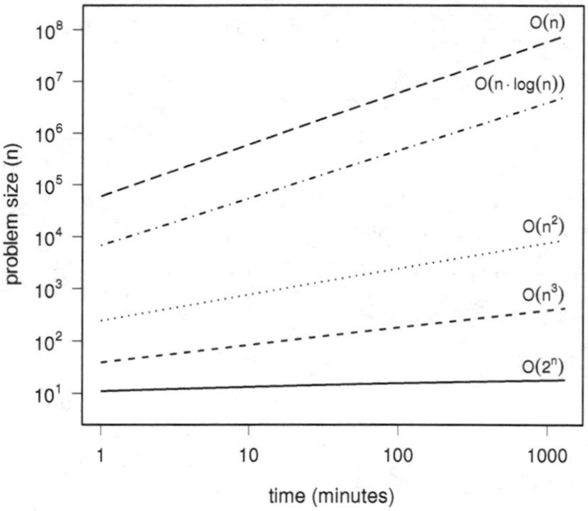

Figure 2.2: Problem size dependence on runtime: $\log_{10}$ scale.

of data be increased for a given calculation time if the computer is 10 times faster? Here, constant prefactors do not play any role. The used relationships are shown directly below the table.

Table 2.2 shows the structural difference between the calculation times n, $n \log n$, n^2, and n^3 on the one side and 2^n on the other side. In the first four cases the number of data increases by a constant factor, depending on the degree of the calculation time polynomial. In the last case the number

Table 2.2: Maximum Number of Data Dependent on the Technology

		Technology
$f(n)$	Old	New (i.e. 10 times faster)
n	p	$10p$
$n \log n$	p	(almost $10)p$
n^2	p	$3.16p$
n^3	p	$2.15p$
2^n	p	$p+3.3$

Note: $\log(10p) = \log p + \log 10 \approx \log p$, $3.16^2 \approx 10$, $2.15^3 \approx 10$, and $2^{3.3} \approx 10$.

of data increases only by a constant summand. This motivates the following definition.

Definition 2.6: Classes of Complexity
Let $f : \mathbb{N} \to \mathbb{R}_0^+$ be a complexity function.

1. f **increases polynomially** (or is called **polynomially constrained**) if there exists a polynomial p with $f = O(p)$.
2. f **increases exponentially** if there exists an $\varepsilon > 0$ with $f = \Omega(2^{n^\varepsilon})$.

If at all possible, algorithms whose runtime is not constrained by a polynomial should be avoided. Unfortunately, some problems are likely not solvable in polynomial time. For example, there is no known algorithm with polynomial runtime that will, given a set of n integers, tell us if there is a (nonempty) subset for which the sum of the integers in the subset is zero. Calculating such a subset is therefore "hard," but given a subset, it is easy to check if it is truly a subset and if its sum is zero. Both checks can be done in polynomial time. Another practical example of a hard problem is finding the shortest round-trip through n cities. This problem is commonly referred to as the traveling salesperson problem.

Notice that complexity is a **random variable**, since $f(n)$ generally depends not only on the number of data n, but also on the (random) ordering of the data. However, since any statement on the complexity of an algorithm should be independent of the input, the distribution of the complexity is often characterized by a location measure, typically by means of the **expected value**. This leads to the so-called **mean complexity**. Additionally, the distribution might be characterized by its **range**, i.e. by its smallest and its largest value (if they exist), which describe the most and the least favorable case. Since the most favorable case is often considered as not so important, only a **worst case analysis** is carried out in such cases.

2.2.1.3 Example: Sorting Algorithms

In the following the complexity of different algorithms for the solution of the same problem is exemplified by the practically very important sorting of n data according to some ordering (usually largest to smallest or vice versa). We will discuss different examples of sorting algorithms. In order to explain their structure step-by-step, we will start with an intuitive algorithm, which is however rarely used in practice.

We use the above complexity characteristics for the comparison of algorithms. The complexity is determined in the worst case as well as in the mean.

Question
Do you know applications of sorting algorithms in statistics?

Example 2.5: Bubble Sort[4]
Method: If only the repeated interchanges of neighboring data elements are permitted as movements of data elements, we can obviously generate a data sequence ordered by ascending values as follows.

We run through the list $a_1, \ldots, a_n$ of data elements considering any two neighboring elements a_i and a_{i+1}, $1 \leq i < n$. If $a_i > a_{i+1}$, then we interchange a_i and a_{i+1}.

After the first pass the largest element has reached its correct position at the right-hand end of the list. Then again, the sequence is passed another time exchanging neighboring elements if necessary. This passing of the sequence is repeated until no interchanges appear anymore, i.e. until all pairs of neighboring elements are correctly ordered. Finally, the list a is sorted in ascending order.

In this sorting algorithm, larger elements obviously have the tendency to rise slowly like air bubbles in water. This analogy induced the name Bubble sort.

Sample example: Apply bubble sort to the vector $a = [15\ 2\ 43\ 17\ 4\ 8\ 47]^T$.

In the first pass, the following interchanges of neighboring elements are carried out:

[4]based on Ottman and Widmayer (1996, pp. 73 – 76)

15	2					
2	15	43	17			
		17	43	4		
			4	43	8	
				8	43	47

After the first pass the ordering of the list a looks as follows: 2, 15, 17, 4, 8, 43, 47.

The second pass delivers the ordering: 2, 15, 4, 8, 17, 43, 47.

The third pass delivers at last: 2, 4, 8, 15, 17, 43, 47,

i.e. the input elements are in ascending order. Another pass would not cause any more interchanges. The bubble sort has successfully ended.

The pseudocode of bubble sort is given in Algorithm 2.1.

Algorithm 2.1 Bubble Sort

Require: n = number of observations; a = list of inputs

$j \leftarrow 1$
while $j > 0$ **do**
    $j \leftarrow 0$ $\{j =$ number of interchanges in the pass$\}$
    for $i = 1$ **to** $n-1$ **do**
        if $a_i > a_{i+1}$ **then**
            $a_i \leftrightarrow a_{i+1}$
            $j \leftarrow j+1$
        end if
    end for
end while

Note that the pseudocodes given in this book are most of the time not algorithms exactly specified in all detail, but can be easily completed to have this property. For example, in Algorithm 2.1 the interchange operation is not specified.

Also note that in Algorithm 2.1 we always pass the whole list, though after the ith pass the i largest elements are already in the correct ordering at the right-hand end. We thus would receive an efficiency improvement by, e.g., inspecting only the positions $1, \ldots, (n-i)+1$ in the ith pass.

Analysis: The estimation of the number of comparisons in the **most favorable** and the **least favorable case** is simple.

If the list a is already sorted in ascending order (**best case**), the `for`-loop of the above algorithm is passed exactly once without any interchanges. Therefore, the minimum number of comparisons is $C_{min}(n) = n-1$ and the minimum number of movements $M_{min}(n) = 0$.

In the least favorable case (**worst case**) for the algorithm bubble sort the list a is initially sorted in descending order. Then, the element with the minimal value advances one position to the left in each pass of the `while`-loop. Therefore, n passes are necessary until it has reached the left-hand end, and until no interchange is needed anymore.

It is easy to see that in this case in the ith pass, $1 \leq i < n$, $(n-i)$ interchanges of neighboring elements are needed, i.e. $3(n-i)$ movements (left element into auxiliary storage, right element to the place of the left element, and contents of auxiliary storage to the place of the right element), and naturally each time $n-i$ comparisons have to be carried out.

Therefore:

$$C_{max} = \sum_{i=1}^{n-1} (n-i) = n(n-1) - \frac{n(n-1)}{2} = \frac{n(n-1)}{2} = \Theta(n^2)$$

and

$$M_{max} = \sum_{i=1}^{n-1} 3(n-i) = \Theta(n^2)$$

One can show that this complexity characterization is also valid in the mean (average case):

$$C_{mean}(n) = M_{mean}(n) = \Theta(n^2).$$

We do not prove this here, since bubble sort is surely an intuitive and popular but very bad elementary sorting method. Only in cases where the list is already nearly sorted are a small number of comparing operations and movements carried out. Moreover, the method is very asymmetric concerning the passing direction. If the original sequence is, e.g., already nearly sorted in the sense that for $a_1, \ldots, a_n$ it is true that $a_i \leq a_{i+1}$, $1 \leq i < n-1$, and a_n is the minimal element, then $n-1$ passes are necessary in order to bring a_n to the left-hand end of the list.

In order to repair this weakness, we could pass the list alternately from left to right and vice versa. This slightly better variant is known as **shakersort**.

The next sorting algorithm is using a different idea. It immediately finds the correct place for a new list element.

Example 2.6: Insertion Sort[5]
Method: Assume that we have already sorted the first i data, i.e.

$$a_1 \leq \ldots \leq a_i, \quad i = 1, \ldots, n-1.$$

[5]based on Wegener (1992, pp. 58 – 59)

Now, a new piece of data a_{i+1} should be arranged in order. For this there are $i+1$ possible positions.

Before a_j there are j positions, behind a_j another $i+1-j$. Therefore, we compare a_{i+1} first with a_j for $j = \lceil (i+1)/2 \rceil$. In that range in which a_{i+1} lies, we continue analogously. This way, $\lceil \log_2(i+1) \rceil$ comparisons are enough to find the correct position of a_{i+1}.

Let, e.g., $i = 7$. Then, because of the continuing split, 3 comparisons are enough to arrange the new piece of data a_{i+1} in order. If, e.g., a_{i+1} is smaller than all a_j, $j = 1, \ldots, i$, then we first compare it with a_4, then with a_2, and finally with a_1.

After we have found the correct position k for a_{i+1}, it is stored in an additional storage place. After shifting the data $a_k, \ldots, a_i$ one position to the right, a_{i+1} takes the position k.

The pseudocode of insertion sort is given in Algorithm 2.2.

Algorithm 2.2 Insertion Sort

Require: n = number of observations; a = list of inputs

 1: **for** $i = 1$ **to** $n-1$ **do**
 2:     $r \leftarrow i$ $\{r = \text{right range limit}\}$
 3:     **if** $a_1 \geq a_{i+1}$ **then**
 4:         $[a_1, \ldots, a_{i+1}] \leftarrow [a_{i+1}, a_1, \ldots, a_i]$ $\{a_{i+1}$ leftmost: NEXT $i\}$
 5:     **else if** $a_i > a_{i+1}$ **then**
 6:         $l \leftarrow 1$ $\{l = \text{left range limit}\}$ $\{a_{i+1}$ not rightmost: Find correct position for $a_{i+1}\}$
 7:         **loop**
 8:             $j \leftarrow \lceil (l+r)/2 \rceil$
 9:             **if** $a_j \leq a_{i+1} \wedge a_{i+1} \leq a_{j+1}$ **then**
10:                 $[a_1, \ldots, a_{i+1}] \leftarrow [a_1, \ldots, a_j, a_{i+1}, a_{j+1}, \ldots, a_i]$
11:                 break loop
12:             **else if** $a_j > a_{i+1}$ **then**
13:                 $r \leftarrow j$
14:             **else**
15:                 $l \leftarrow j$
16:             **end if**
17:         **end loop**
18:     **end if**
19: **end for**
20: **return** a

Note that Algorithm 2.2 assumes that the element a_{i+1} is initially at the right end of the list.

Analysis: In the **worst case** the total number of comparisons is

$$\sum_{i=1}^{n-1} \lceil \log_2(i+1) \rceil = \sum_{i=2}^{n} \lceil \log_2(i) \rceil < \log_2(n!) + n.$$

We will study the term $\log_2(n!)$ in more detail since it appears frequently. Following the **Stirling formula** the ratio of $n!$ and $\sqrt{2\pi}\, n^{n+1/2} e^{-n}$ converges toward 1:

$$\log_2(n!) \approx \log_2(\lfloor \sqrt{2\pi}\, n^{n+1/2} e^{-n} \rfloor) = n \log_2(n) - n \log_2(e) + O(\log_2(n)).$$

Since $\log_2(n) \ll n$ for large n, n is dominated by the term $n \log_2(n)$, and thus $O(\log_2(n))$ gets small relative to the first term. Therefore, the following approximation is very good:

$$\log_2(n!) \approx n \log_2(n) - 1.4427n.$$

Theorem 2.1: Insertion Sort - Worst Case
In the worst case insertion sort approximately needs

$$n \log_2(n) - 0.4427n + O(\log_2(n))$$

comparisons.

Therefore, for comparisons insertion sort is definitely better than bubble sort. However, the data transport is very "expensive," even in the **average case**, for which we consider all $n!$ permutations of the ordered sequence as uniformly probable inputs. For the average case analysis we assume without loss of generality that the ordered sequence has the form $1, \ldots, n$.

Theorem 2.2: Insertion Sort - Average Case
In the average case insertion sort needs $\Theta(n^2)$ interchange operations.

Proof. We count only the operations that move one piece of data one place to the right. There are $(n-1)!$ permutations π with $\pi(i) = j$.

Therefore, a_i ends with probability $\frac{(n-1)!}{n!} = \frac{1}{n}$ at position j assuming uniform distribution of inputs. If $j > i$, a_i has to be moved on average $(j-i)$ times one place to the right. In the average case, the number of movements of a_i to the right is thus

$$\frac{1}{n}(0 + \ldots + 0 + 1 + \ldots + (n-i)) = \frac{(n-i)(n-i+1)}{2n}.$$

Therefore, the total number of movements to the right in the average case is

$$\frac{1}{2n}\sum_{i=1}^{n}(n-i)(n-i+1) = \frac{1}{2n}\sum_{i=1}^{n}(i-1)i = \frac{(n^2-1)}{6}.$$

□

However, even in the worst case $O(n^2)$ operations are sufficient!

From this analysis, we learned something general about the construction of sorting algorithms. On average, we cannot avoid quadratic runtime if the data are only "creeping" in one direction, i.e. are making steps of length 1 in one pass.

Example 2.7: Quick Sort[6]

This sorting algorithm introduced by Hoare (1962) is nowadays one of the most commonly used sorting methods. It is an application of the divide-and-conquer principle to sorting.

Method: At first, one piece of data is chosen, e.g. according to one of the four most widely used variants:

1. Choose the first element of the input list.
2. Choose a position of the input list by means of a random number generator, i.e. choose the position and thus the piece of data randomly.
3. Choose three pieces of data in the list: the first one, the middle one, and the last one (if the number n of data is even, choose, e.g., the left of the two middle ones). Calculate the median of these three pieces of data.
4. Choose three positions randomly and determine the median of the corresponding data.

Then, this first chosen piece of data is brought into its correct position, i.e. the data on its left should not be larger, and on its right, not smaller. Then, quick sort is applied recursively to the two sublists to the left and to the right of the chosen piece of data.

Sublists of length 0 or 1 are already sorted so that the procedure can be stopped. There are implementations in which short lists, e.g. with at most 10 data, are analyzed by means of insertion sort in order to avoid the higher organizational effort of quick sort.

This effort is mainly related to the fact that there are two sublists that both have to be analyzed. We have to decide which sublist should be analyzed first, and we have to put the other sublist on a so-called recursion stack.

[6]based on Wegener (1992, pp. 60 – 63)

Algorithm 2.3 Quick Sort

Require: a = list of inputs; l_0 = leftmost index; r_0 = rightmost index

  1: **if** $r_0 - l_0 \geq 1$ **then**

  2:     $l \leftarrow l_0$

  3:     $i \leftarrow$ RANDOMELEMENT $\in \{l_0, \ldots, r_0\}$

  4:     $r \leftarrow r_0$

  5:     **while** $l \neq r$ **do**

  6:         **while** $(a_l \leq a_i) \wedge (l < i)$ **do**

  7:             $l \leftarrow l + 1$

  8:         **end while**

  9:         **while** $a_r > a_i \wedge r > i$ **do**

10:             $r \leftarrow r - 1$

11:         **end while**

12:         $a_l \leftrightarrow a_r$

13:         **if** $l = i$ **then**

14:             $i \leftarrow r$

15:         **else if** $r = i$ **then**

16:             $i \leftarrow l$

17:         **end if**

18:     **end while**

19:     $a \leftarrow$ Quick sort$(a, l_0, i - 1)$

20:     $a \leftarrow$ Quick sort$(a, i + 1, r_0)$

21: **end if**

22: **return** a

For the **decomposition of the list** into a chosen piece of data and those sublists of data that are not larger and not smaller than the chosen piece of data, respectively, there are again different implementations. We introduce one implementation that also works correctly if some data of the list are equal, and which manages to get along with the minimum number of $n - 1$ comparisons:

1. Assume that the list has positions $1, \ldots, n$ and that a_i is chosen first.

2. Starting from a_1 we search for the first piece of data $a_l > a_i$. This is true for at least $l = i$.

3. Starting from a_n we search for the first piece of data $a_r \leq a_i$. Again, this is true for at least $r = i$.

4. If $l = r = i$, then STOP. Otherwise, interchange a_l and a_r. Notice that the first chosen piece of data a_i could have changed its position now. Therefore,

let $i = r$, if $l = i$, and $i = l$, if $r = i$ otherwise, in order to change the index i along with the corresponding element a_i. Continue the search as in steps 1 and 2, starting at positions l and r. In all cases where a_i is not affected by an interchange, we could start at $l+1$ and $r-1$ in order to reduce the number of comparisons. This is omitted in the pseudocode of Algorithm 2.3 in order to simplify and shorten its representation.

The pseudocode of quick sort is given in Algorithm 2.3.

Note that this is the only exception where we specify a recursive definition of the algorithm, since its implementation is very natural and will typically happen in languages like C. In R, as an example of a statistical programming language, we won't be able to sort even reasonably sized data using a recursive implementation due to R's stack usage restrictions, which stem from its scoping rules. An iterative implementation would involve dealing with stacks that keep information about the subsets of the data that have to be sorted during the next iterations.

The method needs $n-1$ comparisons since a_i is compared with every other piece of data exactly once. The number of interchanges is also at most $n-1$, since every piece of data except a_i is interchanged at most once.

Sample example: Let $n = 13$, $i = 7$, and $a_i = 53$ chosen first. In what follows, the first element chosen is italicized and the elements identified to be interchanged in bold.

15	47	33	**87**	98	17	*53*	76	83	2	53	27	**44**
15	47	33	44	**98**	17	*53*	76	83	2	53	**27**	87
15	47	33	44	27	17	***53***	76	83	2	**53**	98	87
15	47	33	44	27	17	53	**76**	83	2	***53***	98	87
15	47	33	44	27	17	53	***53***	83	**2**	76	98	87
15	47	33	44	27	17	53	2	**83**	***53***	76	98	87
15	47	33	44	27	17	53	2	***53***	83	76	98	87

After this first run, position 9 has received its final piece of data. Quick sort is now called for the position intervals $[1,8]$ and $[10,13]$.

Analysis: In the following we only study the number of comparisons. The number of interchanges cannot be higher.

In **variants (1) and (2)** (see page 27) a_i can be the smallest piece of data. Therefore, in the **worst case** the number of comparisons is

$$C_{max}(n) = C_{max}(n-1) + n - 1 = 1 + 2 + \ldots + (n-1) = \binom{n}{2} = \frac{n(n-1)}{2},$$

since then the chosen element has to be compared with all other pieces of data ($(n-1)$ comparisons) and the next list includes $(n-1)$ elements. Thus, we have found one case with $(n-1),(n-2),\ldots$ comparisons. At least this many comparisons we have to budget.

The maximal number of comparisons is also restricted by $\binom{n}{2}$, since no pair is compared twice. Therefore, in the worst case $\binom{n}{2}$ comparisons are carried out.

For **variants (3) and (4)** (see page 27) the situation is somewhat more favorable. For the determination of the median we need 3 comparisons in the worst case. This median then has to be compared only with the (n-3) remaining data, and the chosen piece of data is at worst the second smallest or the second largest piece of data so that the next list is at most $(n-2)$ elements long. In total, the maximum is not very much changed by this, though.

For these variants we have:

$$C_{max}(n) = C_{max}(n-2)+n = n+(n-2)+(n-4)+\ldots = \Theta(n^2).$$

Indeed, all Quick sort variants need $\Theta(n^2)$ comparisons in the worst case.

For variants (1) and (2) we now derive the **mean number of necessary comparisons**. For variants (3) and (4), see the next example.

First notice that the variants (1) and (3) are essentially different from the variants (2) and (4). In the first case, we have to consider the average over all inputs. In the second case, we can calculate the mean number of comparisons because of the use of random numbers. Having made this clear, we can work jointly on cases (1) and (2).

In the end, the chosen object stands at position j, $1 \le j \le n$, with probability $\frac{1}{n}$. Then, problems of sizes $j-1$ and $n-j$ have to be solved. Therefore,

$$C_{mean}(0) = C_{mean}(1) = 0 \qquad \text{and for} \quad n \ge 2:$$

$$C_{mean}(n) = n-1+\frac{1}{n}\sum_{j=1}^{n}(C_{mean}(j-1)+C_{mean}(n-j)).$$

Exemplarily, we solve this recursion equality abbreviating C_{mean} by C. Obviously,

$$\begin{aligned}
\sum_{j=1}^{n} C(j-1) &= \sum_{j=1}^{n} C(n-j), \text{ hence} \\
C(n) &= n-1+\frac{2}{n}\sum_{j=1}^{n} C(j-1) \text{ and} \\
nC(n) &= n(n-1)+2\sum_{j=1}^{n} C(j-1).
\end{aligned}$$

In this equality the full set of $C(j)$ appears. In order to reduce the number of different C-values to two, we consider the above equation for $n-1$ and subtract this equation from the equation for n:

$$\begin{aligned}
(n-1)C(n-1) &= (n-1)(n-2)+2\sum_{j=1}^{n-1} C(j-1) \\
\Rightarrow nC(n)-(n-1)C(n-1) &= n(n-1)-(n-2)(n-1)+2C(n-1) \\
\Rightarrow nC(n)-(n+1)C(n-1) &= 2(n-1) \\
\Rightarrow \text{(division by } (n+1)n) : \frac{C(n)}{n+1} &= \frac{C(n-1)}{n}+\frac{2(n-1)}{n(n+1)}
\end{aligned}$$

Now, it appears to be natural to study $Z(n) := \frac{C(n)}{n+1}$ instead:

$$\begin{aligned}
Z(n) &= Z(n-1)+\frac{2(n-1)}{n(n+1)} \\
&= Z(n-2)+\frac{2(n-2)}{(n-1)n}+\frac{2(n-1)}{n(n+1)} \\
&= \cdots \\
&= Z(1)+2\sum_{j=2}^{n} \frac{j-1}{j(j+1)}
\end{aligned}$$

Since $C(1)=0$, also $Z(1)=0$. Moreover:

$$\frac{1}{j(j+1)}=\frac{1}{j}-\frac{1}{j+1}.$$

Therefore,

$$\begin{aligned}
Z(n) &= 2\sum_{j=2}^{n}\left(\frac{2}{j+1}-\frac{1}{j}\right) \\
&= 2\left(\sum_{j=3}^{n+1}\frac{2}{j}-\sum_{j=2}^{n}\frac{1}{j}\right) \\
&= 2\left(\sum_{j=3}^{n}\frac{2}{j}+\frac{2}{n+1}-\sum_{j=3}^{n}\frac{1}{j}-\frac{1}{2}\right) \\
&= -1+2\sum_{j=3}^{n}\left(\frac{1}{j}\right)+\frac{4}{n+1}.
\end{aligned}$$

The series $1+\frac{1}{2}+\ldots+\frac{1}{n}$ occurs so often that it has gotten the special name **harmonic series** $H(n)$. Thus,

$$Z(n)=2H(n)-4+\frac{4}{n+1}$$

and therefore

$$C(n) = (n+1)Z(n) = 2(n+1)\left(H(n) - 2 + \frac{2}{n+1}\right) = 2(n+1)H(n) - 4n.$$

Considering **Riemann sums**, for $H(n)$ it is valid that

$$\int_1^{n+1} \frac{1}{x} dx \leq H(n) \leq 1 + \int_1^n \frac{1}{x} dx, \qquad \text{i.e.} \qquad \log(n+1) \leq H(n) \leq 1 + \log(n).$$

One can even show that $(H(n) - \log(n))$ converges. The limit $\gamma \approx 0.57721\ldots$ is called **Euler constant**. Finally, $\log(n) = \log_2(n)\log(2)$.

Overall, we have proved the following theorem:

Theorem 2.3: Quick Sort - Worst and Average Case
In the worst case all Quick sort variants need $\Theta(n^2)$ comparisons. For the variants (1) and (2) the average number of comparisons is:

$$\begin{aligned} C_{mean}(n) &= 2(n+1)\sum_{j=1}^{n} \frac{1}{j} - 4n \\ &\approx 2(n+1)(0.57721 + 0.69315 \cdot \log_2(n)) - 4n \\ &\approx 1.386n\log_2(n) - 2.846n + 1.386\log_2(n) + 1.154. \end{aligned}$$

Example 2.8: Clever Quick Sort[7]
Method: Clever quick sort we call the variants (3) and (4) of quick sort. There, we choose three objects x, y, and z from the input list. Then, the median of these three objects is calculated and used as the dividing object. Obviously, this object only has to be compared with the remaining $n-3$ objects.

Analysis: In order to calculate the median of the three objects x, y, and z, on average $\frac{8}{3}$ comparisons are sufficient and necessary.
As the first comparison, w.l.o.g., the comparison between x and y can be chosen. Let, w.l.o.g., $x > y$. Because of the larger-smaller symmetry we can then choose, w.l.o.g., the comparison between x and z. If $z > x$, then x is the median. Otherwise, z and y also have to be compared.

Without additional information we assume that z falls with equal probability $\frac{1}{3}$ into each of the intervals $(-\infty, y)$, (y, x), and (x, ∞). Hence, we need 3 comparisons with probability $\frac{2}{3}$ and 2 comparisons with probability $\frac{1}{3}$. Therefore, on average we need $\frac{8}{3} = 3\frac{2}{3} + 2\frac{1}{3}$ comparisons. For the first step we thus need on average $n - \frac{1}{3}$ instead of $n-1$ comparisons. In return, the dividing object has the tendency to lie nearer to the median of the whole list. We will

[7]based on Wegener (1991, pp. 8 – 17)

study whether clever quick sort lives up to its name. We calculate the **mean complexity** of this algorithm.

For this we need the probability that the dividing object takes position j in the sorted list. We choose 3 from n objects. For this there are $\binom{n}{3}$ possibilities. The median of these three objects only takes position j if the largest object takes one of the $n-j$ posterior positions, the median object takes position j, and the smallest object takes one of the $j-1$ anterior positions. For this, there are $(j-1)(n-j)$ possibilities. The sought-after probability is therefore $\frac{(j-1)(n-j)}{\binom{n}{3}}$.

Numerical example: Let $n=5$, then for $j=1,\ldots,5$ we have:

$$\frac{(j-1)(n-j)}{\binom{n}{3}} = \frac{(j-1)(5-j)12}{120} = 0, 0.3, 0.4, 0.3, 0.$$

Obviously, here we are not concerned with an equal weighting of positions as for quick sort.

For the average number $C_{mean}(n)$ of comparisons needed by clever quick sort, we then have

$$C_{mean}(0) = C_{mean}(1) = 0, \qquad C_{mean}(2) = 1, \qquad \text{and for } n \geq 3,$$

$$C_{mean}(n) = n - \frac{1}{3} + \binom{n}{3}^{-1} \sum_{j=1}^{n} (j-1)(n-j)(C_{mean}(j-1) + C_{mean}(n-j)).$$

With this recursion equality we could principally calculate the value of $C_{mean}(n)$ for every n. Nevertheless, what we need is an explicit formula for $C_{mean}(n)$. The effort to calculate, e.g., $C_{mean}(100,000)$ would be too high for a recursion even with a computer.

Unfortunately, although the corresponding transformation of the recursion formula is very "artful," it is also very tedious and does not aid in understanding the sorting algorithm. Therefore, the derivation is skipped (see the above literature for a detailed presentation). The result, however, is very remarkable:

Theorem 2.4: Clever Quick Sort - Average Case
The average number of comparisons needed by clever quick sort is

$$C_{mean}(1) = 0,\ C_{mean}(2) = 1,\ C_{mean}(3) = \frac{8}{3},\ C_{mean}(4) = \frac{14}{3},\ C_{mean}(5) = \frac{106}{15}$$

and for $n \geq 6$,

$$\begin{aligned} C_{mean}(n) &= \frac{12}{7}(n+1)\sum_{j=1}^{n-1}\frac{1}{j} - \frac{477}{147}n + \frac{223}{147} + \frac{252}{147n} \\ &\approx 1.188n\log_2(n-1) - 2.255n + 1.188\log_2(n-1) + 2.507. \end{aligned}$$

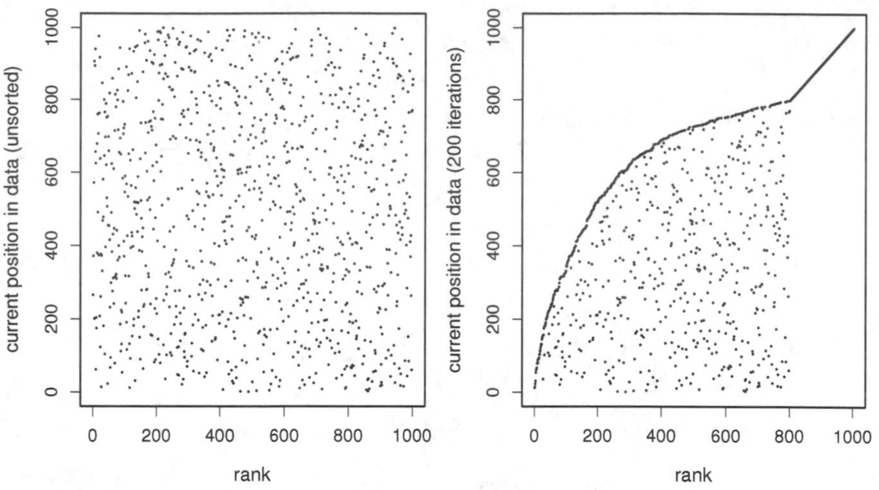

Figure 2.3: Bubble sort. Left: Initialization; Right: Snapshot during the process.

Now it is easy to calculate $C_{mean}(100,000)$ by means of a computer. One can show that clever quick sort asymptotically exceeds the minimum of the expected number of comparisons only by 18.8%. Additionally, since the linear term has a relatively large negative prefactor, clever quick sort is much more efficient than the other introduced methods.

2.2.1.4 Practice and Simulation

Sorting Algorithms on the Internet

There are many simulators for sorting algorithms on the internet, for example

`http://www.ansatt.hig.no/frodeh/algmet/animate.html`.

Inspired by this simulator, Figures 2.3 and 2.4 show snapshots of the process of sorting for bubble sort and for quick sort.

Please note that in Figures 2.3 and 2.4, on the y-axis the current position of the elements with the true rank given at the x-axis is indicated. After sorting, a diagonal should result.

The progress of the different sorting algorithms can also be presented schematically, as illustrated in Figure 2.5. The slow movements of the elements for bubble sort and the block-by-block process for quick sort are obvious.

Outlook: Algorithms

Although the term **algorithm** is very old, it is still very important today. With-

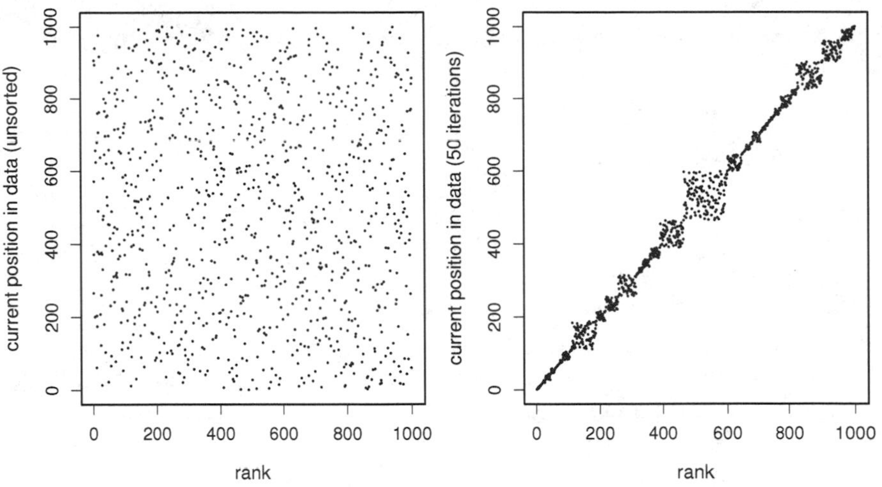

Figure 2.4: Quick sort. Left: Initialization; Right: Snapshot during the process.

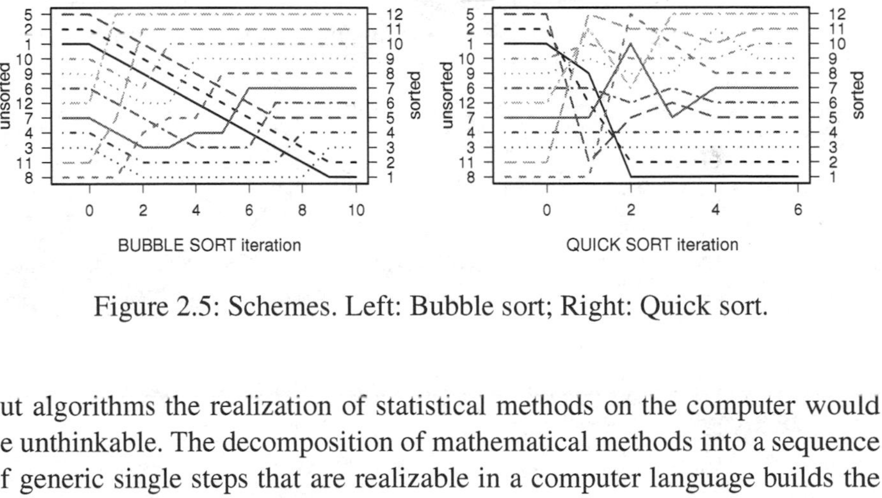

Figure 2.5: Schemes. Left: Bubble sort; Right: Quick sort.

out algorithms the realization of statistical methods on the computer would be unthinkable. The decomposition of mathematical methods into a sequence of generic single steps that are realizable in a computer language builds the basis for today's methodological thinking. However, it should be stressed that here an algorithm is understood as a helping aid for the determination of the solution fixed by the mathematical method. Therefore, for an algorithm, not only the determination of its complexity is important, but also the exactness of the found solution has to be tested. The introduced sorting algorithms always reach the correct solution reliably. However, numerical algorithms often have difficulties with this, dependent on the data situation (cp. the term **condition number** in Sections 2.4.2, 2.4.4.2, and 3.1.2).

The term **Turing machine** introduced in the next section clarifies the decomposition of a mathematical method into generic single steps in a way similar to that of today's machine languages.

2.2.2 *Turing Machines*[8]

2.2.2.1 *Motivation and History: Turing machines*

A Turing machine is an idealized computer named after **A.M. Turing** (1912 - 1954) that specifies the notion of a mechanical or algorithmic calculating machine.

The Turing machine has an external **storage** in the form of an infinitely long tape that is subdivided into single cells each containing one character of a common finite alphabet $\{a_0, a_1, \ldots, a_n\}$.

The machine works in discrete successive **steps**. During each step the information on the tape (one character a_j of the alphabet) is read and processed according to the current machine instruction.

The machine only has a finite number of **instructions**, so-called states, $l_1, l_2, \ldots, l_m$. The input information a_j and the current instruction l_k determine the output information a_{kj}, which overwrites the character a_j. Both a_j and l_k also determine whether the tape is moved to the left, to the right, or not at all. After execution of the instruction the machine takes the new state l_{kj}, and a new step starts.

For given starting information I_a, i.e. a preprinted tape, the Turing machine:

1. May stop after finitely many steps representing the information I_b - in this case the Turing machine is said to be **applicable** to the starting information.
2. May never stop - in this case the Turing machine is said to be **not applicable** to the starting information.

Each Turing machine realizes a fixed **algorithm**. For example, one can construct a Turing machine for the addition of two natural numbers (see below). The larger the problem, the more complex is the corresponding machine.

Turing machines are not physically realized but only used for the exact definition of an algorithm. The Turing machine is a pure "thought experiment," particularly since a computer with infinite storage can never exist. Additionally, as a practical computer the Turing machine would be unusable because of its extremely awkward operation. However, the Turing machine

[8]based on Böhling (1971, pp. 34 – 40, 48 – 49)

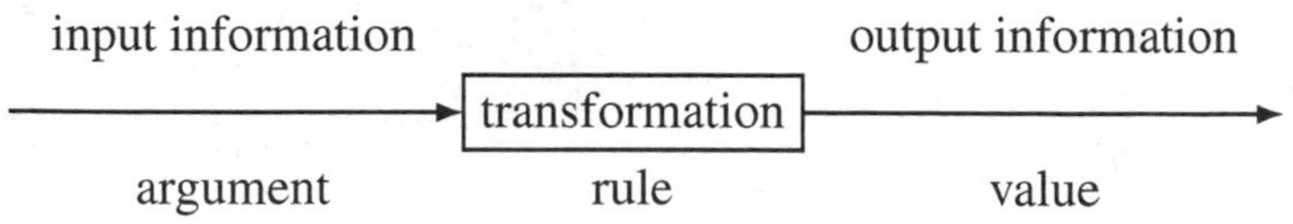

Figure 2.6: Turing machine as character transformation.

is of great theoretical and methodological value for the practical construction of computers and for the conceptual clarification of basic problems in mathematics and logic, especially in the theory of algorithms. Moreover, the Turing machine, a so-called **paper computer**, is a good model for today's computers.

Functional Description

A **Turing machine** (TM) is a mechanism for the effective calculation of a function. Such a calculation can be regarded as a **transformation of character strings**. Each given character string (in the domain of the function) is transformed according to a predefined rule into another character string (in the codomain).

Example 2.9: Turing machine as Character String Transformation
The operation + on the natural numbers can be considered as follows:
Given the alphabets $A = \{1,2,3,\ldots,9,0,+\}$ and $B = \{1,2,3,\ldots,9,0\}$, let f be a mapping $f : A^* \to B^*$, where A^* and B^* are the sets of all words with characters from A and B, respectively (finite character strings over A and B). For example, f should transform $1347+90+1$ ($\in A^*$) into 1438 ($\in B^*$). This process should proceed purely mechanically without any "thinking."

The interpretation of arguments and function values is in the domain of people, not of the mechanism.

In order to specify the mechanism we need:

1. An agreement about the usage of characters for the encoding of objects,
2. a storage medium for these characters, and
3. a facility for the execution of operations on these characters.

The calculation process is illustrated in Figure 2.6.

2.2.2.2 *Theory: Definition*

For a TM, storage medium and operation have a special form. Since character strings are the material to be worked upon, a **tape** can be used as the storage

	+	f	a	K	r	-	b	

Figure 2.7: Form of storage of a Turing machine.

medium. This tape is subdivided into **cells** which represent work areas (see Figure 2.7).

The transformation of character strings is carried out stepwise: The TM always considers exactly **one** cell, the so-called **working cell**. The content of this cell and the state (i.e. the actual instruction) of the TM determine the operation to be performed and the next state.

One operation consists of two suboperations:

1. Change of character in working cell,
2. transfer to a new cell.

The transformation of character strings consists of a finite sequence of operations of this kind.

Let us discuss the **technical construction** and the **operation of a TM** in more detail. Let $A = \{a_0, a_1, \ldots, a_n\}$ be the alphabet from which a valid character string is to be built. a_0 is of special importance. It represents the blank, written as b.

The machine consists of:

1. The control unit CU with:
   a. An input register IR,
   b. an instruction (or command) register CR,
   c. an operation register OR,
   d. a cell selector CS.
2. The tape unit TU with:
   a. A read/write head RW,
   b. the tape guide apparatus TG.
3. The program storage PS.
4. The information storage, the tape.

The purpose of a TM is the planned transformation of words from A^* into words from (w.l.o.g. also) A^*.

Explanations

4: On the **tape** there are the words on A to be worked upon. The tape is decomposed into cells. Into each of these cells "fits" exactly one character from A. A cell is called empty if it contains the character $a_0 =$ b. The tape is arbitrarily extendable to the right. The cells are numerated, starting from zero

	i	a_0	a_{i0}	v_{i0}	l_{i0}
ith instruction	$\vdots$	$\vdots$	$\vdots$	$\vdots$	$\vdots$
	i	a_n	a_{in}	v_{in}	l_{in}

Figure 2.8: TM instruction as quintuple.

at the left. Apart from cells with b, the input only consists of finitely many cells with elements from A.

3: A TM works according to a **Turing program** TP, which consists of finitely many instructions. The instructions are stored in cells of the PS. The cells are numerated starting with 1. Each instruction is uniquely determined by the number of the cell in which it is stored. The number of such cells can be arbitrarily enlarged.

2: The **read/write head** RW reads and writes exactly one cell. Writing means overwriting, writing a b means erasing. The cell underneath the RW is called working cell WC. The symbol $<\ >$ is used to identify the content of a part of the TM. There are three different shift instructions S:

- R directs the cell right to the WC under the RW.
- L directs the cell left to the WC under the RW.
- N (neutral) leaves the cell under the RW as WC.

1: The **cell selector** CS chooses the instruction that is to be executed next. It contains the number i of the cell in which this instruction is stored. In an alphabet with the characters $a_0, \ldots, a_n$ a complete instruction contains $n+1$ implications (subinstructions) of the form:

$< \mathrm{IR} >= a_j \rightarrow$ write a_{ij} into WC and
write v_{ij} (R, N or L) into OR and
load the l_{ij}th instruction into CR.

The ith instruction is briefly written as $n+1$ times a quintuple (see Figure 2.8). In this way, it is fully determined what to do for each possible content of the input register IR.

Execution of a TM

The working of a Turing machine can be summarized as follows: The TM works in steps. Each step consists of five so-called **operating cycles** (see Figure 2.9):

1st cycle: The content of a cell with number i as indicated in the cell selector

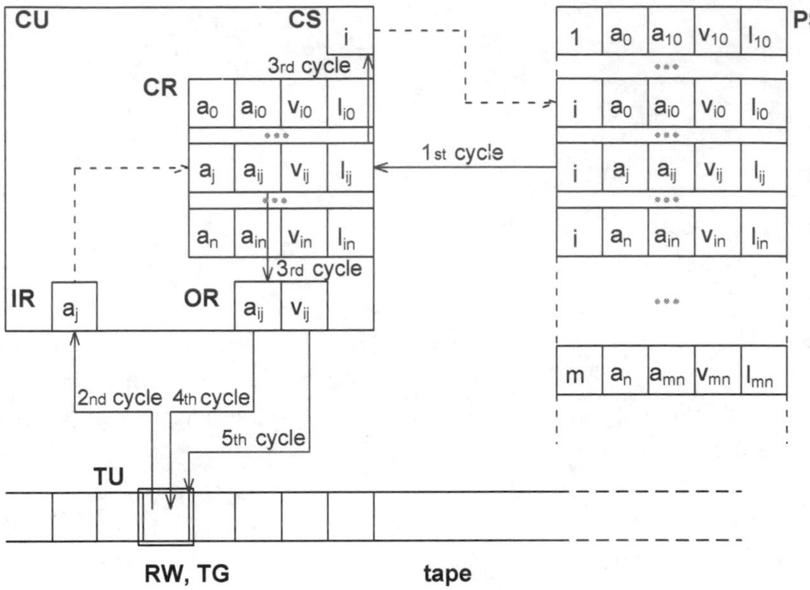

Figure 2.9: Scheme of a Turing machine.

is brought into the command register. The TM takes the state i:

$$(\langle \text{CS} \rangle = i) \longrightarrow (\langle i\text{th cell} \rangle \longrightarrow \text{CR})$$

2nd cycle: The RW reads the character in the WC and the TU gives it to the IR:

$$\langle \text{WC} \rangle \longrightarrow \text{IR}$$

3rd cycle: The CU determines the operation $a_{ij}v_{ij}$ and the next instruction l_{ij} writes the operation into the OR and the number l_{ij} of the next instruction into the CS:

$$[\langle \text{IR} \rangle ; \langle \text{CR} \rangle] \xrightarrow{\text{CU}} [\langle \text{OR} \rangle ; \langle \text{CS} \rangle]$$

4th cycle: The TU takes the character a_{ij} from the OR and the RW writes it into the WC:

$$\langle \text{OR} \rangle \ni a_{ij} \longrightarrow \text{WC}$$

5th cycle: The TU takes the shift character v_{ij} from the OR and the TG executes the shift operation:

$$\langle \text{OR} \rangle \ni v_{ij} \longrightarrow \text{TG}$$

The TM regularly stops iff the content of the CS is zero. The TM stops irregularly, if

1. the zeroth cell of the tape is WC and $< \text{OR} > \ni$ `L` (the TM is blocked).
2. $< \text{CS} >$ is empty.

In the beginning, e.g., $< \text{CS} >= 1$, i.e. the program starts with instruction 1.

The TM as Mathematical Structure
A TM is uniquely determined by the following items:

1. The set of states (= set of instructions) I,
2. the character alphabet A,
3. the Turing program P, where a subinstruction is corresponding to a mapping $I \times A \to A \times S \times I$,
4. the starting state $l_0 \in I$,
5. the set of final states $F \subseteq I$.

Therefore, a TM can be represented by a quintuple: $M = (I,\ A,\ P,\ l_0,\ F)$. Notice that the shift alphabet S is equal to {`R`, `N`, `L`} for all Turing machines. This is the reason why it is not mentioned in the characterization.

Binary Turing machines
An important special case of Turing machines is machines over the binary alphabet {`b`, `I`}. Such Turing machines are particularly similar to today's computers.

TM as a Model for a Computer:
1. Information: The binary TM operates on a binary alphabet $A =$ {`b`, `I`}, and sequences of elements of A are used to encode other objects by adequate interpretation. E.g., in the example program "add" (see below) a natural number n is encoded as $n+1$ times `I`, i.e. a sequence of K times `I` is interpreted as the natural number $K-1$. Today's computers also use a binary alphabet (power, no power) and the characters utilized by the human user are encoded by the machine itself into binary sequences.
2. Storage: The storage of a TM is absolutely comparable with the storage of a computer. It consists of a linear arrangement of single storage spaces. Each space has a fixed address and a variable content, which is determined by the human user by means of a program.
3. Control unit: The control unit controls and organizes the process of the whole activity of the machine.
4. Instructions: The instructions of a TM have the structure of quintuples. Instructions of a computer also have tuple structure. For example, on the ancient IBM System/370 (1976): $AER\ R_1, R_2$ denoted
add (A) "short normalized" (E) two numbers whose addresses are stored in the registers (R) with the numbers R_1 and R_2.

Table 2.3: Turing machine “add”

1	b	b	R	2
1	I	I	R	**0**
2	b	I	R	3
2	I	I	R	2
3	b	b	L	4
3	I	I	R	3
4	b	b	R	**0**
4	I	b	L	5
5	b	b	R	**0**
5	I	b	L	6
6	b	b	R	**0**
6	I	I	L	6

The term **short normalized** corresponds to single-precision normalized floating-point inputs and output (see Section 2.3).

5. Programs: The programming language of a TM is very simple. A program mainly consists of a sequence of instructions that are logically related, building a sensible entity this way. On a computer this corresponds to the programming in so-called **machine languages**.

2.2.2.3 Example: Addition

Example 2.10: Turing machine “add”
Let $A = \{a_0 = \texttt{b}, a_1 = \texttt{I}\}$, let the Turing Program P be given by Table 2.3, and let the content of the input tape be “b I b I b b ...”. The result of the application of the program to this input is given in Table 2.4.

Notice that the position of the read/write head is indicated by a bar over the corresponding character.

In order to characterize the purpose of the different statements in the above TM, please realize that instruction 2 identifies the first summand and instruction 3 the second. Instructions 4 and 5 both erase one I. One I is erased because of the coding of the natural numbers (see above, a zero is represented by one I), the other because of the filling of the gap between the two summands by one I in instruction 2. Instruction 6 stops the program regularly.

In passing, notice that some subinstructions are redundant! **Which?**

Table 2.4: Execution Progress of the Turing machine "add"

$\overline{\texttt{b}}$	I	b	I	b	b	b	...	1 (instruction)
b	$\overline{\texttt{I}}$	b	I	b	b	b	...	2
b	I	$\overline{\texttt{b}}$	I	b	b	b	...	2
b	I	I	$\overline{\texttt{I}}$	b	b	b	...	3
b	I	I	I	$\overline{\texttt{b}}$	b	b	...	3
b	I	I	$\overline{\texttt{I}}$	b	b	b	...	4
b	I	$\overline{\texttt{I}}$	b	b	b	b	...	5
b	$\overline{\texttt{I}}$	b	b	b	b	b	...	6
$\overline{\texttt{b}}$	I	b	b	b	b	b	...	6
b	$\overline{\texttt{I}}$	b	b	b	b	b	...	0

Please check the subinstructions that end with the instruction number 0. The redundant subinstructions were only included for the sake of completeness. In general, they are omitted.

We can summarize the result of the above computation of the TM as follows:

$$P(\overline{\texttt{b}}\texttt{IbI}) = \texttt{b}\overline{\texttt{I}}.$$

More generally, the program provides

$$P(\overline{\texttt{b}}\ \underbrace{\texttt{I}\ldots\texttt{I}}_{(x+1)\text{-times}}\ \texttt{b}\ \underbrace{\texttt{I}\ldots\texttt{I}}_{(y+1)\text{-times}}\ \texttt{bb}\ldots) = \texttt{b}\ \underbrace{\overline{\texttt{I}}\ldots\texttt{I}}_{(x+y+1)\text{-times}}\ .$$

This is the reason why this program is called **add**. Which two numbers were added in the above example?

2.2.2.4 Practice and Simulation

In the internet there are different simulators for Turing machines, which, however, utilize different "programming dialects."

Example 2.11: TM Example 1
See `http://ironphoenix.org/tril/tm/` for a Turing machine simulator. Notice the very different notation of the instructions relative to the above notation, e.g. for the Turing machine "subtraction" in Table 2.5. The instructions are read as follows. The first subinstruction means that if in instruction `1` the input `_` (= `b`) is observed, then the next instruction to be executed is again instruction `1`, a `_` has to be written on tape and the tape has to be shifted to the right. Obviously, this is not a binary machine. The alphabet includes `_`, `1`, `-`, `=`. Moreover, left and right shifts are coded <, >.

Table 2.5: Turing machine "subtraction"

```
1,_ 1,_,>
1,1 1,1,>
1,- 1,-,>
1,= 2,_,<
2,1 3,=,<
2,- H,_,<
3,1 3,1,<
3,- 4,-,<
4,_ 4,_,<
4,1 1,_,>
```

Table 2.6: Turing machine "adding"

```
0 _ _ R 0
0 1 1 R 1
1 _ 1 R 2
1 1 1 R 1
2 _ _ L 3
2 1 1 R 2
3 _ _ L 3
3 1 _ L 4
4 _ _ R stop
4 1 1 L 4
```

Example 2.12: TM Example 2
See `http://www.turing.org.uk/turing/scrapbook/tmjava.html` for another Turing machine simulator. Notice that here the notation of the instructions is similar to our notation above, e.g. for the Turing **adding machine** in Table 2.6.

Outlook: Turing machines
Turing machines are surely a very old idea (1936). Moreover, Turing machines in their pure form are not utilized in practice. Nevertheless, the basic ideas of today's computers can be easily represented by Turing machines. This constitutes their value even today. If you are interested in an actual view on Turing machines you may want to look into, e.g., Wiener et al. (1998).

Summary
In this section we introduced algorithms and Turing machines. **Algorithms** are basic for today's thinking about computer programming. Algorithms are

characterized by their complexity and their accuracy. We introduced complexity measures and discussed the complexities of different algorithms for solving the sorting problem. The discussion of accuracy is postponed to the next sections. **Turing machines** are a purely theoretical vehicle for the realization of algorithms. Nevertheless, Turing machines have astonishing similarities to actual computers. As an example we introduced a binary Turing machine for addition.

2.3 Floating-Point Computations: How Does a Computer Compute?[9]

2.3.1 Motivation and History: Floating-Point Computations

The first mechanical binary programmable computer, the Z1 built by Konrad Zuse in 1938, used 22-bit floating-point numbers. The first commercial computer with floating-point hardware was the Z4 constructed by Zuse in 1942-1945. The IBM 704 followed 10 years later in 1954. In the next decades floating-point hardware was typically optional, and computers with this option were scientific computers. It was only in the 1990s that floating-point units (FPUs) became ubiquitous.[10] They are now an integral part of every modern CPU. However, there are certain applications (embedded devices, for example) where an FPU is still not standard. Often, the functions of the FPU are then emulated in software.

In the next few paragraphs we will introduce the basic principles of arithmetic for **floating-point numbers**. One might think that these principles should hardly be interesting since all modern desktop computers have CPUs that support these basic arithmetic operations in hardware. We hope that after working through this section, it will be clear that knowing the underlying principles of floating-point arithmetic is invaluable if one wants to study error propagation in numerical algorithms.

For brevity and clarity we will focus on so-called **single-precision** floating-point arithmetic. The same ideas apply to the nowadays more common double-precision and multiple-precision floating-point arithmetic as well (see any of the mentioned literature for details). We will also avoid a discussion of **fixed-point arithmetic** which assumes that the decimal point is always at a fixed position. The most common type of fixed-point numbers on computers are the integers. Here there are no digits to the right of the decimal point (cp. Section 6.2).

In fact, as computers have become faster, it is usually best to use floating-

---

[9]Partly based on Knuth (1998, pp. 214 – 223); Wilkinson (1963, pp. 7 – 13).

[10]cp. http://en.wikipedia.org/wiki/Floating_point

point numbers unless the domain is well defined ahead of time and the expected time savings of avoiding the complexity of floating-point operations are necessary.

So what are **floating-point numbers**? The basic idea is to track the location of the decimal point dynamically and adjust it as necessary rather than storing a fixed number of digits to the left and right of the decimal point (or conversely, fix the location of the decimal point). Using this idea, it is possible, given a fixed number of bits, to encode a larger range of numbers than using the fixed-point approach. The representation for decimal numbers is not unlike scientific notation. The advantages of this system become clear when we consider the following example. Say we want to estimate the number of people struck by lightning each year. We know that the probability π of being struck by lightning in a given year is $1.35 \cdot 10^{-6}$ and that the world population N is $6.791 \cdot 10^9$. To calculate the number of people struck by lightning, we need to multiply π by N. If we wanted to do this using a fixed-point representation of the numbers, we would need a number type with at least 18 significant digits. On the other hand, a floating-point type with four significant digits and a suitably sized exponent would suffice to form the product. So floating-point numbers allow us to represent a larger range of numbers using the same number of bits compared to fixed-point types. For this we sacrifice precision at the far end of the representable range of numbers.

Definition 2.7: Floating-Point Numbers
We generally consider numbers with **base** b, **excess** (excess exponent) q, **maximum exponent** $e_{\max}$ and **mantissa length** (= number of represented digits) p. A number x is then represented by a pair (e, f) such that

$$x \approx (e, f) = f \cdot b^{e-q}, \tag{2.1}$$

where $e \in \{0, \ldots, e_{\max}\}$ is the integer-valued **exponent** and f, the **mantissa** or **significand**, is a number between -1 and $+1$ with a fixed number of digits, i.e. $|f| < 1$.

Note that the term **mantissa** is also used in connection with decimal logarithms. There, it denotes the digits after the decimal point. In fact, these digits determine the digit structure of such numbers, whereas the places before the decimal point determine the exponent. As an example, the decimal logarithm of $2.5566 \cdot 10^5$ is equal to $\log_{10}(2.5566) + \log_{10}(10^5) = 0.4077 + 5 = 5.4077$. The digits after the decimal point, i.e. 0.4077, represent the logarithm of 2.5566, i.e. of the term before the exponential part of the number. The digits before the decimal point, i.e. 5, represent the exponent of 10^5.

In the definition, the decimal point appears far left in the, e.g., decimal

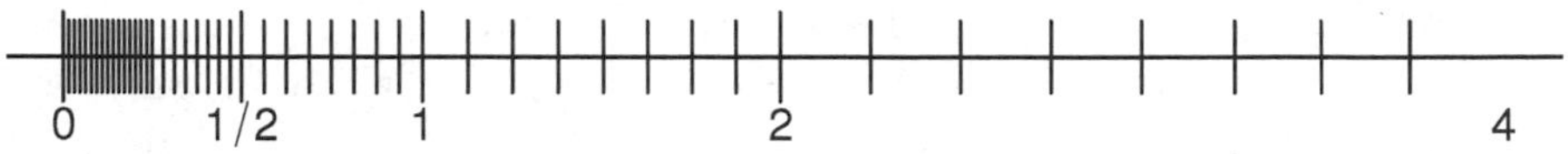

Figure 2.10: Resolution of floating-point numbers. Each dash marks one of the 48 numbers greater than or equal to zero that a ($b = 2$, $p = 4$, $q = 2$, $e_{\max} = 4$)-floating-point number can represent.

representation of f, and since we deal with p-digit numbers, $b^p f$ is an integer so that

$$-b^p < b^p f < b^p. \tag{2.2}$$

While any b is possible, in practice we commonly use $b = 2$ or $b = 10$. The corresponding floating-point numbers are called **binary floating-point numbers** for $b = 2$ and **decimal floating-point numbers** for $b = 10$.

One important thing to note about all floating-point numbers is that their precision decreases with increasing exponent. While the number of digits represented stays constant, the smallest difference between two numbers that can be represented in a given floating-point format depends on the exponent of the numbers. This is illustrated in Figure 2.10, which corresponds to the following example.

Example 2.13: Resolution of Floating-Point Numbers
Let $b = 2$, $p = 4$, $q = 2$, and $e_{\max} = 4$. Then, the corresponding floating-point type can represent the following numbers greater or to equal zero:

$$.b_1b_2b_3b_4 \cdot 2^{e-2},$$

where $e - 2$ has the values $2, 1, 0, -1, -2$ and b_i the values 0 or 1.
Note that $.b_1b_2b_3b_4 \cdot 2^4$ takes on every integer between 0 and 15. Division by $2^2, 2^3, \ldots, 2^6$ leads to the following representable numbers:
$0, \frac{4}{16}, \frac{8}{16}, \ldots, \frac{60}{16}, 0, \frac{2}{16}, \ldots, \frac{30}{16}, 0, \frac{1}{16}, \ldots, \frac{15}{16}, 0, \frac{1}{32}, \ldots, \frac{15}{32}, 0, \frac{1}{64}, \ldots, \frac{15}{64}$.
Because of multiple representations, the following 48 numbers are left representable:
$0, \frac{1}{4}, \frac{2}{4}, \ldots, \frac{15}{4}, \frac{1}{8}, \frac{3}{8}, \ldots, \frac{15}{8}, \frac{1}{16}, \frac{3}{16}, \ldots, \frac{15}{16}, \frac{1}{32}, \frac{3}{32}, \ldots, \frac{15}{32}, \frac{1}{64}, \frac{3}{64}, \ldots, \frac{15}{64}$.

Let us continue the example "being struck by lightning" we gave on page 46: If an excess of $q = 50$ and a 4-digit decimal mantissa are used, we have:

$$N = (60, +0.6791),$$
$$\pi = (43, +0.1350).$$

binary32 floating-point numbers are one type of binary floating-point numbers. Historically they usually have been referred to as **single-precision** floating-point numbers and correspond to the `float` type of C, C++, and Java and the `single` type of Matlab and Pascal.

Single-precision floating-point numbers are made up of 4 bytes, which are filled as follows:

`±eeeeeee`	`efffffff`	`ffffffff`	`ffffffff`

The layout consists of the sign (= 1 bit) and the 8-bit exponent of the number followed by 24 bits for the mantissa, of which only 23 are stored because the highest bit (if $b = 2$) is always set, since all stored numbers are normalized. We will shortly define a **normalized** floating-point number. To simplify the representation, the hexadecimal ($b = 16$) notation is usually used instead of binary notation ($b = 2$).[11]

The excess q is 127 so that exponents in the range of -126 to 127 can be represented because exponent values of 0 and 255 are treated as special. Note that because of the excess we can dispense with the sign of the exponent.

The more common **binary64 floating-point numbers**, often called **double-precision floating-point numbers**, again have $b = 2$ but use 11 bits to encode the exponent and 52 bits for the significand.

Definition 2.8: Normalized Calculation
A floating-point number (e, f) is called **normalized** if the most significant digit in the representation of f is not zero so that

$$\frac{1}{b} \leq |f| < 1; \tag{2.3}$$

or if $f = 0$ and $e = 0$ takes the smallest possible value. In the latter case (e, f) represents zero.

In our example with $0 \leq e < 256$, the only representation of zero is $(e, f) = (0, 0)$. Moreover, the special exponent value $e = 255$ is used to encode **special** values such as $\pm\infty$, not a number (NaN) and others.

For two normalized floating-point numbers, we can decide which one is larger by first comparing their exponents, and second, if both exponents are equal, additionally comparing the mantissas.

Most floating-point programs utilize normalized numbers. Inputs are assumed to be normalized and outputs are always normalized. This convention

[11]Note that these two representations are equivalent since any finite binary number can be exactly represented as a finite hexadecimal number.

hinders us in representing some very small numbers. For example, the value $(e, f) = (0, 0.01)$ cannot be normalized without producing a negative exponent. These numbers are called **subnormal numbers**, but we will not consider them any further because if our algorithms have to deal with them, we lose speed, uniformity, and the capability to relatively easily determine upper limits for the relative errors in floating-point calculations (see Section 2.4).

Since floating-point arithmetic only gives approximate results, in order to distinguish them from the exact operations in the next sections, the following symbols are used for the finite precision floating-point operations addition, subtraction, multiplication, and division with appropriate round-off:

$$\hat{+}, \hat{-}, \hat{*}, \text{ and } \hat{\div}.$$

2.3.2 *Theory: Floating-Point Operations*

2.3.2.1 *Floating-Point Addition*

The following detailed descriptions of the normalized floating-point operations are machine independent. The first (and by far most difficult) algorithm describes the floating-point addition:

$$(e_u, f_u) \hat{+} (e_v, f_v) = (e_w, f_w). \tag{2.4}$$

Algorithm 2.4 Addition of Two Floating-Point Numbers

Require: Base b, excess q, two p-digit normalized floating-point numbers $u = (e_u, f_u)$, $v = (e_v, f_v)$

if $e_u < e_v$ **then**
    swap u and v.
end if
$e_w \leftarrow e_u$
if $e_u - e_v > p$ **then**
    return u
end if
$f_v \leftarrow f_v / b^{e_u - e_v}$
$f_w \leftarrow f_u + f_v$
return Normalized (e_w, f_w)

Adding two floating-point numbers amounts to first choosing the exponent e_w as the maximum of the two exponents e_u and e_v. W.l.o.g. let this maximum be e_u. Then move the decimal point of the mantissa f_v to match the new exponent e_w and add f_u and the adjusted f_v, storing the result in a $2p$-digit temporary mantissa f_w that is reduced to p digits by the normalization step. The double-width accumulator is important to avoid round-off errors. Algorithm 2.4 is a formal description of the just described method. Still, it is missing the normalization step.

Normalization is more difficult because many corner cases have to be dealt with. Let us assume we want to normalize a floating-point number (e, f). Initially, if the number is zero, we can immediately return. Then, if the mantissa is smaller than $1/b$, i.e. the most significant digit is 0, we multiply f by b and subtract one from the exponent e until the first digit is nonzero. Next, we check if the mantissa is larger than 1, and if so, repeatedly divide it by b until it is smaller than 1. Again, we have to adjust the exponent accordingly by adding 1 for each time we divided by b.

Now we must round the result because the mantissa might have more than p digits. If the not-representable rest $(b^p \cdot f) \bmod 1$ of f is smaller than 0.5, we round downward, if it is larger than 0.5, we round upward, and should it be 0.5, we round to the closest p-digit even number. Rounding can lead to situations where $|f| \geq 1$ after rounding. So we have to repeat the check again until we obtain a valid f. Finally, before we return the normalized number, we have to check that the exponent e is not out of range. If it is, we return an error, and otherwise, we have obtained the normalized representation of (e, f). For a formal description see Algorithm 2.5.

Example 2.14: Sample Rounding Examples
Let $b = 10$, $p = 1$, $f = 0.55$. Then, $(b^p \cdot f) \bmod 1 = (10^1 \cdot 0.55) \bmod 1 = 0.5 = r$. Since $b^p \cdot f + r = 10^1 \cdot 0.55 + 0.5 = 6$ is even, we obtain $f + r/b^p = 0.55 + 0.5/10^1 = 0.6$ as the rounded result. The result of this somewhat strange rounding rule is that numbers 5.5 and 6.5 are both rounded to 6 if $b = 10$ and $p = 1$. More generally, the sequence $1.5, 2.5, 3.5, 4.5, 5.5, 6.5, 7.5, 8.5$ is rounded to $2, 2, 4, 4, 6, 6, 8, 8$.

Example 2.15: Addition of Floating-Point Numbers
Algorithm 2.4 for **floating-point addition** is demonstrated by the following examples in 4-digit ($b = 10$, $p = 4$, $q = 10$) floating-point arithmetic. In all examples we want to calculate the sum

$$(e_w, f_w) = w = u \hat{+} v = (e_u, f_u) \hat{+} (e_v, f_v).$$

Let $u = (14, 0.5153)$ and $v = (4, 0.4238)$, then $e_u - e_v = 14 - 4 = 10 > 4 = p$, and thus the result is u.

Next, consider $u = (14, 0.3614)$ and $v = (11, 0.4865)$. We set $e_w = 14$ and shift f_v 3 digits to the right to obtain 0.0004865. The sum of the two mantissa is $f_w = 0.3614 + 0.0004865 = 0.3618865$, which after rounding yields the result $w = (14, 0.3619)$.

If we take $u = (14, 0.6418)$ and $v = (14, 0.7158)$, then $e_w = 14$ and f_v does not need to be shifted. The non-normalized sum of the two mantissa is $f_w = 1.3576$. To normalize it, we need to shift it right by one yielding $e_w = 15$ and $f_w = 0.13576$, which must be rounded to obtain the final result of $w = (15, 0.1358)$.

We can also subtract two numbers using the algorithm. To see this, set $u = (6, 0.6717)$ and $v = (6, -0.6713)$. The exponent of w is $e_w = 6$ and the mantissa is $f_w = 0.6717 + (-0.6713) = 0.0004$. Again, the mantissa is not normalized. We need to shift right by 3 decimal places to obtain the normalized result $w = (3, 0.4000)$.

Finally, let $u = (6, 0.1000)$ and $v = (5, -0.9997)$. $e_w = 6$ and $f_w = f_u + f_v = 0.00003$. Here, the non-normalized mantissa of the result is even smaller than the smallest number representable in the chosen precision of $p = 4$ digits. Normalization then gives $w = (2, 0.3000)$.

Notice that in all the previous examples we assumed that the sum of f_u and f_w was calculated with exact arithmetic. This requires $2p$ significant digits in the worst case. But are all of these digits trustworthy? Reexamine our example of $u \hat{+} v$ with $u = (6, 0.6717)$, and $v = (6, -0.6713)$. The result we obtained was $w = (3, 0.4000)$. But how many of f_w's digits are trustworthy? Only the first digit really conveys any meaning. To see this, let us say the value we wanted to represent with u was $0.67174999 \cdot 10^6$ and the value of v was really $-0.67132122 \cdot 10^6$. Then the true sum is $0.00042877 \cdot 10^6 = 0.42877 \cdot 10^3$. So we have made an error of $0.02877 \cdot 10^3$ in our calculation without even knowing about it. This effect is called **cancellation** and is caused by subtracting two numbers of similar magnitude. If the floating-point representation of the two numbers is exact, no large errors can occur and we speak of **benign cancellation**, but if they are in fact polluted by rounding errors of previous calculations all significant digits can be useless (**catastrophic cancellation**). Sometimes it can help to transform an expression into an algebraically equivalent form to avoid cancellation. We will see an example of this after we have covered multiplication.

Algorithm 2.5 Normalization of a Floating-Point Number

Require: Base b even, excess q, maximum exponent e_{max}, p-digit normalized floating-point number (e, f)

1: **if** $f = 0$ **then**
2:    **return** $(0, 0)$
3: **end if**
4: **while** $|f| < 1/b$ **do** {Shift f left}
5:    $f \leftarrow f \cdot b$;
6:    $e \leftarrow e - 1$;
7: **end while**
8: **repeat**
9:    **while** $|f| \geq 1$ **do** {Shift f right}
10:      $f \leftarrow f/b$;
11:      $e \leftarrow e + 1$;
12:    **end while**
13:    [Round:] $r \leftarrow (b^p \cdot f) \bmod 1$
14:    **if** $r < 0.5$ **then**
15:      $f \leftarrow f - b^{-p} \cdot r$
16:    **else if** $r > 0.5$ **then**
17:      $f \leftarrow f + b^{-p} \cdot (1 - r)$ {$|f|$ now possibly > 1}
18:    **else** {r is exactly 0.5; see Example 2.14}
19:      **if** $b^p \cdot f + r$ is even **then**
20:        $f \leftarrow f + r/b^p$ {$|f|$ now possibly > 1}
21:      **else**
22:        $f \leftarrow f - r/b^p$
23:      **end if**
24:    **end if**
25: **until** $|f| < 1$
26: **if** $e < 1$ **then**
27:    **return** Exponent underflow
28: **end if**
29: **if** $e > e_{max}$ **then**
30:    **return** Exponent overflow
31: **end if**
32: **return** (e, f)

2.3.2.2 Floating-Point Multiplication

After having covered the hard part of normalizing and adding or subtracting floating-point numbers, we are left with the tasks of multiplication and division. Both algorithms are much easier than the algorithm for addition (see Algorithms 2.6 and 2.7).

Algorithm 2.6 Multiplication of Two Floating-Point Numbers

Require: Base b, excess q, two p-digit normalized floating-point numbers $u = (e_u, f_u)$, $v = (e_v, f_v)$

1: $e_w \leftarrow e_u + (e_v - q)$
2: $f_w \leftarrow f_u \cdot f_v$
3: **return** Normalized (e_w, f_w)

Algorithm 2.7 Division of Two Floating-Point Numbers

Require: Base b, excess q, two p-digit normalized floating-point numbers $u = (e_u, f_u)$, $v = (e_v, f_v)$

1: **if** $e_v = 0$ **then**
2:     **return** Division by zero
3: **end if**
4: $e_w \leftarrow e_u - e_v + q + 1$
5: $f_w \leftarrow (b^{-1} \cdot f_u) / f_v$
6: **return** Normalized (e_w, f_w)

In both algorithms the temporary result f_w of the multiplication or division of the mantissa is stored in a so-called double-precision accumulator, that is f_w has $2p$ digits that are truncated to p during the normalization step.

For multiplication, we calculate $e_w = e_u + e_v$ and $f_w = f_u \cdot f_v$. Notice that correction by the excess q is only needed once so that $e_v - q$ has to be added instead of e_v.

We also know that

$$\frac{1}{b^2} \leq |f_u \cdot f_v| < 1$$

unless f_u or f_v is zero. This means that we never have cancellation in multiplication. Combining this with what we know about cancellation from the previous section, we can now give an example where rearranging an expression into an algebraically equivalent one will yield a numerically more stable result.

Consider two floating-point numbers u and v that are exact. We wish to calculate $u^2 - v^2$. If we calculate $u \cdot u$ and $v \cdot v$, we may lose up to p digits of precision and catastrophic cancellation may occur in the final subtraction. On the other hand, if we rewrote $u^2 - v^2 = (u - v) \cdot (u + v)$, we know that we would only observe benign cancellation in the sum and difference and no cancellation in the multiplication. Therefore, we would prefer the second form $((u-v)\cdot(u+v))$ over the first if we wish to calculate $u^2 - v^2$ using floating-point arithmetic.

Example 2.16: Multiplication of Floating-Point Numbers
We wish to calculate

$$u \hat{*} v = (e_u, f_u) \hat{*} (e_v, f_v) = (e_w, f_w) = w$$

in decimal floating-point ($b = 10$, $p = 4$, $q = 10$).

Let $u = (6, 0.9132)$ and $v = (16, 0.5135)$, then $e_w = 6 + 16 - 10 = 12$ and $f_w = 0.46892820$, which is rounded to $(12, 0.4689)$.

If we set $u = (7, 0.1713)$ and $v = (6, 0.1214)$, we obtain $e_w = 7 + 6 - 10 = 3$ and $f_w = 0.1713 \cdot 0.1214 = 0.02079582$. Normalizing yields the answer of $w = (2, 0.2080)$.

For division, we start by calculating $e_w = e_u - e_v$. The double-precision accumulator is first filled by f_u in the first p digits, and zeros in the last p digits. Then, the content of the accumulator is shifted one place to the right and e_w is increased by one accordingly. Finally the accumulator is divided by f_v and normalized to form the p-digit result. If $|f_u| > |f_v|$, then the absolute value of the accumulator automatically lies between $1/b$ and 1, and a normalization is not necessary.

Example 2.17: Division of Floating-Point Numbers
We wish to calculate

$$u \hat{\div} v = (e_u, f_u) \hat{\div} (e_v, f_v) = (e_w, f_w) = w$$

in decimal floating-point ($b = 10$, $p = 4$, $q = 10$).

Let $u = (4, 0.9137) = 0.9137 \times 10^{-6}$ and $v = (8, 0.1312) = 0.1312 \times 10^{-2}$, then $e_w = 4 - 8 + 10 + 1 = 7$ and $f_w = 0.0913\,7000/0.1312 = 0.6964\,17\ldots$. The calculated quotient is then $w = (7, 0.6964) = 0.6964 \times 10^{-3}$.

If we set $u = (14, 0.1235) = 0.1235 \times 10^4$ and $v = (4, 0.9872) = 0.9872 \times 10^{-6}$, then $e_w = 14 - 4 + 10 + 1 = 21$ and $f_w = 0.0123\,5000/0.9872 = 0.0125\,101\ldots$. The calculated quotient is then $w = (20, 0.1251) = 0.1251 \times 10^{10}$.

In all the previous examples, we have never encountered an **exponent over-** or **underflow**. While it is relatively rare that an over- or underflow of the exponent is encountered during addition, the risk of this happening during multiplication is much larger. This stems from the fact that in multiplication the exponents are summed and not increased by at most 1. In the past, it was customary to disable underflow protection in numerical codes due to speed considerations. This should be avoided at all costs because it can lead to subtly incorrect results that are hard to spot. Especially in multiplication, this generally causes a severe loss of precision (indeed, of all significant digits). This can lead to unexpected effects. For example, there are situations in which $(u \hat{*} v) \hat{*} w$ is zero but $u \hat{*} (v \hat{*} w)$ is not (and may even be quite large), since $(u \hat{*} v)$ leads to an exponent underflow, whereas $u \hat{*} (v \hat{*} w)$ can be calculated with valid exponents. Similarly, one can find, e.g., positive numbers a, b, c, d, and y so that

$$(a \hat{*} y \hat{+} b) \hat{\div} (c \hat{*} y \hat{+} d) = 0$$

and

$$(a \hat{+} b \hat{\div} y) \hat{\div} (c \hat{+} d \hat{\div} y) \approx \frac{1}{3} \tag{2.5}$$

if exponent underflow occurs.

In R, this is true, e.g., for $a = 1$, $b = 2$, $c = 3$, $d = 4$, and $y = 10^{308}$. R handles the case of an overflow by setting the value to `Inf` (for infinity) and the case of an underflow by setting the value to zero. The maximal and minimal representable numbers on the current machine are given by the values of `double.xmax` and `double.xmin` in the object `.Machine`. So, for the first expression given above, $a \hat{*} y + b$ is still representable but $c \hat{*} y$ is `Inf`, hence, we get zero as the result of the division. If we choose $y = 10^{309}$ we get a division of an infinitely large number by another infinitely large number and R returns `NaN`, i.e. undefined.

Though it is well-known that floating-point programs do not calculate exactly, such differences as in Equation 2.5 are not expected by most people, particularly since a, b, c, d, and y are all positive. Thus, exponent underflow should be taken seriously and it is negligent to ignore it!

Many of the pitfalls of floating-point numbers are summed up in an excellent article by Goldberg (1991). Although this paper is quite old, its content is as relevant today as it was 20 years ago.

2.3.3 Summary and Outlook

Summary

In this section we introduced floating-point numbers together with their basic arithmetical operations addition, subtraction, multiplication, and division as they are used in today's computers. In particular, we discussed normalization of floating-point numbers and also the effect of cancellation, underflow, and overflow.

Outlook
Though the discussed floating-point representation is already relatively old (IEEE 754-1985; see, e.g., http://en.wikipedia.org/wiki/IEEE_754-1985), it is still standard (IEEE 754-2008; see, e.g. http://en.wikipedia.org/wiki/IEEE_754).

2.4 Precision of Computations: How Exact Does a Computer Compute?

2.4.1 Motivation and History: Rounding Errors in Floating-Point Operations[12]

Floating-point calculations are inexact by nature. One of the main problems of numerical methods is the determination of the accuracy of their results. This causes a certain **credibility gap**: We do not know how much we can believe in the results of the computer.
Newcomers often solve this problem by implicitly believing that the computer is unfailing. They tend to believe that all digits of a result are correct. Some disillusioned computer users, though, have the opposite approach: they permanently fear that their results may be meaningless.

Many mathematicians have tried to analyze specific sequences of floating-point operations on the error size. However, this appeared to be so complicated for realistic problems that plausibility arguments were often used instead.

In fact, here we will not try to undertake a complete study of errors in sequences of floating-point operations. Instead, we will restrict ourselves to some basic considerations on the analysis of errors in floating-point calculations.

Basic for today's view of rounding errors on computers was the work of Wilkinson in the early 1960s (Wilkinson, 1965). He uses a simple but generally useful way to express the error behavior of floating-point operations, namely, the concept of **significant digits**, i.e. of relative errors: If an exact number x is represented in a computer by the approximation $x^* = x(1+\delta)$, then the value $\delta = (x^* - x)/x$ is called **relative error** of the approximation.

Definition 2.9: Absolute and Relative Rounding Error
If $\text{round}(x) = x + \Delta x$, then Δx is called **absolute (rounding) error**. If $\text{round}(x) = x(1+\delta_x)$, then δ_x is called **relative (rounding) error**.

[12] Partly based on Knuth (1998, pp. 229 – 238); Wilkinson (1963, pp. 7 – 13).

The following bounds are valid for absolute and relative errors. Note that the bound for the relative error is independent of x.

Theorem 2.5: Error Bounds for rounding
Consider the error caused by rounding with a p-digit mantissa. Let $b^{e-1} \leq |x| < b^e$, then

$$|\Delta x| \leq \frac{1}{2} b^{e-p}, \tag{2.6}$$

and

$$|\delta_x| = \frac{|\Delta x|}{|x|} \overset{*}{\leq} \frac{|\Delta x|}{b^{e-1} + |\Delta x|} \overset{**}{\leq} \frac{\frac{1}{2} b^{e-p}}{b^{e-1} + \frac{1}{2} b^{e-p}} \overset{***}{\leq} \frac{1}{2} b^{1-p}. \tag{2.7}$$

Proof. Inequality (2.6) is obvious for rounding.
In order to prove inequality (2.7), (*), (**), and (***) have to be proved. (*) is valid if

$$|x| \geq b^{e-1} + |\Delta x|.$$

To show this, we distinguish two cases:
1. $b^{e-1} \leq |x| < b^{e-1} + \frac{1}{2} b^{e-p}$,
2. $|x| \geq b^{e-1} + \frac{1}{2} b^{e-p}$.

In case 1, $|x|$ can be represented as

$$|x| = |\text{round}(x)| + |\Delta x| = b^{e-1} + |\Delta x|,$$

since $(1/2)b^{e-p}$ is the maximal absolute error because of inequality (2.6), and therefore a number $b^{e-1} \leq x < b^{e-1} + \frac{1}{2} b^{e-p}$ would be rounded to b^{e-1} with maximal positive error $\frac{1}{2} b^{e-p}$, and a number $-b^{e-1} - \frac{1}{2} b^{e-p} < x \leq -b^{e-1}$ would be rounded to $-b^{e-1}$ with a negative rounding error between 0 and $-\frac{1}{2} b^{e-p}$.

In case 2,

$$|x| > b^{e-1} + \frac{1}{2} b^{e-p} \geq b^{e-1} + |\Delta x|$$

because of inequality (2.6).

The validity of inequality (**) can be shown by equivalence transformations using Equation 2.6. The validity of (***) can be shown by simple calculations. □

In inequality (2.7), in order to find a bound for the relative error from the bound $\frac{1}{2} b^{e-p}$ for the absolute error, we divided by b^{e-1}, the lower bound for a number to be represented with the same exponent e, leading to the bound $\frac{1}{2} b^{1-p}$.

We can utilize inequality (2.7) to estimate the relative error in normalized floating-point calculations in a simple way. Let us start with assuming that errors in the inputs u, v of calculations can be ignored, meaning that **representation errors of the inputs are neglected**.

Then, for **floating-point addition** one can show that

$$u \hat{+} v = (u+v)(1+\delta_{u+v}).$$

This can be easily derived from Algorithm 2.4. There, we demonstrated that addition is realized by building the exact sum, normalizing to a valid mantissa, and rounding to p digits. Therefore, if, e.g., $f \cdot 2^e$ (or $f \cdot 10^e$) is the normalized **exact sum** of two floating-point numbers, then the absolute value of the absolute representation error is obviously smaller than $\frac{1}{2} \cdot 2^{-p} \cdot 2^e$ (or $\frac{1}{2} \cdot 10^{-p} \cdot 10^e$).

From inequality (2.7) bounds for the relative addition error can be derived.

Corollary 2.1: Bounds for Addition Errors

$$u \hat{+} v = (u+v)(1+\delta), \tag{2.8}$$

$$|\delta| \leq 2^{-p} \quad \text{(binary)}, \tag{2.9}$$

$$|\delta| \leq \frac{1}{2} 10^{1-p} \quad \text{(decimal)}. \tag{2.10}$$

Obviously, the error bound is worse in the decimal system than in the binary system.

Note that in the cases $u = 0$ or $v = 0$, no errors are caused by floating-point addition, and Equality 2.8 is valid for $\delta = 0$: If $v = 0$, then $u \hat{+} v = u$, and if $u = 0$, then $u \hat{+} v = v$.

Thus, ignoring representation errors of inputs, the relative error in the sum is always small. **In all cases in which cancellation appears the relative error is even $\delta = 0$** (see Example 2.15). This fact should be stressed since cancellation is usually associated with accuracy loss (see Section 2.4.2).

Our considerations can be summarized as follows: The calculated sum of u and v is the exact sum of the numbers $u(1+\delta)$ and $v(1+\delta)$, where, e.g., $|\delta| \leq 2^{-p}$ or $|\delta| \leq \frac{1}{2} 10^{1-p}$. For **subtraction** an analogous statement is valid.

Analogously, the calculated **multiplication or division** results are the exact results rounded to p digits.

Corollary 2.2: Bounds for Multiplication and Division Errors
For multiplication it is valid that

$$u \hat{*} v = uv(1+\delta), \tag{2.11}$$
$$|\delta| \le 2^{-p} \qquad \text{(binary)}, \tag{2.12}$$
$$|\delta| \le \frac{1}{2}10^{1-p} \qquad \text{(decimal)}. \tag{2.13}$$

If $v \neq 0$, for division we always have:

$$u \hat{\div} v = (u/v)(1+\delta), \tag{2.14}$$
$$|\delta| \le 2^{-p} \qquad \text{(binary)}, \tag{2.15}$$
$$|\delta| \le \frac{1}{2}10^{1-p} \qquad \text{(decimal)}. \tag{2.16}$$

Note, that the calculated product is the exact product of $u(1+\delta)$ and v, or of u and $v(1+\delta)$, or of $u(1+\delta)^{1/2}$ and $v(1+\delta)^{1/2}$, where, e.g., $|\delta| \le 2^{-p}$ or $|\delta| \le \frac{1}{2}10^{1-p}$. Thus, the factor $(1+\delta)$ can be considered to belong to u or v, or could be distributed to both u and v, depending on what appears to be most convenient.

The calculated ratio is the exact ratio of $u(1+\delta)$ and v, or of u and $v/(1+\delta)$, where, e.g., $|\delta| \le 2^{-p}$ or $|\delta| \le \frac{1}{2}10^{1-p}$.

2.4.2 *Theory: Error Propagation*[13]

Now, we are interested in the error propagation caused by algebraic operations, i.e. to what extent errors already present in the inputs are amplified (increased) or damped (decreased) by floating-point operations.

We consider functions $g : \mathbb{R}^n \to \mathbb{R}$, $y = g(x_1, \ldots, x_n) =: g(x)$, which are differentiable in the relevant region. Let x^* be an approximation of x, then

$$\Delta x_i := x_i^* - x_i, \quad \Delta x := x^* - x$$

are the **absolute approximation errors** and

$$\delta_{x_i} = \Delta x_i / x_i$$

the **relative approximation errors** if $x_i \neq 0$.

The sensitivity of absolute and relative errors to changes in x_i can be measured as follows.

[13]Partly based on Stoer (1972, pp. 8 – 13).

Theorem 2.6: Sensitivity of Absolute and Relative Errors
The absolute error can be approximated as follows:

$$\Delta y := y^* - y = g(x^*) - g(x) \approx \sum_{i=1}^{n} \Delta x_i \frac{\partial g(x)}{\partial x_i}.$$

If $y \neq 0$ and $x_i \neq 0$ for $i = 1, \ldots, n$, then the error propagation formula for relative errors has the form:

$$\delta_y := \frac{\Delta y}{y} \approx \sum_{i=1}^{n} \frac{x_i}{g(x)} \frac{\partial g(x)}{\partial x_i} \delta_{x_i}.$$

Proof. The first-order **Taylor expansion** of g in x^* has the form:

$$\Delta y := y^* - y = g(x^*) - g(x) \approx \sum_{i=1}^{n} (x_i^* - x_i) \frac{\partial g(x)}{\partial x_i} = \sum_{i=1}^{n} \Delta x_i \frac{\partial g(x)}{\partial x_i}.$$

The approximation of the relative error can be derived as follows:

$$\delta_y := \frac{\Delta y}{y} \approx \sum_{i=1}^{n} \frac{\Delta x_i}{g(x)} \frac{\partial g(x)}{\partial x_i} = \sum_{i=1}^{n} \frac{x_i}{g(x)} \frac{\partial g(x)}{\partial x_i} \frac{\Delta x_i}{x_i} = \sum_{i=1}^{n} \frac{x_i}{g(x)} \frac{\partial g(x)}{\partial x_i} \delta_{x_i}.$$

□

Obviously, the proportionality factor $\frac{\partial g(x)}{\partial x_i}$ measures the sensitivity with which y responds to absolute changes Δx_i of x_i. Also, the factor $\frac{x_i}{g} \frac{\partial g}{\partial x_i}$ represents the strength of the reaction of the relative error in y to the relative error in x_i.

Definition 2.10: Condition Numbers
The expressions $\frac{x_i}{g} \frac{\partial g}{\partial x_i}$ are called **amplification factors** or **condition numbers** for the relative errors. They have the advantage that they are independent of the scaling of y and x_i. If absolute values of the condition numbers are large, the problem is called **ill-conditioned**, otherwise **well-conditioned**.

For ill-conditioned problems small relative errors in the inputs x cause large relative errors in the results $y = g(x)$.

The above definition of condition numbers has, however, the disadvantage that it is only sensible for $y \neq 0$ and $x_i \neq 0$. Moreover, for many purposes it is impractical because the condition of an operation is described by more than one number. Therefore, often other definitions for the condition of a problem are in use as well, see, e.g., Sections 2.4.4 and 3.1.2.

Corollary 2.3: Error Propagation for Arithmetical Operations
For the **arithmetical operations** the following error propagation formulas are valid. Let $x \neq 0$ and $y \neq 0$, then

$$\begin{array}{lllll} \text{a)} & g(x,y) := x \cdot y, & \delta_{xy} & \approx & \delta_x + \delta_y, \\ \text{b)} & g(x,y) := x/y, & \delta_{x/y} & \approx & \delta_x - \delta_y, \\ \text{c)} & g(x,y) := x \pm y, & \delta_{x \pm y} & \approx & \frac{x}{x \pm y}\delta_x \pm \frac{y}{x \pm y}\delta_y, \text{ if } x \pm y \neq 0. \end{array}$$

Proof. According to Theorem 2.6 we know:

1. $\delta_{xy} = \frac{x}{xy}\frac{\partial(xy)}{\partial x}\delta_x + \frac{y}{xy}\frac{\partial(xy)}{\partial y}\delta_y = \delta_x + \delta_y$,
2. $\delta_{x/y} = \frac{x}{x/y}\frac{\partial(x/y)}{\partial x}\delta_x + \frac{y}{x/y}\frac{\partial(x/y)}{\partial y}\delta_y = \delta_x + \frac{y^2}{x}\frac{-x}{y^2}\delta_y = \delta_x - \delta_y$, and
3. $\delta_{x \pm y} = \frac{x}{x \pm y}\frac{\partial(x \pm y)}{\partial x}\delta_x + \frac{y}{x \pm y}\frac{\partial(x \pm y)}{\partial y}\delta_y = \frac{x}{x \pm y}\delta_x \pm \frac{y}{x \pm y}\delta_y$.

□

Obviously, multiplication and division are not dangerous operations, since the relative errors of the inputs are not strongly propagated into the result. The same is true for addition if the operands x and y have the same sign, since then the condition numbers $\frac{x}{x+y}, \frac{y}{x+y}$ lie between 0 and 1 and their sum is 1. Therefore:

$$|\delta_{x \pm y}| \leq \max(|\delta_x|, |\delta_y|).$$

This is called **error damping**.

If, however, the addition operands have different signs (which can be rewritten as subtraction), then at least one of the factors $|\frac{x}{x+y}|, |\frac{y}{x+y}|$ is larger than 1, and at least one of the relative errors δ_x, δ_y is amplified. This amplification is especially large if $x \approx -y$, i.e. if cancellation occurs while calculating $x \hat{+} y$.

This leads to the **paradox situation** that cancellation leads to exact results for exact inputs, but makes inexact inputs even more inexact.

As the most important result we should keep in mind the following

Rule of thumb: Addition and subtraction may be dangerous with respect to error propagation in contrast to multiplication and division.

For applied statistics it is dangerous to implement the **standard deviation** of n observations by means of the textbook formula:

$$\sigma = \sqrt{\frac{n\sum_{k=1}^{n} x_k^2 - (\sum_{k=1}^{n} x_k)^2}{n(n-1)}}. \tag{2.17}$$

This often leads to a negative argument value under the square root because of cancellation. In Section 2.4.4 better algorithms for the calculation of the standard deviation are introduced.

2.4.3 *Theory: Axiomatic Problems*[14]

One consequence of the inexactness of floating-point operations is that some laws of arithmetical operations are not valid anymore. Let us discuss the commutative and associative laws here. Let us start with good news.

Theorem 2.7: Commutative Laws
The **commutative law of addition** and the **commutative law of multiplication** are valid, i.e.

$$u \hat{+} v = v \hat{+} u, \tag{2.18}$$

as well as

$$u \hat{*} v = v \hat{*} u. \tag{2.19}$$

Proof. For Equality 2.18: $u \hat{+} v = \text{round}(u+v) = \text{round}(v+u) = v \hat{+} u$. □

Thus, commutativity is valid despite the inexactness of the floating-point operations. However, for the corresponding associative laws properties are much more problematic.

The **associative law of addition** fails, i.e. generally

$$(u \hat{+} v) \hat{+} w \neq u \hat{+} (v \hat{+} w) \qquad \text{for most } u, v, w. \tag{2.20}$$

The corresponding difference can indeed be extremely large.

Example 2.18: Associative Law of Addition

$$\begin{aligned}(11111113.\hat{+}(-11111111.))\hat{+}7.5111111 &= 2.0000000\hat{+}7.5111111\\ &= 9.5111111;\\ 11111113.\hat{+}((-11111111.)\hat{+}7.5111111) &= 11111113.\hat{+}(-11111103.)\\ &= 10.00000.\end{aligned}$$

Therefore, the absolute value of the relative error in the associative law of addition is around 5% here.

Thus, programmers should be extremely careful, since mathematical expressions like $a_1 + a_2 + a_3$ or $\sum_{k=1}^{n} a_k$ implicitly assume the validity of associativity. **For numerical reasons, it is therefore important to differentiate between different mathematically equivalent formulations of a calculation method**. We have seen examples of this in the previous section when we hinted at the problem of **cancellation**.

[14]Partly based on Knuth (1998, pp. 229 – 238).

As an example for a typical **error propagation** consider the **associative law of multiplication**.

Theorem 2.8: Associative Law of Multiplication
$(u \hat{*} v) \hat{*} w$ is generally unequal to $u \hat{*} (v \hat{*} w)$. However, the situation is much better than that for the associative law of addition, since

$$(u \hat{*} v) \hat{*} w = ((uv)(1+\delta_1)) \hat{*} w = uvw(1+\delta_1)(1+\delta_2),$$
$$u \hat{*} (v \hat{*} w) = u \hat{*} ((vw)(1+\delta_3)) = uvw(1+\delta_3)(1+\delta_4)$$

for certain δ_1, δ_2, δ_3, δ_4 if no exponent underflow or overflow appears, where $|\delta_j| \leq \frac{1}{2}b^{1-p}$ for every j. Therefore,

$$\frac{(u \hat{*} v) \hat{*} w}{u \hat{*} (v \hat{*} w)} = \frac{(1+\delta_1)(1+\delta_2)}{(1+\delta_3)(1+\delta_4)} \approx 1+\delta,$$

where

$$|\delta| < \frac{2b^{1-p}}{(1-\frac{1}{2}b^{1-p})^2}. \tag{2.21}$$

Proof. The last inequality can be derived from the following expression neglecting all second-order terms in the numerator:

$$\frac{(1+\delta_1)(1+\delta_2)}{(1+\delta_3)(1+\delta_4)} = 1 + \frac{(1+\delta_1)(1+\delta_2)-(1+\delta_3)(1+\delta_4)}{(1+\delta_3)(1+\delta_4)}$$
$$\approx 1 + \frac{(1+\delta_1+\delta_2)-(1+\delta_3+\delta_4)}{(1+\delta_3)(1+\delta_4)} = 1+\delta.$$

Finally, for δ, we know that

$$|\delta| = \left|\frac{(1+\delta_1+\delta_2)-(1+\delta_3+\delta_4)}{(1+\delta_3)(1+\delta_4)}\right| < \frac{4\max_{i\in\{1,\dots,4\}}(|\delta_i|)}{(1-\max_{i\in\{3,4\}}(|\delta_i|))^2}$$
$$\leq \frac{2b^{1-p}}{(1-\frac{1}{2}b^{1-p})^2}.$$ □

Let us now discuss the relative size of errors in the different situations discussed so far.

Definition 2.11: Unit in the Last Place (ulp)
The number b^{1-p} is called **ulp**, meaning **unit in the last place** of the mantissa.

The unit in the last place is, thus, the minimal spacing between floating-point numbers in their chosen representation. Imagine we have a floating-point number with base $b = 10$ and $p = 3$ digits in the mantissa, say $0.314 \cdot 10^2$. Because we only have 3 digits, the nearest larger number that we can represent is obviously $0.315 \cdot 10^2$. This number differs from $0.314 \cdot 10^2$ by one unit in the last place. Any real number r between $0.314 \cdot 10^2$ and $0.315 \cdot 10^2$ can at best be represented by one of these two numbers. If r is actually $\pi \cdot 10^1$, then $31.415926\ldots$ is best represented by $0.314 \cdot 10^2$, and the absolute rounding error is $(\pi \cdot 10^1 - 0.314 \cdot 10^2) = 0.015926\ldots$. In the worst case, the real number $0.3145 \cdot 10^2$ would have the absolute rounding error 0.05. Thus, the maximum absolute rounding error occurs when r is halfway between two representable numbers, leading to the rounding error 0.5 units in the last place. This obviously relates to the general bound $0.5b^{e-p}$ in Formula 2.6 for absolute errors with $e = 2$. In inequality (2.7), in order to find a bound for the relative error, we divided by b^{e-1}, the lower bound for a number to be represented with the same exponent e (in our case $b^{e-1} = 10^1$), leading to the bound $0.5b^{1-p}$. With Definition 2.11, this leads to the following result:

Corollary 2.4: Errors Expressed in ulp
Floating-point numbers are correct up to **0.5 ulp** (see Section 2.4.1). The calculation of $u \cdot v \cdot w$ by two floating-point multiplications using Algorithm 2.6 is correct up to approximately **1 ulp** (neglecting all second-order terms). Moreover, the associative law of multiplication is true up to an error of **2 ulp**.

We have thus shown that $(u \,\hat{*}\, v) \,\hat{*}\, w$ is approximately equal to $u \,\hat{*}\, (v \,\hat{*}\, w)$, except if exponent overflow or underflow appears. This **phenomenon of approximate equality** will be discussed more precisely in what follows.

Programmers using floating-point arithmetic rarely test whether two calculated values are exactly equal, since that is very inaccurate because of accumulated rounding errors. For example, if textbooks state that x_n has a limit for $n \to \infty$ for an **iteration** $x_{n+1} = f(x_n)$, it will not necessarily be successful to wait until $x_{n+1} = x_n$ exactly for some n. The sequence x_n might, e.g., be periodical with a longer period because of the rounding of intermediate results.

A more adequate method would be to wait until $|x_{n+k} - x_n| \leq \Delta$ for a certain k and a certain fixed small number Δ. However, since we generally do not a priori know the size of x_n, it is even better to wait until

$$|x_{n+k} - x_n| \leq \delta |x_n|, \tag{2.22}$$

because it is much simpler to choose a sensible δ. The relation (2.22) can be thought of as another way to express that x_{n+k} and x_n are approximately equal. So, for floating-point operations a relation "approximately equal" is much more useful than the traditional equality.

Therefore, we introduce new operations for **floating-point comparison**, aiming at the assessment of the relative difference between the floating-point numbers.

Definition 2.12: Floating-Point Comparison
Let b be the base and the normalized floating-point numbers $u = f_u \cdot b^{e_u}$, $v = f_v \cdot b^{e_v}$, $1/b \leq |f_u|, |f_v| < 1$. Then we define four different floating-point relations:

u **is surely smaller than** v**:**

$$u \prec v\ (\delta) \quad \text{iff} \quad v - u > \delta \max(b^{e_u}, b^{e_v})$$

u **is approximately equal to** v**:**

$$u \sim v\ (\delta) \quad \text{iff} \quad |v - u| \leq \delta \max(b^{e_u}, b^{e_v})$$

u **is essentially equal to** v**:**

$$u \approx v\ (\delta) \quad \text{iff} \quad |v - u| \leq \delta \min(b^{e_u}, b^{e_v})$$

u **is surely greater than** v**:**

$$u \succ v\ (\delta) \quad \text{iff} \quad u - v > \delta \max(b^{e_u}, b^{e_v})$$

These definitions are applicable to both non-normalized and normalized values. Notice that exactly one of the conditions $u \prec v$, $u \sim v$, or $u \succ v$ is always true for every given pair of values u and v, and that the relation $u \approx v$ is somewhat stronger than $u \sim v$. All these relations are specified for a positive real number δ, measuring the degree of approximation.

The above definitions can be best illustrated by means of so-called **neighborhoods** $N(u) = \{x \mid |x - u| \leq \delta b^{e_u}\}$ associated with a floating-point number u. Obviously, $N(u)$ represents a set of values neighboring u based on the exponent of the floating-point representation of u.

Theorem 2.9: Comparison by Neighborhoods
The four floating-point relations can be expressed by means of the neighborhoods associated with floating-point numbers:
$u \prec v$ iff $N(u) < v$ and $u < N(v)$ (element-wise),
$u \sim v$ iff $u \in N(v)$ or $v \in N(u)$,
$u \succ v$ iff $u > N(v)$ and $N(u) > v$ (element-wise),
$u \approx v$ iff $u \in N(v)$ and $v \in N(u)$.

Proof. The first statement is proved as follows:
$x \in N(u)$ is equivalent to $|x - u| \leq \delta b^{e_u}$ and $y \in N(v)$ to $|y - v| \leq \delta b^{e_v}$.

Element-wise $N(u) < v$ and $u < N(v)$ are thus equivalent to
$v - u = |v - u| > \delta b^{e_u}$ and $v - u = |v - u| > \delta b^{e_v}$.
The other statements can be proved analogously. □

Note that the exponents of u and v do not have to be equal. Thus, the neighborhoods associated with u and v do not have to have the same size. We assume that the parameter δ, measuring the degree of approximation, is a constant. A more complete notation would indicate the dependence of $N(u)$ on δ.

Example 2.19: Inequality in Floating-Point Comparison
Let $v = 0.10 \cdot 10^1 = 1$, $u = 0.1 \cdot 10^0 = 0.1$, and $\delta = 0.05$, then:
$v - u = |v - u| = 0.9$ and $\delta \max(b^{e_u}, b^{e_v}) = 0.05 \cdot \max(1, 10) = 0.5$.
Therefore, $u \prec v\ (\delta)$ and not $u \sim v\ (\delta)$. Moreover,
$N(u) = \{x | \, |x - u| \leq \delta b^{e_u} = 0.05\} = [0.05, 0.15]$ and
$N(v) = \{x | \, |x - v| \leq \delta b^{e_v} = 0.5\} = [0.5, 1.5]$.
Obviously, $N(u) < v$ and $u < N(v)$.

Example 2.20: Approximate Equality in Floating-Point Comparison
Let $u = 0.10 \cdot 10^1 = 1$, $v = 0.6 \cdot 10^0 = 0.6$, and $\delta = 0.05$, then:
$\delta \max(b^{e_u}, b^{e_v}) = 0.05 \cdot \max(10, 1) = 0.5$ and
$\delta \min(b^{e_u}, b^{e_v}) = 0.05 \cdot \min(10, 1) = 0.05$,
thus $u \sim v(\delta)$, and not $u \approx v(\delta)$.

Example 2.21: Essential Equality in Floating-Point Comparison
Let $u = 0.10 \cdot 10^1 = 1$, $v = 0.95 \cdot 10^0 = 0.95$, and $\delta = 0.05$, then:
$\delta \max(b^{e_u}, b^{e_v}) = 0.05 \cdot \max(10, 1) = 0.5$ and
$\delta \min(b^{e_u}, b^{e_v}) = 0.05 \cdot \min(10, 1) = 0.05$,
thus $u \approx v(\delta)$.

We will show that floating-point operations are actually not mathematically exact, but their result is at least essentially equal to the exact result. Obviously, it suffices to show this result for the rounding operation.

Theorem 2.10: Essential Equality of Rounding
Let $\delta_0 = b^{1-p}$, i.e. one **ulp**. Then:

$$x \approx \mathrm{round}(x) \left(\frac{1}{2} \delta_0 \right). \tag{2.23}$$

Proof. From inequality (2.7) we derive the inequality:
$|x - \mathrm{round}(x)| = |\Delta x| \leq \frac{1}{2}\delta_0 \min(|x|, |\mathrm{round}(x)|) \leq \frac{1}{2}\delta_0 \min(b^{e_x}, b^{e_{\mathrm{round}(x)}})$;
since if $|\mathrm{round}(x)| > |x|$, then inequality (2.7) suffices,

and if $|\text{round}(x)| < |x|$, then at least $|\text{round}(x)| \geq b^{e_x - 1}$, and therefore $\frac{|\Delta x|}{|\text{round}(x)|} \leq \frac{|\Delta x|}{b^{e_x-1}} \leq \frac{\frac{1}{2}b^{e_x-p}}{b^{e_x-1}} = \frac{1}{2}b^{1-p}$.
This directly leads to the statement. □

Similar to Theorem 2.10, one can show that $u \mathbin{\hat{+}} v \approx (u+v)(\frac{1}{2}\delta_0)$ and so forth for the other floating-point operations.

Moreover, essential equality is also true for the associative law of multiplication.

Theorem 2.11: Essential Equality for the Associative Law of Multiplication

$$(u \mathbin{\hat{*}} v) \mathbin{\hat{*}} w \approx u \mathbin{\hat{*}} (v \mathbin{\hat{*}} w)(\delta) \tag{2.24}$$

if $\delta \geq \frac{2\delta_0}{(1-\frac{1}{2}\delta_0)^2}$.

Proof. With the approximate associative law of multiplication (Theorem 2.8) we can show that

$$|(u \mathbin{\hat{*}} v) \mathbin{\hat{*}} w - u \mathbin{\hat{*}} (v \mathbin{\hat{*}} w)| < \frac{2\delta_0}{(1-\frac{1}{2}\delta_0)^2}|u \mathbin{\hat{*}} (v \mathbin{\hat{*}} w)|,$$

and an analogous inequality is valid if $(u \mathbin{\hat{*}} v) \mathbin{\hat{*}} w$ and $u \mathbin{\hat{*}} (v \mathbin{\hat{*}} w)$ are interchanged. □

Example 2.22: Essential Equality for the Associative Law of Multiplication
If $b = 10$ and $p = 8$, then $\delta = 0.00000021$ can be used in the previous theorem, since $\delta_0 = b^{1-p} = 10^{-7}$.

Therefore, the associative law of multiplication (and also all other not exactly valid arithmetical laws) is at least essentially valid, though with a larger uncertainty δ than for the arithmetical operations. Obviously, for the associative law of addition this uncertainty is very high, as exemplified in the beginning of this subsection.

The relations $\prec$, $\sim$, $\succ$, and $\approx$ are especially useful in numerical algorithms where convergence should be checked. Therefore, we will meet these terms again with iterative optimization methods in Chapter 4.

2.4.4 Example: Calculation of the Empirical Variance[15]

2.4.4.1 Algorithms

The problem of numerical calculation of the sample variance of n data $\{x_i\}_{i=1}^n$ unfortunately appears to be trivial only at first sight. Quite the contrary, it is

[15]Partly based on Chan et al. (1983).

relatively difficult, in particular if n is large and the variance small. Let us start with the definition of the two most well-known algorithms for the calculation of the sample variance.

Definition 2.13: Standard Algorithms for the Sample Variance
The sample variance is given by S/n or $S/(n-1)$, according to its use, where S is the sum of squares of the deviations from the mean:

$$S = \sum_{i=1}^{n} (x_i - \bar{x})^2, \tag{2.25}$$

where

$$\bar{x} = \frac{1}{n} \sum_{i=1}^{n} x_i. \tag{2.26}$$

Formulas (2.25) and (2.26) define a direct algorithm for the calculation of S. In what follows, this algorithm is called **standard two-pass algorithm**, since it needs two runs through the data.

In order to avoid two runs through the data to calculate S in Formula 2.25, it is common to represent S in the following form:

$$S = \sum_{i=1}^{n} x_i^2 - \frac{1}{n} \left(\sum_{i=1}^{n} x_i \right)^2. \tag{2.27}$$

Since this form is often proposed in statistics textbooks, it will be called **text-book one-pass algorithm** in what follows.

A two-pass algorithm may not be desirable in many applications, e.g. if the sample is too large to be stored as a whole in the main storage, or if the variance should be calculated dynamically, i.e. step by step during data collection.

Unfortunately, although the one-pass algorithm (2.27) is mathematically equivalent to Formulas 2.25 and 2.26, numerically it is disastrous. This is because the values of $\sum x_i^2$ and $(\sum x_i)^2$ become very large in practice and are calculated with rounding errors. If the variance is very small, nearly all digits will be canceled by the subtraction in algorithm (2.27). Large **cancellation**, however, leads, as demonstrated in Section 2.4.2, to amplification of representation errors so that the calculated S possibly shows a very large relative error, potentially even leading to a negative value, which paradoxically can be seen as a blessing since it clearly uncovers the rounding error problem.

In order to avoid such difficulties, numerous other algorithms were developed that get along with one run through the data. We will concentrate on

updating algorithms in Youngs and Cramer (1971) as well as the **pairwise algorithm** in Chan et al. (1979).

Definition 2.14: Updating Algorithm of Youngs and Cramer
Let T_{ij}, M_{ij}, and S_{ij} be the sum, the mean, and the sum of squares of the data from x_i to x_j:

$$T_{ij} = \sum_{k=i}^{j} x_k, \qquad M_{ij} = \frac{1}{(j-i+1)} T_{ij}, \qquad S_{ij} = \sum_{k=i}^{j} (x_k - M_{ij})^2.$$

Then, the following formulas define the **updating algorithm of Youngs and Cramer**:

$$T_{1,j} = T_{1,j-1} + x_j, \tag{2.28}$$

$$S_{1,j} = S_{1,j-1} + \frac{1}{j(j-1)} (jx_j - T_{1,j})^2 \tag{2.29}$$

with $T_{1,1} = x_1$ and $S_{1,1} = 0$.

Theorem 2.12: Updating Algorithm of Youngs and Cramer
The updating algorithm of Youngs and Cramer calculates the sample variance, i.e. $S = S_{1,n}$.

Proof.

$$\begin{aligned}
\sum_{k=1}^{j} \left(x_k - \frac{1}{j} T_{1j} \right)^2 &= \sum_{k=1}^{j} \left(x_k - \frac{1}{j} (T_{1,j-1} + x_j) \right)^2 \\
&= \sum_{k=1}^{j} \left(\left(x_k - \frac{1}{j-1} T_{1,j-1} \right) + \left(\frac{1}{j(j-1)} T_{1,j-1} - \frac{1}{j} x_j \right) \right)^2 \\
&\quad \left[\text{since } \frac{1}{j} = \frac{1}{j-1} - \frac{1}{j(j-1)} \right] \\
&= \sum_{k=1}^{j-1} \left(x_k - \frac{1}{j-1} T_{1,j-1} \right)^2 + \left(x_j - \frac{1}{j-1} T_{1,j-1} \right)^2 \\
&\quad + 2 \sum_{k=1}^{j} \left(x_k - \frac{1}{j-1} T_{1,j-1} \right) \left(\frac{1}{j(j-1)} T_{1,j-1} - \frac{1}{j} x_j \right) \\
&\quad + j \left(\frac{1}{j(j-1)} T_{1,j-1} - \frac{1}{j} x_j \right)^2
\end{aligned}$$

$$
\begin{aligned}
&= \sum_{k=1}^{j-1}\left(x_k - \frac{1}{j-1}T_{1,j-1}\right)^2 + \left(x_j - \frac{1}{j-1}T_{1,j-1}\right)^2\left(1-\frac{2}{j}\right) \\
&\quad + j\left(\frac{1}{j(j-1)}T_{1,j-1} - \frac{1}{j}x_j\right)^2 \\
&\quad \left[\text{since } \sum_{k=1}^{j-1}\left(x_k - \frac{1}{j-1}T_{1,j-1}\right) = 0\right] \\
&= S_{1,j-1} + \left(x_j - \frac{1}{j-1}(T_{1j}-x_j)\right)^2\left(1-\frac{2}{j}+\frac{1}{j}\right) \\
&= S_{1,j-1} + \frac{1}{(j-1)^2}(jx_j - T_{1j})^2\,\frac{j-1}{j}. \qquad \square
\end{aligned}
$$

This algorithm is more stable than the textbook algorithm. Please note in particular that $S = S_{1,n}$ is calculated by a sum of non-negative numbers.

The updating formulas 2.28 and 2.29 can be generalized in order to allow for a combination of two samples of arbitrary size.

Theorem 2.13: Combination of two samples
Suppose that we have two samples $\{x_i\}_{i=1}^{m}$ and $\{x_i\}_{i=m+1}^{m+n}$ with

$$
T_{1,m} = \sum_{i=1}^{m} x_i, \qquad T_{m+1,m+n} = \sum_{i=m+1}^{m+n} x_i,
$$

$$
S_{1,m} = \sum_{i=1}^{m}(x_i - \frac{1}{m}T_{1,m})^2, \qquad S_{m+1,m+n} = \sum_{i=m+1}^{m+n}(x_i - \frac{1}{n}T_{m+1,m+n})^2.
$$

For the combination of all data to one sample of the size $m+n$ let

$$
T_{1,m+n} = T_{1,m} + T_{m+1,m+n}, \tag{2.30}
$$

$$
S_{1,m+n} = S_{1,m} + S_{m+1,m+n} + \frac{m}{n(m+n)}\left(\frac{n}{m}T_{1,m} - T_{m+1,m+n}\right)^2. \tag{2.31}
$$

Then $S = S_{1,m+n}$.

Proof. 1. Special case $m = j-1$, $n = 1$. From Formula 2.31 it follows that

$$
\begin{aligned}
S_{1,j} &= S_{1,j-1} + \frac{j-1}{j}\left(\frac{1}{j-1}T_{1,j-1} - x_j\right)^2 \\
&= S_{1,j-1} + \frac{j-1}{j}\left(\frac{1}{j-1}(T_{1j}-x_j) - x_j\right)^2 \\
&= S_{1,j-1} + \frac{1}{j(j-1)}(T_{1j} - x_j - (j-1)x_j)^2.
\end{aligned}
$$

Obviously, in this case Formula 2.31 is equivalent to Formula 2.29.

2. In general:

$$\sum_{i=1}^{m+n}\left(x_i-\frac{1}{m+n}T_{1,m+n}\right)^2=\sum_{i=1}^{m+n}\left(x_i-\frac{1}{m+n}(T_{1,m}+T_{m+1,m+n})\right)^2$$

$$=\sum_{i=1}^{m}\left(\left(x_i-\frac{1}{m}T_{1,m}\right)+\frac{1}{m+n}\left(\frac{n}{m}T_{1,m}-T_{m+1,m+n}\right)\right)^2+$$

$$\sum_{i=m+1}^{m+n}\left(\left(x_i-\frac{1}{n}T_{m+1,m+n}\right)+\frac{1}{m+n}\left(\frac{m}{n}T_{m+1,m+n}-T_{1,m}\right)\right)^2$$

$$=\sum_{i=1}^{m}\left(x_i-\frac{1}{m}T_{1,m}\right)^2+\sum_{i=m+1}^{m+n}\left(x_i-\frac{1}{n}T_{m+1,m+n}\right)^2+$$

$$\frac{1}{(m+n)^2}\left(\frac{n}{m}T_{1,m}-T_{m+1,m+n}\right)^2\left(m+\frac{m^2}{n}\right)$$

$$\left[\text{since }\sum_{i=1}^{m}\left(x_i-\frac{1}{m}T_{1,m}\right)=0\text{ and }\sum_{i=m+1}^{m+n}\left(x_i-\frac{1}{n}T_{m+1,m+n}\right)=0\right]$$

$$=\sum_{i=1}^{m}\left(x_i-\frac{1}{m}T_{1,m}\right)^2+\sum_{i=m+1}^{m+n}\left(x_i-\frac{1}{n}T_{m+1,m+n}\right)^2+$$

$$\frac{m}{n(m+n)}\left(\frac{n}{m}T_{1,m}-T_{m+1,m+n}\right)^2.\quad\square$$

Definition 2.15: Pairwise Algorithm
If $m=n$, the above formula for the combination of two samples simplifies to

$$S=S_{1,2m}=S_{1,m}+S_{m+1,2m}+\frac{1}{2m}(T_{1,m}-T_{m+1,2m})^2 \qquad (2.32)$$

and defines the **pairwise algorithm** for the calculation of the sample variance.

There are other possibilities to increase accuracy of the calculated S. For data with a large mean value $\bar{x}$ experience shows that, e.g., essential accuracy gains can be achieved by approximately shifting by the mean $\bar{x}$ prior to the calculation of S. Even very rough approximations of the mean appear to be useful so that a **two-pass algorithm** for the preceding calculation of $\bar{x}$ is unnecessary (see Section 2.4.4.3).

Moreover, if the shift is actually carried out by the correct mean before applying the textbook algorithm (2.27), we will get the so-called corrected two-pass algorithm:

Definition 2.16: Corrected Two-Pass Algorithm
The **corrected two-pass algorithm** is defined by

$$S = \sum_{i=1}^{n}(x_i - \bar{x})^2 - \frac{1}{n}\left(\sum_{i=1}^{n}(x_i - \bar{x})\right)^2. \tag{2.33}$$

Note that the first term represents the **two-pass algorithm** (2.25), and the second term would be exactly zero for exact calculations. However, from experience the second term is a very good approximation to the error in the first term in practice. Moreover, note that this algorithm generally does not lead to cancellation since the correcting term is most of the time very much smaller than the first term.

2.4.4.2 Condition and Error Bounds

Originally, algorithms for the calculation of the variance were only assessed on the basis of empirical studies. However, in the end of the 1970s more exact error bounds were found for many algorithms.

Chan and Lewis (1978) first derived a **condition number κ of a sample $\{x_i\}$ for variance calculation**. This condition number κ measures the sensitivity of S for a given data set, i.e. if relative errors δ exist in x_i, then the relative error in S is bounded by $\kappa\delta$.

On computers, a maximum relative error of κu in S will appear for a **machine precision** $u = 0.5 \cdot \text{ulp}$ measuring the size of representation errors in the data (see Formulas 2.9 and 2.10 in Section 2.4.1).

Attention: This statement is independent of the used algorithm in contrast to the algorithm specific condition numbers derived in Section 2.4.2.

The value κu can thus be used as an **indicator for the accuracy of different algorithms** if error bounds that only depend on κ, u and n can be derived for the individual algorithms.

For the derivation of another algorithm-independent condition number see Section 3.1.2 on the linear least squares method.

Definition 2.17: Condition Number of Variance Calculation
Based on the L2 norm $\|\boldsymbol{x}\|_2^2 = \sum_{i=1}^n x_i^2$ of the data, the **condition number** of the calculation of S is given by:

$$\kappa = \frac{\|\boldsymbol{x}\|_2}{\sqrt{S}} = \sqrt{1 + \frac{\bar{x}^2 n}{S}}. \tag{2.34}$$

The latter equality can be shown as follows:

$$S + \bar{x}^2 n = \sum_{i=1}^{n}(x_i - \bar{x})^2 + \bar{x}^2 n = \sum_{i=1}^{n} x_i^2 - \bar{x}^2 n + \bar{x}^2 n = \|\boldsymbol{x}\|_2^2. \tag{2.35}$$

If S is small and $\bar{x} \neq 0$, then we get the useful approximation

$$\kappa \approx \bar{x}\sqrt{\frac{n}{S}} = \frac{\bar{x}}{\sqrt{S/n}}, \tag{2.36}$$

i.e. κ is approximately the mean divided by the standard deviation. Obviously, $\kappa \geq 1$, and in many situations κ is very large.

Attention: κ depends on the value S it is a condition number for. Thus, κ cannot be exactly calculated before the calculation of S. Therefore, expressions containing κ are only used for comparison of different algorithms.

Table 2.7 shows the error bounds given in Chan et al. (1983) for all variance algorithms discussed up to now. These are bounds for the relative error $|(S - S^*)/S|$ in the calculated value S^*. Small constant prefactors were neglected for the sake of clarity. The shown terms dominate the error bounds if the relative error is smaller than one. Note that the error analysis of the **corrected two-pass algorithm** will be discussed below (see page 78).

Table 2.7: Error Bounds for $|(S - S^*)/S|$ in the Calculated Value S^*

Algorithm	Error bound
1. textbook	$n\kappa^2 u$
2. textbook with pairwise summation	$\kappa^2 u \log_2 n$
3. two-pass	$nu + n^2\kappa^2 u^2$
4. two-pass with pairwise summation	$u \log_2 n + (\kappa u \log_2 n)^2$
5. corrected two-pass	$nu + n^3\kappa^2 u^3$
6. corrected two-pass with pairwise summation	$u \log_2 n + \kappa^2 u^3 \log_2^3 n$
7. Youngs and Cramer updating	$n\kappa u$
8. Youngs and Cramer updating pairwise	$\kappa u \log_2 n$

Notice that the bounds for the textbook algorithm (see 1 and 2) can be directly compared with the bounds of the Youngs and Cramer algorithm (7 and 8). The former depend on κ^2, whereas the latter only on κ, all other terms the same. Thus, the Youngs and Cramer algorithm is much more stable than the textbook algorithm.

Simulation Study

In an empirical study for the assessment of the bounds in Table 2.7 data were generated by means of a random number generator (cp. Chapter 6) for normally distributed data with expected value 1 and different variances σ^2 with $1 \geq \sigma^2 \geq 10^{-26}$. For these choices of expected value and variance $\kappa \approx 1/\sigma$ (see Equation 2.36). In all cases the results are mean relative errors averaged over 100 repetitions. Double-precision floating-point arithmetic was used with a **machine precision** of $u \approx 10^{-16}$. The "correct" result was assumed to be the result calculated by the software R. Figure 2.11 shows the results. The relative errors are marked by unfilled (for $n = 4$) and filled (for $n = 1024$) circles. In each plot $\kappa = 1/\sigma$ is given on the abscissa and the relative error in S on the ordinate. To allow for comparisons, the bounds from Table 2.7 can also be found in Figure 2.11. In each plot, assuming a prefactor of 10 each, the lower dashed curve corresponds to the error bound for $n = 4$ data and the upper solid curve to $n = 1024$. Obviously, the experimental results confirm the general form of the error bounds in Table 2.7. Note that the textbook algorithm gives acceptable results (i.e. a relative error smaller than 10^{-2}) only until $\kappa = 10^7$, whereas the results of the Youngs and Cramer updating algorithm are acceptable at least until $\kappa = 10^{13}$. This reflects the squared and linear dependency of the bound on κ, respectively. Further note that the bounds are somewhat higher for $n = 1024$ than for $n = 4$, whereas the realized relative errors are slightly lower for $n = 1024$. The pairwise procedure only has a small effect on the realized errors.

2.4.4.3 Shifted Data

Let us now discuss the effect of a suitable shift of the original data on the condition number κ. We will see that a shift with approximations of the mean can reduce the condition number drastically.

If we replace the original data $\{x_i\}$ by shifted data

$$\tilde{x}_i = x_i - d \tag{2.37}$$

for a fixed shift value d, then the new data have mean $\bar{x} - d$ and S is preserved (assuming that the $\tilde{x}_i$ are exactly calculated). In practice, data with non-vanishing mean are often shifted by some a priori estimate of the mean before calculating S. This generally improves the accuracy of the calculated S.

We will now analyze this improvement by studying the dependence of the condition number on the shift. Bounds for $\tilde{\kappa}$, the condition number of the shifted data, will be derived for different choices of the shift d. $\tilde{\kappa}$ could then be inserted for κ in the bounds in Table 2.7 to receive error bounds for each of the algorithms for shifted data.

Lemma 2.2: Condition of shifted data
Let $\tilde{\kappa}$ be the condition number of the shifted data $\tilde{x}_i = x_i - d$, $d \in \mathbb{R}$. Then

$$\tilde{\kappa}^2 = 1 + \frac{n}{S}(\bar{x} - d)^2, \tag{2.38}$$

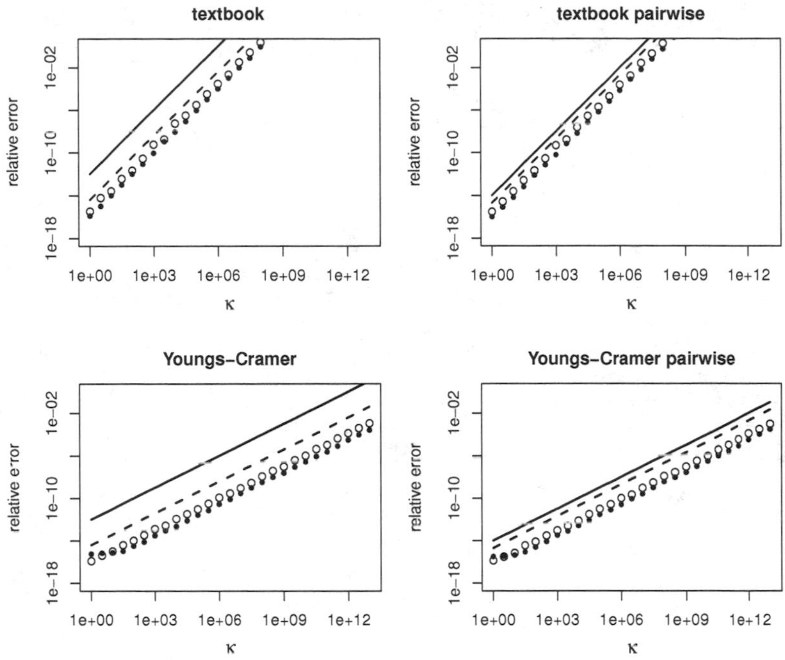

Figure 2.11: Error bounds for variance algorithms: Lower dashed curve corresponds to the error bound for $n = 4$ data and the upper solid curve to $n = 1024$. Errors of the empirical study are marked by unfilled (for $n = 4$) and filled (for $n = 1024$) circles.

$$\tilde{\kappa} < \kappa \Leftrightarrow |d - \bar{x}| < |\bar{x}| \Leftrightarrow \begin{cases} 0 < d < 2\bar{x} & \text{if } \bar{x} > 0 \\ 2\bar{x} < d < 0 & \text{if } \bar{x} < 0, \end{cases} \tag{2.39}$$

$$\tilde{\kappa} = 1 \text{ if } d = \bar{x}. \tag{2.40}$$

Proof. (2.38) is true by definition of the condition number (Definition 2.17), (2.39) by comparing (2.38) for shifted and non-shifted data, and (2.40) by insertion of $d = \bar{x}$ in (2.38). □

Obviously, the data are perfectly conditioned for the calculation of the sample variance if $d = \bar{x}$. Additionally, $d = 0$ is the only sensible value in cases where $\bar{x} = 0$.

In practice, we cannot exactly guess $\bar{x}$ without a special run through the data, i.e. without the **two-pass** algorithm, but we can find a rough estimate without large computational burden. This will be discussed in the following theorems.

Theorem 2.14: Shift by One Observation
Let us assume that we shift by any single original observation.
If $d = x_j$ for any $j \in \{1,\dots,n\}$, then

$$\tilde{\kappa}^2 \leq 1+n. \tag{2.41}$$

If $d = x_j$ is randomly chosen from $\{x_i\}$, then

$$\mathrm{E}(\tilde{\kappa}^2) = 2 \tag{2.42}$$

independently of n and S, and

$$\tilde{\kappa}^2 < 1+k \tag{2.43}$$

with a probability of at least $1-1/k$ for $1 \leq k \leq n$.

Proof. (2.41): If $d = x_j$ for any $j \in \{1,\dots,n\}$, then $\min_i(x_i) \leq d \leq \max_i(x_i)$, and thus $(\bar{x}-d)^2 \leq \sum_i (\bar{x}-x_i)^2 = S$. From (2.38) we get the statement.
(2.42): We know that $\mathrm{E}[(\bar{x}-d)^2] = \mathrm{E}[(\bar{x}-x_j)^2] = S/n$. By (2.38) this leads to the statement.
(2.43): For fixed k, $1 \leq k \leq n$, the inequality $(\bar{x}-x_j)^2 \geq kS/n$ can be valid for maximally n/k values j. Otherwise, we would have $\sum_j (\bar{x}-x_j)^2 > (n/k)(kS/n) = S$. If we thus choose x_j uniformly random, then $(\bar{x}-x_j)^2 < kS/n$ at least with a probability of $(n-n/k)/n = 1-1/k$. The statement follows from (2.38). □

The bound in inequality (2.41) might be sufficient for moderately large n. However, we never get equality in inequality (2.41), and approximate equality only if

$$(\bar{x}-x_j)^2 \approx \sum_i (\bar{x}-x_i)^2,$$

i.e. only if x_j is substantially more distant from $\bar{x}$ than all the other x_i.
Property (2.42) is a probabilistic refinement of inequality (2.41) stating that the expected value is much lower than $1+n$. Note that inequality (2.42) is even independent of the distribution of the $\{x_i\}$. By a prior permutation of the data and by choosing x_j as the first element of the resulting list, we achieve sampling from the assumed uniform distribution of the $\{x_i\}$. This is naturally not possible if the data have a fixed starting value.
From property (2.43), we learn, e.g., that $\tilde{\kappa}^2 < 11$ with probability ≥ 0.90. This is, again, independent of the sample size $n \geq 10$ and S if the shift value x_j is drawn uniformly random from the original sample.

Let us now switch to the consideration of more than one original observation for the determination of the shift d.

Theorem 2.15: Shift by the Mean of Observations
Let $d = \bar{x}_p = \sum_j x_j/p$ be the mean of any p original data values, $p \ll n$. We assume that p is sufficiently small so that rounding errors in the calculation of $\bar{x}_p$ do not play any role. In particular, let $\kappa p u < 1$, where u denotes the machine precision. Then, the condition number of this shift is bounded as follows:

$$\tilde{\kappa}^2 \leq 1 + \frac{n}{p}. \tag{2.44}$$

Let $d = fl(\bar{x}) = \frac{1}{n}\sum_{i=1}^n x_i(1+\xi_i)$ be the floating-point approximation of the overall mean. Then

$$\tilde{\kappa}^2 \leq 1 + n^2\kappa^2 u^2. \tag{2.45}$$

Proof. (2.44): We use the following special case of the **Cauchy inequality**:

$$|\boldsymbol{x}^T\boldsymbol{y}|^2 \leq \|\boldsymbol{x}\|_2^2\|\boldsymbol{y}\|_2^2,$$

where $\boldsymbol{y} = (1/p \cdots 1/p)^T$ is a p-vector with $\|\boldsymbol{y}\|_2^2 = 1/p$.

Therefore,

$$\begin{aligned}
\tilde{\kappa}^2 &= 1 + \frac{n}{S}(\bar{x}-\bar{x}_p)^2 = 1 + \frac{n}{S}\left(\frac{1}{p}\sum_{j=1}^{p}(\bar{x}-x_j)\right)^2 \\
&\leq 1 + \frac{n}{Sp}\sum_{j=1}^{p}(\bar{x}-x_j)^2 \leq 1 + \frac{n}{p}.
\end{aligned} \tag{2.46}$$

(2.45): If the shift is really carried out by the computed mean, rounding errors generally cannot be ignored. Instead of $\bar{x}$, its floating-point approximation

$$fl(\bar{x}) = \frac{1}{n}\sum_{i=1}^{n} x_i(1+\xi_i) \tag{2.47}$$

is calculated, where for floating-point addition the ξ_i values are bounded by

$$|\xi_i| \leq nu \tag{2.48}$$

if only linear terms are taken into account and prefactors are ignored (cp. (2.8)–(2.10) and (2.14)–(2.16) in Section 2.4.1).

Therefore, $\tilde{\kappa}^2$ is bounded by

$$\begin{aligned}
\tilde{\kappa}^2 &= 1 + \frac{n}{S}(\bar{x} - fl(\bar{x}))^2 = 1 + \frac{1}{nS}\left(\sum_{i=1}^{n} x_i\xi_i\right)^2 \\
&\leq 1 + \frac{1}{nS}\|\boldsymbol{x}\|_2^2\|\boldsymbol{\xi}\|_2^2 \qquad \text{(by the Cauchy inequality)} \\
&= 1 + \frac{1}{n}\kappa^2\|\boldsymbol{\xi}\|_2^2 \leq 1 + \kappa^2\|\boldsymbol{\xi}\|_\infty^2.
\end{aligned} \tag{2.49}$$

Here, we have used Definition 2.34 and the general inequality

$$\|\xi\|_2^2 \leq n\|\xi\|_\infty^2,$$

where $\|\xi\|_\infty = \max_i |\xi_i|$. With (2.48) we can write (2.49) as

$$\tilde{\kappa}^2 \leq 1 + n^2\kappa^2 u^2. \qquad \square$$

For $p = 1$, inequality (2.44) leads to inequality (2.41). Note that this kind of shift, as simple as the idea is, can lead to a very large improvement of accuracy. Moreover, note that because of the dependence on κ, the bound (2.45) can be worse than the bounds for more primitive estimates of d. This is really relevant in practice, since we can construct examples where the calculated mean does not even lie between $\min(x_i)$ and $\max(x_i)$, and therefore $(\bar{x} - fl(\bar{x}))^2 > \max_i(\bar{x} - x_i)^2$. In such cases it is better to shift by one single data point and not by the calculated mean.

Naturally, a shift by the mean might also be undesired for reasons of efficiency since we need an additional run through the data. Nevertheless, if a two-pass algorithm is acceptable and $n^2\kappa^2u^2$ is small (e.g. < 1), then this shift followed by a one-pass algorithm delivers a very reliable method for the calculation of S.

The **corrected two-pass algorithm** (2.33) is of this form. It consists of the textbook algorithm with data shifted by $fl(\bar{x})$. Its error bound $nu(1 + n^2\kappa^2u^2)$ easily results from inequality (2.45) and the bound of the textbook algorithm in Table 2.7.

Other one-pass algorithms can also be used in combination with a shift by the calculated mean. However, if a good shift can be found so that $\tilde{\kappa} \approx 1$, then all one-pass algorithms are essentially equivalent with an error bound nu (or $u\log_2 n$ with pairwise summation). Since the textbook algorithm is the fastest one-pass algorithm, because it only needs n multiplications and $2n$ additions in contrast to, e.g., $4n$ multiplications and $3n$ additions for the updating algorithm, it is recommended here except in rare cases.

2.4.4.4 An Overall Approach

The results of the preceding sections deliver a basis for an intelligent choice of an algorithm for the most accurate calculation of the sample variance. As often in realized approaches, not just one of the possible algorithms is used for the solution of a problem. Instead, the available algorithms are combined in a clever way. This is demonstrated here for the calculation of the sample variance following Chan et al. (1983).

There is at least one situation where the textbook algorithm (2.27) can be recommended as it is, namely, if the original data consist of only integer numbers that are small enough to guarantee that an overflow cannot appear. Then, Formula 2.27 should be used with integer addition. In this case, rounding errors do not appear except in the last step, in which the two sums are combined by subtraction and a division by n is carried out.

For non-integer numbers we first have to decide whether we want to use a one-pass or a two-pass algorithm. If all data fit into the (high-speed) main storage and we are not interested in dynamic updates for every new observation, then, probably, a two-pass algorithm is acceptable and the **corrected two-pass algorithm** (2.33) is recommended. If n is large and high speed is needed, then pairwise summation should be considered.

If a **one-pass algorithm** has to be used, then the first step should be to shift the data as well as possible, e.g. by an x_j (see Section 2.4.4.3). Then, an adequate one-pass algorithm should be chosen. For this, we first have to estimate $\tilde{\kappa}$, the condition number of the shifted data, e.g. by means of the formulas in Section 2.4.4.3. If $n\tilde{\kappa}^2 u$, the error bound for the textbook algorithm, is smaller than the desired relative accuracy, then the textbook algorithm should be applied to the shifted data. If this bound is too large, then, to be on the safe side, a slower algorithm should be used. Again, the dependency on n could be reduced by pairwise summation. The dependency on $\tilde{\kappa}$ can be reduced by means of an **updating algorithm**. The use of the updating algorithm with pairwise summation could reduce both dependencies.

If n is a multiple of 2, pairwise summation can be easily implemented. For general n more (in particular human) work is needed making the algorithm less attractive.

The just described decision tree is presented in Figure 2.12.

2.4.5 *Practice and Simulation*

Early Stopping

Unfortunately, instead of checking approximate equality (see Section 2.4.3) the restriction of the number of iteration steps is recently adopted with increased frequency as the method of choice to control the convergence of a sequence to its limit (so-called **early stopping**). This choice does not make too much sense, as one can see by means of a sequence simulator. [16] Figure 2.13 shows the correspondence of the sequences $(-1)^n/n$ and $(-1)^n/n^2$ with their limit 0 after 50 iterations. Obviously, convergence speed is different for the two sequences, and thus the correspondence to the limit is different.

[16]For example by using the simulator in `http://www.flashandmath.com/mathlets/calc/sequences/sequences.html`.

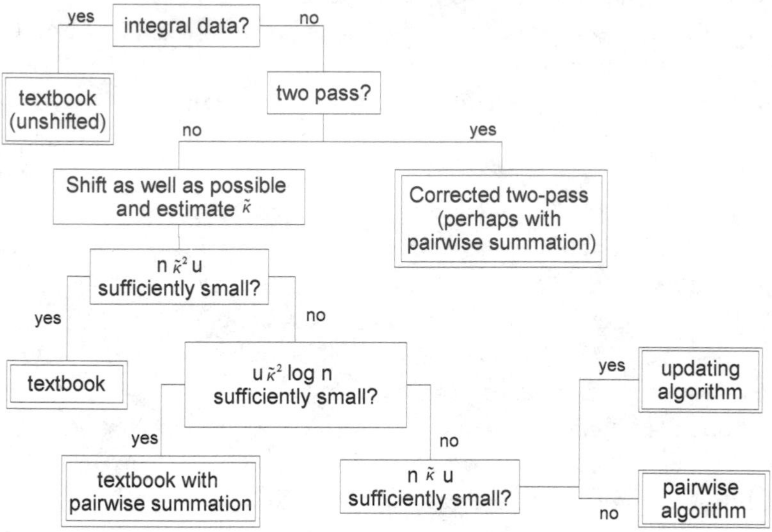

Figure 2.12: Choice of an algorithm for the calculation of the sample mean.

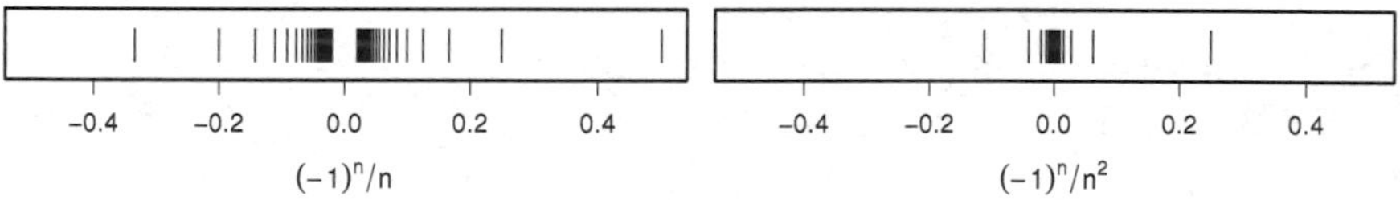

Figure 2.13: Sequences: Convergence after 50 iterations.

Analogous properties are valid for series. We studied the series $\sum_{k=1}^{n}(-1)^{k+1}/k$ and $\sum_{k=1}^{n}(-1)^{k+1}/k^2$ with the limits $\log(2) = 0.69315$ and $\pi^2/12 = 0.82163$ after $n = 50$ iterations (see Figure 2.14).

Iterative methods will also be studied for nonlinear optimization in Chapter 4. There, convergence speed, which is decisive for the goodness of limit approximation after n iterations, is formally introduced.

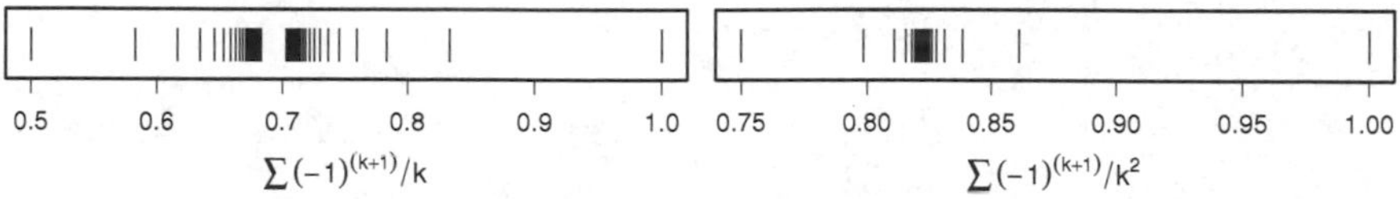

Figure 2.14: Series: Convergence after 50 iterations.

2.4.6 *Summary and Outlook*

Summary

In this section we discussed the accuracy of floating-point operations. We derived error bounds for rounding, the basic arithmetical operations, and the associative law of multiplication. We gave an example for a case where the associative law of addition is very wrong using floating-point operations. We introduced condition numbers for error propagation of arithmetical operations, and we introduced new approximate comparison operations that appear to be more adequate for floating-point numbers than exact comparisons. Finally, we discussed the accuracy of various methods for the calculation of the empirical variance.

Outlook

Recently, the incremental calculation of mean, variance, and other statistical moments has undergone something of a revival, since it is needed in so-called **data stream mining**. In data streams so many data are observed that not all can be stored at the same time. Therefore, e.g., only sufficient statistics of these data are stored and analyzed. These sufficient statistics have to be updated successively (see, e.g., Gaber et al., 2005).

2.5 Implementation in R

2.5.1 *Sorting*

The core function for sorting numerical vectors in R is `sort.int` which is contained in the base package. Depending on the method requested, either a quick Sort variant (Singleton, 1969) that uses a randomly chosen pivot element is used or a shell sort (Sedgewick, 1986) is employed if partial sorting is requested. It is interesting to note that for **classed objects**, that is objects that have a `class` attribute, `sort` is implemented using the `order` command. This is useful because `rank` can also be implemented as an order operation. If no ties are present, it holds that `rank(x) == order(order(x))` since the first application of `order` returns the ordering of the elements in `x` and outer `order` operation then returns the position of each element in the ordered set that corresponds to the rank of the observation.

2.5.2 *Floating-Point Numbers*

R does not have a single-precision data type. In R, all numbers are stored in double-precision format. The functions `as.single` and `single` are identical to the functions `as.double` and `double` except that they set the attribute

`Csingle` used in the `.Fortran` and `.C` interfaces, and they are only intended to be used in that context.

2.5.3 (Co-)Variance

Checking the `var` function in R, reveals that the actual calculation is performed by a `.Internal` C function named `cov`. Looking at the source file `src/main/names.c` in the R source code, we see that `cov` is actually implemented in a C function named `do_cov` located in the file `src/main/cov.c`. This function performs quite a bit of error checking and `NA` handling. Finally, the calculation is performed by the function `cov_complete1` in the same file. It uses a simple two-pass algorithm for the computation.

2.6 Conclusion

In this chapter we discussed the fundamental questions for statistical computing, namely:

- **What** can a computer compute?
- **How** does a computer compute?
- **How exact** does a computer compute?

As a contribution to the **What**-question we introduced the basic term of this book, namely **algorithm**, and the **Turing machine** as a theoretical vehicle to implement algorithms and as a near relative, i.e. one of the forefathers, of modern computers.

To answer the **How**-question we introduced **floating-point numbers** and operations as being standard in modern computers. We discussed the pros and cons of this number representation, and as the main result of our discussion of the **How exact**-question we saw that not even basic laws of arithmetical operations are valid for floating-point numbers with important implications on the validity of computer calculations.

We introduced the two main properties of algorithms: **Complexity** and **accuracy**. We discussed these properties by means of two standard examples: Complexity of sorting algorithms, as utilized for quantile determination, and accuracy of the calculation of the empirical variance. We introduced measures for complexity and accuracy, namely **Landau symbols** and **conditions numbers**, respectively. Both properties, complexity as well as accuracy, will be further discussed in the next chapters, e.g. complexity with multivariate optimization methods (see Section 4.4) and accuracy with the verification procedure in Chapter 3.

Indeed, all that we discussed in this chapter should be understood as basic foundations for higher-order characterizations of statistical algorithms in the next chapters, without which the understanding of these characterizations and algorithms cannot be complete.

2.7 Exercises

Exercise 2.2.1: Implement the Euclidean algorithm to calculate the GCD of two whole numbers. Try to be as efficient as possible in your implementation.

Exercise 2.2.2: Devise an algorithm to calculate the least common multiple (LCM) of two whole numbers. Implement and test your algorithm.

Hint
You can reuse your GCD algorithm implementation.

Exercise 2.2.3: Implement a recursive function to calculate arbitrary elements of the Fibonacci sequence. Recall that the Fibonacci sequence is defined as

$$f_{fib}(n) = \begin{cases} n & \text{if } n < 2 \\ f_{fib}(n-1) + f_{fib}(n-2) & \text{else} \end{cases}.$$

Measure the runtime of your implementation for different n (say $n = 2, 3, 5, 8, 30$). What do you observe?

Exercise 2.2.4: Design a more efficient algorithm to calculate an arbitrary Fibonacci number. Implement it and compare its runtime to the naive recursive implementation from the previous exercise. What do you observe?

Exercise 2.2.5: Consider the Algorithm 2.8 to calculate the nth power of a matrix $\boldsymbol{A}$. Is this algorithm free of errors?

Give the exact number of matrix multiplications ($\times$) that are performed as a function of b and o where b is the number of digits n has as a binary number and o is the number of ones in the binary representation of n. Example: If $n = 5_{10} = 101_2$, then $b = 3$ and $o = 2$. For $n = 11_{10} = 1011_2$, $b = 4$ and $o = 3$.

Represent the exact runtime using the Landau symbols.

Exercise 2.2.6: Execute Algorithm 2.8 for

$$A = \begin{Bmatrix} 1 & 1 \\ 1 & 0 \end{Bmatrix}$$

and $n = 6$. What do you observe?

Exercise 2.2.7: Implement the sorting algorithms

Algorithm 2.8 Matrix Exponentiation

Require: $\boldsymbol{A} \in \mathbb{R}^{k \times k}$, $n \in \mathbb{N}$
Ensure: $\boldsymbol{B} = \boldsymbol{A}^n = \boldsymbol{A} \times \boldsymbol{A} \times \cdots \times \boldsymbol{A}$

```
B ← I_k
while n ≠ 0 do
  if n is odd then
    B ← B × A
    n ← n − 1
  else
    A ← A × A
    n ← n/2
  end if
end while
```

Note: $\boldsymbol{I}_k$ is the k-dimensional identity matrix.

- Bubble sort
- Insertion sort
- Quick sort
- Clever quick sort

and test them on a series of problems including lists of length 1, already sorted lists, and lists containing ties.

Exercise 2.2.8: Write a simulator for a generic Turing machine over the alphabet A with the instruction set I. The simulator should accept a Turing program P and initial instruction L_0 as well as the final states $F \subset I$. Test your simulator using several small example programs. Take care to test different alphabets and instruction sets.

Exercise 2.2.9: Write a Turing program that multiplies two numbers and test it.

Exercise 2.3.1: Find two example floating-point numbers ($b = 10$, $p = 4$) u and v where

$$u^2 - v^2 \neq (u - v) \cdot (u + v).$$

What is the maximum error you can observe?

Exercise 2.3.2: Find the normalized floating-point representation for 0.0029, 0.029, 0.29, 2.9, 29, 290 with base $b = 10$, excess $q = 3$, mantissa $p = 2$ and maximal exponent $e_{\max} = 6$.

Exercise 2.3.3: Find the normalized floating-point representation for -4.2 with $b = 2$, $p = 24$, $q = 127$, and $e_{\text{max}} = 255$. Specify the mantissa, the exponent, and the sign explicitly. This corresponds to **single-precision** floating-point numbers according to the IEEE754 standard.

Example
6.125_{10} corresponds to $110.001_2 = 1.10001 \times 2^2$. Hence, the following values can be derived:

mantissa: 10001000000000000000000 (leading 1 is **not** stored)

exponent: $2 + q = 129_{10} = 10000001_2$

sign: $+$ (represented as 0).

Exercise 2.3.4: Find floating-point numbers u, v, w such that

$$0 = (u \hat{\times} v) \hat{\times} w \qquad\qquad 0 \neq u \hat{\times} (v \hat{\times} w)$$

with $b = 2$, $p = 24$, $q = 127$ and $e_{\text{max}} = 255$.

Exercise 2.3.5: In R or a similar software, compare the result of $0.1 + 0.2$ with 0.3 as in `0.1 + 0.2 == 0.3`. What is the difference between $0.1 + 0.2$ and 0.3? Why is there any difference? Specify all three numbers in floating-point representation and explain any difference you observe.

Exercise 2.3.6: Devise and implement an smart algorithm to sum up n floating point numbers. Try to avoid cancellation and other effects that could reduce the precision.

Hint
The order in which the elements are summed is important.

Exercise 2.4.1:

1. Implement the following one-pass algorithms to calculate the sample variance:
   (a) The textbook algorithm.
   (b) The textbook algorithm with sensible shift.
   (c) The textbook algorithm with pairwise summation.
   (d) The Youngs-Cramer algorithm.
   (e) The Youngs-Cramer algorithm with sensible shift.
   (f) The Youngs-Cramer algorithm with pairwise summation.
2. Simulate and visualize the error propagation for the algorithms in (a) and compare them with each other. For this, use data from different distributions (normal, uniform, ...). Assume that the "true" variance is the value

produced by the programming language you use (though this value might not always be true!).

Chapter 3

Verification

3.1 Motivation and History[1]

Systematic verification of the results of numerical algorithms is one of the most important and at the same time one of the most neglected tasks in the development of such algorithms. On the one hand, there is the well-established field of software engineering that studies how to design and verify large software systems so that they adhere to formal specifications (see, e.g., Sommerville (2010)). On the other hand, the verification of the exactness of numerical results is obviously restricted to problems for which the correct solution is well-known a priori. Moreover, in order to be able to verify the results in the general case, there is need for such correct solutions for all degrees of difficulty. For this, one has to fully understand the numerical problem to be solved, and there has to be a general theory for the generation of test problems with known solutions. For this purpose, a systematic approach is necessary, i.e. some sort of an **experimental design of testing**. Testing sporadic examples will likely show a completely distorted image. In particular, the well-established practice of testing new algorithms on standard problems from literature does not in any way assess the general capabilities of the algorithm under test.

This chapter demonstrates how a general verification procedure can be constructed. In order to be able to rigorously understand the problem to be solved for being able to assess the difficulty of data situations for this problem, we concentrate on probably the most used model in statistics, the **linear model** $y = X\beta + \epsilon$. For its popularity alone, it should be of particular importance to analyze the numerical problems in the estimation of unknown coefficients.

For the general verification of least squares solutions $\hat{\beta}$ for such models, it would be useful to cover the entire space of possible inputs, namely of the

[1] Partly based on Weihs (1977), Weihs (2009).

(influential) factor matrix $\boldsymbol{X}$ and the **response vector** y, with as little examples as possible. Additionally, it would be helpful to construct **representative factor matrices and response vectors** for which the accuracy of the least squares solution can be calculated easily, e.g., for which, in the ideal case, **the least squares solution can be exactly computed**. We will demonstrate how to generate singular test matrices with full column rank where the generalized inverses can be exactly computed, and how representative response vectors can be constructed. However, the reader should be aware that in his or her real applications other types of factor matrices might appear, and that no test system can generate all possible types of such matrices. Nevertheless, our system demonstrates the most important **principles of the verification of computer algorithms** by means of **Linear Least Squares (LLS) solvers**.

In order to formalize the representativeness of test cases we use characterization measures for the difficulty of a data situation regarding the LLS problem. As such measures we utilize **condition numbers** of factor matrices and the angle between the response vector and its projection (in Section 3.1.2). This way, we can control the difficulty of the LLS problem. Moreover, we concentrate on some select **LLS solvers** (see Section 3.2.1). The **construction of numerically favorable representative test matrices and representative response vectors** is demonstrated in Section 3.2.2. Here, the focus is on the exact determination of the least squares solution and on the as full as possible coverage of the relevant problems. For the latter, we construct not too specific test examples with representative condition numbers. Finally, a full **verification method** is proposed (Section 3.3.1) and **verification results** (Section 3.3.2) for the introduced LLS solvers are discussed.

3.1.1 Preliminaries

In this subsection, we will give notation, definitions, and results from matrix theory that we assume to be known. In the following sections we will not explicitly refer to this subsection when using these facts. All other theory needed will be cited or proved.

Notation 3.1: Matrix Properties
Let $L(m,n)$ be the set of all real-valued $m \times n$ matrices. Then any $\boldsymbol{X} \in L(m,n)$ defines a linear map from $\mathbb{R}^n$ to $\mathbb{R}^m$, hence the name L for this set. If $\boldsymbol{X}$ is such a matrix ($\boldsymbol{X} \in L(m,n)$), then

- $\boldsymbol{X}^T$ is the **transpose** of $\boldsymbol{X}$,
- $\text{im}(\boldsymbol{X})$ is the **image** of $\boldsymbol{X}$,
- $\ker(\boldsymbol{X})$ is the **kernel** of $\boldsymbol{X}$,

– $\det(\boldsymbol{X})$ is the **determinant** of $\boldsymbol{X}$,

– $\operatorname{rank}(\boldsymbol{X})$ is the **rank** of $\boldsymbol{X}$.

Definition 3.1: Special Matrices
We need the following special matrices in this chapter.

– The **identity matrix** $\in L(n,n)$ is called $\boldsymbol{I}$ or $\boldsymbol{I}_n$.

– Let $\boldsymbol{P}_{i,j}$ be the matrix generated by the interchange of the ith and the jth row or column of $\boldsymbol{I}$. $\boldsymbol{P}_{i,j}$ is called **elementary permutation matrix**. Multiplication from the right by $\boldsymbol{P}_{i,j}$ interchanges the columns i and j of a matrix $\boldsymbol{X}$, multiplication from the left interchanges rows i and j. A product $\boldsymbol{P}$ of elementary permutation matrices is called **permutation matrix**. Check that

$$\boldsymbol{P}_{i,j}^{-1} = \boldsymbol{P}_{i,j}, \quad \boldsymbol{P}^{-1} = \boldsymbol{P}^T.$$

– We call $\boldsymbol{X} = \operatorname{diag}(d_1,\ldots,d_n)$ a **diagonal matrix** with diagonal elements $d_1,\ldots,d_n$, $d_i \neq 0$ iff $\boldsymbol{X}_{ij} = 0$ for $i \neq j$.

We call a matrix $\boldsymbol{X}$

– **upper triangular** iff $\boldsymbol{X}_{ij} = 0$ for $i > j$,

– **lower triangular** iff $\boldsymbol{X}_{ij} = 0$ for $i < j$,

– **orthogonal** iff $\boldsymbol{X}^T\boldsymbol{X} = \operatorname{diag}(d_1,\ldots,d_n)$,

– **orthonormal** iff $\boldsymbol{X}^T\boldsymbol{X} = I$.

Definition 3.2: Moore-Penrose Generalized Inverse
For any $\boldsymbol{X} \in L(m,n)$ there is a unique matrix called the **Moore-Penrose generalized inverse** $\boldsymbol{X}^+ \in L(n,m)$ with the following four properties:

1. $\boldsymbol{X}\boldsymbol{X}^+\boldsymbol{X} = \boldsymbol{X}$
2. $\boldsymbol{X}^+\boldsymbol{X}\boldsymbol{X}^+ = \boldsymbol{X}^+$
3. $(\boldsymbol{X}\boldsymbol{X}^+)^T = \boldsymbol{X}\boldsymbol{X}^+$
4. $(\boldsymbol{X}^+\boldsymbol{X})^T = \boldsymbol{X}^+\boldsymbol{X}$

If $\boldsymbol{X}$ has maximum column rank n, then $\boldsymbol{X}^+ = (\boldsymbol{X}^T\boldsymbol{X})^{-1}\boldsymbol{X}^T$. This generalized inverse has the **minimum-norm least squares property**, i.e.
$\hat{\beta} := \boldsymbol{X}^+\boldsymbol{y}$ approximately solves the equation system $\boldsymbol{X}\beta = \boldsymbol{y}$ with the properties:

(1.) $\|\boldsymbol{y} - \boldsymbol{X}\hat{\beta}\|_2 \leq \|\boldsymbol{y} - \boldsymbol{X}\beta\|_2$ for all β (**least squares property**),

(1.) $\|\hat{\beta}\|_2 \leq \|\beta\|_2$ for all β with equality in (1) (**minimum-norm property**).

Proposition 3.1: Image and Kernel of Generalized Inverses
Image and kernel of a matrix are, in a way, complementary, since
$\operatorname{im}(\boldsymbol{X})^\perp = \ker(\boldsymbol{X}^T)$, where $\operatorname{im}(\boldsymbol{X})^\perp$, the **orthogonal complement**, stands

for all vectors orthogonal to $\operatorname{im}(\boldsymbol{X})$. Moreover,
$\ker(\boldsymbol{X}^T) = \ker(\boldsymbol{X}^+)$ and $\ker(\boldsymbol{X}) = \ker((\boldsymbol{X}^+)^T)$ as well as
$\operatorname{im}(\boldsymbol{X}^T) = \operatorname{im}(\boldsymbol{X}^+)$ and $\operatorname{im}(\boldsymbol{X}) = \operatorname{im}((\boldsymbol{X}^+)^T)$.

Definition 3.3: Matrix Norm
Given $\boldsymbol{X} \in L(m,n)$, we call a function
$\|\cdot\| : L(m,n) \to \mathbb{R}^+$ a **matrix norm** iff:

1. For $\boldsymbol{X} \in L(m,n)$, $\|\boldsymbol{X}\| \geq 0$ and $\|\boldsymbol{X}\| = 0$ iff $\boldsymbol{X} = O$, the null matrix.
2. For any scalar $a \in \mathbb{R}$, $\|a\boldsymbol{X}\| = |a|\|\boldsymbol{X}\|$.
3. For $\boldsymbol{X}_1, \boldsymbol{X}_2 \in L(m,n)$, $\|\boldsymbol{X}_1 + \boldsymbol{X}_2\| \leq \|\boldsymbol{X}_1\| + \|\boldsymbol{X}_2\|$.
4. For all $\boldsymbol{X}_1, \boldsymbol{X}_2$ for which the matrix product is defined, in addition $\|\boldsymbol{X}_1\boldsymbol{X}_2\| \leq \|\boldsymbol{X}_1\|\|\boldsymbol{X}_2\|$.

A matrix norm is called **consistent with a vector norm** iff

$$\|\boldsymbol{X}\boldsymbol{\beta}\| \leq \|\boldsymbol{X}\|\|\boldsymbol{\beta}\|.$$

In particular

$$\begin{aligned}\|\boldsymbol{X}\|_2 &:= \max_{\|\boldsymbol{\beta}\|_2=1} \|\boldsymbol{X}\boldsymbol{\beta}\|_2 \\ &= \max\{\sqrt{\lambda_i} \mid \lambda_i \text{ is eigenvalue of } \boldsymbol{X}\boldsymbol{X}^T\}\end{aligned}$$

is called the **spectral norm** of matrix $\boldsymbol{X}$ and

$$\|\boldsymbol{X}\|_F := \left(\sum_{i=1}^{m}\sum_{j=1}^{n} |x_{ij}|^2\right)^{0.5}$$

is called **Frobenius norm** of $\boldsymbol{X} \in L(m,n)$.

Let $\boldsymbol{X} \in L(m,n)$ and $r := \operatorname{rank}(\boldsymbol{X})$. Since $\operatorname{rank}(\boldsymbol{X}\boldsymbol{X}^T) = \operatorname{rank}(\boldsymbol{X})$, the eigenvalues $\lambda_j(\boldsymbol{X}\boldsymbol{X}^T)$, $j = 1, \ldots, m$, of $\boldsymbol{X}\boldsymbol{X}^T$ can be ordered so that

$$\lambda_1(\boldsymbol{X}\boldsymbol{X}^T) \geq \ldots \geq \lambda_r(\boldsymbol{X}\boldsymbol{X}^T) > \lambda_{r+1}(\boldsymbol{X}\boldsymbol{X}^T) = \ldots = \lambda_m(\boldsymbol{X}\boldsymbol{X}^T) = 0.$$

The $s_j(\boldsymbol{X}) := \sqrt{\lambda_j(\boldsymbol{X}\boldsymbol{X}^T)}$, $j = 1, \ldots, r$, are called **singular values** of $\boldsymbol{X}$. The positive eigenvalues of $\boldsymbol{X}\boldsymbol{X}^T$ and $\boldsymbol{X}^T\boldsymbol{X}$ are equal so that we can use $\boldsymbol{X}^T\boldsymbol{X}$ instead of $\boldsymbol{X}\boldsymbol{X}^T$ in the definition of singular values. Obviously, $\|\boldsymbol{X}\|_2 = s_1(\boldsymbol{X})$, and one can show that $\|\boldsymbol{X}^+\|_2 = s_r(\boldsymbol{X})^{-1}$.

One can show that the spectral matrix norm $\|\cdot\|_2$ and the Euclidean vector norm $\|\cdot\|_2$ are consistent. This motivates the index 2 of the spectral matrix norm once more. For orthonormal matrices $\|\boldsymbol{Q}\|_2 = 1$, since $\boldsymbol{Q}^T\boldsymbol{Q} = I$.
For the Frobenius norm one can show:
$\|\boldsymbol{X}\|_F^2 = \sum_{j=1}^r s_j^2(\boldsymbol{X})$ and $\|\boldsymbol{X}^+\|_F^2 = \sum_{j=1}^r (s_j^2(\boldsymbol{X}))^{-1}$.

3.1.2 *The Condition of the Linear Least Squares Problem*

Let us now start with the introduction of characterization measures for the difficulty of a data situation regarding the LLS problem. As such measures we will propose **condition numbers** of factor matrices and the angle between the response vector and its projection. For the verification of linear least squares solutions we are looking for a procedure to adequately cover the entire space of possible inputs, namely of the **factor matrices** and the **response vectors**, with as few examples as possible. For this, we will characterize the dependency of the accuracy of the least squares solution on the factor matrix $\boldsymbol{X}$ as well as on the response vector $\boldsymbol{y}$. Let us start, however, with the exact definition of the problem.

Definition 3.4: Linear Least Squares Problem, LLS Problem
Let $\boldsymbol{X} \in L(m,n)$ and $\boldsymbol{y} \in \mathbb{R}^m$. Then, $\hat{\boldsymbol{\beta}} \in \mathbb{R}^n$ with

$$\|\boldsymbol{y} - \boldsymbol{X}\hat{\boldsymbol{\beta}}\|_2 = \min_{\boldsymbol{\beta} \in \mathbb{R}^n} \|\boldsymbol{y} - \boldsymbol{X}\boldsymbol{\beta}\|_2$$

is called the **Linear Least Squares solution (LLS solution)** of the linear model $\boldsymbol{y} = \boldsymbol{X}\boldsymbol{\beta}$. Note that for the sake of simplicity the statistics notation $\boldsymbol{y} = \boldsymbol{X}\boldsymbol{\beta} + \boldsymbol{\epsilon}$ with error term $\boldsymbol{\epsilon}$ is not used in this chapter.

Note that $\hat{\boldsymbol{\beta}} := \boldsymbol{X}^+\boldsymbol{y}$ is a **minimum-norm LLS solution**. In what follows, we want to compute $\hat{\boldsymbol{\beta}} := \boldsymbol{X}^+\boldsymbol{y}$ either directly, e.g. by means of the normal equations (see (3.1)), or indirectly via the computation of the generalized inverse $\boldsymbol{X}^+$. Note also that the normal equations implicitly compute the generalized inverse, since $\boldsymbol{X}^+ = (\boldsymbol{X}^T\boldsymbol{X})^{-1}\boldsymbol{X}^T$ for matrices $\boldsymbol{X}$ with maximum column rank.

In practice, the factor matrix $\boldsymbol{X} \in L(m,n)$ as well as the response vector $\boldsymbol{y} \in \mathbb{R}^m$ of the LLS problem are not exactly known in general, e.g. because the measurement or the representation precision is finite. So it is desirable to find a measure for the sensitivity of an LLS problem to disturbances in the data $\boldsymbol{X}$, $\boldsymbol{y}$. The value of such a measure should be varied in the verification procedure in order to simulate different degrees of difficulty of the LLS problem. For this, condition numbers are useful.

Definition 3.5: Condition Numbers for the LLS Problem
If the elements of $\boldsymbol{X}$, $\boldsymbol{y}$ contain relative errors of maximum size δ, then let the relative error in the LLS solution $\hat{\boldsymbol{\beta}}$ be constrained by $f(\kappa)\delta$, where $f(\kappa)$ is a function of κ. Such measures κ are called **condition numbers** for the LLS problem.

An LLS problem is called **well-conditioned** iff the corresponding condition number is small.

Thus, a function of the condition number κ for the LLS problem serves as a multiplier of the relative input error in the bound of the relative error of the LLS solution.

What well-conditioned really means, i.e. how large a condition number may be, somewhat depends on the accuracy one wants to achieve and on the LLS solver used. In the literature bounds like $f(\kappa) < 1$ are in use. We will discuss whether this is reasonable.

In what follows, we will explain that the following characteristics of the matrix $\boldsymbol{X}$ may serve as condition numbers for the LLS problem:

Definition 3.6: Special Condition Numbers
Let $\boldsymbol{X} \in L(m,n)$ and $r := \text{rank}(\boldsymbol{X})$. Any $K(\boldsymbol{X}) := \|\boldsymbol{X}\|\|\boldsymbol{X}^+\|$ is called a **condition number** of the matrix $\boldsymbol{X}$ for the LLS problem.

Note that the matrix norm can be varied in this definition. $K_2(\boldsymbol{X}) := \|\boldsymbol{X}\|_2\|\boldsymbol{X}^+\|_2 = s_1(\boldsymbol{X})/s_r(\boldsymbol{X})$ is called **spectral condition number** of $\boldsymbol{X}$.

It holds that $K_2(\boldsymbol{X}\boldsymbol{X}^T) = K_2(\boldsymbol{X}^T\boldsymbol{X}) = K_2^2(\boldsymbol{X})$ since if $\boldsymbol{X}\boldsymbol{X}^T\beta = \lambda_i\beta$, then $\boldsymbol{X}\boldsymbol{X}^T\boldsymbol{X}\boldsymbol{X}^T\beta = \boldsymbol{X}\boldsymbol{X}^T\lambda_i\beta = \lambda_i^2\beta$.

$K_F(\boldsymbol{X}) := \|\boldsymbol{X}\|_F\|\boldsymbol{X}^+\|_F$ is called **F-condition number** of $\boldsymbol{X} \in L(m,n)$.

The following theorem gives a relationship between $K_F(\boldsymbol{X})$ and $K_2(\boldsymbol{X})$ and explains why $K_F(\boldsymbol{X})$ is also an LLS condition number if $K_2(\boldsymbol{X})$ is.

Theorem 3.1: Relations between Condition Numbers
$K_2(\boldsymbol{X}) \le K_F(\boldsymbol{X}) \le \min(m,n)K_2(\boldsymbol{X})$.

Proof. Let $r := \text{rank}(\boldsymbol{X}) \le \min(m,n)$. Since $\|\boldsymbol{X}\|_F^2 = \sum_{j=1}^r s_j^2(\boldsymbol{X})$ and $\|\boldsymbol{X}^+\|_F^2 = \sum_{j=1}^r (s_j^2(\boldsymbol{X}))^{-1}$, it holds that $s_1(\boldsymbol{X}) \le \|\boldsymbol{X}\|_F$, $(s_r(\boldsymbol{X}))^{-1} \le \|\boldsymbol{X}^+\|_F$, $\|\boldsymbol{X}\|_F \le \sqrt{r}s_1(\boldsymbol{X})$, and $\|\boldsymbol{X}^+\|_F \le \sqrt{r}(s_r(\boldsymbol{X}))^{-1}$.

Therefore, the theorem follows from $K_2(\boldsymbol{X}) = s_1(\boldsymbol{X})/s_r(\boldsymbol{X})$. □

Obviously, $K_F(\boldsymbol{X})$ is of the same magnitude as $K_2(\boldsymbol{X})$ for arbitrary $\boldsymbol{X} \in L(m,n)$. This motivates the equivalent usefulness of the two characteristics as condition numbers.

For ease of notation, the index 2 in K_2 is left out in what follows, making the spectral condition number the standard condition number.

Notice that the LLS problem can generally have more than one solution. $\boldsymbol{X}^+\boldsymbol{y}$ is a minimum-norm solution. A **unique LLS solution** exists, iff $\boldsymbol{X}$ has full column rank, since only then $\boldsymbol{X}^T\boldsymbol{X}$ can be inverted, and the so-called **normal equations** $\boldsymbol{X}^T\boldsymbol{X}\hat{\beta} = \boldsymbol{X}^T\boldsymbol{y}$ (see below) can be solved uniquely.

From now on, all results presented apply solely to LLS problems with **factor matrices with full column rank**. Thus, in the following let:

$$\boldsymbol{X} \in L(m,n), \quad m \geq n, \quad \text{rank}(\boldsymbol{X}) = n.$$

Normal Equations
Historically, Gauss's normal equations are the first proposed LLS solver. However, the numerical problems in the solution of the LLS problem by the normal equations became obvious early after the introduction of computers. It was realized that the **normal equations**

$$\boldsymbol{X}^T\boldsymbol{X}\hat{\boldsymbol{\beta}} = \boldsymbol{X}^T\boldsymbol{y} \tag{3.1}$$

(cp. Proposition 4.3) are rarely adequate for the solution of not very well-conditioned LLS problems on a computer, because the condition of the normal equations is extremely bad for not well-conditioned LLS problems since

$$K(\boldsymbol{X}^T\boldsymbol{X}) = K^2(\boldsymbol{X}).$$

Alternatives
In the beginning of the 1960s so-called **orthogonalization methods** were proposed for the solution of LLS problems (Golub, 1965; Bauer, 1965).

This leads to a decomposition $\boldsymbol{X} = \boldsymbol{Q}\boldsymbol{R}$, where $\boldsymbol{Q}^T\boldsymbol{Q} = \boldsymbol{I}$ and $\boldsymbol{R}$ is a **nonsingular** upper triangle matrix. By this, the normal equations $\boldsymbol{X}^T\boldsymbol{X}\hat{\boldsymbol{\beta}} = \boldsymbol{X}^T\boldsymbol{y}$ are transformed to $\boldsymbol{R}^T\boldsymbol{Q}^T\boldsymbol{Q}\boldsymbol{R}\hat{\boldsymbol{\beta}} = \boldsymbol{R}^T\boldsymbol{Q}^T\boldsymbol{y}$. Thus,

$$\boldsymbol{R}\hat{\boldsymbol{\beta}} = \boldsymbol{Q}^T\boldsymbol{y} \tag{3.2}$$

and since $\boldsymbol{Q}^T\boldsymbol{Q} = \boldsymbol{I}$ it holds that

$$K(\boldsymbol{X}) = K(\boldsymbol{Q}\boldsymbol{R}) = K(\boldsymbol{R}).$$

This leads to the hope to be able to avoid the square of the condition number by means of orthogonalization. Unfortunately, not much later this hope proved to be deceptive, since already Golub and Wilkinson (1966) published a general upper bound for the absolute error in the LLS solution of a "disturbed linear model", which contains the factor $K^2(\boldsymbol{X})$:

Theorem 3.2: Upper Bound for the Absolute Error in Least Squares Estimation
Let $\boldsymbol{X}, \boldsymbol{dX} \in L(m,n)$, $m \geq n$, rank$(\boldsymbol{X}) = n$ and $\boldsymbol{y}, \boldsymbol{dy} \in \mathbb{R}^m$, as well as $\|\boldsymbol{X}\|_2 = 1, \|\boldsymbol{y}\|_2 = 1$ and $\|\boldsymbol{dX}\|_2, \|\boldsymbol{dy}\|_2 \leq \delta, 0 < \delta \ll 1$.
Moreover, let $\hat{\boldsymbol{\beta}}$ be the LLS solution of $\boldsymbol{y} = \boldsymbol{X}\boldsymbol{\beta}$, $r_0 := \boldsymbol{y} - \boldsymbol{X}\hat{\boldsymbol{\beta}}$ and $\hat{\boldsymbol{\beta}}_{\boldsymbol{d}}$ the LLS solution of the disturbed model $\boldsymbol{y} + \boldsymbol{dy} = (\boldsymbol{X} + \boldsymbol{dX})\boldsymbol{\beta}$. Then:

$$\|\hat{\boldsymbol{\beta}}_{\boldsymbol{d}} - \hat{\boldsymbol{\beta}}\|_2 \leq \delta(K^2(\boldsymbol{X})\|r_0\|_2 + K(\boldsymbol{X})\|\hat{\boldsymbol{\beta}}\|_2 + K(\boldsymbol{X})) + O(\delta^2). \tag{3.3}$$

Proof. See Golub and Wilkinson (1966, pp. 143 – 144). □

Notice that here the idea is to study the effect of round-off errors caused by the numerical solution of the LLS problem as if they were induced by disturbed data. Since, indeed, it could be shown that the methods of Golub and Bauer caused disturbances $\boldsymbol{dX}$ and $\boldsymbol{dy}$ with

$$\|\boldsymbol{dX}\|_2 \leq \delta\|\boldsymbol{X}\|_2,\ \|\boldsymbol{dy}\|_2 \leq \delta\|\boldsymbol{y}\|_2 \quad \text{with (generally)} \quad 0 < \delta \ll 1$$

(see Wilkinson, 1965, pp. 153–160 and Björck, 1967, p. 20), the bound (3.3) really delivers an upper bound for the absolute error in the computed LLS solution when an orthogonalization method is used. Thus, for these methods $K^2(\boldsymbol{X})$ is relevant in the bound (3.3) for not small residuals $\boldsymbol{r}_0$.

This, at the first time, gave an idea about the dependency of the absolute error in the LLS solution on the input error and the spectral condition number, explaining that $K(\boldsymbol{X})$ is a condition number for the LLS problem. However, Golub and Wilkinson only dealt with the special case $\|\boldsymbol{X}\|_2 = 1, \|\boldsymbol{y}\|_2 = 1$ and did not deliver a lower bound for the absolute error in the LLS solution of a disturbed linear model. Thus, hope remained that the real error is heavily overestimated by (3.3).

Upper and Lower Bounds for the Absolute Error

In 1975 van der Sluis delivered **lower bounds** for the error in the LLS solution of a disturbed linear model **that can occur in the worst case** (he argued that without specific information about the factor matrix, the selected method, etc., the worst case has to be assumed realistic!). Unfortunately, the results of van der Sluis did not give reason for rejoicing since they show that the upper bounds of Golub and Wilkinson are realistic, at least in the case $\|\boldsymbol{X}\|_2 = 1$:

Theorem 3.3: Bounds for the Absolute Error in Least Squares Estimation
Let $\boldsymbol{X}, \boldsymbol{dX} \in L(m,n)$, $m \geq n$, $\text{rank}(\boldsymbol{X}) = n$, and $\boldsymbol{y}, \boldsymbol{dy} \in \mathbb{R}^m$, as well as

$$\|\boldsymbol{dX}\|_2 \leq \delta\|\boldsymbol{X}\|_2, \quad \|\boldsymbol{dy}\|_2 \leq \delta\|\boldsymbol{y}\|_2 \quad \text{and} \quad \mu := \delta\frac{s_1(\boldsymbol{X})}{s_n(\boldsymbol{X})} = \delta K(\boldsymbol{X}) < 1.$$

Furthermore, let $\hat{\beta}$ be the LLS solution of the system $\boldsymbol{y} = \boldsymbol{X}\beta$, let $\boldsymbol{r}_0 := \boldsymbol{y} - \boldsymbol{X}\hat{\beta}$ be the corresponding residual, and let $\boldsymbol{d}\hat{\beta}$ chosen so that $(\hat{\beta} + \boldsymbol{d}\hat{\beta})$ is the LLS solution of the disturbed system $\boldsymbol{y} + \boldsymbol{dy} = (\boldsymbol{X} + \boldsymbol{dX})(\beta + \boldsymbol{d}\beta)$. Then:

1. For every pair $(\boldsymbol{X}, \boldsymbol{y})$ and any kind of disturbance $(\boldsymbol{dX}, \boldsymbol{dy})$ it is valid that

$$\|\boldsymbol{d}\hat{\beta}\|_2 \leq \frac{\delta}{s_n(\boldsymbol{X})}\left[\frac{s_1(\boldsymbol{X})\|\boldsymbol{r}_0\|_2}{s_n(\boldsymbol{X})(1-\mu^2)} + \frac{s_1(\boldsymbol{X})\|\hat{\beta}\|_2}{1-\mu} + \frac{\|\boldsymbol{y}\|_2}{1-\mu}\right].$$

2. For every pair $(\boldsymbol{X}, \boldsymbol{y})$ there is a disturbance $(\boldsymbol{dX}, \boldsymbol{dy})$ so that

$$\|\boldsymbol{d}\hat{\boldsymbol{\beta}}\|_2 \geq \frac{\delta}{s_n(\boldsymbol{X})}\left[\frac{s_1(\boldsymbol{X})\|\boldsymbol{r}_0\|_2}{s_n(\boldsymbol{X})(1-\mu^2)} + \frac{\|\boldsymbol{y}\|_2}{1-\mu^2}\right].$$

3. For every pair $(\boldsymbol{X}, \boldsymbol{y})$ there is a "disturbance" $(\boldsymbol{dX}, \boldsymbol{dy})$ so that

$$\|\boldsymbol{d}\hat{\boldsymbol{\beta}}\|_2 \geq \frac{\delta}{s_n(\boldsymbol{X})}\left[s_1(\boldsymbol{X})\|\hat{\boldsymbol{\beta}}\|_2 + \|\boldsymbol{y}\|_2\right].$$

Proof. See van der Sluis (1975, pp. 246 – 248, Theorem 4.3). □

Notice that $K^2(\boldsymbol{X})$ appears in the upper bound (1) as well as in the first lower bound (2) if $\|\boldsymbol{X}\|_2 = 1$. Thus, the term $K^2(\boldsymbol{X})$ appears to be realistic for not small residuals $\boldsymbol{r}_0$, at least in the worst case. But how realistic is the worst case? Shouldn't one expect the worst case to be exceptional so that in the normal case one could hope for a more favorable result? Moreover, what will happen when $\|\boldsymbol{X}\|_2 \neq 1$? We will study such questions by our verification procedure.

Bounds for Relative Errors

Notice also that the original definition of a condition number for the LLS problem is a statement about the relative error in the LLS solution. Thus, we will restate the van der Sluis bounds for relative errors.

Corollary 3.1: Bounds for Relative Errors

Under the conditions of Theorem 3.3 the relative error in the LLS solution has the bounds

$$\begin{aligned}
\frac{\|\boldsymbol{d}\hat{\boldsymbol{\beta}}\|_2}{\|\hat{\boldsymbol{\beta}}\|_2} &\leq \delta K(\boldsymbol{X})\left[K(\boldsymbol{X})\frac{\tan\phi(\boldsymbol{y})}{1-\mu^2} + \frac{1}{1-\mu}\left(1 + \frac{1}{\cos\phi(\boldsymbol{y})}\right)\right], \\
\frac{\|\boldsymbol{d}\hat{\boldsymbol{\beta}}\|_2}{\|\hat{\boldsymbol{\beta}}\|_2} &\geq \frac{\delta}{s_n(\boldsymbol{X})(1-\mu^2)}\left[\frac{s_1(\boldsymbol{X})\|\boldsymbol{r}_0\|_2}{s_n(\boldsymbol{X})\|\hat{\boldsymbol{\beta}}\|_2} + \frac{\|\boldsymbol{y}\|_2}{\|\hat{\boldsymbol{\beta}}\|_2}\right], \\
\frac{\|\boldsymbol{d}\hat{\boldsymbol{\beta}}\|_2}{\|\hat{\boldsymbol{\beta}}\|_2} &\geq \frac{\delta}{s_n(\boldsymbol{X})}\left[s_1(\boldsymbol{X}) + \frac{\|\boldsymbol{y}\|_2}{\|\hat{\boldsymbol{\beta}}\|_2}\right],
\end{aligned}$$

where $\phi(\boldsymbol{y})$ is the angle between $\boldsymbol{y}$ and $\boldsymbol{y}_0 := \boldsymbol{X}\hat{\boldsymbol{\beta}}$.

Proof. 1.) Upper bound:

$$\begin{aligned}
\frac{\|d\hat{\beta}\|_2}{\|\hat{\beta}\|_2} &\leq \frac{\delta}{s_n(X)}\left[\frac{s_1(X)\|r_0\|_2}{s_n(X)\|\hat{\beta}\|_2(1-\mu^2)} + \frac{s_1(X)}{1-\mu} + \frac{\|y\|_2}{\|\hat{\beta}\|_2(1-\mu)}\right] \\
&= \delta K(X)\left[\|X^+\|\frac{\|r_0\|}{\|\hat{\beta}\|(1-\mu^2)} + \frac{1}{1-\mu}\right] + \delta\|X^+\|\frac{\|y\|}{\|\hat{\beta}\|(1-\mu)} \\
&\leq \delta K(X)\left[K(X)\frac{\|r_0\|}{\|y_0\|(1-\mu^2)} + \frac{1}{1-\mu} + \frac{\|y\|}{\|y_0\|(1-\mu)}\right], \\
&\quad \text{since } \|y_0\| = \|X\hat{\beta}\| \leq \|X\|\|\hat{\beta}\|, \\
&= \delta K(X)\left[K(X)\frac{\tan\phi(y)}{1-\mu^2} + \frac{1}{1-\mu}\left(1+\frac{1}{\cos\phi(y)}\right)\right],
\end{aligned}$$

where the last equality is to be proven.
r_0 is orthogonal to y_0 ($r_0 \perp y_0$), since

$$\begin{aligned}
r_0^T y_0 &= (y - X\hat{\beta})^T X\hat{\beta} = y^T XX^+ y - y^T X^{+T} X^T XX^+ y \\
&= y^T XX^+ y - y^T XX^+ XX^+ y = 0.
\end{aligned}$$

Then the relevant trigonometrical functions can be written as

$$\begin{aligned}
\cos\phi(y) &:= \frac{y_0^T y}{\|y_0\|_2\|y\|_2} = \frac{y_0^T(y_0+r_0)}{\|y_0\|_2\|y\|_2} \overset{r_0\perp y_0}{=} \frac{\|y_0\|_2^2}{\|y_0\|_2\|y\|_2} = \frac{\|y_0\|_2}{\|y\|_2}, \\
\cos^2\phi(y) &= \frac{\|y_0\|_2^2}{\|y\|_2^2} = \frac{\|y_0\|_2^2}{\|y_0\|_2^2 + \|r_0\|_2^2},
\end{aligned}$$

since $\|y\|_2^2 = (y_0+r_0)^T(y_0+r_0) = \|y_0\|_2^2 + \|r_0\|_2^2$, and

$$\begin{aligned}
\sin^2\phi(y) &= 1 - \cos^2\phi(y) = \frac{\|r_0\|_2^2}{\|y_0\|_2^2 + \|r_0\|_2^2} = \frac{\|r_0\|_2^2}{\|y\|_2^2}, \\
\tan\phi(y) &= \frac{\sin\phi(y)}{\cos\phi(y)} = \frac{\|r_0\|_2}{\|y_0\|_2}.
\end{aligned}$$

This proves the last equality in the above upper bound for the relative error in the LLS solution.

2.) The lower bounds are obvious. □

Notice that the derivation of the upper bound with tangent and cosine terms cannot be generalized to the lower bounds since the derived upper bound is even larger than the van der Sluis bound.

Relevance
According to Theorem 3.3, (1) and (2), an amplification factor

$$K^2(\boldsymbol{X})\frac{\|\boldsymbol{r}_0\|_2}{1-\mu^2}$$

of the relative input error δ is not unrealistic if $1 = \|\boldsymbol{X}\|_2 = s_1(\boldsymbol{X})$. Otherwise, the amplification factor

$$\frac{s_1(\boldsymbol{X})}{s_n^2(\boldsymbol{X})}\frac{\|\boldsymbol{r}_0\|_2}{1-\mu^2} = \frac{K(\boldsymbol{X})}{s_n(\boldsymbol{X})}\frac{\|\boldsymbol{r}_0\|_2}{1-\mu^2}$$

is to be expected.

From the upper bound of the relative output error it can be deduced that $K^2(\boldsymbol{X})\frac{\tan\phi(\boldsymbol{y})}{(1-\mu^2)}$ is an amplification factor for δ. On the one hand, $\frac{\tan\phi(\boldsymbol{y})}{(1-\mu^2)}$ is especially large when the angle $\phi(\boldsymbol{y})$ is near 90° and the factor $(1-\mu^2)$ is near 0, i.e., $\mu := \delta K(\boldsymbol{X})$ is near 1, i.e. the condition number $K(\boldsymbol{X})$ nearly compensates the relative input error δ. On the other hand, $\frac{\tan\phi(\boldsymbol{y})}{(1-\mu^2)}$ is especially small when the angle $\phi(\boldsymbol{y})$ is near 0° and the factor $(1-\mu^2)$ is near 1, i.e., $\mu := \delta K(\boldsymbol{X})$ is near 0, i.e. the condition number $K(\boldsymbol{X})$ is small and cannot compensate the relative input error δ. In the latter case, the factor $K^2(\boldsymbol{X})$ can even be reduced by $\frac{\tan\phi(\boldsymbol{y})}{(1-\mu^2)}$. Thus, for small angles and small condition numbers there might be a chance that the factor $K^2(\boldsymbol{X})$ can be avoided.

If the factor matrix $\boldsymbol{X}$ has maximum column rank, the results of van der Sluis show that $K(\boldsymbol{X})$ is a condition number for the LLS problem. Unfortunately, the lower bounds reveal the importance of the terms of the upper bound. Nevertheless, the results of van der Sluis do not indicate how realistic the bounds are in practice. The intention of this chapter is thus to analyze by means of adequate verification problems, how realistic it is to assume a dependence of the error in the LLS solution on $K^2(\boldsymbol{X})$ for different types of matrices, response vectors, and LLS solvers.

3.1.3 *Summary*

In this section we introduced the spectral and the Frobenius condition numbers $K(\boldsymbol{X})$ and $K_F(\boldsymbol{X})$ and explained why they can act as characterizations of the difficulty of the LLS problem $\|\boldsymbol{y}-\boldsymbol{X}\hat{\beta}\|_2 = \min_{\beta\in\mathbb{R}^n}\|\boldsymbol{y}-\boldsymbol{X}\beta\|_2$. We stated that for the relative error in the LLS solution $\hat{\beta}$ the square of the condition numbers have to be expected as an amplification factor for the relative error in the data. We explained why this can be avoided for small condition numbers and small angles between the response vector and its projection on

$\mathrm{im}(\boldsymbol{X})$. As a consequence, in the following we are looking for test matrices $\boldsymbol{X}$ with specified condition numbers and for response vectors $\boldsymbol{y}$ with specified angles to their projections.

3.2 Theory

3.2.1 LLS Solvers

Before we describe the verification procedure, let us first introduce the LLS solvers used for the verification of the bounds. Notice that not the solvers are intended to be compared primarily. It rather should be estimated how realistic the dependency of the relative error in the LLS solutions on the square of the condition number is in practice. However, in order to realize differences between different LLS solvers concerning precision of solutions, different types of solvers will be used: the normal equations will be compared to an orthogonalization method and to a method that can easily adapt the LLS solution to new data.

The most common method to calculate the least squares solution is the **normal equations**. It is known, however, that this method is numerically problematic already for not very badly conditioned LLS problems because the condition number of $\boldsymbol{X}^T\boldsymbol{X}$ is equal to the square of the condition number of $\boldsymbol{X}$. This resulted in the development of many alternative methods trying to avoid this problem. In the following, only two of these methods are introduced: at first the Gram-Schmidt method, an orthogonalization process, and then the method of Greville that turns out to be particularly useful in the construction of test matrices. There are certainly far more methods to construct the LLS solution, which are not discussed here, cp., e.g., Lawson and Hanson (1974).

3.2.1.1 Gram-Schmidt Method

The Gram-Schmidt (GS) orthogonalization process produces a so-called **full-rank decomposition (frd)**, which substantially simplifies the computation of the generalized inverse $\boldsymbol{X}^+$ of $\boldsymbol{X}$.

Definition 3.7: Full-Rank Decomposition
Let $\boldsymbol{X} \in L(m,n)$, $m \geq n$, $\mathrm{rank}(\boldsymbol{X}) = r > 0$. A decomposition of $\boldsymbol{X}$ so that

$$\boldsymbol{X} = \boldsymbol{B}\boldsymbol{C} \quad \text{with} \quad \boldsymbol{B} \in L(m,r),\ \boldsymbol{C} \in L(r,n),\ \mathrm{rank}(B) = \mathrm{rank}(C) = r \quad (3.4)$$

is called **full-rank decomposition (frd)** of $\boldsymbol{X}$.

Obviously, matrix $\boldsymbol{B}$ has maximum column rank, and matrix $\boldsymbol{C}$ has maximum row rank. This motivates the term full-rank decomposition.

Theorem 3.4: Generalized Inverse for frd
If $\boldsymbol{X} = \boldsymbol{B}\boldsymbol{C}$ is an frd of $\boldsymbol{X}$, then

$$\boldsymbol{X}^+ = \boldsymbol{C}^T(\boldsymbol{C}\boldsymbol{C}^T)^{-1}(\boldsymbol{B}^T\boldsymbol{B})^{-1}\boldsymbol{B}^T. \tag{3.5}$$

Proof. Since $\boldsymbol{C}$ has maximum row rank, $(\boldsymbol{C}\boldsymbol{C}^T)^{-1}$ exists. Since $\boldsymbol{B}$ has maximum column rank, $(\boldsymbol{B}^T\boldsymbol{B})^{-1}$ exists. Thus, only the conditions for a generalized inverse have to be checked.

$$\begin{aligned}
\boldsymbol{X}\boldsymbol{X}^+\boldsymbol{X} &= \boldsymbol{B}\boldsymbol{C}\boldsymbol{C}^T(\boldsymbol{C}\boldsymbol{C}^T)^{-1}(\boldsymbol{B}^T\boldsymbol{B})^{-1}\boldsymbol{B}^T\boldsymbol{B}\boldsymbol{C} = \boldsymbol{B}\boldsymbol{C} = \boldsymbol{X},\\
\boldsymbol{X}^+\boldsymbol{X}\boldsymbol{X}^+ &= \boldsymbol{C}^T(\boldsymbol{C}\boldsymbol{C}^T)^{-1}(\boldsymbol{B}^T\boldsymbol{B})^{-1}\boldsymbol{B}^T\boldsymbol{B}\boldsymbol{C}\boldsymbol{C}^T(\boldsymbol{C}\boldsymbol{C}^T)^{-1}(\boldsymbol{B}^T\boldsymbol{B})^{-1}\boldsymbol{B}^T\\
&= \boldsymbol{C}^T(\boldsymbol{C}\boldsymbol{C}^T)^{-1}(\boldsymbol{B}^T\boldsymbol{B})^{-1}\boldsymbol{B}^T = \boldsymbol{X}^+,\\
\boldsymbol{X}\boldsymbol{X}^+ &= \boldsymbol{B}\boldsymbol{C}\boldsymbol{C}^T(\boldsymbol{C}\boldsymbol{C}^T)^{-1}(\boldsymbol{B}^T\boldsymbol{B})^{-1}\boldsymbol{B}^T\\
&= \boldsymbol{B}(\boldsymbol{B}^T\boldsymbol{B})^{-1}\boldsymbol{B}^T \quad \text{symmetric},\\
\boldsymbol{X}^+\boldsymbol{X} &= \boldsymbol{C}^T(\boldsymbol{C}\boldsymbol{C}^T)^{-1}(\boldsymbol{B}^T\boldsymbol{B})^{-1}\boldsymbol{B}^T\boldsymbol{B}\boldsymbol{C}\\
&= \boldsymbol{C}^T(\boldsymbol{C}\boldsymbol{C}^T)^{-1}\boldsymbol{C} \quad \text{symmetric}.
\end{aligned}$$

□

In fact, the Gram-Schmidt orthogonalization process produces a special frd of a matrix $\boldsymbol{X}$, namely a so-called **triangular decomposition** since the matrix C is upper triangular. This means that at least one of the matrices $\boldsymbol{B}, \boldsymbol{C}$ is a triangular matrix. Apparently, for the determination of the frd the **correct assignment of the column rank** r of the corresponding matrix $\boldsymbol{X}$ becomes critical; thus, if r is assigned incorrectly, then the detected decomposition of $\boldsymbol{X}$ cannot be an frd and Theorem 3.4 is not applicable. Therefore, methods that deal with triangular decompositions generally use so-called **pivot strategies** for the rank assignment, i.e. adequate **column or row re-orderings**. This implies that instead of the frd of the matrix $\boldsymbol{X}$ we need to construct the frd of

$$\tilde{\boldsymbol{X}} = \boldsymbol{P}_1\boldsymbol{X}\boldsymbol{P}_2, \text{ where} \tag{3.6}$$

$\boldsymbol{P}_1 \in L(m,m)$ and $\boldsymbol{P}_2 \in L(n,n)$ are permutation matrices.

Corollary 3.2: Generalized Inverses for frd with Permutations
If $\tilde{\boldsymbol{X}} = \boldsymbol{B}\boldsymbol{C}$ is an frd of $\tilde{\boldsymbol{X}}$, then $\boldsymbol{B}\boldsymbol{C} = \tilde{\boldsymbol{X}} = \boldsymbol{P}_1\boldsymbol{X}\boldsymbol{P}_2$.

Since $\boldsymbol{P}_i^T = \boldsymbol{P}_i^{-1}$, $i = 1,2$, it follows that $\boldsymbol{X} = \boldsymbol{P}_1^T\tilde{\boldsymbol{X}}\boldsymbol{P}_2^T$, and therefore because of Theorem 3.4:

$$\boldsymbol{X}^+ = \boldsymbol{P}_2\tilde{\boldsymbol{X}}^+\boldsymbol{P}_1 = \boldsymbol{P}_2\boldsymbol{C}^T(\boldsymbol{C}\boldsymbol{C}^T)^{-1}(\boldsymbol{B}^T\boldsymbol{B})^{-1}\boldsymbol{B}^T\boldsymbol{P}_1. \tag{3.7}$$

Proof.

$$\begin{aligned}
\boldsymbol{X}\boldsymbol{X}^+\boldsymbol{X} &= P_1^T\tilde{\boldsymbol{X}}P_2^T P_2\tilde{\boldsymbol{X}}^+P_1P_1^T\tilde{\boldsymbol{X}}P_2^T \\
&= P_1^T\tilde{\boldsymbol{X}}\tilde{\boldsymbol{X}}^+\tilde{\boldsymbol{X}}P_2^T = P_1^T\tilde{\boldsymbol{X}}P_2^T = \boldsymbol{X}, \qquad (3.8)\\
\boldsymbol{X}^+\boldsymbol{X}\boldsymbol{X}^+ &= P_2\tilde{\boldsymbol{X}}^+P_1P_1^T\tilde{\boldsymbol{X}}P_2^T P_2\tilde{\boldsymbol{X}}^+P_1 \\
&= P_2\tilde{\boldsymbol{X}}^+\tilde{\boldsymbol{X}}\tilde{\boldsymbol{X}}^+P_1 = P_2\tilde{\boldsymbol{X}}^+P_1 = \boldsymbol{X}^+, \qquad (3.9)\\
\boldsymbol{X}\boldsymbol{X}^+ &= P_1^T\tilde{\boldsymbol{X}}P_2^T P_2\tilde{\boldsymbol{X}}^+P_1 = P_1^T\tilde{\boldsymbol{X}}\tilde{\boldsymbol{X}}^+P_1 \quad \text{symmetric}, \qquad (3.10)\\
\boldsymbol{X}^+\boldsymbol{X} &= P_2\tilde{\boldsymbol{X}}^+P_1P_1^T\tilde{\boldsymbol{X}}P_2^T = P_2\tilde{\boldsymbol{X}}^+\tilde{\boldsymbol{X}}P_2^T \quad \text{symmetric.} \quad \square
\end{aligned}$$

In the Gram-Schmidt orthogonalization process the **linear independent column vectors of $\boldsymbol{X}$ are orthogonalized**. For the classical description of the Gram-Schmidt process (GS process) for the orthogonalization of k linear independent vectors $\boldsymbol{x}_1^{(1)}, \ldots, \boldsymbol{x}_k^{(1)}$, see Algorithm 3.1.

Algorithm 3.1 Gram-Schmidt Orthogonalization (GS1)

Require: $\boldsymbol{X} = [\boldsymbol{x}_1^{(1)} \ldots \boldsymbol{x}_k^{(1)}]$ a column representation of $\boldsymbol{X}$

$\boldsymbol{q}_1 \leftarrow \boldsymbol{x}_1^{(1)}$
$d_1 \leftarrow \boldsymbol{q}_1^T\boldsymbol{q}_1$
for $s = 2$ **to** k **do**
  $\boldsymbol{q}_s \leftarrow \boldsymbol{x}_s^{(1)}$
  for $i = 1$ **to** $s-1$ **do**
    $r_{is} \leftarrow (\boldsymbol{x}_s^{(1)})^T\boldsymbol{q}_i/d_i$
    $\boldsymbol{q}_s \leftarrow \boldsymbol{q}_s - r_{is}\boldsymbol{q}_i$
  end for
  $d_s \leftarrow \boldsymbol{q}_s^T\boldsymbol{q}_s$
end for

The following proposition interprets the GS process geometrically.

Proposition 3.2: Geometrical Interpretation of Gram-Schmidt Orthogonalization 1
Geometrically, Algorithm 3.1 can be interpreted as follows: In step s, $s = 2, \ldots, k$, transform the vector s so that it is orthogonal to each of the $(s-1)$ vectors already orthogonalized.

Proof. In step s, we have

$$\boldsymbol{q}_j^T\boldsymbol{q}_s = \boldsymbol{q}_j^T\left(\boldsymbol{x}_s^{(1)} - \sum_{i=1}^{s-1}\frac{\boldsymbol{q}_i^T\boldsymbol{x}_s^{(1)}}{\boldsymbol{q}_i^T\boldsymbol{q}_i}\boldsymbol{q}_i\right) = \boldsymbol{q}_j^T\boldsymbol{x}_s^{(1)} - \boldsymbol{q}_j^T\boldsymbol{x}_s^{(1)} = 0$$

for $j = 1, \ldots, s-1$, since $\boldsymbol{q}_j^T \boldsymbol{q}_i = 0$ for all $j \neq i$. □

An obvious variant of the GS process transforms all vectors $s+1, \ldots, k$ in step s so that they are orthogonal to vector s. This variant is called GS2 (see Algorithm 3.2).

Algorithm 3.2 Gram-Schmidt Orthogonalization (GS2)

Require: $X = [\boldsymbol{x}_1^{(1)} \ldots \boldsymbol{x}_k^{(1)}]$

for $s = 1$ **to** $k-1$ **do**
  $\boldsymbol{q}_s \leftarrow \boldsymbol{x}_s^{(s)}$
  $d_s \leftarrow \boldsymbol{q}_s^T \boldsymbol{q}_s$
  for $i = s+1$ **to** k **do**
    $r_{si} \leftarrow (\boldsymbol{x}_i^{(1)})^T \boldsymbol{q}_s / d_s$
    $\boldsymbol{x}_i^{(s+1)} \leftarrow \boldsymbol{x}_i^{(s)} - r_{si} \boldsymbol{q}_s$
  end for
end for
$\boldsymbol{q}_k \leftarrow \boldsymbol{x}_k^{(k)}$
$d_k \leftarrow \boldsymbol{q}_k^T \boldsymbol{q}_k$

Again, the geometrical interpretation helps to understand the steps of the algorithm.

Proposition 3.3: Geometrical Interpretation of Gram-Schmidt Orthogonalization 2
In step s, $s = 1, \ldots, k-1$, algorithm GS2 transforms the vectors $s+1, \ldots, k$ so that they are orthogonal to vector s.

Proof. For $i = s+1, \ldots, k$:

$$\begin{aligned}
\boldsymbol{x}_i^{(s+1)^T} \boldsymbol{q}_s &= \left(\boldsymbol{x}_i^{(s)} - \frac{\boldsymbol{x}_i^{(1)^T} \boldsymbol{q}_s}{d_s} \boldsymbol{q}_s \right)^T \boldsymbol{q}_s \\
&= (\boldsymbol{x}_i^{(s)} - \boldsymbol{x}_i^{(1)})^T \boldsymbol{q}_s \quad \text{since } d_s := \boldsymbol{q}_s^T \boldsymbol{q}_s \qquad \text{(a)} \\
&= (-r_{s-1,i}\boldsymbol{q}_{s-1} - \ldots - r_{1,i}\boldsymbol{q}_1)^T \boldsymbol{q}_s \\
&= 0
\end{aligned}$$

since $\boldsymbol{q}_s$ is orthogonal to all $\boldsymbol{q}_j$, $j = 1, \ldots, s-1$. □

Obviously, the number of operations is equal in the two algorithms. However, algorithm GS2 possibly needs more storage since

1. in algorithm GS1 $\boldsymbol{x}_s^{(1)}$ is not needed anymore in step $(s+1)$, and thus $\boldsymbol{q}_s$ can take the place of $\boldsymbol{x}_s^{(1)}$ if $\boldsymbol{x}_s^{(1)}$ is not needed otherwise, and
2. in algorithm GS2 $\boldsymbol{x}_i^{(j)}$, $j < i$, cannot take the place of $\boldsymbol{x}_i^{(1)}$ since $\boldsymbol{x}_i^{(1)}$ is also needed in the next steps.

However, algorithm GS2 can be revised so that this disadvantage is removed. Notice that from (a) in the proof of Proposition 3.3 it holds that

$$(\boldsymbol{x}_i^{(s)})^T \boldsymbol{q}_s = (\boldsymbol{x}_i^{(1)})^T \boldsymbol{q}_s.$$

This leads to the so-called modified Gram-Schmidt process (MGSprocess) given in Algorithm 3.3.

Algorithm 3.3 Modified Gram-Schmidt Orthogonalization (MGS)

Require: $\boldsymbol{X} = [\boldsymbol{x}_1^{(1)} \dots \boldsymbol{x}_k^{(1)}]$

for $s = 1$ **to** $k-1$ **do**
    $\boldsymbol{q}_s \leftarrow \boldsymbol{x}_s^{(s)}$
    $d_s \leftarrow \boldsymbol{q}_s^T \boldsymbol{q}_s$
    for $i = s+1$ **to** k **do**
        $r'_{si} \leftarrow (\boldsymbol{x}_i^{(s)})^T \boldsymbol{q}_s / d_s$
        $\boldsymbol{x}_i^{(s+1)} \leftarrow \boldsymbol{x}_i^{(s)} - r'_{si} \boldsymbol{q}_s$
    end for
end for
$\boldsymbol{q}_k \leftarrow \boldsymbol{x}_k^{(k)}$
$d_k \leftarrow \boldsymbol{q}_k^T \boldsymbol{q}_k$

Obviously, in this algorithm $\boldsymbol{x}_i^{(s)}$ can take the storage of $\boldsymbol{x}_i^{(1)}$ if the values of $\boldsymbol{x}_i^{(1)}$ are not needed otherwise.

The classical GS process is known to be numerically unstable. This was already indicated by results of Rice (1966). These results also suggested that the MGS process is numerically substantially more stable than the GS process. However, this process seems to have been used considerably earlier instead of the classical GS process (cp. Björck, 1967, p. 20 and Peters and Wilkinson, 1970, p. 313).

The reason why the MGS process is numerically more stable than the GS process, obviously, has to be related to the fact that $(\boldsymbol{x}_i^{(s)})^T \boldsymbol{q}_s$ is used instead of $(\boldsymbol{x}_i^{(1)})^T \boldsymbol{q}_s$ in the orthogonalization.

Indeed, the usage of $(\boldsymbol{x}_i^{(1)})^T \boldsymbol{q}_s$ is dangerous since the two terms in the scalar product lie $(s-1)$ orthogonalization steps apart and the goodness of orthogonality becomes worse for larger s because of round-off errors in $\boldsymbol{q}_s$. In contrast, in the MGS process, $\boldsymbol{x}_i^{(s)}$, $i = s+1, \ldots, k$, and $\boldsymbol{q}_s = \boldsymbol{x}_s^{(s)}$ are both computed as results of the orthogonalization with regard to $\boldsymbol{q}_{s-1}$. Thus, the orthogonalization optimally applies to the results of the previous orthogonalization steps. Because of the numerical problems of algorithms GS1 and GS2, we only concentrate on algorithm MGS in the verification of numerical stability of the construction of the generalized inverse $\boldsymbol{X}^+$ used for the construction of the LLS solution.

As already mentioned in the beginning of this paragraph, the MGS process constructs a special frd of the factor matrix $\boldsymbol{X} \in L(m,n)$, namely a so-called triangular decomposition. In order to find the correct rank of $\boldsymbol{X}$ a so-called **pivot strategy** (Peters and Wilkinson, 1970, p. 313) is applied ensuring the column with maximum norm is processed next (see Algorithm 3.4).

Algorithm 3.4 Pivot Strategy (PS)

Require: $\boldsymbol{X} = [\boldsymbol{x}_1^{(1)} \ldots \boldsymbol{x}_n^{(1)}]$

In each step s of Algorithm 3.3 apply a so-called **pivot-strategy**:
for $s = 1$ **to** $n-1$ **do**
    $i_0 \leftarrow \arg\max_{i=s,\ldots,n} \|\boldsymbol{x}_i^{(s)}\|_2$ (take the 1st such index)
    if $s \neq i_0$ **then** interchange columns s and i_0 (use the same column numbering $i = s, \ldots, n$ after the interchange as before)
    Orthogonalize as in Algorithm 3.3 storing $u_{si} \leftarrow r'_{si}$ and $d_s \leftarrow \boldsymbol{q}_s^T \boldsymbol{q}_s$.
end for
$\boldsymbol{Q} \leftarrow [\boldsymbol{q}_1\ \boldsymbol{q}_2\ \ldots\ \boldsymbol{q}_n]$ as in Algorithm 3.3 with $\boldsymbol{Q}^T\boldsymbol{Q} = \operatorname{diag}(d_1, d_2, \ldots, d_n)$.

Notice that $d_i > 0,\ i = 1, \ldots, n$, iff $\operatorname{rank}(\boldsymbol{X}) = n$. The resulting triangular decomposition and the corresponding generalized inverse are given in the next corollary.

Corollary 3.3: MGS Generalized Inverse
Let $\boldsymbol{X} \in L(m,n)$ be a factor matrix with $\operatorname{rank}(\boldsymbol{X}) = n$, $\boldsymbol{Q}$ the matrix defined in Algorithm 3.4, and $\boldsymbol{U} = (u_{si})$, where $u_{si},\ 1 \le s \le n, s+1 \le i \le n$, are defined as in Algorithm 3.4, $u_{ss} := 1,\ 1 \le s \le n$, and $u_{si} = 0,\ 1 \le i \le n,\ 1 \le s \le i-1$. Then, $\boldsymbol{Q}^T\boldsymbol{Q} = \operatorname{diag}(d_1, d_2, \ldots, d_n) =: \boldsymbol{D}$ and $\boldsymbol{U}$ is an upper triangular matrix with maximum row rank, thus, being invertible. Moreover, Algorithm 3.4 applies column permutations to the matrix $\boldsymbol{X}$ implicitly defining a transformed matrix $\tilde{\boldsymbol{X}}$. $\tilde{\boldsymbol{X}} = \boldsymbol{Q}\boldsymbol{U}$ is an frd of $\tilde{\boldsymbol{X}}$ and

$$\tilde{\boldsymbol{X}}^+ = \boldsymbol{U}^{-1}\boldsymbol{D}^{-1}\boldsymbol{Q}^T, \tag{3.11}$$

as well as

$$X^+ = P_2 U^{-1} D^{-1} Q^T. \tag{3.12}$$

Proof.

$$\tilde{X}^+ = U^T (UU^T)^{-1} (Q^T Q)^{-1} Q^T = U^T (UU^T)^{-1} D^{-1} Q^T = U^{-1} D^{-1} Q^T$$

by Theorem 3.4. Since only column interchanges by a permutation matrix P_2 were carried out, it follows from Corollary 3.2 that

$$X^+ = P_2 U^{-1} D^{-1} Q^T. \qquad \square$$

This completes the computation of X^+ by means of the MGS process. With this, the LLS solution can be written as

$$\hat{\beta} = P_2 U^{-1} D^{-1} Q^T y. \tag{3.13}$$

Note that this is the solution of the equation system

$$D U P_2^T \hat{\beta} = Q^T y, \tag{3.14}$$

which is often solved in practice after orthogonalization according to Equation (3.2).

3.2.1.2 Greville's Method

Unfortunately, by using the MGS process one nice property of the GS process is lost, namely the easy adaptation of the LLS solution to new data. With the GS process a new variable can easily be added and orthogonalized by the standard process. With the MGS process, however, a new variable has to be treated individually, i.e. in special steps. Moreover, all three Gram-Schmidt-processes are not able to easily adapt the result to new observations, i.e. to new rows of the matrix X.

For the method of Greville (1960) all such adaptations are no problem. This method constructs the generalized inverse of the matrix X column by column or row by row. For example, in step j the generalized inverse of X_j := (first j columns of X) or $X_{(j)}$:= (first j rows of X) is calculated.

Theorem 3.5: Greville's Method

Column version: Let $X = [x_1 \ldots x_n]$ be a column representation of $X \in L(m,n)$ with maximum column rank and let

$$X_1 := [x_1], \quad X_j := [X_{j-1}\, x_j], \quad j = 2, \ldots, n,$$
$$d_j := X_{j-1}^+ x_j, \quad c_j := x_j - X_{j-1} d_j \text{, and } b_j^T := c_j^+ = (c_j^T c_j)^{-1} c_j^T.$$

Then

$$X_1^+ = (x_1^T x_1)^{-1} x_1^T, \quad X_j^+ = [X_{j-1}\, x_j]^+ = \begin{bmatrix} X_{j-1}^+ - d_j b_j^T \\ b_j^T \end{bmatrix}, \; j = 2, \ldots, n.$$

Row version: Let $X = \begin{bmatrix} x_1^T \\ \vdots \\ x_m^T \end{bmatrix}$ be a row representation of $X \in L(m,n)$ with maximum row rank and let

$$X_{(1)} := [x_1^T], \quad X_{(j)} := \begin{bmatrix} X_{(j-1)} \\ x_j^T \end{bmatrix}, \quad j = 2, \ldots, m,$$

$$d_j^T := x_j^T X_{(j-1)}^+, \quad c_j^T := x_j^T - d_j^T X_{(j-1)} \text{ and } b_j := c_j^{T+} = (c_j^T c_j)^{-1} c_j.$$

Then

$$\begin{aligned} X_{(1)}^+ &= (x_1^T x_1)^{-1} x_1, \\ X_{(j)}^+ &= \begin{bmatrix} X_{(j-1)} \\ x_j^T \end{bmatrix}^+ = \begin{bmatrix} X_{j-1}^+ - b_j d_j^T & b_j \end{bmatrix}, \; j = 2, \ldots, n. \end{aligned}$$

Proof. Verification of the properties of the generalized inverse (see Ben-Israel and Greville, 2003, pp. 263 – 265, and Kishi, 1964, pp. 344 – 350). □

Theorem 3.5 apparently provides two construction rules for the calculation of X^+ recursively by X_j^+ resp. $X_{(j)}^+$. In both rules, the actual generalized inverse is adapted to the new column (or row) and complemented by an additional row (or column).

Obviously, one important condition is $c_j \neq 0 \Leftrightarrow x_j$ is not a linear combination of the columns of X_{j-1}. Otherwise, $(c_j^T c_j)^{-1}$ and thus b_j would not be defined. Analogously, this is true for the row version. As we assume maximum column rank and maximum row rank, c_j resp. c_j^T can never be zero. Thus, the case $c_j = 0$ can be ignored, and the variants of Greville's theorem dealing with the case $c_j = 0$ are not relevant here. Note, however, that small $c_j \neq 0$ might cause numerical trouble.

Having constructed the generalized inverse, obviously, the LLS solution can be generated by $\hat{\beta} = X^+ y$.

3.2.1.3 *Summary*

In this subsection we have introduced three LLS solvers: the normal equations, variants of the Gram-Schmidt (GS) process, and variants of Greville's method. The normal equations are known to be numerically unstable. For the GS process we derived a numerically stable variant, the MGS process. Greville's method has the important advantage that it can easily adapt to new data, variables as well as observations. Moreover, this method will be prominently used to generate test matrices as described in the next subsection.

3.2.2 *Test Data Generation*

The accuracy of the LLS solutions computed by the methods introduced in the previous subsection can, naturally, only be studied precisely if the exact LLS solutions of the studied problems are well-known. Notice that all methods presented in this chapter, including the normal equations, can be seen as first computing the generalized inverse $\boldsymbol{X}^+$ of the factor matrix $\boldsymbol{X}$, which is then used to compute the LLS solution $\boldsymbol{X}^+\boldsymbol{y}$. Therefore, in the following, test matrices will be introduced, the generalized inverse of which is not only known, but also can be computed without any round-off errors. This way, if the response vector $\boldsymbol{y}$ can be exactly represented, the exact LLS solution can be computed.

Let us begin with a short remark. In the following, test matrices are developed for the statistical model $\boldsymbol{y} = \boldsymbol{X}\boldsymbol{\beta} + \boldsymbol{\epsilon}$, where no structure is imposed on $\boldsymbol{X}$. In particular, no constant $\boldsymbol{1}$-column is assumed to be included in $\boldsymbol{X}$. One may argue that such a column, indicating the constant in the model, is standard for most statistical models. Such models might be written as $y_i = \beta_0 + \boldsymbol{\beta}^T\boldsymbol{x}_i + \varepsilon_i$, $i = 1,\ldots,m$. It is standard to assume $\mathrm{E}(\varepsilon_i) = 0$. If we now switch to mean-centered variables $\tilde{y}_i := y_i - \mathrm{E}(y_i)$ and $\tilde{\boldsymbol{x}}_i := \boldsymbol{x}_i - \mathrm{E}(\boldsymbol{x}_i)$, then the corresponding model would look like $\tilde{y}_i = \tilde{\boldsymbol{\beta}}^T\tilde{\boldsymbol{x}}_i + \tilde{\varepsilon}_i$, i.e. the constant term is eliminated. Naturally, in such models the columns of the factor matrices and the responses have to be mean-centered. The usage of the following test matrices in such cases will be discussed in the Exercises 3.3.6 to 3.3.8.

3.2.2.1 *Nonsingular Test Matrices*

Often, there is the misunderstanding that only in computations with very large data sets numerical errors have a chance to accumulate. In what follows, though, we will show that it is enough to focus on very well-chosen small scale data sets to simulate situations where large numerical errors can be observed.

In Zielke (1974) three types of nonsingular test matrices were described, the elements of which can be chosen so that not only the test matrix itself, but also its inverse is integer and easily constructed. In the following, we will restrict ourselves to only one of these types of test matrices.

Proposition 3.4: Nonsingular Zielke Matrices
Let

$$\boldsymbol{X}_Z(\boldsymbol{Z}_1,\ldots,\boldsymbol{Z}_{m-1},n,p) := \begin{bmatrix} \boldsymbol{Z}_1+\boldsymbol{I}_p & \boldsymbol{Z}_2+\boldsymbol{I}_p & \cdots & \boldsymbol{Z}_{m-1}+\boldsymbol{I}_p & \boldsymbol{Z}_1 \\ \vdots & & \cdot^{\cdot^{\cdot}} & \boldsymbol{Z}_{m-1} & \cdots \\ \vdots & \boldsymbol{Z}_2+\boldsymbol{I}_p & \cdot^{\cdot^{\cdot}} & & \cdots \\ & & & & \\ \boldsymbol{Z}_1+\boldsymbol{I}_p & \boldsymbol{Z}_2 & \cdots & \boldsymbol{Z}_{m-1} & \boldsymbol{Z}_1 \\ \boldsymbol{Z}_1 & \boldsymbol{Z}_2 & \cdots & \boldsymbol{Z}_{m-1} & \boldsymbol{Z}_1-\boldsymbol{I}_p \end{bmatrix},$$

where $\boldsymbol{Z}_i \in L(p,p)$ is an integer matrix and $1 \leq i \leq m-1$, $mp = n$, $m \geq 3$. Then, the corresponding inverse has the form

$$\boldsymbol{X}_Z(\boldsymbol{Z}_1,\ldots,\boldsymbol{Z}_{m-1},n,p)^{-1} :=$$

$$\begin{bmatrix} -\boldsymbol{Z}_{m-1} & \boldsymbol{Z}_{m-1}-\boldsymbol{Z}_{m-2} & \cdots & \boldsymbol{Z}_3-\boldsymbol{Z}_2 & \boldsymbol{Z}_2-\boldsymbol{Z}_1+\boldsymbol{I}_p & \boldsymbol{Z}_1 \\ & & & \boldsymbol{I}_p & -\boldsymbol{I}_p & \\ (0) & & \cdot^{\cdot^{\cdot}} & -\boldsymbol{I}_p & & \\ & \boldsymbol{I}_p & \cdot^{\cdot^{\cdot}} & & & (0) \\ \boldsymbol{I}_p & -\boldsymbol{I}_p & & & & \\ \boldsymbol{Z}_{m-1} & \boldsymbol{Z}_{m-2}-\boldsymbol{Z}_{m-1} & \cdots & \boldsymbol{Z}_2-\boldsymbol{Z}_3 & \boldsymbol{Z}_1-\boldsymbol{Z}_2 & -\boldsymbol{Z}_1-\boldsymbol{I}_p \end{bmatrix}.$$

Unfortunately, the construction of these matrices does not appear to be obvious at first glance. Therefore, we will describe the construction in some detail.

Construction of matrix $\boldsymbol{X}_Z$: Choose the first and the last row of submatrices. Then, add such rows from below, until overall m such rows are built, adding $\boldsymbol{I}_p$ to the last row, first in the first and the last column of submatrices, and then also in column 2, etc. Notice that in all rows, except the last, the last element is $\boldsymbol{Z}_1$.

Construction of matrix $\boldsymbol{X}_Z^{-1}$: The inverse $\boldsymbol{X}_Z^{-1}$ is built column-wise, while the matrix $\boldsymbol{X}_Z$ was built row-wise: Choose column 1 of submatrices and the last column m. Column 1 consists of $-\boldsymbol{Z}_{m-1}$ as the first, $\boldsymbol{Z}_{m-1}$ as the last and $\boldsymbol{I}_p$ as the last but one element. The last column consists of $\boldsymbol{Z}_1$ as the first and $-\boldsymbol{Z}_1 - \boldsymbol{I}_p$ as the last element. All other elements are equal to zero. Then choose one by one the columns $j = m-1,\ldots,m-(m-2) = 2$ beginning at the right-hand side. These columns consist of $(\boldsymbol{Z}_{m-j+1} - \boldsymbol{Z}_{m-j})$ as the first and $-(\boldsymbol{Z}_{m-j+1} - \boldsymbol{Z}_{m-j})$ as the last element. All other elements are equal to zero, except adding $\boldsymbol{I}_p$ to the $(m-j)$th and subtracting it from the $(m-j+1)$th element.

Example 3.1: Zielke Matrices and Their Inverses

1. Let $Z_i = 998$, $i = 1,2$, $m = 3$. Then

$$\boldsymbol{X}_Z(998,998,3,1) = \begin{bmatrix} 999 & 999 & 998 \\ 999 & 998 & 998 \\ 998 & 998 & 997 \end{bmatrix}$$

and

$$\boldsymbol{X}_Z(998,998,3,1)^{-1} = \begin{bmatrix} -998 & 1 & 998 \\ 1 & -1 & 0 \\ 998 & 0 & -999 \end{bmatrix}.$$

2. Let $\boldsymbol{Z}_i = \boldsymbol{Z} := \begin{bmatrix} 998 & 0 \\ 0 & 0 \end{bmatrix}$, $i = 1,2$, $m = 3$, $p = 2$. Then

$$\boldsymbol{X}_Z(\boldsymbol{Z},\boldsymbol{Z},6,2) = \begin{bmatrix} 999 & 0 & 999 & 0 & 998 & 0 \\ 0 & 1 & 0 & 1 & 0 & 0 \\ 999 & 0 & 998 & 0 & 998 & 0 \\ 0 & 1 & 0 & 0 & 0 & 0 \\ 998 & 0 & 998 & 0 & 997 & 0 \\ 0 & 0 & 0 & 0 & 0 & -1 \end{bmatrix},$$

$$\boldsymbol{X}_Z(\boldsymbol{Z},\boldsymbol{Z},6,2)^{-1} = \begin{bmatrix} -998 & 0 & 1 & 0 & 998 & 0 \\ 0 & 0 & 0 & 1 & 0 & 0 \\ 1 & 0 & -1 & 0 & 0 & 0 \\ 0 & 1 & 0 & -1 & 0 & 0 \\ 998 & 0 & 0 & 0 & -999 & 0 \\ 0 & 0 & 0 & 0 & 0 & -1 \end{bmatrix}.$$

3. Let $Z_1 = 998$, $Z_i = 0$, $i = 2,3,4$, $m = 5$. Then

$$\boldsymbol{X}_Z(998,0,0,0,5,1) = \begin{bmatrix} 999 & 1 & 1 & 1 & 998 \\ 999 & 1 & 1 & 0 & 998 \\ 999 & 1 & 0 & 0 & 998 \\ 999 & 0 & 0 & 0 & 998 \\ 998 & 0 & 0 & 0 & 997 \end{bmatrix},$$

$$\boldsymbol{X}_Z(998,0,0,0,5,1)^{-1} = \begin{bmatrix} 0 & 0 & 0 & -997 & 998 \\ 0 & 0 & 1 & -1 & 0 \\ 0 & 1 & -1 & 0 & 0 \\ 1 & -1 & 0 & 0 & 0 \\ 0 & 0 & 0 & 998 & -999 \end{bmatrix}.$$

Let us recall the aim we started from. We wanted to control the matrix condition, i.e. we wanted to be able to construct matrices with a prefixed magnitude of the condition number.

Zielke (1974, p. 34) wrote that "unless they are not only used to expose gross failures in the algorithm, test matrices should have the worst possible condition, i.e. a high condition number. Then it is possible to test the quality of a method with respect to error propagation avoiding enormous computational costs".

This raises the question how test matrices with high condition can be constructed. **This is generally neither possible nor sensible using random numbers for all matrix entries!** In contrast, however, we can control the condition of the above $\boldsymbol{X}_Z(Z_1,\ldots,Z_{m-1},n,1)$ by means of only the free parameters $Z_1,\ldots,Z_{m-1} \in \mathbb{R}$. It can be shown (Zielke, 1974, p. 47) that

$$K_F(\boldsymbol{X}_Z(Z_1,\ldots,Z_{n-1},n,1)) \geq 2nZ^2,$$

if $|Z_i| \geq Z \gg 1$, $i = 1,\ldots,n-1$, are integers. So the F-condition number increases with the square of the smallest free parameter Z and linearly with the rank n. Thus, even for small ranks n one can get test matrices with very high condition numbers without having to set the free parameters to very high values. For instance:

$$K_F(\boldsymbol{X}_Z(10^3,10^3,3,1)) \geq 6 \cdot 10^6,$$
$$K_F(\boldsymbol{X}_Z(10^3,10^3,10^3,10^3,5,1)) \geq 10 \cdot 10^6 = 10^7,$$
$$K_F(\boldsymbol{X}_Z(10^5,10^5,10^5,10^5,5,1)) \geq 10^{11}.$$

3.2.2.2 Singular Test Matrices

Zielke (1986) gave an overview of singular test matrices. He proved (in Theorem 4) that the generalized inverse of a singular integer matrix $\boldsymbol{X}$ has to be a non-integer if $\boldsymbol{X}$ does not result from a nonsingular integer matrix by adding zero rows or columns and $\boldsymbol{X}$ is not the zero matrix. Therefore, if one is interested in exact least squares solutions, then, obviously, the best one could reach is generalized inverses only containing non-integers exactly representable on the computer with a rather short mantissa. This has been realized in Weihs (1977, 2009) in a very general manner.

The general procedure is as follows. By means of **row deletion in matrices with full row rank** based on Zielke's nonsingular matrices, singular test matrices can be successfully constructed with the desired property, i.e. elements of the generalized inverse are integers or exactly representable non-integers. In order to prove this property a converse of Greville's method is used. Test matrices with full column rank can then be created by transposition. Unfortunately, directly dealing with columns leads to identical rows, i.e. to undesired identical observations.

Let us first construct generalized inverses of matrices after row deletions and then apply these results to special Zielke matrices in order to guarantee exactly representable inverses of test matrices.

Notation 3.2: Matrix with Deleted Columns and Rows
The matrix constructed by deletion of columns $j_1 \neq \ldots \neq j_p$ and rows $i_1 \neq \ldots \neq i_q$ from $\boldsymbol{X}$ is denoted by $\boldsymbol{X}_{(j_1,\ldots,j_p;i_1,\ldots,i_q)}$.

Theorem 3.6: Converse of Greville's Theorem: Deletion of One Row
Let $\boldsymbol{X}_j \in L(j,n)$ be of maximum row rank, $\boldsymbol{X}_{j-1} := \boldsymbol{X}_{j(;i)}$, $1 \leq i \leq j$, $\boldsymbol{x}_i^T :=$ (ith row of $\boldsymbol{X}_j$) and $\boldsymbol{b}_i :=$ (ith column of $\boldsymbol{X}_j^+$). Then

$$\boldsymbol{X}_{j-1}^+ = \left(\boldsymbol{I}_n - \frac{\boldsymbol{b}_i \boldsymbol{b}_i^T}{\boldsymbol{b}_i^T \boldsymbol{b}_i}\right) \boldsymbol{X}_{j\ (i)}^+ .$$

Proof.

1. Let first $i = j$. We use the row version of Greville's Theorem 3.5 to construct the generalized inverse of $\boldsymbol{X}_j \in L(j,n)$ from the generalized inverse of $\boldsymbol{X}_{j-1} := \boldsymbol{X}_{j(;j)}$:
   a. Let $\boldsymbol{d}_j^T := \boldsymbol{x}_j^T \boldsymbol{X}_{j-1}^+$, $\boldsymbol{c}_j^T := \boldsymbol{x}_j^T - \boldsymbol{d}_j^T \boldsymbol{X}_{j-1}$. Thus,
   $\boldsymbol{c}_j = \boldsymbol{x}_j - (\boldsymbol{X}_{j-1}^+ \boldsymbol{X}_{j-1})^T \boldsymbol{x}_j = \boldsymbol{x}_j - \boldsymbol{X}_{j-1}^+ \boldsymbol{X}_{j-1} \boldsymbol{x}_j$.
   As mentioned after Greville's Theorem 3.5,
   $\boldsymbol{c}_j = \boldsymbol{0} \Leftrightarrow \boldsymbol{x}_j^T$ linearly dependent of the rows of $\boldsymbol{X}_{j-1}$.

   b. Greville's theorem 3.5 implies:

   $$\begin{bmatrix} \boldsymbol{X}_{j-1} \\ \boldsymbol{x}_j^T \end{bmatrix}^+ = \begin{bmatrix} \boldsymbol{X}_{j-1}^+ - \boldsymbol{b}_j \boldsymbol{d}_j^T & \boldsymbol{b}_j \end{bmatrix} = \begin{bmatrix} (\boldsymbol{I} - \boldsymbol{b}_j \boldsymbol{x}_j^T) \boldsymbol{X}_{j-1}^+ & \boldsymbol{b}_j \end{bmatrix},$$

   where $\boldsymbol{b}_j := \boldsymbol{c}_j^{T+}$.

   c. Because $\boldsymbol{X}_j$ has maximum row rank, $\boldsymbol{x}_j^T$ is linearly independent of the rows of $\boldsymbol{X}_{j-1} \overset{(a)}{\Rightarrow} \boldsymbol{c}_j \neq \boldsymbol{0} \Rightarrow \boldsymbol{b}_j = \boldsymbol{c}_j^{T+} \neq \boldsymbol{0} \Rightarrow \boldsymbol{X}_{j-1} \boldsymbol{b}_j = \boldsymbol{X}_{j-1} \boldsymbol{c}_j / (\boldsymbol{c}_j^T \boldsymbol{c}_j)$.
   Moreover, $\boldsymbol{X}_{j-1} \boldsymbol{c}_j = \boldsymbol{X}_{j-1} (\boldsymbol{I} - \boldsymbol{X}_{j-1}^+ \boldsymbol{X}_{j-1}) \boldsymbol{x}_j = \boldsymbol{0}$.
   Thus, $\boldsymbol{X}_{j-1} \boldsymbol{b}_j = \boldsymbol{0}$. For $i = j$, the statement of the theorem then follows

from the following argument:

$$\begin{aligned}
X_{j-1}b_j = 0 \quad &\Leftrightarrow \quad b_j \in \ker(X_{j-1}) = \ker(X_{j-1}^{T+}) \\
&\Leftrightarrow \quad X_{j-1}^{T+} b_j = 0 \\
&\Leftrightarrow \quad b_j^T X_{j-1}^{+} = 0 \\
\Rightarrow \left(I - \frac{b_j b_j^T}{b_j^T b_j}\right) X_j^+ \quad &\overset{(b)}{=} \quad \left(I - \frac{b_j b_j^T}{b_j^T b_j}\right) [(I - b_j x_j^T) X_{j-1}^+ \quad b_j] \\
&= \quad [X_{j-1}^+ - b_j x_j^T X_{j-1}^+ + b_j x_j^T X_{j-1}^+ \quad 0] \\
&= \quad [X_{j-1}^+ \quad 0] .
\end{aligned}$$

2. Now, let i be general. Then let $C_j := PX_j := \begin{bmatrix} X_{j-1} \\ x_i^T \end{bmatrix}$, where P is the following product of elementary permutation matrices: $P := P_{j-1,j} \cdot \ldots \cdot P_{i,i+1}$. Therefore: $P^{-1} = P^T$. Moreover, let $C_{j-1} := C_{j(;j)} = X_{j-1}$. (1) implies

$$X_{j-1}^+ = C_{j-1}^+ - \left(I_n - \frac{b_i b_i^T}{b_i^T b_i}\right) C_j^+{}_{(j)} .$$

Then, the statement of the theorem follows from

$$C_j^+{}_{(j)} = (PX_j)_{(j)}^+ = (X_j^+ P^T)_{(j)} = X_j^+{}_{(i)} . \qquad \square$$

For a converse of the column version of Greville's method see Fletcher (1969).

Notation 3.3: Weighting Matrices
Matrices W so that $X_{j-k}^+ = W\, X_j^+{}_{(i_1,\ldots,i_k)}$ are called **weighting matrices**.

This theorem obviously constructs the generalized inverse of a matrix after deletion of row i by deleting column i in the generalized inverse of the original matrix and by left-multiplying this matrix by a weighting matrix.

Obviously, it is not enough to delete only one row, since this would lead to too special test matrices. However, a corresponding statement can also be derived for the deletion of more than one row.

Corollary 3.4: Converse of Greville's Theorem: Deletion of More Than One Row
Let $X_j \in L(j,n)$ be of maximum row rank and $X_{j-k} := X_{j(;i_1,\ldots,i_k)}$, $1 \leq i_1,\ldots,i_k \leq j$, $1 \leq k < j$, $b_i^{(k)} :=$ (ith column of X_j^+ after k deletions), $1 \leq$

$i \leq j-k, 0 \leq k < j$. Notice that $\boldsymbol{b}_i^{(0)}$ is the ith column of $\boldsymbol{X}_j^+$ itself. Then

$$\boldsymbol{X}_{j-k}^+ = \left(\boldsymbol{I}_n - \frac{\boldsymbol{b}_{i_k}^{(k-1)} \boldsymbol{b}_{i_k}^{(k-1)^T}}{\boldsymbol{b}_{i_k}^{(k-1)^T} \boldsymbol{b}_{i_k}^{(k-1)}} - \cdots - \frac{\boldsymbol{b}_{i_1}^{(0)} \boldsymbol{b}_{i_1}^{(0)^T}}{\boldsymbol{b}_{i_1}^{(0)^T} \boldsymbol{b}_{i_1}^{(0)}} \right) \boldsymbol{X}_{j\ (i_1,\ldots,i_k)}^+ .$$

Proof.

1. Since $\boldsymbol{X}_j$ has maximum row rank, all row-deleted versions of $\boldsymbol{X}_j$ also have maximum row rank. It follows by induction from Theorem 3.6 that

$$\boldsymbol{X}_{j-k}^+ = \left(\boldsymbol{I}_n - \frac{\boldsymbol{b}_{i_k}^{(k-1)} \boldsymbol{b}_{i_k}^{(k-1)^T}}{\boldsymbol{b}_{i_k}^{(k-1)^T} \boldsymbol{b}_{i_k}^{(k-1)}} \right) \cdots \left(\boldsymbol{I}_n - \frac{\boldsymbol{b}_{i_1}^{(0)} \boldsymbol{b}_{i_1}^{(0)^T}}{\boldsymbol{b}_{i_1}^{(0)^T} \boldsymbol{b}_{i_1}^{(0)}} \right) \boldsymbol{X}_{j\ (i_1,\ldots,i_k)}^+ .$$

2. We will show that $\boldsymbol{b}_{i_p}^{(p-1)^T} \boldsymbol{b}_{i_q}^{(q-1)} = 0$ for $p \neq q$.
Proof by induction: Let w.l.o.g. $p > q$.
   a. Let $p = q+1$, then by Theorem 3.6

$$\boldsymbol{b}_{i_p}^{(p-1)^T} \boldsymbol{b}_{i_q}^{(q-1)} = \boldsymbol{b}_{i_p}^{(q)^T} \boldsymbol{b}_{i_q}^{(q-1)} = \boldsymbol{b}_{i_p}^{(q-1)^T} \left(\boldsymbol{I} - \frac{\boldsymbol{b}_{i_q}^{(q-1)} \boldsymbol{b}_{i_q}^{(q-1)^T}}{\boldsymbol{b}_{i_q}^{(q-1)^T} \boldsymbol{b}_{i_q}^{(q-1)}} \right) \boldsymbol{b}_{i_q}^{(q-1)} = 0.$$

   b. If the statement is valid for $p = q+1, \ldots, q+s-1$, then:

$$\begin{aligned}
\boldsymbol{b}_{i_{q+s}}^{(q+s-1)^T} \boldsymbol{b}_{i_q}^{(q-1)} &= \boldsymbol{b}_{i_{q+s}}^{(q+s-2)^T} \left(I - \frac{\boldsymbol{b}_{i_{q+s-1}}^{(q+s-2)} \boldsymbol{b}_{i_{q+s-1}}^{(q+s-2)^T}}{\boldsymbol{b}_{i_{q+s-1}}^{(q+s-2)^T} \boldsymbol{b}_{i_{q+s-1}}^{(q+s-2)}} \right) \boldsymbol{b}_{i_q}^{(q-1)} \\
&= \boldsymbol{b}_{i_{q+s}}^{(q+s-2)^T} \boldsymbol{b}_{i_q}^{(q-1)}, \text{ since } \boldsymbol{b}_{i_{q+s-1}}^{(q+s-2)^T} \boldsymbol{b}_{i_q}^{(q-1)} = 0 \\
&= \ldots \\
&= \boldsymbol{b}_{i_{q+s}}^{(q)^T} \boldsymbol{b}_{i_q}^{(q-1)} \\
&= 0 \quad \text{as in (a).}
\end{aligned}$$

3. The statement follows from (1) and (2) □

Obviously, the construction of the generalized inverse after deletion of k rows is generalized from the corresponding construction after deletion of one row. Note, however, that the different weighting matrices are based on the columns of the generalized inverse after $0, 1, \ldots, k-1$ deletions, meaning that the generalized inverse has to be constructed stepwise:

- Construct the weighting matrix $\left(I_n - \frac{b_{i_1}^{(0)} b_{i_1}^{(0)T}}{b_{i_1}^{(0)T} b_{i_1}^{(0)}}\right)$ for the deletion of one row based on the original generalized inverse, i.e. take $b_{i_1}^{(0)}$ from the original generalized inverse.
- Construct the weighting matrix for the deletion of two rows $\left(I_n - \frac{b_{i_1}^{(0)} b_{i_1}^{(0)T}}{b_{i_1}^{(0)T} b_{i_1}^{(0)}} - \frac{b_{i_2}^{(1)} b_{i_2}^{(1)T}}{b_{i_2}^{(1)T} b_{i_2}^{(1)}}\right)$ by application of the weighting matrix for the deletion of one row from the previous step to $b_{i_2}^{(0)}$, i.e. by taking $b_{i_2}^{(1)} = \left(I_n - \frac{b_{i_1}^{(0)} b_{i_1}^{(0)T}}{b_{i_1}^{(0)T} b_{i_1}^{(0)}}\right) b_{i_2}^{(0)}$.
- ...
- Construct $b_{i_k}^{(k-1)} = \left(I_n - \frac{b_{i_1}^{(0)} b_{i_1}^{(0)T}}{b_{i_1}^{(0)T} b_{i_1}^{(0)}} - \cdots - \frac{b_{i_{(k-1)}}^{(k-2)} b_{i_{(k-1)}}^{(k-2)T}}{b_{i_{(k-1)}}^{(k-2)T} b_{i_{(k-1)}}^{(k-2)}}\right) b_{i_k}^{(0)}$.
- Construct $X_{j-k}^{+} = \left(I_n - \frac{b_{i_1}^{(0)} b_{i_1}^{(0)T}}{b_{i_1}^{(0)T} b_{i_1}^{(0)}} - \cdots - \frac{b_{i_k}^{(k-1)} b_{i_k}^{(k-1)T}}{b_{i_k}^{(k-1)T} b_{i_k}^{(k-1)}}\right) X_{j\ (i_1,\ldots,i_k)}^{+}$.

Notice that the weighting matrix in Corollary 3.4 is numerically much more stable than the corresponding form in (1) in the proof of the corollary, since the orthogonalities are theoretically utilized and not just numerically expected.

In order to guarantee that the generalized inverses of the Zielke matrices $X_Z(Z_1,\ldots,Z_{m-1},n,p)$ are exactly representable on a computer after row deletions we now concentrate on matrices with the property:

$$Z_{i+1} = Z_i + I_p, \quad 1 \le i \le m-2 \tag{3.15}$$

(cp. Weihs, 1977, pp. 93 – 94). These test matrices have only one free parameter matrix $Z = Z_1$ and are denoted by $X_{Z1}(Z,n,p)$:

$$X_{Z1}(Z,n,p) = \begin{bmatrix} Z+I_p & Z+2I_p & \cdots & Z+(m-1)I_p & Z \\ Z+I_p & Z+2I_p & \cdots & Z+(m-2)I_p & Z \\ \vdots & \vdots & & \vdots & \vdots \\ Z+I_p & Z+2I_p & \cdots & Z+(m-2)I_p & Z \\ Z+I_p & Z+I_p & \cdots & Z+(m-2)I_p & Z \\ Z & Z+I_p & \cdots & Z+(m-2)I_p & Z-I_p \end{bmatrix},$$

$$
\boldsymbol{X}_{Z1}(\boldsymbol{Z},n,p)^{-1} =
\begin{bmatrix}
-\boldsymbol{Z}-(m-2)\boldsymbol{I}_p & \boldsymbol{I}_p & \cdots & \boldsymbol{I}_p & 2\boldsymbol{I}_p & \boldsymbol{Z} \\
 & & & \boldsymbol{I}_p & -\boldsymbol{I}_p & \\
 & \textbf{(0)} & & .\cdot\ ^{\cdot} & -\boldsymbol{I}_p & \\
 & & \boldsymbol{I}_p & \cdot\ ^{\cdot} & & \textbf{(0)} \\
 & \boldsymbol{I}_p & -\boldsymbol{I}_p & & & \\
\boldsymbol{Z}+(m-2)\boldsymbol{I}_p & -\boldsymbol{I}_p & \cdots & -\boldsymbol{I}_p & -\boldsymbol{I}_p & -\boldsymbol{Z}-\boldsymbol{I}_p
\end{bmatrix}
$$

with $m := n/p \geq 3$.

Example 3.2: One Parameter Zielke Matrix and its Inverse
Let $Z = 998$, $n = m = 3$. Then

$$
\boldsymbol{X}_{Z1}(998,3,1) =
\begin{bmatrix}
999 & 1000 & 998 \\
999 & 999 & 998 \\
998 & 999 & 997
\end{bmatrix}
$$

and

$$
\boldsymbol{X}_{Z1}(998,3,1)^{-1} =
\begin{bmatrix}
-999 & 2 & 998 \\
1 & -1 & 0 \\
999 & -1 & -999
\end{bmatrix}.
$$

Let us apply Theorem 3.6 to these matrices, Z being an integer.

Corollary 3.5: Deletion of One Row in Zielke Matrices
Let $\boldsymbol{X}_n := \boldsymbol{X}_{Z1}(Z,n,1)$, $Z \in \mathbb{Z}$, and $\boldsymbol{X}_{n-1} := \boldsymbol{X}_{n(;i)}$ with $1 < i < n-1$. Then

$$
\overset{(n-i+1)\text{th column}}{\downarrow}
$$

$$
\boldsymbol{X}_{n-1}^{+} =
\begin{bmatrix}
0.75 & 0 & \cdots & 0 & -0.25 & 0.25 & 0 & \cdots & 0 & 0.25 \\
 & 1 & & & & & & & & \\
\textbf{(0)} & & \ddots & & & & \textbf{(0)} & & & \\
 & & & 1 & & & & & & \\
-0.25 & 0 & \cdots & 0 & 0.75 & 0.25 & 0 & \cdots & 0 & 0.25 \\
0.25 & 0 & \cdots & 0 & 0.25 & 0.75 & 0 & \cdots & 0 & -0.25 \\
 & & & & & & 1 & & & \\
 & \textbf{(0)} & & & & & & \ddots & & \textbf{(0)} \\
 & & & & & & & & 1 & \\
0.25 & 0 & \cdots & 0 & 0.25 & -0.25 & 0 & \cdots & 0 & 0.75
\end{bmatrix}
\boldsymbol{X}_{n(i)}^{+}
$$

Proof. Obviously,

$$
\overset{(n-i+1)\text{th column}}{\downarrow}
$$

$$
\boldsymbol{b}_i^{(0)} = \begin{bmatrix} 1 & 0 & \cdots & 0 & 1 & -1 & 0 & \cdots & 0 & -1 \end{bmatrix}^T
$$

with $\boldsymbol{b}_i^{(0)T}\boldsymbol{b}_i^{(0)} = 4$ and thus

$$\frac{\boldsymbol{b}_i^{(0)}}{\boldsymbol{b}_i^{(0)T}\boldsymbol{b}_i^{(0)}} = \begin{bmatrix}0.25 & 0 & \cdots & 0 & 0.25 & -0.25 & 0 & \cdots & 0 & -0.25\end{bmatrix}^T.$$

The weighting matrix is derived by multiplication with $\boldsymbol{b}_i^{(0)T}$ and subtraction from the identity. □

Obviously, the weighting matrix and thus the computed generalized inverse of the original matrix after the deletion of one row can be represented exactly on a computer since all elements in the weighting matrix are multiples of powers of 2 if $1 < i < n-1$. Unfortunately, this result cannot easily be generalized to deletions of more than one row. Consider, e.g., the following case:

Corollary 3.6: Deletion of Two Succeeding Rows in Zielke Matrices
Let $\boldsymbol{X}_n := \boldsymbol{X}_{Z1}(Z,n,1)$, $Z \in \mathbb{Z}$, $\boldsymbol{X}_{n-1} := \boldsymbol{X}_{n(;i)}$, and $\boldsymbol{X}_{n-2} := X_{n(;i,i+1)}$, $1 < i$, $i+1 < n-1$. Then

$$\boldsymbol{X}_{n-2}^+ =$$

$$\begin{array}{c} \\ (n-i+1)\text{th column} \\ \downarrow \end{array}$$

$$\begin{bmatrix}
3/5 & 0 & \cdots & 0 & -1/5 & 0 & 1/5 & 0 & \cdots & 0 & 2/5 \\
 & 1 & & & & & & & & & \\
(\mathbf{0}) & & \ddots & & & & & (\mathbf{0}) & & & \\
 & & & 1 & & & & & & & \\
-1/5 & 0 & \cdots & 0 & 11/15 & 1/3 & -1/15 & 0 & \cdots & 0 & 1/5 \\
0 & 0 & \cdots & 0 & 1/3 & 1/3 & 1/3 & 0 & \cdots & 0 & 0 \\
1/5 & 0 & \cdots & 0 & -1/15 & 1/3 & 11/15 & 0 & \cdots & 0 & -1/5 \\
 & & & & & & & 1 & & & \\
 & & & & (\mathbf{0}) & & & & \ddots & & (\mathbf{0}) \\
 & & & & & & & & & 1 & \\
2/5 & 0 & \cdots & 0 & 1/5 & 0 & -1/5 & 0 & \cdots & 0 & 3/5
\end{bmatrix} \boldsymbol{X}_{n\ (i,i+1)}^+$$

Proof. By Corollary 3.5,

$$
\boldsymbol{b}_{i+1}^{(1)} =
$$

$$
\begin{bmatrix}
0.75 & 0 & \cdots & 0 & -0.25 & 0.25 & 0 & \cdots & 0 & 0.25 \\
 & 1 & & & & & & & & \\
(\mathbf{0}) & & \ddots & & & & (\mathbf{0}) & & & \\
 & & & 1 & & & & & & \\
-0.25 & 0 & \cdots & 0 & 0.75 & 0.25 & 0 & \cdots & 0 & 0.25 \\
0.25 & 0 & \cdots & 0 & 0.25 & 0.75 & 0 & \cdots & 0 & -0.25 \\
 & & & & & & 1 & & & \\
 & & & (\mathbf{0}) & & & & \ddots & & (\mathbf{0}) \\
 & & & & & & & & 1 & \\
0.25 & 0 & \cdots & 0 & 0.25 & -0.25 & 0 & \cdots & 0 & 0.75
\end{bmatrix}
\begin{bmatrix} 1 \\ 0 \\ \vdots \\ 0 \\ 1 \\ -1 \\ 0 \\ 0 \\ \vdots \\ 0 \\ -1 \end{bmatrix}
=
\begin{bmatrix} 3/4 \\ 0 \\ \vdots \\ 0 \\ 1 \\ -5/4 \\ 1/4 \\ 0 \\ \vdots \\ 0 \\ -3/4 \end{bmatrix}
$$

with

$$
\boldsymbol{b}_{i+1}^{(1)T} \boldsymbol{b}_{i+1}^{(1)} = \frac{60}{16} = \frac{15}{4}
$$

and

$$
\frac{\boldsymbol{b}_{i+1}^{(1)}}{\boldsymbol{b}_{i+1}^{(1)T} \boldsymbol{b}_{i+1}^{(1)}} = \left[\tfrac{1}{5} \quad 0 \quad \cdots \quad 0 \quad \tfrac{4}{15} \quad -\tfrac{1}{3} \quad \underset{\substack{\uparrow \\ (n-i+1)\text{th place}}}{\tfrac{1}{15}} \quad 0 \quad \cdots \quad 0 \quad -\tfrac{1}{5}\right]^T .
$$

Then, by Corollary 3.4, the weighting matrix has the stated form, since

$$
\left(I_n - \frac{\boldsymbol{b}_i^{(0)} \boldsymbol{b}_i^{(0)^T}}{\boldsymbol{b}_i^{(0)^T} \boldsymbol{b}_i^{(0)}} \right) - \frac{\boldsymbol{b}_{i+1}^{(1)} \boldsymbol{b}_{i+1}^{(1)^T}}{\boldsymbol{b}_{i+1}^{(1)^T} \boldsymbol{b}_{i+1}^{(1)}} =
$$

$$
\begin{bmatrix}
0.75 & 0 & \cdots & 0 & -0.25 & 0.25 & 0 & \cdots & 0 & 0.25 \\
 & 1 & & & & & & & & \\
(\mathbf{0}) & & \ddots & & & & (\mathbf{0}) & & & \\
 & & & 1 & & & & & & \\
-0.25 & 0 & \cdots & 0 & 0.75 & 0.25 & 0 & \cdots & 0 & 0.25 \\
0.25 & 0 & \cdots & 0 & 0.25 & 0.75 & 0 & \cdots & 0 & -0.25 \\
 & & & & & & 1 & & & \\
 & & & (\mathbf{0}) & & & & \ddots & & (\mathbf{0}) \\
 & & & & & & & & 1 & \\
0.25 & 0 & \cdots & 0 & 0.25 & -0.25 & 0 & \cdots & 0 & 0.75
\end{bmatrix}
-
\begin{bmatrix} 1/5 \\ 0 \\ \vdots \\ 0 \\ 4/15 \\ -1/3 \\ 1/15 \\ 0 \\ \vdots \\ 0 \\ -1/5 \end{bmatrix}
\begin{bmatrix} 3/4 \\ 0 \\ \vdots \\ 0 \\ 1 \\ -5/4 \\ 1/4 \\ 0 \\ \vdots \\ 0 \\ -3/4 \end{bmatrix}^T
$$

□

Obviously, in the case of the deletion of two successive rows in

$\boldsymbol{X}_{Z1}(Z,n,1)$ the weighting matrix is not exactly representable with standard floating-point representation since, e.g., $1/3$ is not. Also the deletion of more than two successive rows does not lead to a convenient result (see Exercise 3.2.3). Therefore, the deletion of successive rows in $\boldsymbol{X}_{Z1}(Z,n,1)$ will not be admitted.

Notice: Therefore, not more than $\lfloor (n-2)/2 \rfloor$ rows are admitted to be deleted in $\boldsymbol{X}_{Z1}(Z,n,1)$.

In the case of the deletion of non-successive rows in $\boldsymbol{X}_{Z1}(Z,n,1)$ one can show:

Corollary 3.7: Deletion of More Than One Row in Zielke Matrices

Let $\boldsymbol{X}_n := \boldsymbol{X}_{Z1}(Z,n,1),\, Z \in \mathbb{Z}$, and $\boldsymbol{X}_{n-k} := \boldsymbol{X}_{n(;i_1,i_2,\ldots,i_k)}$, $1 < i_1 < \ldots < i_k < n-1$, $|i_p - i_q| > 1$ for $p \neq q$. Then, with $m := 2(k+1)$, $\tilde{m} := \frac{k}{m}$:

$\boldsymbol{X}^+_{n-k} =$

$$\begin{bmatrix}
1-\tilde{m} & 0 & \cdots & 0 & -1/m & 1/m & \cdots & -1/m & 1/m & 0 & \cdots & 0 & \tilde{m} \\
 & 1 & & & & & & & & & & & \\
(0) & & \ddots & & & & & (0) & & & & & \\
 & & & 1 & & & & & & & & & \\
-1/m & 0 & \cdots & 0 & 1-\tilde{m} & \tilde{m} & \cdots & 1/m & -1/m & 0 & \cdots & 0 & 1/m \\
1/m & 0 & \cdots & 0 & \tilde{m} & 1-\tilde{m} & \cdots & -1/m & 1/m & 0 & \cdots & 0 & -1/m \\
 & & & & \cdots & \cdots & \cdots & \cdots & \cdots & & & & \\
-1/m & 0 & \cdots & 0 & 1/m & -1/m & \cdots & 1-\tilde{m} & \tilde{m} & 0 & \cdots & 0 & 1/m \\
1/m & 0 & \cdots & 0 & -1/m & 1/m & \cdots & \tilde{m} & 1-\tilde{m} & 0 & \cdots & 0 & -1/m \\
 & & & & & & & & & 1 & & & \\
 & & & & & (0) & & & & & \ddots & & (0) \\
 & & & & & & & & & & & 1 & \\
\tilde{m} & 0 & \cdots & 0 & 1/m & -1/m & \cdots & 1/m & -1/m & 0 & \cdots & 0 & 1-\tilde{m}
\end{bmatrix} \boldsymbol{X}^+_{n\;(i_1,\ldots,i_k)}$$

$\uparrow$ $(n-i_k+1)$th $\qquad$ $\uparrow$ $(n-i_1+1)$th column

Notice: There are k pairs of columns with $\left(-1/m \quad 1/m\right)$ in the first row.

Proof. Induction is started by Corollary 3.6.

Then, assume that the i_kth column of the original inverse looks as follows after $(k-1)$ steps:

$$b_{i_k}^{(k-1)} =$$

$$\begin{bmatrix}
1-\frac{k-1}{2k} & 0 & \dots & 0 & -\frac{1}{2k} & \frac{1}{2k} & \dots & -\frac{1}{2k} & \frac{1}{2k} & 0 & \dots & 0 & \frac{k-1}{2k} \\
 & 1 & & & & & & & & & & & \\
(0) & & \ddots & & & & & (0) & & & & & \\
 & & & 1 & & & & & & & & & \\
-\frac{1}{2k} & 0 & \dots & 0 & 1-\frac{k-1}{2k} & \frac{k-1}{2k} & \dots & \frac{1}{2k} & -\frac{1}{2k} & 0 & \dots & 0 & \frac{1}{2k} \\
\frac{1}{2k} & 0 & \dots & 0 & \frac{k-1}{2k} & 1-\frac{k-1}{2k} & \dots & -\frac{1}{2k} & \frac{1}{2k} & 0 & \dots & 0 & -\frac{1}{2k} \\
 & & & & \dots & \dots & \dots & \dots & \dots & \dots & & & \\
-\frac{1}{2k} & 0 & \dots & 0 & \frac{1}{2k} & -\frac{1}{2k} & \dots & 1-\frac{k-1}{2k} & \frac{k-1}{2k} & 0 & \dots & 0 & \frac{1}{2k} \\
\frac{1}{2k} & 0 & \dots & 0 & -\frac{1}{2k} & \frac{1}{2k} & \dots & \frac{k-1}{2k} & 1-\frac{k-1}{2k} & 0 & \dots & 0 & -\frac{1}{2k} \\
 & & & & & & & & & 1 & & & \\
 & & & & & (0) & & & & & \ddots & & (0) \\
 & & & & & & & & & & & 1 & \\
\frac{k-1}{2k} & 0 & \dots & 0 & \frac{1}{2k} & -\frac{1}{2k} & \dots & \frac{1}{2k} & -\frac{1}{2k} & 0 & \dots & 0 & 1-\frac{k-1}{2k}
\end{bmatrix}
\begin{bmatrix} 1 \\ (0) \\ 1 \\ -1 \\ (0) \\ 0 \\ 0 \\ \vdots \\ 0 \\ 0 \\ (0) \\ -1 \end{bmatrix}
=
\begin{bmatrix} 1/k \\ (0) \\ 1 \\ -1 \\ (0) \\ -1/k \\ 1/k \\ \vdots \\ -1/k \\ 1/k \\ (0) \\ -1/k \end{bmatrix}$$

Therefore, $\boldsymbol{b}_{i_k}^{(k-1)^T} \boldsymbol{b}_{i_k}^{(k-1)} = 2 + \frac{2k}{k^2} = \frac{m}{k} = \frac{1}{\tilde{m}}$ and thus,

$$\frac{\boldsymbol{b}_{i_k}^{(k-1)}}{\boldsymbol{b}_{i_k}^{(k-1)^T} \boldsymbol{b}_{i_k}^{(k-1)}} =$$

$$\begin{bmatrix} \frac{1}{m} & 0 & \cdots & 0 & \underset{\substack{\uparrow \\ (n-i_k)\text{th}}}{\tilde{m}} & -\tilde{m} & 0 & \cdots & 0 & \underset{\substack{\uparrow \\ (n-i_{k-1})\text{th}}}{-\frac{1}{m}} & \frac{1}{m} & 0 & \cdots & 0 & \underset{\substack{\uparrow \\ (n-i_1)\text{th place}}}{-\frac{1}{m}} & \frac{1}{m} & 0 & \cdots & 0 & -\frac{1}{m} \end{bmatrix}^T.$$

It follows that

$$\boldsymbol{W} = \left(\boldsymbol{I}_n - \frac{\boldsymbol{b}_{i_{k-1}}^{(k-2)} \boldsymbol{b}_{i_{k-1}}^{(k-2)^T}}{\boldsymbol{b}_{i_{k-1}}^{(k-2)^T} \boldsymbol{b}_{i_{k-1}}^{(k-2)}} - \cdots - \frac{\boldsymbol{b}_{i_1}^{(0)} \boldsymbol{b}_{i_1}^{(0)^T}}{\boldsymbol{b}_{i_1}^{(0)^T} \boldsymbol{b}_{i_1}^{(0)}} \right) - \frac{\boldsymbol{b}_{i_k}^{(k-1)} \boldsymbol{b}_{i_k}^{(k-1)^T}}{\boldsymbol{b}_{i_k}^{(k-1)^T} \boldsymbol{b}_{i_k}^{(k-1)}}$$

has the stated form (see Table 3.1). □

Obviously, it is important here that $\frac{1}{m} = \frac{1}{(2k+2)}$ can be computed exactly. But this is only true for $k = 2^i - 1$, $i \in \mathbb{N}$, since then $m = 2^{i+1}$. From this, another restriction follows for row deletion in $\boldsymbol{X}_{Z1}(Z, n, 1)$. Indeed, if $n = 3$, no row may be deleted as $1 < i < n - 1 = 2$ would be required. If $n = 4$, only the second row is permitted to be deleted. Even for $n = 5$ only one row can be deleted, namely row 2 or 3. One should keep in mind that because of the condition $k = 2^i - 1$, $i \in \mathbb{N}$, only $k = 1, 3, 7, 15, \ldots$ non-successive rows may

Table 3.1: Weighting matrix W in proof of Corollary 3.7

$$W = \begin{bmatrix}
1-\frac{k-1}{2k} & 0 & \dots & 0 & -\frac{1}{2k} & \frac{1}{2k} & \dots & -\frac{1}{2k} & \frac{1}{2k} & 0 & \dots & 0 & \frac{k-1}{2k} \\
 & 1 & & & & & & & & & & & \\
(0) & & \ddots & & & & & (0) & & & & & \\
 & & & 1 & & & & & & & & & \\
-\frac{1}{2k} & 0 & \dots & 0 & 1-\frac{k-1}{2k} & \frac{k-1}{2k} & \dots & \frac{1}{2k} & -\frac{1}{2k} & 0 & \dots & 0 & \frac{1}{2k} \\
\frac{1}{2k} & 0 & \dots & 0 & \frac{k-1}{2k} & 1-\frac{k-1}{2k} & \dots & -\frac{1}{2k} & \frac{1}{2k} & 0 & \dots & 0 & -\frac{1}{2k} \\
 & & & & \dots & \dots & \dots & \dots & \dots & \dots & & & \\
-\frac{1}{2k} & 0 & \dots & 0 & \frac{1}{2k} & -\frac{1}{2k} & \dots & 1-\frac{k-1}{2k} & \frac{k-1}{2k} & 0 & \dots & 0 & \frac{1}{2k} \\
\frac{1}{2k} & 0 & \dots & 0 & -\frac{1}{2k} & \frac{1}{2k} & \dots & \frac{k-1}{2k} & 1-\frac{k-1}{2k} & 0 & \dots & 0 & -\frac{1}{2k} \\
 & & & & & & & & & 1 & & & \\
 & & & & & (0) & & & & & \ddots & & (0) \\
 & & & & & & & & & & & 1 & \\
\frac{k-1}{2k} & 0 & \dots & 0 & \frac{1}{2k} & -\frac{1}{2k} & \dots & \frac{1}{2k} & -\frac{1}{2k} & 0 & \dots & 0 & 1-\frac{k-1}{2k}
\end{bmatrix} - \begin{bmatrix}
\frac{1}{2(k+1)} \\ (0) \\ \frac{k}{2(k+1)} \\ \frac{-k}{2(k+1)} \\ (0) \\ \frac{-1}{2(k+1)} \\ \frac{1}{2(k+1)} \\ \vdots \\ \frac{-1}{2(k+1)} \\ \frac{1}{2(k+1)} \\ (0) \\ \frac{-1}{2(k+1)}
\end{bmatrix} \begin{bmatrix}
1/k \\ (0) \\ 1 \\ -1 \\ (0) \\ -1/k \\ 1/k \\ \vdots \\ -1/k \\ 1/k \\ (0) \\ -1/k
\end{bmatrix}^T$$

be deleted. For this, matrices of the type $\boldsymbol{X}_{Z1}(Z,n,1)$ with $n = 4, 8, 16, 32, \ldots$ rows are needed.

Example 3.3: Zielke Matrices and their (Generalized) Inverses
Let $n = m = 4$ and $Z = 998$. Then

$$\boldsymbol{X}_{Z1}(998,4,1) = \begin{bmatrix} 999 & 1000 & 1001 & 998 \\ 999 & 1000 & 1000 & 998 \\ 999 & 999 & 1000 & 998 \\ 998 & 999 & 1000 & 997 \end{bmatrix},$$

$$\boldsymbol{X}_{Z1}(998,4,1)^{-1} = \begin{bmatrix} -1000 & 1 & 2 & 998 \\ 0 & 1 & -1 & 0 \\ 1 & -1 & 0 & 0 \\ 1000 & -1 & -1 & -999 \end{bmatrix},$$

and

$$\boldsymbol{X}_{Z1}(998,4,1)_{(;2)} = \begin{bmatrix} 999 & 1000 & 1001 & 998 \\ 999 & 999 & 1000 & 998 \\ 998 & 999 & 1000 & 997 \end{bmatrix},$$

$\boldsymbol{X}_{Z1}(998,4,1)^{+}_{(;2)} =$

$$\begin{bmatrix} 0.75 & -0.25 & 0.25 & 0.25 \\ -0.25 & 0.75 & 0.25 & 0.25 \\ 0.25 & 0.25 & 0.75 & -0.25 \\ 0.25 & 0.25 & -0.25 & 0.75 \end{bmatrix} \begin{bmatrix} -1000 & 2 & 998 \\ 0 & -1 & 0 \\ 1 & 0 & 0 \\ 1000 & -1 & -999 \end{bmatrix}$$

$$= \begin{bmatrix} -499.75 & 1.5 & 498.75 \\ 500.25 & -1.5 & -499.25 \\ -499.25 & 0.5 & 499.25 \\ 499.75 & -0.5 & -499.75 \end{bmatrix}.$$

Check that the latter matrix is really the generalized inverse of $\boldsymbol{X}_{Z1}(998,4,1)_{(;2)}$ (see Exercise 3.2.4).

Up to now, only the case $p = 1$ was treated. In the case $p > 1$ there are more possibilities for row deletions with exactly representable generalized inverses (Weihs, 1977, p. 98).

Note that by transposing the above matrices, test matrices with full column rank with 1, 3, 7, and 15 degrees of freedom and $n = 4, 8, 16$, and 32 observations, respectively, are generated. So with the above procedure sensible degrees of freedom for applications in statistics are automatically attained.

Even adding linearly dependent rows to the created test matrices with full column rank could be realized in such a way that the corresponding generalized inverses are exactly representable (Weihs, 1977, p. 101).

Overall, we have succeeded in constructing general (at least with respect to condition) integer test matrices with exactly representable generalized inverses. With these matrices it appears possible to cover the space $L(m,n)$ of factor matrices in a way adequate for the LLS problem, namely by choosing matrices with condition numbers covering a whole range.

3.2.2.3 Response Vectors

In addition to test matrices, the verification algorithm also requires the response vectors for which the LLS solvers are to be tested. Response vectors should be constructed to be as general and as exactly representable as possible. The generality of the response vectors $\boldsymbol{y}$ is defined by the generality of the angle $\phi(\boldsymbol{y})$ between $\boldsymbol{y}$ and $\text{im}(\boldsymbol{X})$ as explained by Corollary 3.1 in Section 3.1.2.

There, we showed that the vectors $\boldsymbol{y}, \boldsymbol{y}_0$, and $\boldsymbol{r}_0$ build an orthogonal decomposition since $\boldsymbol{y} = \boldsymbol{y}_0 + \boldsymbol{r}_0$ with $\boldsymbol{y}_0 \in \text{im}(\boldsymbol{X})$ and $\boldsymbol{y}_0 \perp \boldsymbol{r}_0$. $\boldsymbol{y}_0$ is the projection of $\boldsymbol{y}$ on $\text{im}(\boldsymbol{X})$ and $\boldsymbol{r}_0$ the corresponding residual.

If we vary the length of the vector $\boldsymbol{r}_0$ leaving the projection $\boldsymbol{y}_0$ fixed, we can construct arbitrary angles between the vector $\boldsymbol{y} := \boldsymbol{y}_0 + r$ (with $\boldsymbol{r} \perp \boldsymbol{y}_0$) and $\text{im}(\boldsymbol{X})$. To understand this, we just have to recall the representations of sine, cosine, and tangent derived in Corollary 3.1 in Section 3.1.2:

$$\cos\phi(\boldsymbol{y}) = \frac{\|\boldsymbol{y}_0\|_2}{\|\boldsymbol{y}\|_2} = \frac{\|\boldsymbol{y}_0\|_2}{\|\boldsymbol{y}_0 + \boldsymbol{r}\|_2},$$

$$\sin\phi(\boldsymbol{y}) = \frac{\|\boldsymbol{r}\|_2}{\|\boldsymbol{y}\|_2} = \frac{\|\boldsymbol{r}\|_2}{\|\boldsymbol{y}_0 + \boldsymbol{r}\|_2}, \quad \text{and}$$

$$\tan\phi(\boldsymbol{y}) = \frac{\|\boldsymbol{r}\|_2}{\|\boldsymbol{y}_0\|_2}.$$

Obviously, the bigger $\|\boldsymbol{r}\|$, the lower $\cos\phi(\boldsymbol{y})$ and the bigger $\tan\phi(\boldsymbol{y})$.

In order to study the dependency of the precision of the computed LLS solutions on the angle between response vector and $\text{im}(\boldsymbol{X})$, in the following a method for the construction of arbitrary small, medium, and large angles will be proposed.

For this, let us first study the relationship between the sizes of an angle and its tangent. Let

$\tan\phi_j := 2^{-21}, 2^{-19}, 2^{-3}, 2^{-1}, 2, 2^3, 2^{19}, 2^{21},\ j = 1,\ldots,8$. Then,

$\phi_j \cong 5\cdot 10^{-7}, 2\cdot 10^{-6}, 0.12, 0.46, 1.1, 1.45, 1.570794, 1.570796$ (in rad), or

$\phi_j \triangleq (3\cdot 10^{-5})^o, (10^{-4})^o, 1.8^o, 26.6^o, 63.4^o, 82.9^o, 89.99989^o, 89.99997^o$.

For given $\boldsymbol{y}_0$, $\boldsymbol{r}_0$ with "not too different L2 norms" the following algorithm chooses response vectors $\boldsymbol{y}_k$, $k = 1,\ldots,4$, so that $\phi(\boldsymbol{y}_k) \in [\phi_{2k-1}, \phi_{2k})$:

Algorithm 3.5 Generation of Response Vectors (RVs)

Require: $\boldsymbol{y}_0, \boldsymbol{r}_0$
```
1: i ← −21
2: for j = 19 to − 19 by − 2 do
3:     if ||r0||2 ≥ 2^j ||y0||2 then {i ← j; goto OUT }
4: end for
5: OUT: c ← (2^(−21−i), 2^(−3−i), 2^(1−i), 2^(19−i))
6: for k = 1 to 4 do
7:     yk ← y0 + ck r0
8: end for
```

Obviously, this way the greatest $i \in \{19, 17, \ldots, 1, -1, \ldots, -19\}$ is chosen so that $\|\boldsymbol{r}_0\|_2 \geq 2^i\|\boldsymbol{y}_0\|_2$, using $i := -21$, if no such i exists. Then we set $\boldsymbol{y}_k := \boldsymbol{y}_0 + c_k\boldsymbol{r}_0 =: \boldsymbol{y}_0 + \boldsymbol{r}_k$ with $c_k := 2^{-21-i}, 2^{-3-i}, 2^{1-i}, 2^{19-i}$, $k = 1,\ldots,4$.

This implies that if the norms of $\boldsymbol{y}_0$ and $\boldsymbol{r}_0$ are not too different,

$$c_k 2^i \leq c_k \frac{\|\boldsymbol{r}_0\|_2}{\|\boldsymbol{y}_0\|_2} = \tan\phi(\boldsymbol{y}_k) < c_k 2^{i+2},$$

i.e. $\phi(\boldsymbol{y}_k) \in [\phi_{2k-1}, \phi_{2k})$ if $i > -21$. That means we are guaranteed one very small angle in the interval $[(3\cdot 10^{-5})^o, (10^{-4})^o]$, one medium-small angle in $[1.8^o, 26.6^o]$, one medium-large angle in $[63.4^o, 82.9^o]$, and one very large angle in $[89.99989^o, 89.99997^o]$.

Now, in order to generate $4\cdot S$ response vectors with prefixed angle magnitudes for LLS problems with test matrices $\boldsymbol{X}$ generated as in Section 3.2.2.2, the generalized inverse of which is exactly representable, the following generation process is run S times.

Generation of Response Vectors

1. Choose an arbitrary exactly representable (e.g. integer) vector $\boldsymbol{y} \in \mathbb{R}^m$, using a random number generator (see Chapter 6).

2. Compute the exact LLS solution $\hat{\boldsymbol{\beta}} := \boldsymbol{X}^+\boldsymbol{y}$ of the model $\boldsymbol{y} = \boldsymbol{X}\boldsymbol{\beta}$ by means of the exact generalized inverse $\boldsymbol{X}^+$, the exact projection $\boldsymbol{y}_0 := \boldsymbol{X}\hat{\boldsymbol{\beta}}$, and the exact residual $\boldsymbol{r}_0 := \boldsymbol{y} - \boldsymbol{y}_0$.
3. If $\boldsymbol{r}_0 \neq \boldsymbol{0}$, choose $\boldsymbol{y}_1, \ldots, \boldsymbol{y}_4$ from Algorithm 3.5 as response vectors, otherwise, choose $\boldsymbol{y} = \boldsymbol{y}_0$.

Notice that the LLS solutions corresponding to $\boldsymbol{y}_1, \ldots, \boldsymbol{y}_4$ are identical since $\boldsymbol{y}_i = \boldsymbol{y}_0 + \boldsymbol{r}_i, i = 1, \ldots, 4$, with $\boldsymbol{y}_0^T\boldsymbol{r} = 0$ and $\mathrm{im}(\boldsymbol{X})^\perp = \ker(\boldsymbol{X}^T) = \ker(\boldsymbol{X}^+)$, i.e. $\boldsymbol{X}^+\boldsymbol{y}_i = \boldsymbol{X}^+(\boldsymbol{y}_0 + \boldsymbol{r}_i) = \boldsymbol{X}^+\boldsymbol{y}_0 = \boldsymbol{X}^+\boldsymbol{y}$.

$\mathrm{im}(\boldsymbol{X})^\perp = \ker(\boldsymbol{X}^T)$ says that all vectors in the kernel of $\boldsymbol{X}^T$ are orthogonal to all vectors in the image of $\boldsymbol{X}$ (Ben-Israel and Greville, 2003, p. 12). $\ker(\boldsymbol{X}^T) = \ker(\boldsymbol{X}^+)$ follows from the fact that $\boldsymbol{X}^+ = (\boldsymbol{X}^T\boldsymbol{X})^{-1}\boldsymbol{X}^T$ in case of maximum column rank. Moreover, notice that in order to guarantee exact $\boldsymbol{y}_1, \ldots, \boldsymbol{y}_4$, one should restrict the size of $\boldsymbol{y}$, e.g. by $\|\boldsymbol{y}\|_\infty \leq 1000$.

3.2.2.4 Summary

In this subsection full column rank integer test matrices are constructed with generalized inverses that are exactly representable on a computer. Moreover, these test matrices can easily be used to construct test matrices with known magnitude of condition numbers. Very big conditions can be generated without similarly big matrix entries. Additionally, a method is proposed to construct response vectors with arbitrary angles with $\mathrm{im}(\boldsymbol{X})$ in order to be able to verify the dependency between the error in the LLS solution and this angle. Overall, this subsection gives methods for the generation of all test data needed for the verification procedure.

3.3 Practice and Simulation

3.3.1 Verification Procedure

Based on the test matrices and the response vectors in Sections 3.2.2.2 and 3.2.2.3, the following verification procedure is proposed to assess the LLS problem solvers in Section 3.2.1.

1. Specify **invertible test matrices** of the type $\boldsymbol{X}_{Z1}(Z, n, 1)$ with $n = 4, 8, 16, 32, 64$. Choose the free integer number Z to generate test matrices with specified magnitudes of the condition numbers, assuming the **machine precision** is $1.1 \cdot 10^{-16}$ for double-precision arithmetic [2]. In order to base verification on all relevant magnitudes of condition number, we set $Z = 2^0, 2^1, \ldots, 2^{20} \approx 10^{12}$.

[2] see `http://en.wikipedia.org/wiki/Machine_epsilon`

2. Delete $k = 1,3,7,15,31$ rows from these test matrices and use the exactly representable inverses of $\boldsymbol{X}_{Z1}(Z,n,1)$ to construct exact generalized inverses of the generated matrices with full row rank. Transpose these matrices in order to obtain **test matrices with full column rank** and their exact generalized inverses.
3. Generate S integer starting response vectors $\boldsymbol{y}$ and corresponding quadruples $\boldsymbol{y}_1 \cdots \boldsymbol{y}_4$ of **response vectors** for each of the chosen test matrices by means of Algorithm 3.5. Calculate the **exact LLS solutions** by multiplying the exact generalized inverses by the integer response vector $\boldsymbol{y}$. Notice that the LLS solutions are equal to $\boldsymbol{X}^+\boldsymbol{y}$ for integer $\boldsymbol{y}$ for all $\boldsymbol{y}_1 \cdots \boldsymbol{y}_4$ (see Section 3.2.2.3).
4. Apply the MGS algorithm for **Gram-Schmidt orthogonalization**, the column version of **Greville's method** and the **normal equations method** to the generated test problems. Record the accuracy of the results and the condition number of the full column rank test matrix.
5. **Compare** the results. How does the accuracy depend on the condition number of the full column rank test matrix? How does the accuracy change for different angles of the response vectors? Which ranking of the methods does arise?

The **accuracy of the results** can be characterized, e.g., by the mean value and the standard deviation of the relative errors $\|d\hat{\boldsymbol{\beta}}_1\|_2/\|\hat{\boldsymbol{\beta}}_1\|_2, \ldots,$ $\|d\hat{\boldsymbol{\beta}}_4\|_2/\|\hat{\boldsymbol{\beta}}_4\|_2$ of the LLS solutions over the S repetitions for a specified angle magnitude. In what follows, the index k corresponds to the angle magnitude, i.e. $\|d\hat{\boldsymbol{\beta}}_k\|_2/\|\hat{\boldsymbol{\beta}}_k\|_2$ corresponds to the LLS solution $\hat{\boldsymbol{\beta}}_k := \boldsymbol{X}^+\boldsymbol{y}_k$.

Under the conditions of Theorem 3.3 the relative error of the LLS solution was shown (Corollary 3.1) to have the upper bound

$$\begin{aligned}\frac{\|d\hat{\boldsymbol{\beta}}_k\|_2}{\|\hat{\boldsymbol{\beta}}_k\|_2} &\leq \frac{\delta}{s_n(\boldsymbol{X})}\left[\frac{s_1(\boldsymbol{X})\|\boldsymbol{r}_k\|_2}{s_n(\boldsymbol{X})\|\hat{\boldsymbol{\beta}}_k\|_2(1-\mu^2)} + \frac{s_1(\boldsymbol{X})}{1-\mu} + \frac{\|\boldsymbol{y}_k\|_2}{\|\hat{\boldsymbol{\beta}}_k\|_2(1-\mu)}\right] \\ &\leq \delta K(\boldsymbol{X})\left[K(\boldsymbol{X})\frac{\tan\phi(\boldsymbol{y}_k)}{1-\mu^2} + \frac{1}{1-\mu}\left(1+\frac{1}{\cos\phi(\boldsymbol{y}_k)}\right)\right], \quad (3.16)\end{aligned}$$

where $n = \text{rank}(\boldsymbol{X})$, $\mu := \delta K(\boldsymbol{X}) < 1$, and $\phi(\boldsymbol{y}_k)$ is the angle between $\boldsymbol{y}_k$ and $\boldsymbol{y}_0$.

From this bound it has to be suspected (as mentioned before) that $\delta K(\boldsymbol{X})^2$ appears as an amplification factor of $\tan\phi(\boldsymbol{y}_k)$ in the relative error in the LLS solution. However, given the findings of van der Sluis, the remaining question is whether the worst case lower bounds are rather tight and realistic. Thus, in the **comparison of the results** of the different LLS solvers we will compare the percentages of the results that have even lower relative errors in the LLS solutions than expected by the lower bounds of van der Sluis. For such results one can show:

$$\begin{aligned}\frac{\|d\hat{\beta}_k\|_2}{\|\hat{\beta}_k\|_2} &\leq \frac{\delta}{s_n(X)(1-\mu^2)}\left[\frac{s_1(X)\|r_k\|_2}{s_n(X)\|\hat{\beta}_k\|_2}+\frac{\|y_k\|_2}{\|\hat{\beta}_k\|_2}\right] \\ &\leq \delta\frac{K(X)}{1-\mu^2}\left[K(X)\tan\phi(y_k)+\frac{1}{\cos\phi(y_k)}\right],\end{aligned}$$

$$\frac{\|d\hat{\beta}_k\|_2}{\|\hat{\beta}_k\|_2} \leq \frac{\delta}{s_n(X)}\left[s_1(X)+\frac{\|y_k\|_2}{\|\hat{\beta}_k\|_2}\right] \leq \delta K(X)\left[1+\frac{1}{\cos\phi(y_k)}\right].$$

Correspondingly, we define so-called **solution classes** S_1, S_2 by all LLS solutions satisfying

$$S_1: \frac{\|d\hat{\beta}_k\|_2}{\|\hat{\beta}_k\|_2} \leq \delta\frac{K_F(X)}{1-\mu^2}\left[K_F(X)\tan\phi(y_k)+\frac{1}{\cos\phi(y_k)}\right], \quad (3.17)$$

$$S_2: \frac{\|d\hat{\beta}_k\|_2}{\|\hat{\beta}_k\|_2} \leq \delta K_F(X)\left[1+\frac{1}{\cos\phi(y_k)}\right]. \quad (3.18)$$

In Formulas 3.17 and 3.18 the minimum angle for all repetitions of $y_1,\ldots,y_4$ is chosen, respectively. In Formula 3.16 the maximum angle is chosen.

Notice that the **percentage of the LLS solutions in solution class** S_2 provides information about the probability that the reviewed methods for the LLS solution depend only linearly on $K(X)$.

3.3.2 Verification Results

Let us start the discussion of the results by an example realization of the procedure.

Example 3.4: Example for a Simulation Step
In order to be able to present the example in a relatively compact form, we use $m=8$, $Z=2^{12}=4096$. Then, the test matrix and its generalized inverse

look as follows:

$$X = \begin{bmatrix} 4097 & 4097 & 4097 & 4097 & 4096 \\ 4098 & 4098 & 4098 & 4097 & 4097 \\ 4099 & 4099 & 4099 & 4098 & 4098 \\ 4100 & 4100 & 4099 & 4099 & 4099 \\ 4101 & 4101 & 4100 & 4100 & 4100 \\ 4102 & 4101 & 4101 & 4101 & 4101 \\ 4103 & 4102 & 4102 & 4102 & 4102 \\ 4096 & 4096 & 4096 & 4096 & 4095 \end{bmatrix},$$

$$X^+ =$$

$$\begin{bmatrix} -1025\frac{3}{8} & 1025\frac{3}{8} & -1025\frac{3}{8} & 1025\frac{3}{8} & -1025\frac{3}{8} & 1025\frac{7}{8} & -1024\frac{7}{8} & 1025\frac{3}{8} \\ \frac{1}{2} & -\frac{1}{2} & \frac{1}{2} & 0 & 1 & -1 & 0 & -\frac{1}{2} \\ \frac{1}{2} & 0 & 1 & -1 & 0 & -\frac{1}{2} & \frac{1}{2} & -\frac{1}{2} \\ 1 & -1 & 0 & -\frac{1}{2} & \frac{1}{2} & -\frac{1}{2} & \frac{1}{2} & 0 \\ 1023\frac{5}{8} & -1024\frac{1}{8} & 1024\frac{1}{8} & -1024\frac{1}{8} & 1024\frac{1}{8} & -1024\frac{1}{8} & 1024\frac{1}{8} & -1024\frac{5}{8} \end{bmatrix}.$$

The corresponding condition number is $K_F = 106263986$, i.e. $K_F \approx 10^8$. The originally generated response vector y is

$$y = \begin{bmatrix} 299 & -30 & -446 & 217 & 270 & 13 & -794 & 255 \end{bmatrix}^T.$$

The chosen response vectors in one step of our simulation have the form:

$$\begin{aligned} Y &= [y_1 \; y_2 \; y_3 \; y_4] \\ &= \begin{bmatrix} 136.25016 & 176.9375 & 787.25 & 170655880 \\ -97.24994 & -80.4375 & 171.75 & 70516639 \\ -378.75006 & -395.5625 & -647.75 & -70517115 \\ 384.24984 & 342.4375 & -284.75 & -175373952 \\ 102.75016 & 144.5625 & 771.75 & 175374439 \\ -249.74975 & -184.0625 & 801.25 & 275513094 \\ -531.25025 & -596.9375 & -1582.25 & -275513875 \\ 417.74984 & 377.0625 & -233.25 & -170655326 \end{bmatrix}. \end{aligned}$$

Remember that the corresponding exact LLS solutions for all y_1, y_2, y_3, y_4 are identical, namely,

$$\hat{\beta} = \begin{bmatrix} 1,154,181.8 & 71.0 & -1044.5 & -48.0 & -1,153,441.8 \end{bmatrix}^T.$$

Using this LLS solution, the projection of y on $\text{im}(X)$ can be shown to be

$$y_0 = \begin{bmatrix} 136\frac{1}{4} & -97\frac{1}{4} & -378\frac{3}{4} & 384\frac{1}{4} & 102\frac{3}{4} & -249\frac{3}{4} & -531\frac{1}{4} & 417\frac{3}{4} \end{bmatrix}^T.$$

Table 3.2: LLS Solutions for the Different Solvers and Response Vectors

normal equation	Greville		MGS	
y_1, y_2, y_3, y_4	y_1, y_2, y_3	y_4	y_1, y_2, y_3	y_4
1211302.70	1224443.35	1225129.05	1154181.7	1154181.43
43.15	49.70	49.38	71.0	71.00
-1072.36	-1103.06	-1103.41	-1044.5	-1044.50
-75.85	-72.52	-72.84	-48.0	-48.00
-1210493.07	-1223616.10	-1224300.97	-1153441.7	-1153441.43

After applying the normal equations, Greville's method, and MGS to estimate these solutions, we found that all methods resulted in (nearly) the same solutions for the four different response vectors with the exception that Greville and MGS produced a somewhat different solution for y_4 (see Table 3.2).

Note that MGS (y_1, y_2, y_3) looks nearly the same as the exact solution. However, there are more deviations in digits after the decimal point not indicated because of rounding. The corresponding relative errors are
0.04947613, 0.06085740, 0.06145133, 3.313175e-10, 2.760719e-07.

Obviously, the errors of the solutions produced by the normal equations and Greville's method are much bigger than the errors by MGS. Moreover, the largest angle between the response vector and its projection is the most problematic concerning solution accuracy.

Results of the Verification Procedure

Let us now consider the results of the above specified complete verification procedure. We used the minimum number of replications ($S = 1$) and $\delta = 5 \cdot 10^{-14}$, explained by $\delta = 2 \cdot n^{3/2} \cdot \textbf{machine precision} = 2 \cdot n^{3/2} \cdot 1.1 \cdot 10^{-16}$, proposed by Björck (1967) for MGS, where n is taken to be somewhat bigger than the maximum number of columns used for testing (namely 33).

First, we illustrate the dependency of the relative error on the condition number using log-scale on both axes in order to show the magnitudes (see Figures 3.1 and 3.2). Moreover, we compare the realized relative errors to the bounds 3.16, 3.17, and 3.18 for the MGS results.

Obviously, the upper bound of relative errors valid for MGS is not valid for the other methods in the case of very small angles between the response vector and its projection (y_1, see Figure 3.1, left). When angles get somewhat larger (y_2), the bound becomes valid for the other methods and too big for MGS (see Figure 3.1, right). Moreover, note that already for medium-large angles (y_3), the upper bound and the first lower bound for the worst case are bigger than all relative errors (see Figure 3.2, left). The second lower bound,

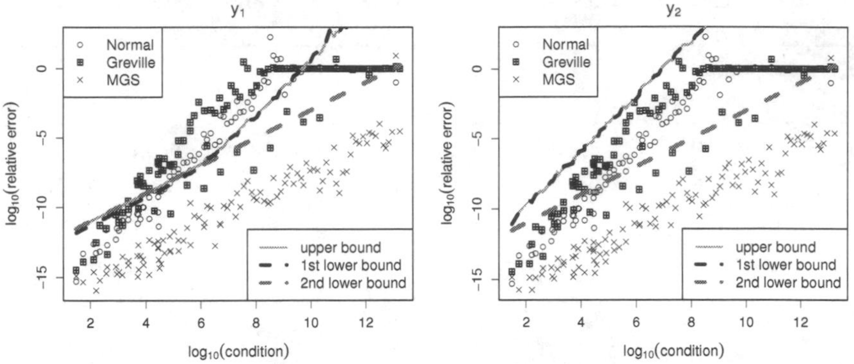

Figure 3.1: Relative errors for different test matrices and very and medium small angles of response vectors ($\boldsymbol{y}_1$ left, $\boldsymbol{y}_2$ right).

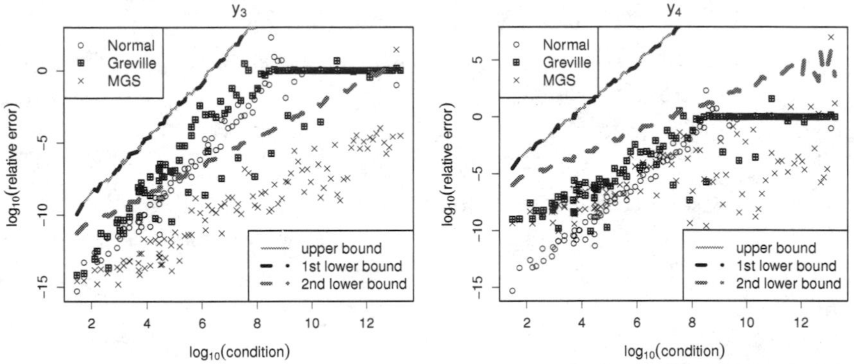

Figure 3.2: Relative errors for different test matrices and medium and very large angles of response vectors ($\boldsymbol{y}_3$ left, $\boldsymbol{y}_4$ right).

however, is only bigger than (nearly) all relative errors for very large angles ($\boldsymbol{y}_4$).

Also notice that the plots indicate that the notion of a "small condition number" might be different for the different LLS solvers. This might be fixed by means of a prespecified maximum magnitude of relative errors allowed. For example, if the relative error should not be greater than 10^{-5}, all condition numbers up to 10^{12} are acceptable for MGS, except for very large angles for which $K(\boldsymbol{X})$ should be $< 10^6$. For the other methods $K(\boldsymbol{X}) < 10^6$ should be valid in any case (see Figures 3.1 and 3.2).

Inspired by the plots we try linear models for the dependency of $\log_{10}$(relative error) on $\log_{10}$(condition number). The results are presented in Table 3.3. Note that, in an ideal world, the results should be identical for

Table 3.3: Regression Results: $\log_{10}$(Relative Error) $\sim$ $\log_{10}$(Condition)

Method	Angle Type	Increment	Slope	Fit (R^2)
Normal equations	very small (y_1)	-16.4	1.69	0.92
Greville	very small (y_1)	-14.1	1.39	0.73
MGS	very small (y_1)	-17.2	0.97	0.94
Normal equations	medium-small (y_2)	-16.4	1.69	0.92
Greville	medium-small (y_2)	-14.1	1.40	0.73
MGS	medium-small (y_2)	-17.2	0.97	0.93
Normal equations	medium-large (y_3)	-16.4	1.69	0.92
Greville	medium-large (y_3)	-14.1	1.39	0.73
MGS	medium-large (y_3)	-16.3	0.92	0.90
Normal equations	very large (y_4)	-16.4	1.69	0.92
Greville	very large (y_4)	-10.4	0.93	0.66
MGS	very large (y_4)	-9.2	0.48	0.38

the different angles since the model coefficients should be identical. However, in practice, for MGS the fit of the linear model gets very bad for very large angles. This can also be verified by the right plot in Figure 3.2. Obviously, for such angles any model is inadequate because of the big variance in the results. Moreover, for Greville's method the fit is never very good. This is also obvious from all plots in Figures 3.1 and 3.2 since there are, in a way, two branches of resulting relative error sizes. Notice that for Greville's method and the normal equations the fit of the linear model is only based on those data where the relative error is smaller than 0.99. For larger values, the linear relationship is broken since the relative error is, in a way, automatically restricted by 1. This is particularly obvious for Greville's method.

Considering the percentages of relative errors greater than the three bounds (see Table 3.4), obviously, with MGS all relative errors are lower than all the bounds. On the other hand, the second lower bound is problematic for the normal equations and Greville's method and all angles but the biggest. The barplots in Figure 3.3 illustrate the dependence of the percentages of relative errors that are larger than the bounds on the angle between the response vector and its projection.

In summary, a linear dependency of $\log_{10}$(relative error) on $\log_{10}$(condition) appears to be not unrealistic. This leads to the following type of relationship

Table 3.4: Percentages of (Relative Errors > Error Bounds)

Normal equations	y_1	y_2	y_3	y_4
upper bound	37.143	0.000	0.000	0.000
lower bound 1	41.905	0.000	0.000	0.000
lower bound 2	66.667	66.667	60.000	0.952
Greville				
upper bound	49.524	0.000	0.000	0.000
lower bound 1	50.476	0.000	0.000	0.000
lower bound 2	76.190	76.190	68.571	0.952
MGS				
upper bound	0.000	0.000	0.000	0.000
lower bound 1	0.000	0.000	0.000	0.000
lower bound 2	0.952	0.952	0.952	0.952

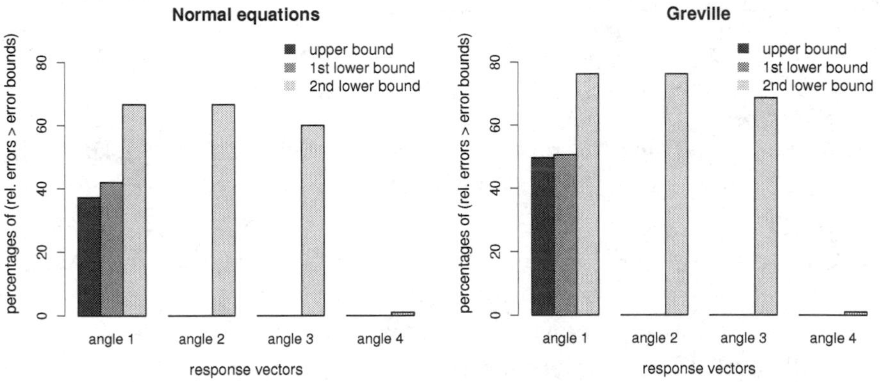

Figure 3.3: Barplot of percentages of (relative errors > error bounds) for the normal equations (left) and Greville's method (right).

between the relative error itself and the condition number:

$$\text{relative error} = const_1 \cdot (\text{condition})^{const_2} = 10^{\text{increment}} \cdot (\text{condition})^{\text{slope}},$$

where increment and slope are the corresponding parameters in the dependency of $\log_{10}$(relative error) on $\log_{10}$(condition). Taking the two methods MGS and normal equations and the angles where the fit of the linear model is satisfactory, the increment is near -17 and the slope is near 1 for MGS and near 1.75 for the normal equations. Thus, for MGS there is no indication for a quadratic dependency on the condition number in contrast to the normal equations.

Moreover, the upper bound of van der Sluis generally appears to be much too pessimistic for MGS (for which δ was fixed), and even the lower bounds are most of the time not exceeded in the case of MGS.

This completes our discussion of the results of the proposed verification procedure. Notice that we have not included repetitions in the procedure so that the computation of means and standard deviations did not make sense. See Exercise 3.3.5 for a verification procedure with repetitions.

3.3.3 Summary

In this section we described a complete verification procedure for LLS solvers based on exactly representable solutions. We described how singular test matrices with different magnitudes of condition numbers can be generated as well as general response vectors with different magnitudes of angle $\phi(\boldsymbol{y})$ between $\boldsymbol{y}$ and $\text{im}(\boldsymbol{X})$. Also, we discussed the results of one example implementation of this procedure. We found that modern orthogonalization methods like MGS only lead to linear dependence of the error rate on the condition number of the test matrix, thus fulfilling the original idea of these methods in practice. In contrast, Greville's method and the normal equations have severe problems for condition numbers greater than, say, 10^6.

3.4 Implementation in R

While it is certainly possible to solve linear least squares problems in R using only basic matrix algebra routines (cf. Exercises 3.2.6 – 3.2.8) this is not advisable. R provides a multitude of functions that can solve an LLS problem given the data. These are `lm` and `aov`, which fit a classical linear model and an analysis of variance model, respectively, to a given data set. Both routines are a high-level interface to the low-level `lm.fit` function which solves an LLS problem for a given design matrix $\boldsymbol{X}$ and response vector $\boldsymbol{y}$ using a QR decomposition. The actual implementation is provided by the `dqrls` function from LINPACK, a `Fortran` library of linear algebra subroutines. For more details on the history of LINPACK see Section 8.2.

Why would a user want to rely on `lm` instead of say a custom implementation of one of the algorithms presented in this chapter? In fact, the LINPACK routines are highly optimized and considered to be stable and bug-free. The actual QR decomposition via Householder transformations (Householder, 1958) is implemented in the LINPACK routine `dqrdc` which was implemented in 1978 and has withstood the test of time. The other reason to use `lm` is that it provides a lot of convenience. There are lots of utility functions that build on `lm`, it is easy to test for significant effects in the model, and last but not least, there is quite a bit of diagnostic output.

If you are working with a data set that is huge, you may run into difficulties because of memory constraints. In these cases the biglm (Lumley, 2011) package may provide a solution. The provided `biglm` function is almost a drop-in replacement for `lm` and uses an iterative algorithm that requires fewer resources (Miller, 1992).

3.5 Conclusion

This chapter derived and exemplified a general procedure for the verification of LLS solvers.

Unfortunately, the literature about systematic verification of algorithms for the LLS problem is very thin. Zielke was one of the few authors who worked on this problem in a more general way.

During the last years it became modern to take the same small number of real data sets for verification of LLS solvers. From our point of view, this is a pity since it rarely leads to a realistic overview of the capabilities of the solvers. The condition of the data sets is not controlled and an over-specialization to these data is to be expected. Particularly often utilized is the UCI Machine Learning Repository[3] with only 24 different data sets recommended to be solved by LLS regression.

Notice that there are also well-working generalized inverse generators on the internet, e.g. with `http://www.bluebit.gr/matrix-calculator/default.aspx`. You might want to test it with our singular test matrices.

3.6 Exercises

Exercise 3.2.1: Program a generator for nonsingular Zielke test matrices of the type $\boldsymbol{X}_{Z1}(Z,n,1)$.

Exercise 3.2.2: Program a generator for singular Zielke test matrices of the type

$$\boldsymbol{X}_{n-k} := \boldsymbol{X}_{n(;i_1,i_2,\ldots,i_k)},$$

$1 < i_1 < \ldots < i_k < n-1$, $|i_p - i_q| > 1$ for $p \neq q, k = 1,3,7,15$, based on non-

[3] `http://archive.ics.uci.edu/ml/`, accessed May 9, 2013

singular Zielke test matrices of the type $\boldsymbol{X} = \boldsymbol{X}_{Z1}(Z,n,1)$ with $n = 4, 8, 16, 32$ (see Exercise 3.2.1).

Exercise 3.2.3: Demonstrate that for the deletion of three successive rows in $\boldsymbol{X}_{Z1}(Z,n,1)$ the weighting matrix is not exactly representable on a computer.

Exercise 3.2.4: Check that the generalized inverse of

$$\boldsymbol{X}_{Z1}(998,4,1)_{(;2)} = \begin{bmatrix} 999 & 1000 & 1001 & 998 \\ 999 & 999 & 1000 & 998 \\ 998 & 999 & 1000 & 997 \end{bmatrix}$$

is

$$\boldsymbol{X}_{Z1}(998,4,1)^{+}_{(;2)} = \begin{bmatrix} -499.75 & 1.5 & 498.75 \\ 500.25 & -1.5 & -499.25 \\ -499.25 & 0.5 & 499.25 \\ 499.75 & -0.5 & -499.75 \end{bmatrix}.$$

Exercise 3.2.5: Program the proposed quadruples of response vectors.

Exercise 3.2.6: Program the normal equations. Test with singular Zielke test matrices and the proposed quadruples of response vectors.

Exercise 3.2.7: Program the column version of Greville's method and the corresponding construction of the LLS solution. Test with singular Zielke test matrices and the proposed quadruples of response vectors.

Exercise 3.2.8: Program the MGS algorithm together with the pivot strategy and the corresponding construction of the LLS solution. Test with singular Zielke test matrices and the proposed quadruples of response vectors.

Exercise 3.3.1: Program the proposed verification procedure in the replication case $S = 10$ (see Exercises 3.2.2 and 3.2.6–3.2.8). Program mean and standard deviation of the repetitions. Test with singular Zielke test matrices and the proposed quadruples of response vectors.

Exercise 3.3.2: Program a plot of relative errors vs. condition numbers, both on $\log_{10}$ scale.

Exercise 3.3.3: Program the Givens orthogonalization algorithm, another method for the construction of the generalized inverse using a QR decomposition. The pseudocode of the algorithm is given in Algorithm 3.6. The algorithm constructs the generalized inverse using successive rotations of the columns of the original matrix.

Algorithm 3.6 Givens Orthogonalization

Require: $\boldsymbol{X} = [\boldsymbol{x}_1^{(1)} \dots \boldsymbol{x}_n^{(1)}]$

$\boldsymbol{Q} \leftarrow \boldsymbol{I}_m$
for $j = 1$ **to** n **do**
  for $i = (j+1)$ **to** m **do**
    if $\boldsymbol{X}_{jj} = 0$ and $\boldsymbol{X}_{ij} = 0$ **then**
      $\boldsymbol{J} \leftarrow \boldsymbol{I}_2$
    else
      if $|\boldsymbol{X}_{jj}| > |\boldsymbol{X}_{ij}|$ **then**
        $a \leftarrow 1/\sqrt{1+(\boldsymbol{X}_{ij}/\boldsymbol{X}_{jj})^2}$
        $b \leftarrow t \cdot a$
      else
        $b \leftarrow 1/\sqrt{1+(\boldsymbol{X}_{jj}/\boldsymbol{X}_{ij})^2}$
        $a \leftarrow t \cdot b$
      end if
      $\boldsymbol{J} \leftarrow \begin{bmatrix} b & a \\ -a & b \end{bmatrix}$
    end if
    $\boldsymbol{X}_{\{j,i\},\cdot} \leftarrow \boldsymbol{J}\boldsymbol{X}_{\{j,i\},\cdot} = \boldsymbol{J}[\boldsymbol{X}_j \ \ \boldsymbol{X}_i]$
    $\boldsymbol{Q}_{\{j,i\},\cdot} \leftarrow \boldsymbol{J}\boldsymbol{Q}_{\{j,i\},\cdot}$
  end for
end for
return $\boldsymbol{Q}$ and $\boldsymbol{X}$ as upper triangular matrix $\boldsymbol{R}$.

Exercise 3.3.4: Determine the runtime complexity of the Givens algorithm as best you can.

Exercise 3.3.5: Compare the MGS algorithm and the Givens algorithm (see Exercise 3.2.8 and 3.3.3) using the verification procedure from Exercise 3.3.1). Use $S = 10$ repetitions for the response vectors and fixed n, Z for the test matrices.

Exercise 3.3.6: Consider the model $\boldsymbol{y} = \boldsymbol{X}\boldsymbol{\beta}$ for mean-centered variables $y, x_j, j = 1, \dots, n$. Adapt step 2 of the verification procedure to this situation in the following way:

2a. Delete $k = 1, 3, 7, 15, 31$ rows from these test matrices and use the exactly representable inverses of $\boldsymbol{X}_{Z1}(Z, n, 1)$ to construct exact generalized inverses of the generated matrices with full row rank. Use the exact generalized inverses as **test matrices with full column rank** and the original

Zielke matrices with rows deleted as the corresponding exact generalized inverses.

Attention: Do not transpose the matrices, since the generalized inverses of row deleted Zielke matrices automatically have full column rank.

Show that the columns of these test matrices are (at least nearly) mean-centered, i.e. that the arithmetical means of the columns are (at least nearly) zero. Hint: Consider the structure of the generalized inverses of the nonsingular Zielke matrices, show that column means are (at least nearly) zero, and then generalize to the singular case. Prove the generalization to the singular case for the deletion of $k = 1$ row.

Exercise 3.3.7: Consider the model $\boldsymbol{y} = \boldsymbol{X}\boldsymbol{\beta}$ for mean-centered variables $y, x_j, j = 1, \ldots, n$. Adapt step 3 of the verification procedure to this situation in the following way:

3a. Generate S integer starting response vectors $\boldsymbol{y}$, mean center them, and generate corresponding quadruples $\boldsymbol{y}_1 - \boldsymbol{y}_4$ of **response vectors** for each of the chosen test matrices by means of Algorithm 3.5. Calculate the **exact LLS solutions** by multiplying the exact generalized inverses by the exact mean-centered response vectors $\boldsymbol{y} - \bar{\boldsymbol{y}}$.

Show that

1. $\boldsymbol{y} - \bar{\boldsymbol{y}}$ is exactly representable, if $\boldsymbol{y}$ is integer and $n = 4, 8, 16, 32, 64$.
2. $\boldsymbol{y}_1 - \boldsymbol{y}_4$ has mean zero if $\boldsymbol{y}$ does.

Exercise 3.3.8: Carry out the verification procedure with steps 2a and 3a and $S = 10$ (cp. Exercise 3.3.1). Compare the results for the different LLS solvers.

Chapter 4

Iteration

In the previous chapter we studied the properties of the linear least squares (LLS) problem. For it, we could derive a closed-form analytic solution. When we used this solution to derive an algorithm to solve the LLS, numerical instabilities surfaced and different algorithms that try to mitigate these problems were introduced. In this chapter we will study solution strategies for problems where no closed-form analytic solution is known. In fact, there may even be no closed-form solution.

We will therefore resort to methods that improve an initial (approximate) solution in each iteration of the algorithm. Hence, all the methods presented in this chapter are, at their core, methods that, given an (approximate) solution, return a new, improved solution. We then **iterate** these until we reach either a fixed-point or some other termination criterion is reached. This idea is a powerful general concept. Instead of trying to solve a problem in one big step, we can develop a, usually simpler, method that only improves a given solution. By iteration this method will then reach something akin to a locally optimal solution. While we will focus on classical optimization problems in this chapter, this concept can be applied to a much broader set of problems.

We saw a very similar idea in Chapter 2 when we looked at sorting. There we ended up dividing the input into smaller chunks that are easier to sort and then recursively applied the procedure to the resulting smaller data sets (see Algorithm 2.3 (Quick Sort) for details). This is not the type of iteration we will look at in this chapter. In contrast to the defined upper limit of iteration steps in Chapter 2—once we run out of data in one of the partitions, no further recursion is possible—the methods in this chapter have no fixed limit on the number of iterations required. We will see examples where these methods never converge. When developing iterative algorithms we have to pay special attention to the termination criterion.

4.1 Motivation

Recall that the LLS is defined as follows. Given a suitable factor matrix $\boldsymbol{X} \in L(m,n)$ and response vector $\boldsymbol{y} \in \mathbb{R}^m$, find $\boldsymbol{\beta} \in \mathbb{R}^n$ such that

$$f_{LLS}(\boldsymbol{\beta}) = \|\boldsymbol{y} - \boldsymbol{X}\boldsymbol{\beta}\|_2 \tag{4.1}$$

becomes minimal. We saw that this problem is relatively easy to solve. The main obstacle are numerical issues when the condition number of $\boldsymbol{X}$ becomes large. But what happens if we replace the square in the Euclidean norm with the absolute value? That is, we seek a $\boldsymbol{\beta}$ that minimizes

$$f_{LAS}(\boldsymbol{\beta}) = \|\boldsymbol{y} - \boldsymbol{X}\boldsymbol{\beta}\|_1. \tag{4.2}$$

This is a much harder problem to solve. To see this, let us simplify our problem for a minute. Assume that X has exactly one column ($n = 1$), i.e. we have a regression model with one regressor and no intercept term. This reduces equation (4.2) to

$$f_{1D-LAS}(\beta) = \sum_{i=1}^{m} |y_i - \beta x_i|, \tag{4.3}$$

which has at least m non-differentiable points $\beta_i = y_i/x_i$, $i = 1$ to m.

The previous example is but one of many problems that arise in statistics for which it is hard or even impossible to derive a closed-form solution. In these cases methods described in this chapter can usually be applied to obtain a good, sometimes even optimal, solution to the problem. They all hinge on the following idea. Given some initial solution $\beta^{(k)}$, derive a new solution $\beta^{(k+1)}$, so that $f(\beta^{(k+1)})$ is better than $f(\beta^{(k)})$. Now repeat the previous step using the new solution to derive a series of continually improving solutions $\beta^{(k+2)}, \beta^{(k+3)}, \ldots$. Terminate once the current solution is "good enough" or no progress is made.

4.2 Preliminaries

Before we start looking at methods to derive a sequence of improving βs, let us fix some notation and give some well-known results from the literature. In this chapter we will only cover optimization problems as they arise in statistics and these problems can in most cases be solved efficiently. So let us first formalize what type of optimization problem we will consider in this chapter.

Definition 4.1: Unconstrained Optimization Problem
Given a function $f\colon \mathbb{R}^n \to \mathbb{R}$, an **unconstrained optimization problem** is given by

$$\underset{\boldsymbol{\beta} \in \mathbb{R}^n}{\text{minimize}}\ f(\boldsymbol{\beta}).$$

We will call $\operatorname{dom} f$, the domain of f, the **parameter space** and any element β of the parameter space will often be called a **parameter vector** or just a **parameter**. Similarly, the image of f, $\operatorname{im} f$, is called the **objective space** or sometimes also the **decision space**.

We call β^* the **global minimum** of f if for all $\beta \in \mathbb{R}^n$ $f(\beta^*) \leq f(\beta)$.

Note that we make only one assumption about f, namely, that the function is a mapping from $\mathbb{R}^n$ to $\mathbb{R}$. We do not assume that the function is convex, continuous, or has any other special properties. In order to simplify the notation, we assume all problems to be minimization problems. Should we in fact want to maximize f, we can solve the equivalent problem of minimizing $\tilde{f}(\beta) := -f(\beta)$. Also note that in the numerical optimization literature, which includes most of the referenced material, β is often called x. Because we use x to denote our data in the examples, we have opted to use β throughout this chapter to denote the parameter with regard to which we wish to optimize. This is a natural choice, as most statistical optimization problems arise in the context of estimation of a parameter often called β seeking a best estimate by either minimizing some loss or maximizing a likelihood.

Notice that our definition excludes all types of combinatorial optimization problems. They usually require very specialized methods or can, through relaxation, be reformulated as a continuous optimization problem. We will see an example of this "trick" when we cover linear programming type problems.

But before we get to that, let us start by looking at one of the simplest possible cases covered by Definition 4.1. In the multivariate case, i.e. in the case $\beta \in \mathbb{R}^n$, $n > 1$, we will first concentrate on the well-known convex functions, which we will define here and give some properties without proof.

Definition 4.2: Convex Function

A function f from an interval I (or more general from a convex subset $\mathscr{S}$ of a real vector space) into $\mathbb{R}$ is called **convex**, iff for all $\beta_1, \beta_2 \in I$ (or $\mathscr{S}$) and $\nu \in (0,1)$

$$f(\nu\beta_1 + (1-\nu)\beta_2) \leq \nu f(\beta_1) + (1-\nu)f(\beta_2).$$

If even

$$f(\nu\beta_1 + (1-\nu)\beta_2) < \nu f(\beta_1) + (1-\nu)f(\beta_2),$$

then f is called **strictly convex**.

A function f from an interval I (or more general from a convex subset $\mathscr{S}$ of a real vector space) into $\mathbb{R}$ is called **concave**, iff for all $\beta_1, \beta_2 \in I$ (or $\mathscr{S}$) and $\nu \in (0,1)$

$$f(\nu\beta_1 + (1-\nu)\beta_2) \geq \nu f(\beta_1) + (1-\nu)f(\beta_2).$$

Graphically, this definition means: the function values between two arguments β_1 and β_2 lie below or at the line connecting the two function values of β_1 and β_2.

The following properties and transformation rules for convex functions are well-known and only repeated for the sake of completeness. Particularly the first two properties are the reason for first elaborating on convex functions.

Proposition 4.1: Properties of Convex Functions

1. If $f\colon \mathbb{R}^n \to \mathbb{R}$ is convex on a nonempty, convex set $\mathscr{S} \subset \mathbb{R}^n$, then every local minimum of function f is a global minimum.
2. If f is strictly convex, then the global minimum of f is uniquely determined.
3. If $f \in C^1(\mathbb{R}^n)$ and $\mathscr{S} \subset \mathbb{R}^n$ are convex and nonempty, then

$$f \text{ convex on } \mathscr{S} \Leftrightarrow \forall \beta_1, \beta_2 \in \mathscr{S}\colon f(\beta_2) \geq f(\beta_1) + \nabla f(\beta_1)^T(\beta_2 - \beta_1).$$

4. If $f \in C^2(\mathbb{R}^n)$ and $\mathscr{S} \subset \mathbb{R}^n$ open, convex and nonempty, then

$$f \text{ convex on } \mathscr{S} \Leftrightarrow \forall \boldsymbol{\beta} \in \mathscr{S}\colon \nabla^2 f(\boldsymbol{\beta}) \text{ is positive semidefinite.}$$

Here $C^k(\mathbb{R}^n), k \in \{1,2\}$, denotes the set of all functions $f\colon \mathbb{R}^n \to \mathbb{R}$ that are k times continuously differentiable.

Proposition 4.2: Transformation Rules for Convex Functions

1. If $f\colon \mathbb{R}^n \to \mathbb{R}$ is convex, $\boldsymbol{A} \in L(n,m)$, and $\boldsymbol{b} \in \mathbb{R}^n$, then

$$g(\boldsymbol{\beta}) := f(\boldsymbol{A}\boldsymbol{\beta} + \boldsymbol{b})$$

   is convex for $\boldsymbol{\beta} \in \mathbb{R}^m$.
2. If $f_1, f_2\colon \mathbb{R}^n \to \mathbb{R}$ are convex functions, then

$$g(\boldsymbol{\beta}) := \max\{f_1(\boldsymbol{\beta}), f_2(\boldsymbol{\beta})\} \quad \text{and} \quad h(\boldsymbol{\beta}) := f_1(\boldsymbol{\beta}) + f_2(\boldsymbol{\beta})$$

   are also convex functions.
3. If $f_1\colon \mathbb{R}^n \to \mathbb{R}$ is convex and $f_2\colon \mathbb{R} \to \mathbb{R}$ is convex and non-decreasing, then

$$g(\boldsymbol{\beta}) := f_2(f_1(\boldsymbol{\beta}))$$

   is also convex.
4. If $f_1\colon \mathbb{R}^n \to \mathbb{R}$ is concave and $f_2\colon \mathbb{R} \to \mathbb{R}$ is convex and non-increasing, then

$$g(\boldsymbol{\beta}) := f_2(f_1(\boldsymbol{\beta}))$$

   is convex.

Definition 4.3: Multivariate Differentiation
Let $f(\beta): \mathbb{R}^n \to \mathbb{R}$ be a twice differentiable function. Then,

- $\nabla f(\beta^{(k)}) := \left[\frac{\partial f}{\partial \beta}\Big|_{\beta^{(k)}}\right]^T = \left[\frac{\partial f}{\partial \beta_1}\Big|_{\beta^{(k)}} \cdots \frac{\partial f}{\partial \beta_n}\Big|_{\beta^{(k)}}\right]^T$ is called **gradient (vector)** of f in $\beta^{(k)}$, where $\left[\frac{\partial f}{\partial \beta}\Big|_{\beta^{(k)}}\right]$ is the row vector of partial derivatives of f with respect to the elements β_i of β after insertion of the elements of $\beta^{(k)}$,
- $\nabla^2 f(\beta^{(k)}) := \boldsymbol{H}(\beta^{(k)}) := \frac{\partial^2 f}{\partial \beta \partial \beta^T}\Big|_{\beta^{(k)}} = \left[\frac{\partial^2 f}{\partial \beta_i \partial \beta_j}\right]_{ij}\Big|_{\beta^{(k)}}$ is called **Hessian (matrix)** of f in $\beta^{(k)}$, where $\left[\frac{\partial^2 f}{\partial \beta_i \partial \beta_j}\right]_{ij}\Big|_{\beta^{(k)}}$ is the matrix of the second partial derivatives of f with respect to the elements β_i, β_j of β after insertion of the elements of $\beta^{(k)}$.

Let $f(\beta): \mathbb{R}^n \to \mathbb{R}^m$ be a once differentiable function. Let $f = [f_1 \dots f_m]^T$, where $f_i, i = 1, \dots, m$, are the component functions of f. Then,

$$\boldsymbol{J}_f(\beta^{(k)}) := \begin{bmatrix} \frac{\partial f_1}{\partial \beta}\Big|_{\beta^{(k)}} \\ \vdots \\ \frac{\partial f_m}{\partial \beta}\Big|_{\beta^{(k)}} \end{bmatrix} \in L(m,n)$$

is called **Jacobi matrix** or **Jacobian** of f in $\beta^{(k)}$.

Let us now review some of the rules for vector/matrix differentiation.

Proposition 4.3: Derivatives in Matrix Form

- If $f(\beta) = \boldsymbol{a}^T \beta$, then $\nabla f(\beta) = \boldsymbol{a}$.
- If $f(\beta) = \beta^T \boldsymbol{X} \beta$ and $\boldsymbol{X} = \boldsymbol{X}^T$, then $\nabla f(\beta) = 2\boldsymbol{X}\beta$.
- If $S(\beta) = (\boldsymbol{y} - \boldsymbol{X}\beta)^T(\boldsymbol{y} - \boldsymbol{X}\beta)$, then $\nabla S(\beta) = -2\boldsymbol{y}^T\boldsymbol{X} + 2\boldsymbol{X}^T\boldsymbol{X}\beta$. By setting this to zero, we get the **Normal equations** (see Section 3.1.2): $\boldsymbol{X}^T\boldsymbol{X}\beta = \boldsymbol{y}^T\boldsymbol{X}$
- If $f(\beta) = \boldsymbol{X}\beta$, then $\boldsymbol{J}_f(\beta) = \boldsymbol{X}$.

4.3 Univariate Optimization

Instead of diving into the deep end and directly tackling the more exotic problems covered by our definition of an unconstrained optimization problem, let us add some restrictions for a moment. First, we will assume, that our function only has one parameter ($n = 1$) and that all valid values for β lie in the closed interval $[\beta_{\text{lower}}, \beta_{\text{upper}}]$. Our problem therefore reduces to

$$\underset{\beta_{\text{lower}} \leq \beta \leq \beta_{\text{upper}}}{\text{minimize}} \quad f(\beta). \tag{4.4}$$

An example from introductory statistics is the following problem. Given m observations $x_1, \ldots, x_m$, find

$$\underset{\beta}{\text{minimize}} \ f(\beta) = \sum_{i=1}^{m} (x_i - \beta)^2.$$

The solution is of course the sample mean $\bar{x}$. (Proof: Calculate f' and solve $f'(\beta) = 0$ for β).

Just as we did in the introduction, we can replace the L2 norm with the $L1$ norm and get

$$\underset{\beta}{\text{minimize}} \ f(\beta) = \sum_{i=1}^{m} |x_i - \beta|. \tag{4.5}$$

The solution to this optimization problem is the median. We know how to efficiently calculate the median from Chapter 2 using sorting algorithms[1], but what if we did not know this "trick"? We cannot give the derivative of f in a closed-form because for $\beta = x_i 1$ it is undefined. We could resort to the theory of subdifferentials, which would at least give us a set of subderivatives for the points $\beta = x_i$. But this would not bring us closer to finding a closed-form solution to the optimization problem at hand.

What we do know, however, is that β^* is bounded by $\min(x_1, \ldots, x_m)$ and $\max(x_1, \ldots, x_m)$ and that f is convex. Therefore, β^* exists. All we need to do is find it. Instead of sorting the x_i, we will search the space of all admissible β. How does this fit into our general scheme of iterating to find a sequence of improving solutions? Let us start with an initial guess β_{best} and the upper and lower bounds

$$\beta_{\text{lower}} := \min(x_1, \ldots, x_m) \quad \text{and} \quad \beta_{\text{upper}} := \max(x_1, \ldots, x_m).$$

Now we need a way to find a new solution $\beta^{(1)}$ so that $f(\beta^{(1)}) \leq f(\beta_{\text{best}})$. Say we pick a random new $\beta_{\text{candidate}} \in (\beta_{\text{lower}}, \beta_{\text{upper}})$. Then either of two things can happen: $f(\beta_{\text{candidate}}) < f(\beta_{\text{best}})$ or $f(\beta_{\text{candidate}}) \geq f(\beta_{\text{best}})$. If $f(\beta_{\text{candidate}}) < f(\beta_{\text{best}})$ we have found our $\beta^{(1)}$ and we can also decrease the size of our search interval by noting that since we know f to be convex, we can replace either β_{lower} with β_{best} or β_{upper} with β_{best} depending on whether

[1]It should be noted, that we can do even better than the lower bound of $O(m \log m)$ from Chapter 2 by using a so-called **selection algorithm**. In fact, there are selection algorithms for finding the kth smallest observation with a worst case runtime of $O(m)$ (Blum et al., 1973).

Algorithm 4.1 Generic Univariate Section Search

Require: $f\colon [\beta_{\text{lower}}, \beta_{\text{upper}}] \to \mathbb{R}$

$\beta_{\text{best}} \leftarrow$ uniformly distributed random number $\in (\beta_{\text{lower}}, \beta_{\text{upper}})$
while not finished **do**
  $\beta_{\text{candidate}} \leftarrow$ new candidate solution
  if $f(\beta_{\text{candidate}}) < f(\beta_{\text{best}})$ **then**
    $\beta_{\text{best}} \leftrightarrow \beta_{\text{candidate}}$
  end if
  $\beta_{\text{lower}} \leftarrow \begin{cases} \beta_{\text{lower}} & \text{, if } \beta_{\text{best}} < \beta_{\text{candidate}} \\ \beta_{\text{candidate}} & \text{, else} \end{cases}$
  $\beta_{\text{upper}} \leftarrow \begin{cases} \beta_{\text{candidate}} & \text{, if } \beta_{\text{best}} < \beta_{\text{candidate}} \\ \beta_{\text{upper}} & \text{, else} \end{cases}$
end while
return $\frac{1}{2}(\beta_{\text{lower}} + \beta_{\text{upper}})$

$\beta^{(1)} > \beta_{\text{best}}$ or not. If, on the other hand, our candidate solution $\beta_{\text{candidate}}$ is worse than β_{best}, we reject it. We can, however, still shrink the search interval using this information using similar logic as above.

If we formalize the above idea into a concrete algorithm, we obtain Algorithm 4.1. Looking at the body of the main loop, it might seem strange that we exchange β_{best} and $\beta_{\text{candidate}}$ if $\beta_{\text{candidate}}$ is the better solution, but it makes the rest of the algorithm substantially shorter if we do not explicitly deal with the case that $\beta_{\text{candidate}}$ is not better than β_{best}, and instead exchange them and pretend that we had generated a better solution. After the exchange, we are certain that

$$f(\beta_{\text{best}}) \leq f(\beta_{\text{candidate}}) \leq \max(f(\beta_{\text{lower}}), f(\beta_{\text{upper}}))$$

(since the function f is convex in $[\beta_{\text{lower}}, \beta_{\text{upper}}]$), but

$$f(\beta_{\text{candidate}}) \leq \min(f(\beta_{\text{lower}}), f(\beta_{\text{upper}}))$$

is not true in general. In fact, the candidate solution will often not be better than both boundary solutions. Upon closer inspection we also see that two important pieces are missing from the algorithm description. First, we have neglected to specify how the next candidate solution is chosen, and second, the algorithm has no explicit stop criterion.

Before we look at an example run of the above algorithm, we will try to derive a good stop criterion. A first idea would be to stop if the width of the search interval is smaller than some given ε, that is, $|\beta_{\text{lower}} - \beta_{\text{upper}}| < \varepsilon$.

While conceptually simple, this criterion has the drawback that the value of ε must depend on the location of the true optimum β^*. To see this, consider a convex function $f\colon [0, 2^{10}] \to \mathbb{R}$. If f has its minimum close to 0, then we can choose ε close to the machine precision, but if the minimum is close to 2^{10}, we need to choose ε at least 2^{10} times larger because we cannot achieve any smaller ε without $\beta_{\text{lower}} = \beta_{\text{upper}}$ because of the loss of precision of double-precision floating-point numbers as we move away from 0 (see Example 2.13). So, although in theory the above criterion would be desirable because it would bound the absolute error we make, it turns out to be impractical because of the numerics involved.

Instead we will bound the relative error by stopping only if

$$\frac{|\beta_{\text{lower}} - \beta_{\text{upper}}|}{|\beta_{\text{best}}|} < \varepsilon.$$

Here we are approximating the absolute error by $|\beta_{\text{lower}} - \beta_{\text{upper}}|$ and the magnitude of the true value by our best solution so far. But also this definition has one final flaw. In the special case that $\beta_{\text{best}} = 0$, we never terminate. A small additive term, often 1, is therefore added to $|\beta_{\text{best}}|$ to bound the denominator away from zero. This results in the final stop criterion

$$\frac{|\beta_{\text{lower}} - \beta_{\text{upper}}|}{1 + |\beta_{\text{best}}|} < \tau$$

or

$$|\beta_{\text{lower}} - \beta_{\text{upper}}| < \tau(1 + |\beta_{\text{best}}|). \tag{4.6}$$

One more thing to note about the algorithm is the return value. Instead of returning the last value of β_{best}, the center of the last search interval is returned. There are arguments for both return values. On the one hand β_{best} is the best observed function value, so it is a natural candidate. On the other hand, we know that β^* lies somewhere between β_{lower} and β_{upper}, and therefore $\frac{1}{2}(\beta_{\text{lower}} + \beta_{\text{upper}})$ is our best guess.

Example 4.1: Approximation of Median
Let us come back to our motivating example of finding the median, or any quantile for that matter, of m observations. In Figure 4.1 we can see four iterations of Algorithm 4.1 optimizing the problem in Formula 4.5 for a small toy data set. New candidate solutions $\beta_{\text{candidate}}$ were picked at random from the search interval. While we see that the search progresses nicely, there must be a better way to choose new candidate solutions. This will be the final missing piece for our first iterative refinement algorithm.

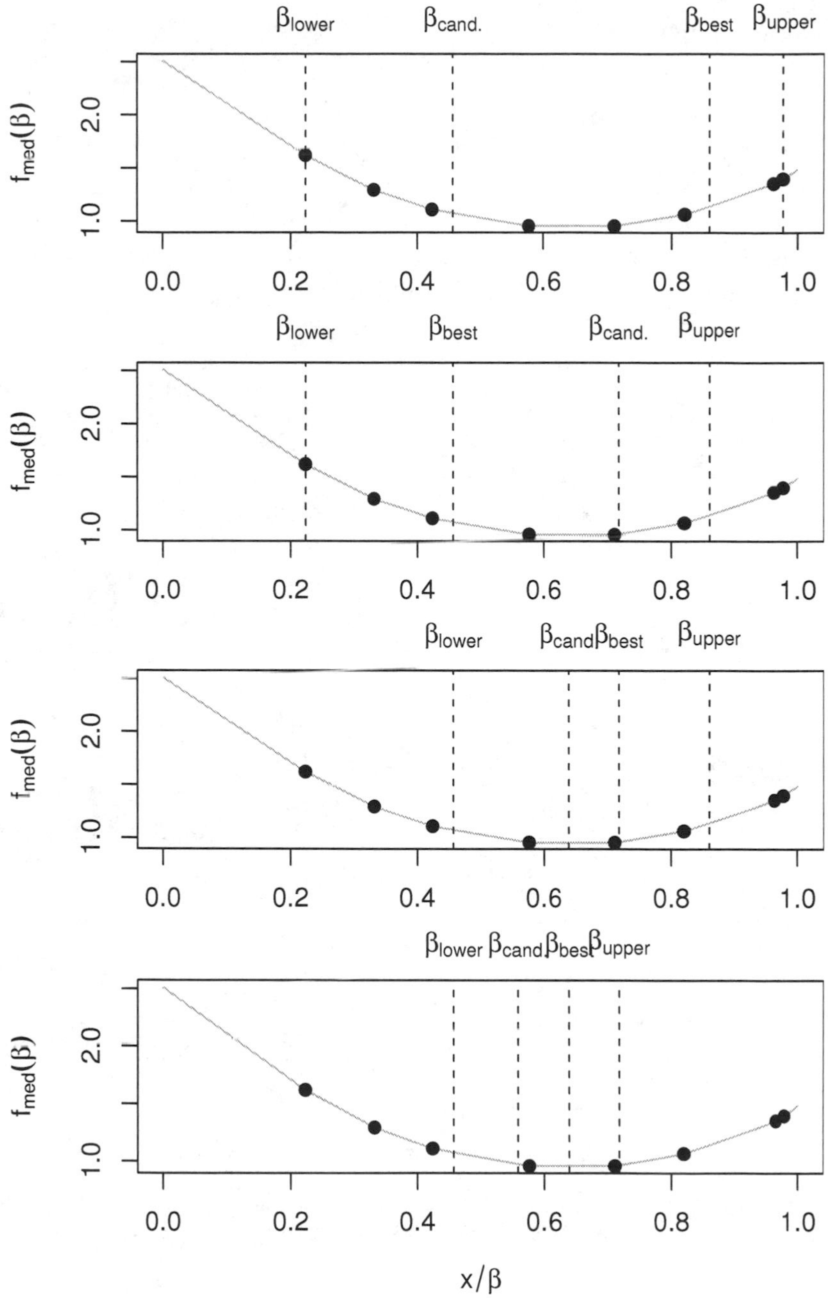

Figure 4.1: Example of applying Algorithm 4.1 with random selection of a new search point. Points denote the observed x_i. We can see that from top (first iteration) to bottom (fourth and last iteration) the size of the search interval which must contain the optimal value steadily decreases.

4.3.1 Golden Section Search

The last missing piece for a high quality variation of Algorithm 4.1 is a method to choose a new candidate $\beta_{\text{candidate}}$. We could go about this randomly as done in the previous example, but that would not be efficient in general. What we really want to do is choose $\beta_{\text{candidate}}$ such that we maximize the minimal reduction in search space that we might obtain because ultimately we need to shrink our search interval in order to terminate with any confidence. By that observation alone it is clear that $\beta_{\text{candidate}}$ should lie between β_{lower} and β_{best} if $\beta_{\text{best}} - \beta_{\text{lower}}$ is larger than $\beta_{\text{upper}} - \beta_{\text{best}}$ and between β_{best} and β_{upper} otherwise. This makes sure that no matter what the outcome, we shrink the larger of the two intervals $(\beta_{\text{lower}}, \beta_{\text{best}})$, $(\beta_{\text{best}}, \beta_{\text{upper}})$.

Of all the thinkable strategies that abide by this restriction there are two prominent examples of which we will discuss one in detail. Both were first described by Kiefer (1953). The first one assumes that we have a fixed budget of k function evaluations. It is called **Fibonacci search** and will not be discussed further here because for most practical problems we do not know in advance how many function evaluations we are willing to perform. Instead we will discuss the other common strategy which is called **Golden Section search**. It is applicable in cases where the number of function evaluations is not known a priori. Golden Section search picks the candidate points $\beta_{\text{candidate}}$ in such a way that the newly induced search interval is **self-similar** to the previous search interval.

Let us assume for a moment, that $\beta_{\text{lower}} < \beta_{\text{best}} < \beta_{\text{candidate}} < \beta_{\text{upper}}$, then we want to choose $\beta_{\text{candidate}}$ such that

$$|\beta_{\text{lower}} - \beta_{\text{candidate}}| = |\beta_{\text{best}} - \beta_{\text{upper}}|.$$

The length of the new search interval is then the same, regardless of whether $f(\beta_{\text{best}}) \leq f(\beta_{\text{candidate}})$ or $f(\beta_{\text{candidate}}) < f(\beta_{\text{best}})$. Now define:

$$\begin{aligned}
a &:= |\beta_{\text{lower}} - \beta_{\text{upper}}|, \\
b &:= |\beta_{\text{lower}} - \beta_{\text{candidate}}| = |\beta_{\text{best}} - \beta_{\text{upper}}|, \\
c &:= |\beta_{\text{lower}} - \beta_{\text{best}}| = |\beta_{\text{candidate}} - \beta_{\text{upper}}|.
\end{aligned}$$

Then a is the length of the current search interval. In the next iteration b will be the length of the search interval, and then c will be the length of the search interval in the iteration after that. We also have $a = b + c$. This allows us to give a precise definition of our vague notion of self-similarity in terms of these three lengths. We want the quotient of successive search intervals to stay constant:

$$\frac{b}{a} = \frac{c}{b} = \phi < 1.$$

This implies that the algorithm will converge linearly because the width of the search interval will decrease by a constant factor in each iteration. Note that this guarantee holds regardless of the function f, which we are minimizing!

Remembering that $a = b + c$ and using the identity

$$\phi^2 = \frac{b}{a}\frac{c}{b} = \frac{c}{a}$$

we obtain

$$\begin{aligned} & a = b + c \\ \Rightarrow\quad & \frac{a}{a} = \frac{b}{a} + \frac{c}{a} \\ \Rightarrow\quad & 1 = \phi + \phi^2 \\ \Rightarrow\quad & 0 = \phi^2 + \phi - 1 \end{aligned}$$

which gives us

$$\phi = \frac{\sqrt{5} - 1}{2}$$

as the only positive solution to the quadratic equation above. This ratio appears in many branches of mathematics and is called the **Golden Section** and lends its name to the associated search algorithm. Using the constraints we derived for the lengths of successive search intervals, we are now able to calculate our initial β_{best} given only β_{lower} and β_{upper}

$$\beta_{\text{best}} = \beta_{\text{upper}} - \phi(\beta_{\text{upper}} - \beta_{\text{lower}}) = \beta_{\text{lower}} + (1 - \phi)(\beta_{\text{upper}} - \beta_{\text{lower}}).$$

Using the length constraint ($a = b + c$), we can derive that

$$\beta_{\text{candidate}} = \beta_{\text{lower}} + (\beta_{\text{upper}} - \beta_{\text{best}}).$$

Substituting these results and the stop criterion from Equation 4.6 into Algorithm 4.1, we obtain the final **Golden Section Search** Algorithm 4.2. Note that while we have assumed that $\beta_{\text{best}} < \beta_{\text{candidate}}$ for the derivation of ϕ, this is the case when we initialize β_{best} in the algorithm. The relation used to calculate the new $\beta_{\text{candidate}}$ holds regardless of whether $\beta_{\text{best}} < \beta_{\text{candidate}}$ or $\beta_{\text{candidate}} < \beta_{\text{best}}$.

4.3.2 Convergence

In the previous section we casually observed that the quotient of the width of consecutive search intervals is constant and smaller than 1. From this we

Algorithm 4.2 Golden Section Search

Require: $f\colon [\beta_{\text{lower}}, \beta_{\text{upper}}] \to \mathbb{R}$, desired precision τ

$\phi \leftarrow \frac{\sqrt{5}-1}{2}$

$\beta_{\text{best}} \leftarrow \beta_{\text{upper}} - \phi(\beta_{\text{upper}} - \beta_{\text{lower}})$

while $|\beta_{\text{upper}} - \beta_{\text{lower}}| \geq \tau(1 + |\beta_{\text{best}}|)$ **do**

  $\beta_{\text{candidate}} \leftarrow \beta_{\text{lower}} + (\beta_{\text{upper}} - \beta_{\text{best}})$

  if $f(\beta_{\text{candidate}}) < f(\beta_{\text{best}})$ **then**

    $\beta_{\text{best}} \leftrightarrow \beta_{\text{candidate}}$

  end if

  $\beta_{\text{lower}} = \begin{cases} \beta_{\text{lower}} & \text{, if } \beta_{\text{best}} < \beta_{\text{candidate}} \\ \beta_{\text{candidate}} & \text{, else} \end{cases}$

  $\beta_{\text{upper}} = \begin{cases} \beta_{\text{candidate}} & \text{, if } \beta_{\text{best}} < \beta_{\text{candidate}} \\ \beta_{\text{upper}} & \text{, else} \end{cases}$

end while

return $\frac{1}{2}(\beta_{\text{lower}} + \beta_{\text{upper}})$

concluded that the algorithm converges linearly. In this section we will come back to this observation and lay the foundation for a rigorous characterization of the speed of convergence. For this it is helpful to begin with a coarse characterization of the speed of convergence.

Definition 4.4: Order of Convergence
A convergent sequence $\{a^{(k)}\}$ with limit a^* is said to **converge with order** p iff

$$0 \leq \limsup_{k\to\infty} \frac{|a^{(k+1)} - a^*|}{|a^{(k)} - a^*|^p} < \infty.$$

Notice that $\limsup_{k\to\infty} x^{(k)}$ stands for the greatest limit of a convergent subsequence of $\{x^{(k)}\}$. Thus, the series itself might not converge as a whole. The above definition uses the convention that $0/0$ is defined to be finite. Let us illustrate the previous definition with two simple examples:

1. The sequence $a^{(k)} = c^k$ where $0 < c < 1$ converges to zero with order 1 because $\frac{a^{(k+1)}}{a^{(k)}} = c$.

2. The sequence $a^{(k)} = c^{2^k}$ where again $0 < c < 1$ also converges to zero but with order $p = 2$:

$$\frac{a^{(k+1)}}{{a^{(k)}}^2} = \frac{c^{2^{k+1}}}{(c^{2^k})^2} = 1.$$

Most algorithms we will encounter have an order of convergence of one, so it makes sense to further characterize this special case. This is achieved using the next definition.

Definition 4.5: Rate of Convergence
A sequence $\{a^{(k)}\}$ that converges to a^* is said to converge **linearly** iff

$$\lim_{k\to\infty} \frac{|a^{(k+1)} - a^*|}{|a^{(k)} - a^*|} = \rho$$

and $\rho \in (0,1)$. If $\rho = 1$, the sequence is said to converge **sublinearly** and if $\rho = 0$, the sequence converges **superlinearly**.

Notice that the Golden Section search converges linearly in that the lengths of its search intervals converge linearly to zero with $\rho = \phi$.

We will now turn our attention to other optimization strategies that achieve a faster rate of convergence than the Golden Section search.

4.3.3 Faster Methods

In order to speedup our search we will have to add additional assumptions about the function we want to minimize. If, for example, we assume that f is smooth in some sense, then it is plausible that a search strategy might use this information to speedup the search. Most of these techniques are based around the idea of fitting a smooth curve, usually a low-order polynomial, to the last few intermediate solutions, and then using information about that curve to guide the search.

One of the simplest approaches to implement such a strategy is to reuse the general framework from the previous section. However, instead of using the rule from the Golden Section search to choose a new candidate solution $\beta_{\text{candidate}}$, we will fit a quadratic function through the three points β_{lower}, β_{best} and β_{upper} and use its minimum, which we can determine analytically, as the new candidate solution. The three points $[\beta_{\text{lower}}\ \ f(\beta_{\text{lower}})]^T$, $[\beta_{\text{best}}\ \ f(\beta_{\text{best}})]^T$, and $[\beta_{\text{upper}}\ \ f(\beta_{\text{upper}})]^T$ uniquely define a quadratic function $y = qx^2 + px + r$, assuming that they are not collinear. We can find the coefficients of this quadratic equation by solving the following system of linear equations:

$$\begin{aligned} f(\beta_{\text{lower}}) &= q\beta_{\text{lower}}^2 + p\beta_{\text{lower}} + r \\ f(\beta_{\text{best}}) &= q\beta_{\text{best}}^2 + p\beta_{\text{best}} + r \\ f(\beta_{\text{upper}}) &= q\beta_{\text{upper}}^2 + p\beta_{\text{upper}} + r. \end{aligned}$$

This can be rewritten in matrix form as

$$\begin{bmatrix} f(\beta_{\text{lower}}) \\ f(\beta_{\text{best}}) \\ f(\beta_{\text{upper}}) \end{bmatrix} = \begin{bmatrix} \beta_{\text{lower}}^2 & \beta_{\text{lower}} & 1 \\ \beta_{\text{best}}^2 & \beta_{\text{best}} & 1 \\ \beta_{\text{upper}}^2 & \beta_{\text{upper}} & 1 \end{bmatrix} \begin{bmatrix} p \\ q \\ r \end{bmatrix}$$

which gives us

$$\begin{bmatrix} \beta_{\text{lower}}^2 & \beta_{\text{lower}} & 1 \\ \beta_{\text{best}}^2 & \beta_{\text{best}} & 1 \\ \beta_{\text{upper}}^2 & \beta_{\text{upper}} & 1 \end{bmatrix}^{-1} \begin{bmatrix} f(\beta_{\text{lower}}) \\ f(\beta_{\text{best}}) \\ f(\beta_{\text{upper}}) \end{bmatrix} = \begin{bmatrix} p \\ q \\ r \end{bmatrix}.$$

Solving for the inverse and using simple algebraic manipulations as well as the fact that the minimum of the quadratic function is at $x = \frac{q}{2p}$, it is easy to show that

$$\frac{1}{2}\frac{f(\beta_{\text{upper}})(\beta_{\text{lower}}^2 - \beta_{\text{best}}^2) + f(\beta_{\text{best}})(\beta_{\text{upper}}^2 - \beta_{\text{lower}}^2) + f(\beta_{\text{lower}})(\beta_{\text{best}}^2 - \beta_{\text{upper}}^2)}{f(\beta_{\text{upper}})(\beta_{\text{lower}} - \beta_{\text{best}}) + f(\beta_{\text{best}})(\beta_{\text{upper}} - \beta_{\text{lower}}) + f(\beta_{\text{lower}})(\beta_{\text{best}} - \beta_{\text{upper}})}$$

is the global minimum of the quadratic function that passes through the points $[\beta_{\text{lower}} \;\; f(\beta_{\text{lower}})]^T$, $[\beta_{\text{best}} \;\; f(\beta_{\text{best}})]^T$, and $[\beta_{\text{upper}} \;\; f(\beta_{\text{upper}})]^T$. Using this formula and Algorithm 4.1 as a template we arrive at Algorithm 4.3.

This quadratic interpolation search is not as robust as the Golden Section search but converges superlinearly under mild regularity conditions (Brent, 1973). This lack of robustness leads to the three extra termination conditions in the main loop of the algorithm. These are triggered when

1. The algorithm chose the same candidate point as in the previous iteration ($\beta'_{\text{candidate}} = \beta_{\text{candidate}}$). No matter how many more iterations are performed, the algorithm will not make any progress because β_{lower}, β_{upper} and β_{best} will remain unchanged.

2. The algorithm chose a new candidate point not inside the search interval $[\beta_{\text{lower}}, \beta_{\text{upper}}]$. This can happen for several reasons. The most likely is that the function has no (local) minimum in the search interval and the quadratic function therefore (correctly) predicts a stationary point outside of the search interval. In the extreme case, the quadratic approximation has degenerated into a linear function and has no finite minimum ($\beta_{\text{candidate}} = \pm\infty$).

There are a number of other corner cases that a production quality version of the algorithm should deal with. These have been omitted here. In practice these drawbacks have led to the development of a combined strategy that switches between Golden Section search and quadratic interpolation. The idea is to use quadratic interpolation for fast convergence if possible and Golden Section search as a robust fallback if necessary. This is called

Algorithm 4.3 Parabolic Interpolation Search

Require: $f\colon [\beta_{\text{lower}}, \beta_{\text{upper}}] \to \mathbb{R}$, desired precision τ

$\beta_{\text{best}} \leftarrow \frac{1}{2}(\beta_{\text{lower}} + \beta_{\text{upper}})$

$\beta_{\text{candidate}} \leftarrow \frac{1}{2}\frac{f(\beta_{\text{upper}})(\beta_{\text{lower}}^2-\beta_{\text{best}}^2)+f(\beta_{\text{best}})(\beta_{\text{upper}}^2-\beta_{\text{lower}}^2)+f(\beta_{\text{lower}})(\beta_{\text{best}}^2-\beta_{\text{upper}}^2)}{f(\beta_{\text{upper}})(\beta_{\text{lower}}-\beta_{\text{best}})+f(\beta_{\text{best}})(\beta_{\text{upper}}-\beta_{\text{lower}})+f(\beta_{\text{lower}})(\beta_{\text{best}}-\beta_{\text{upper}})}$

while $|\beta_{\text{upper}} - \beta_{\text{lower}}| \geq \tau(1 + |\beta_{\text{best}}|)$ **do**

  if $f(\beta_{\text{candidate}}) < f(\beta_{\text{best}})$ **then**

    $\beta_{\text{candidate}} \leftrightarrow \beta_{\text{best}}$

  end if

  $\beta_{\text{lower}} = \begin{cases} \beta_{\text{lower}} & \text{, if } \beta_{\text{best}} < \beta_{\text{candidate}} \\ \beta_{\text{candidate}} & \text{, else} \end{cases}$

  $\beta_{\text{upper}} = \begin{cases} \beta_{\text{candidate}} & \text{, if } \beta_{\text{best}} < \beta_{\text{candidate}} \\ \beta_{\text{upper}} & \text{, else} \end{cases}$

  $\beta'_{\text{candidate}} \leftarrow \beta_{\text{candidate}}$

  $\beta_{\text{candidate}} \leftarrow \frac{1}{2}\frac{f(\beta_{\text{upper}})(\beta_{\text{lower}}^2-\beta_{\text{best}}^2)+f(\beta_{\text{best}})(\beta_{\text{upper}}^2-\beta_{\text{lower}}^2)+f(\beta_{\text{lower}})(\beta_{\text{best}}^2-\beta_{\text{upper}}^2)}{f(\beta_{\text{upper}})(\beta_{\text{lower}}-\beta_{\text{best}})+f(\beta_{\text{best}})(\beta_{\text{upper}}-\beta_{\text{lower}})+f(\beta_{\text{lower}})(\beta_{\text{best}}-\beta_{\text{upper}})}$

  if $\beta_{\text{candidate}} = \beta'_{\text{candidate}}$ or $\beta_{\text{candidate}} \leq \beta_{\text{lower}}$ or $\beta_{\text{candidate}} \geq \beta_{\text{upper}}$ **then**

    return β_{best}

  end if

end while

return β_{best}

Brent's Method and is described in detail in Brent (1973), including the rules that govern when to use which candidate generation routine. The reason this method works so well is that as we get closer to the minimum all twice differentiable functions tend to look like a quadratic function inside the search interval.

Simulation: Comparison of Univariate Optimization Methods

To judge just how much faster the more advanced methods are, let us compare them on our initial problem of estimating the median of m observations. Here we will fix $m = 1001$ and vary the width of the interval into which the observations fall between 1 and 10^8. For each interval width 100 random data sets are generated and each algorithm is run on these, recording both the relative error of the returned solution and the number of function evaluations used. These results are shown in Table 4.1. We can see that the added complexity of Golden Section search is well worth it, saving on average about one-third of the function evaluations compared to Random-Selection. Brent's method is even faster, beating the naive Random-Selection by a factor of five and Golden Section search by a factor of at least three. On the other hand, Brent's

Table 4.1: Comparison of Random-Selection, Golden Section, and Brent's Searches on the Computation of the Median of Data with Different Ranges; # FE = Number of Function Evaluations

Interval	Random-Selection		Golden Section		Brent	
width	rel. Error	# FE	rel. Error	# FE	rel. Error	# FE
1×10^{0}	1.02×10^{-7}	127.5	7.83×10^{-8}	85.4	3.96×10^{-3}	15.9
1×10^{1}	2.29×10^{-8}	134.2	3.49×10^{-8}	93.7	1.06×10^{-4}	18.7
1×10^{2}	3.45×10^{-9}	145.0	2.54×10^{-9}	99.6	3.64×10^{-5}	21.5
1×10^{3}	1.27×10^{-9}	147.1	1.24×10^{-9}	101.0	7.73×10^{-7}	24.9
1×10^{4}	1.09×10^{-9}	148.7	1.14×10^{-9}	102.2	1.28×10^{-7}	27.6
1×10^{5}	1.45×10^{-9}	148.7	1.00×10^{-9}	101.1	1.48×10^{-8}	30.0
1×10^{6}	1.14×10^{-9}	148.5	9.99×10^{-10}	101.7	5.87×10^{-9}	31.0
1×10^{7}	1.30×10^{-9}	145.9	1.05×10^{-9}	101.3	4.56×10^{-9}	31.2
1×10^{8}	1.28×10^{-9}	144.8	1.09×10^{-9}	101.1	4.28×10^{-9}	31.6

method achieves a considerably higher relative error for the narrow search intervals. This is likely due to the somewhat different termination criterion used by this algorithm. Nevertheless, in practice one would always favor the more complex but much faster method by Brent. In the next section we will discuss a method that is even faster. However, we have to make some strong assumptions about the function we are minimizing. Only this additional information can speedup the optimization process.

4.3.4 Newton's Method

If our function f that we wish to minimize is twice differentiable and we know the first and second derivatives, we can use that information to speedup the convergence rate of our search even further. To do this we will again fit a parabola to approximate the function locally and then use its apex as the new candidate solution. But instead of using all three points to determine the coefficients of the quadratic equation, we use only the best solution found so far and the associated function value as well as the first and second derivative. We now need to find a quadratic function whose value and first two derivatives are equal to those of our function f in the current solution $\beta_{\text{best}}^{(k)}$. By Taylor's theorem that quadratic function is given by

$$q(\beta) = f(\beta_{\text{best}}^{(k)}) + f'(\beta_{\text{best}}^{(k)})(\beta - \beta_{\text{best}}^{(k)}) + \frac{1}{2}f''(\beta_{\text{best}}^{(k)})(\beta - \beta_{\text{best}}^{(k)})^2. \tag{4.7}$$

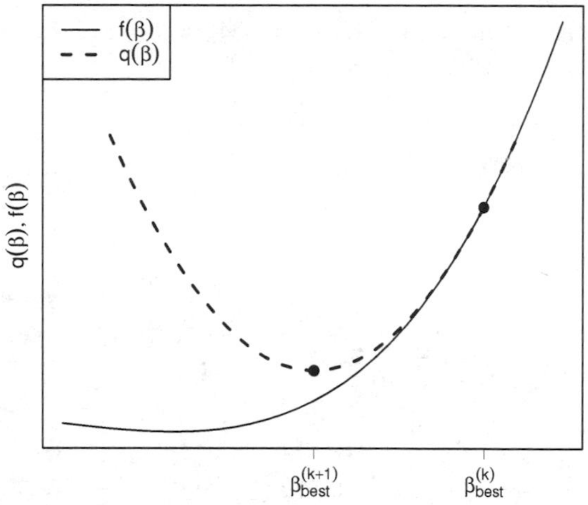

Figure 4.2: Newton method for optimization.

Algorithm 4.4 Univariate Newton Method

Require: Twice differentiable $f\colon \mathbb{R} \to \mathbb{R}$, tolerance on gradient τ, initial solution $\beta^{(1)}$

1: $k \leftarrow 1$
2: $\beta_{\text{best}}^{(k)} \leftarrow \beta^{(1)}$
3: **while** $|f'(\beta_{\text{best}}^{(k)})| \geq \tau$ **do**
4: $\quad \beta_{\text{best}}^{(k+1)} \leftarrow \beta_{\text{best}}^{(k)} - \frac{f'(\beta_{\text{best}}^{(k)})}{f''(\beta_{\text{best}}^{(k)})}$
5: $\quad k \leftarrow k+1$
6: **end while**
7: **return** $\beta_{\text{best}}^{(k)}$

To derive its minimum we need to solve

$$0 = q'(\beta) = f'(\beta_{\text{best}}^{(k)}) + f''(\beta_{\text{best}}^{(k)})(\beta - \beta_{\text{best}}^{(k)})$$

for β to arrive at

$$\beta_{\text{best}}^{(k+1)} = \beta_{\text{best}}^{(k)} - \frac{f'(\beta_{\text{best}}^{(k)})}{f''(\beta_{\text{best}}^{(k)})}$$

for the next candidate solution $\beta_{\text{best}}^{(k+1)}$ (see Figure 4.2).

Note that instead of a steadily shrinking interval $[\beta_{\text{lower}}, \beta_{\text{upper}}]$ we now construct a sequence of intermediate solutions $\{\beta_{\text{best}}^{(k)}\}$. This simplifies the algorithm because we do not need to track the current search bounds, but it

Table 4.2: First 10 Iterations of Newton's Method when Minimizing $F(\beta) = \beta^3$ Starting from $\beta = 10$

k	$\beta_{\text{best}}^{(k)}$	$f(\beta_{\text{best}}^{(k)})$	$f'(\beta_{\text{best}}^{(k)})$	$f''(\beta_{\text{best}}^{(k)})$
0	10.000	1000.000	300.000	60.000
1	5.000	125.000	75.000	30.000
2	2.500	15.625	18.750	15.000
3	1.250	1.953	4.688	7.500
4	0.625	0.244	1.172	3.750
5	0.312	0.031	0.293	1.875
6	0.156	0.004	0.073	0.938
7	0.078	0.000	0.018	0.469
8	0.039	0.000	0.005	0.234
9	0.020	0.000	0.001	0.117
10	0.010	0.000	0.000	0.059

also means that we cannot use the stop criterion based on the relative error estimate. Instead, we will terminate once the first derivative is close enough to zero. When we combine all of these parts, we obtain Algorithm 4.4.

There are several key differences between this algorithm and the previous univariate optimization strategies we have looked at. This algorithm searches on the whole real line for a minimum instead of in a given interval and only terminates when the current solution is sufficiently close to a stationary point (i.e. $f'(\beta) \approx 0$). This also means that the returned solution might be a saddle point instead of a minimum if the function is not convex.

Example 4.2: Newton Method Applied to $f(\beta) = \beta^3$
To illustrate the saddle point problematic, we will manually solve

$$\text{minimize}\, f(\beta) = \beta^3$$

using Newton's method. Table 4.2 lists the first few iterations of the algorithm. We see that the algorithm quickly converges to the stationary point $\beta = 0$. Just how fast Newton's method is will be discussed later.

Another key difference to the other univariate methods is that the choice of the next intermediate solution does not depend on the current function value. In fact, we can recast Newton's method as an algorithm to find the root

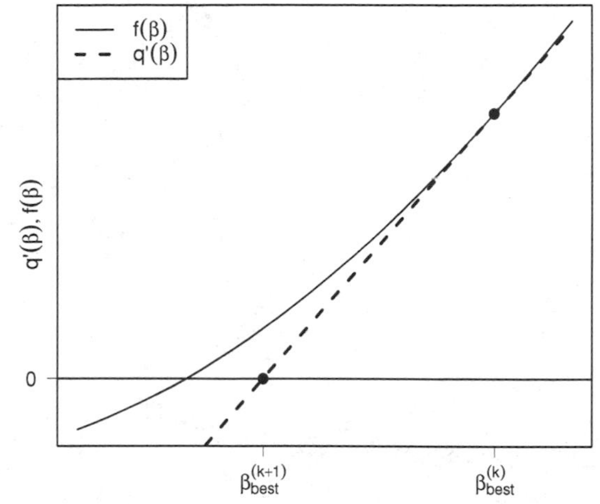

Figure 4.3: Newton method for root finding.

of a function $g(\beta) := f'(\beta)$. The update rule then reduces to

$$\beta_{\text{best}}^{(k+1)} = \beta_{\text{best}}^{(k)} - \frac{g(\beta_{\text{best}}^{(k)})}{g'(\beta_{\text{best}}^{(k)})}.$$

We can again interpret this update rule in the context of function approximation. $\beta_{\text{best}}^{(k+1)}$ is now the intercept of the linear function that passes through the point $[\beta_{\text{best}}^{(k)}\ g(\beta_{\text{best}}^{(k)})]$ and has slope $g'(\beta_{\text{best}}^{(k)})$ (see Figure 4.3).

With this we now have enough insight to prove, under some conditions, that Newton's method has convergence order 2 and is therefore substantially faster than any of the methods we have seen in previous sections. This increase in speed is bought at the expense of robustness and generality. Newton's method can only be applied to a small subset of the functions that Golden Section search can solve.

Theorem 4.1: Convergence Order of Newton's Method
Let $g\colon \mathbb{R} \to \mathbb{R}$ be a convex function that is twice continuous differentiable and let β^* exist with $g(\beta^*) = 0$ and $g'(\beta^*) \neq 0$. If the starting value $\beta^{(0)}$ is close enough to β^*, then the sequence $\{\beta^{(k)}\}$ with

$$\beta^{(k+1)} = \beta^{(k)} - \frac{g(\beta^{(k)})}{g'(\beta^{(k)})}$$

converges with at least order 2 and its limit is β^*.

Proof. For any point β close to β^* there exist r_1 and r_2 such that

$$|g''(\beta)| < r_1 \quad \text{and} \quad |g'(\beta)| > r_2.$$

Then it follows that

$$\begin{aligned}
\beta^{(k+1)} - \beta^* &= \beta^{(k)} - \beta^* - \frac{g(\beta^{(k)})}{g'(\beta^{(k)})} \\
&= -\frac{g(\beta^{(k)}) + g'(\beta^{(k)})(\beta^* - \beta^{(k)})}{g'(\beta^{(k)})} \\
&= -\frac{g(\beta^{(k)}) - g(\beta^*) + g'(\beta^{(k)})(\beta^* - \beta^{(k)})}{g'(\beta^{(k)})}.
\end{aligned}$$

Here we used the fact that $g(\beta^*) = 0$ in the last step. Using Taylor's theorem, we see that the nominator is zero except for a second-order term, so we get

$$\beta^{(k+1)} - \beta^* = -\frac{1}{2}\frac{g''(\beta')}{g'(\beta^{(k)})}(\beta^{(k)} - \beta^*)^2$$

for some β' between β^* and $\beta^{(k)}$ and we therefore have in a neighborhood of β^*

$$|\beta^{(k+1)} - \beta^*| \leq \frac{r_1}{2r_2}|\beta^{(k)} - \beta^*|^2$$

from which we conclude that Newton's method has at least convergence order 2. □

While this result highlights the strong point of the Newton method, fast convergence when we are close to the minimum, it also hints at one of the limitations of the algorithm. If our initial solution is far away from the minimum, we need not converge quickly. In fact, we need not converge at all!

We will now illustrate the usefulness of Newton's method using a simple example.

Example 4.3: Calculation of Quantiles
If we want to find the α quantile of some distribution not knowing the quantile function in analytical form, then we need to resort to numerical methods. Here, Newton's method proves to be particularly powerful if the distribution is continuous and we know its cumulative distribution function F and probability density function $f = F'$. What we seek then is an x such that

$$\alpha = F(x) = \int_{-\infty}^{x} f(t)\,\mathrm{d}t.$$

It is easy to solve the above equation using Newton's method by viewing this as a root finding problem. That is, our current solution will be updated in each generation to

$$x^{(k+1)} = x^{(k)} - \frac{F(x^{(k)}) - \alpha}{f(x^{(k)})}.$$

Here we use the root finding interpretation of Newton's method to formulate the update rule. It is nevertheless instructive to also explicitly write down the underlying optimization problem, which we can recover by integrating over the derivative F from the update rule

$$\underset{x}{\text{minimize}} \int_{-\infty}^{x} (F(t) - \alpha)\,\mathrm{d}t. \tag{4.8}$$

To see this, let $G(t)$ be the antiderivative of $F(t) - \alpha$. Then

$$\int_{-\infty}^{x} (F(t) - \alpha)\,\mathrm{d}t = G(x) - G(-\infty)$$

and by definition,

$$\frac{\mathrm{d}}{\mathrm{d}x} \int_{-\infty}^{x} (F(t) - \alpha)\,\mathrm{d}t = \frac{\mathrm{d}}{\mathrm{d}x}(G(x) - G(-\infty)) = F(x) - \alpha.$$

The reason why it is instructive to also look at the minimization problem that we are implicitly solving is because we know that Newton's method only works reasonably well if the objective function is convex. Luckily we can infer this from the fact that its first derivative F is monotonic because it is a cumulative distribution function and monotonicity of the first derivative is a sufficient condition for convexity in the case of univariate functions. Using the above method, we can, e.g., derive quantiles for the standard normal distribution. Here we used a tolerance of $\tau = 10^{-8}$ for the stop criterion and a starting value of $x_0 = 0$. The results are shown in Table 4.3. We see that the algorithm converges rather quickly even for extreme quantiles, and that the absolute error of the returned solution is quite low for all values of α. This is surprising, given that the objective functions, which are depicted in Figure 4.4, are asymmetric for all $\alpha \neq 0.5$ and get successively flatter around the optimum for smaller values of α. Later in this chapter we will see that an optimization algorithm that purely relies on first derivatives would have a hard time with such an objective function. Here Newton's method profits from the additional curvature information from the second derivative.

So far we have focused exclusively on unimodal problems, that is, problems where any minimum we find will be the global minimum. If we have a

Table 4.3: Results of the Quantile Calculations Using Newton's Method

Quantile α	Final Value x	Error $\lvert x-x^*\rvert$	# of Iterations
0.00001	-4.2648908	1.678657×10^{-13}	14
0.00010	-3.7190165	4.440892×10^{-16}	12
0.00100	-3.0902323	4.440892×10^{-16}	10
0.01000	-2.3263479	4.440892×10^{-16}	8
0.05000	-1.6448536	2.220446×10^{-16}	6
0.10000	-1.2815516	1.854072×10^{-13}	5
0.25000	-0.6744898	1.065814×10^{-14}	4
0.50000	0.0000000	$0.000000 \times 10^{+00}$	0

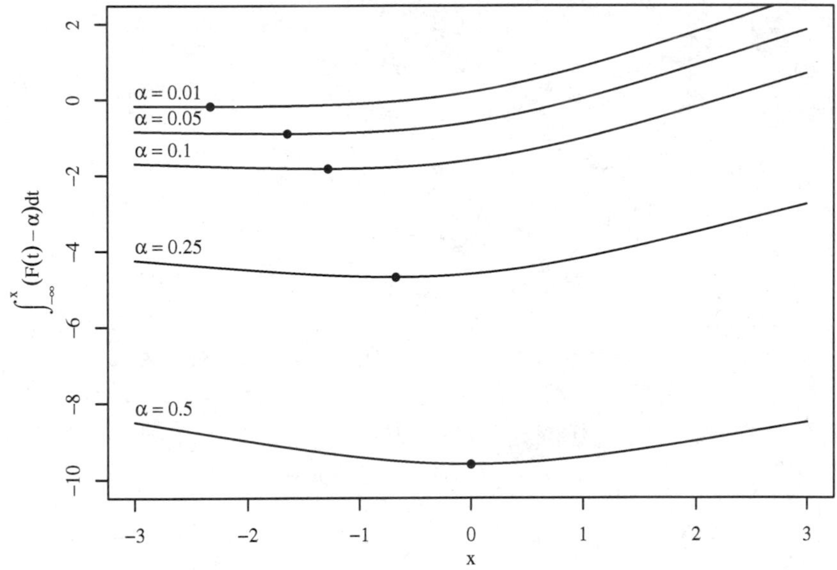

Figure 4.4: Objective function for the quantile problem from Formula 4.8. The location of the minimum is indicated by the points.

multimodal function, we can still try to apply the methods we have learned about, but we need to be aware of some caveats. All interval-based search methods will converge to a local minimum if one exists within the initial interval. The Newton method will also generally converge to a stationary point of the function. This does not have to be an extremum, and even if it is an extremum it may be a maximum and not a minimum. Therefore, one has to be careful when basing further calculations on the result of such an optimization without first confirming that a minimum was indeed found. If the function is highly multimodal, it is often advisable to perform several runs with different starting points or starting intervals, depending on the method used.

4.3.5 Summary

In this section we introduced basic methods for univariate optimization. We started with the discussion of a generic method, and then introduced more and more special methods requiring more and more assumptions. We discussed the ideas and properties of interval-based methods like the Golden Section search, the quadratic interpolation search, and Brent's method. Moreover, with Newton's method we introduced a first method only constructing a sequence of intermediate solutions. In particular, we discussed the convergence speed of these methods.

4.4 Multivariate Optimization [2]

4.4.1 Convex Problems

In the previous section we restricted ourselves to univariate optimization problems. But what if our problem at hand is multivariate, that is, our $\boldsymbol{\beta}$ is not a real value but comes from $\mathbb{R}^n$? Methods to deal with these types of problems will be the topic of this section. Instead of starting with the most general type of optimization problem and then slowly adding restrictions to derive better algorithms, we will defer solving general multivariate optimization problems to the next section and initially only focus on **convex problems**. So we will be solving problems of the form

$$\underset{\boldsymbol{\beta} \in \mathbb{R}^n}{\text{minimize}} \quad f(\boldsymbol{\beta}),$$

where $f\colon \mathbb{R}^n \to \mathbb{R}$ is a convex function. Why do we want f to be convex and not just unimodal as in the univariate case? Convexity makes the job of finding the global minimum easy enough because any local minimum we find will be the global minimum (see Section 4.2). At first it may seem like this restriction to convex functions might in practice be a hindrance but it turns out that many "interesting" statistical optimization problems are indeed convex or may be transformed in a way that they are convex. The properties of convex functions and which transformations preserve convexity were covered in Section 4.2.

[2] Partly based on Judge et al. (1981, pp. 951 – 969).

4.4.1.1 *General Structure of Iterative Algorithms*

In the univariate case most of the algorithms we discussed were based on the idea of iteratively reducing the size of a search interval. In the multivariate case, most times a starting value $\beta^{(1)}$ is updated in each iteration step into a better solution, as in the Newton method. This is illustrated both in Algorithm 4.5 and as a flowchart in Figure 4.5. Notice how we transform the problem of solving a multivariate optimization problem into a problem of solving successive univariate optimization problems by restricting the search for a "better" solution in the kth iteration to a direction $\boldsymbol{d}^{(k)}$. $\boldsymbol{d}^{(k)}$ is called **search direction** and v **step length**.

Algorithm 4.5 General Iterative Algorithm

Require: Function $f\colon \mathbb{R}^n \to \mathbb{R}$, initial value $\beta^{(1)}$
1: $k \leftarrow 1$
2: **while** not done **do**
3: Find a better solution $\beta^{(k+1)}$ based on the current solution $\beta^{(k)}$ along the line $\beta^{(k)} + v\boldsymbol{d}^{(k)}$.
4: $k \leftarrow k+1$
5: **end while**
6: **return** $\beta^{(k)}$

The algorithms we will discuss only differ in their stop criteria and in how they find a better solution $\beta^{(k+1)}$ based on the currently best solution $\beta^{(k)}$ by local search.

This local improvement step, given the temporary solution $\beta^{(k)}$ and the search direction $\boldsymbol{d}^{(k)}$, is often called **line search**. Methods for performing such a line search will be introduced when we discuss the different strategies for determining search directions. Note that in general we do not need to find the best v to minimize $\beta^{(k)} + v\boldsymbol{d}^{(k)}$, often times all that is needed is a sufficient improvement over $\beta^{(k)}$.

First, we will introduce five different stop criteria and four different methods to determine the search direction $\boldsymbol{d}^{(k)}$.

Stop Criteria

Ideally, an iterative algorithm should stop when no further improvement of the objective function f can be achieved. More importantly, we want the point where the algorithm terminates to be identical with the global optimum. Let us qualify these requests in two ways. First, we already discussed in Section 2.4 that exact equality is not to be expected on computers. Therefore, we introduced the notions of approximate and essential equality, which means

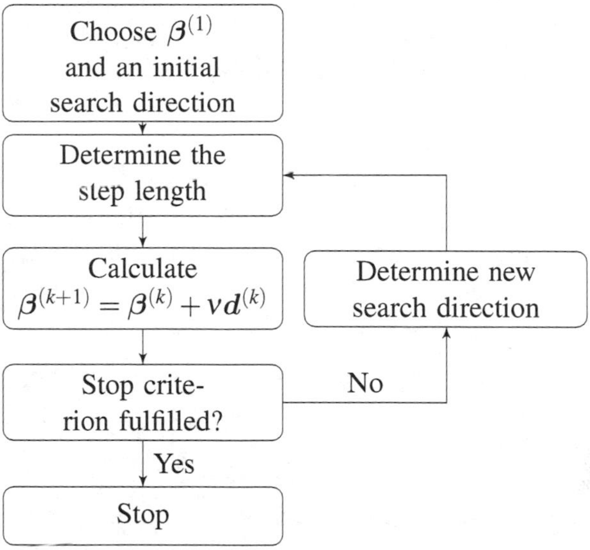

Figure 4.5: Flow chart for iterative optimization methods.

that the best we can hope for is essential equality of the found and the true optimum. Moreover, in practice we relax this criterion even more and terminate if one of the following **stop criteria** is fulfilled for a certain predefined small $\varepsilon > 0$ and a positive integer number p:

1. $\|\beta^{(k+p)} - \beta^{(k)}\|_2 < \varepsilon$,
2. $f(\beta^{(k)}) - f(\beta^{(k+p)}) < \varepsilon$,
3. $\|\nabla f(\beta^{(k)})\|_2 < \varepsilon$.
4. A prefixed upper bound for the number of iterations is reached.
5. A prefixed upper bound for the computation time is reached.

Why do we need five stop criteria in practice?

- Condition 1 does not guarantee termination, e.g. if the minimum is not unique.
- Conditions 2 and 3 might never be valid, e.g. if no minimum exists or the algorithm cannot find a minimum for some reason.
- Condition 3 is only applicable if f has a gradient and ∇f is known.
- Conditions 4 and 5 limit the number of function evaluations we perform and the amount of computing time we spend on the optimization.

Note that conditions 4. and 5. are solely safeguards to ensure that the algorithm always terminates. The returned solution will usually be far away from the global minimum in the parameter space. It is therefore best to always report the values of the first three criteria to detect nonconvergence.

If our function f is not convex, we can still apply many of the algorithms that follow. Of course in such a case the returned solution does not have to correspond to the global minimum and in fact will often not be the global minimum. Many of the algorithms might not even terminate, so the stop criteria 4 and 5 from the above list become crucial. It is often advisable to rerun the optimization with different initial solutions $\beta^{(1)}$ and to compare the returned solutions.

4.4.1.2 Coordinate Descent

A very old method to minimize a convex function f is the **coordinate descent method**. Here, the objective function f is minimized in one of the n coordinate directions in each iteration step. Therefore, in iteration step k the following univariate and unimodal optimization problem in ν is solved:

$$\underset{\nu \in \mathbb{R}}{\text{minimize}} \quad \tilde{f}^{(k)}(\nu) := f(\beta^{(k)} + \nu \boldsymbol{e}^{(k)}), \tag{4.9}$$

where $\boldsymbol{e}^{(k)}$ denotes the kth unit vector in coordinate direction. As the new approximation for the global minimum $\beta^{(k+1)} = \beta^{(k)} + \nu^* \boldsymbol{e}^{(k)}$ is used, ν^* being the above minimum. The reason this method is called **coordinate descent** is that in each iteration we search in one of the coordinate directions. The method terminates if no improvement can be found in any coordinate direction. The solution found this way is taken as the final approximation of the global minimum β^*.

What is left is to specify how to solve the inner (univariate) minimization problem and how the $\boldsymbol{e}^{(k)}$ should be chosen.

In order to find the global minimum ν^* of the function $\tilde{f}^{(k)}(\nu) := f(\beta^{(k)} + \nu \boldsymbol{e}^{(k)})$, we first determine ν_{lower} and ν_{upper} so that $\nu^* \in [\nu_{\text{lower}}, \nu_{\text{upper}}]$. Since f is convex, $\tilde{f}^{(k)}$ is also convex and therefore either $\nu_{\text{lower}} = 0$ or $\nu_{\text{upper}} = 0$. If $\tilde{f}^{(k)}(\varepsilon) < \tilde{f}^{(k)}(0)$ for a small $\varepsilon > 0$, choose $\nu_{\text{upper}} = 0$ and reduce ν_{lower} until $\tilde{f}^{(k)}(\nu_{\text{lower}}) > \tilde{f}^{(k)}(\nu_{\text{lower}} + \varepsilon)$ for a small $\varepsilon > 0$. Then, $\nu_{\text{lower}} \leq \nu^* \leq \nu_{\text{upper}}$. By reversing the inequalities we arrive at analogous rules in case $\tilde{f}^{(k)}(\varepsilon) > \tilde{f}^{(k)}(0)$.

The found search interval $[\nu_{\text{lower}}, \nu_{\text{upper}}]$ can then be shrunken by one of the univariate methods from the previous section, e.g., Golden Section Search or quadratic interpolation search, in order to determine ν^*.

Often, the inner minimization along one of the coordinate directions is terminated before reaching the minimum but after reducing $\tilde{f}^{(k)}$ sufficiently. Although this often leads to more iteration steps, in each iteration less func-

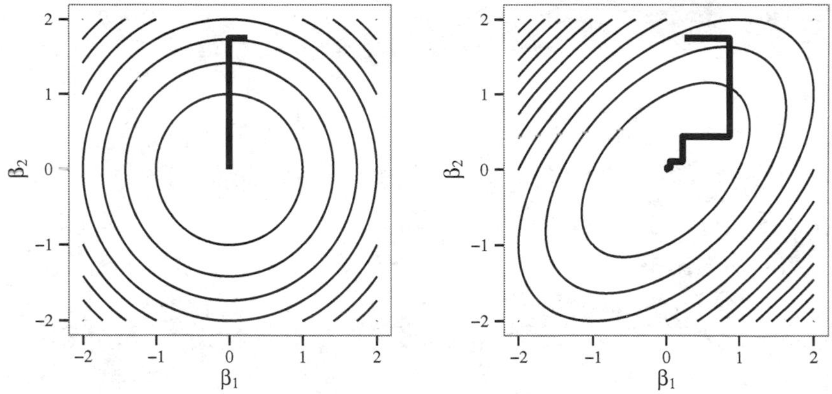

Figure 4.6: Examples: well-conditioned problem (left) and ill-conditioned problem (right) for coordinate descent.

tion evaluations are performed so that the overall number of function evaluations is usually lower than in the case where in each **line search** the minimum of the auxiliary function $\tilde{f}^{(k)}$ is approximated as exactly as possible. Note that this is the first example for a line search method and it should be clear why these are called line searches because the search is restricted to a line in the parameter space of f.

The final missing piece is the choice of $\mathbf{e}^{(k)}$ in the kth iteration. A robust and often used method to choose the $e^{(k)}$ is

$$\mathbf{e}^{(k)} := \mathbf{e}_{((k \bmod n)+1)}.$$

Here, e_i denotes the ith unity vector, i.e. the vector, whose only nonnull element is the ith entry, which is 1. With this method, it is sufficient to stop if there is no improvement after n steps.

Alternatively, in each step one could choose a random coordinate or one could choose a permutation of the coordinate vectors after every n steps and then search the coordinate directions according to this permutation. In practice, it is not very important which variant is chosen. For a theoretical analysis the first mentioned deterministic choice of $\mathbf{e}^{(k)}$ is more convenient than the two randomized variants.

Not all problems are equally well suited for the coordinate descent method (see Figure 4.6). The coordinate descent method works well on separable functions but becomes less attractive if the function is not separable.

Definition 4.6: Separable Function
A function $f\colon \mathbb{R}^n \to \mathbb{R}$ is called **separable**, iff for each coordinate direction e_i

$$\forall \beta_1, \beta_2 \in \mathbb{R}^n\colon\ \beta_1 + v_1^* e_i = \beta_2 + v_2^* e_i,$$

where v_1^* and v_2^* are the minima of (4.9) for β_1 and β_2.

Obviously, separability means that a function can be minimized separately in each of its arguments in order to achieve its joint minimum. The setting of the ith element of β^* is then independent of the value of the other elements of β^*. Thus, the problem can be divided into n univariate problems that can be solved individually.

4.4.1.3 Gradient Descent

If our function f to be minimized is at least once differentiable, then one can choose a better descent direction than a coordinate vector. Since f is differentiable, we can use the gradient ∇f of f as a search direction because we know that the gradient always points into the direction of steepest ascent. Thus, in each iteration the following problem is solved:

$$\underset{v \in \mathbb{R}^+}{\text{minimize}} \quad \tilde{f}(v) := f(\beta^{(k)} - v \nabla f(\beta^{(k)})).$$

Again, v^* can be found, e.g., via Golden Section search, quadratic interpolation, or using a more advanced line search method. In the kth iteration we search for a $\beta^{(k+1)}$ on the **line**

$$S := \{\beta^{(k)} - v \nabla f(\beta^{(k)}) \mid v \in \mathbb{R}^+\}$$

minimizing f on S. Therefore, again we have a **line search** in each iteration step.

The pseudocode of the steepest descent method is given in Algorithm 4.6.

Algorithm 4.6 Steepest Descent Method

Require: $f\colon \mathbb{R}^n \to \mathbb{R}$, gradient $\nabla f\colon \mathbb{R}^n \to \mathbb{R}^n$, initial solution β

```
while not done do
    Δβ ← ∇f(β)
    g(ν) := f(β − νΔβ)
    ν ← linesearch(g)
    β ← β − νΔβ
end while
```

Convergence of the Gradient Descent

We now show that the gradient descent method linearly converges for a special kind of functions, namely the quadratic functions.

Theorem 4.2: Convergence of Steepest Descent

Let

$$f(\boldsymbol{\beta}) = \frac{1}{2}\boldsymbol{\beta}^T \boldsymbol{Q}\boldsymbol{\beta} + \boldsymbol{q}^T \boldsymbol{\beta},$$

where $\boldsymbol{Q} \in \mathbb{R}^{n\times n}$ is a symmetrical positive definite matrix with eigenvalues $\lambda_1 \geq \lambda_2 \geq \cdots \geq \lambda_n$, $\boldsymbol{q} \in \mathbb{R}^n$. Then, the steepest descent method linearly converges for f with a convergence rate

$$\rho \leq \left(\frac{\lambda_1/\lambda_n - 1}{\lambda_1/\lambda_n + 1}\right)^2.$$

Proof. The gradient of f is obviously given by $\nabla f(\boldsymbol{\beta}) = \boldsymbol{Q}\boldsymbol{\beta} + \boldsymbol{q}$, and thus for the minimum $\boldsymbol{\beta}^*$ it is true that

$$\boldsymbol{\beta}^* = -\boldsymbol{Q}^{-1}\boldsymbol{q} \text{ and}$$
$$f(\boldsymbol{\beta}^*) = f(-\boldsymbol{Q}^{-1}\boldsymbol{q}) = -\frac{1}{2}\boldsymbol{q}^T \boldsymbol{Q}^{-1}\boldsymbol{q}.$$

Moreover, $\boldsymbol{d}(\boldsymbol{\beta}) = -\boldsymbol{Q}\boldsymbol{\beta} - \boldsymbol{q}$ is the steepest descent direction in each iteration. Thus, in each iteration the function

$$f(\boldsymbol{\beta} + v\boldsymbol{d}(\boldsymbol{\beta}))$$
$$= \frac{1}{2}(\boldsymbol{\beta} + v\boldsymbol{d}(\boldsymbol{\beta}))^T \boldsymbol{Q}(\boldsymbol{\beta} + v\boldsymbol{d}(\boldsymbol{\beta})) + \boldsymbol{q}^T(\boldsymbol{\beta} + v\boldsymbol{d}(\boldsymbol{\beta}))$$
$$= f(\boldsymbol{\beta}) + \frac{1}{2}(v\boldsymbol{d}(\boldsymbol{\beta})^T \boldsymbol{Q}\boldsymbol{\beta} + \boldsymbol{\beta}^T \boldsymbol{Q} v\boldsymbol{d}(\boldsymbol{\beta})) + \boldsymbol{q}^T v\boldsymbol{d}(\boldsymbol{\beta}) + \frac{1}{2}v^2 \boldsymbol{d}(\boldsymbol{\beta})^T \boldsymbol{Q}\boldsymbol{d}(\boldsymbol{\beta})$$
$$= f(\boldsymbol{\beta}) + v\boldsymbol{d}(\boldsymbol{\beta})^T \boldsymbol{Q}\boldsymbol{\beta} + v\boldsymbol{d}(\boldsymbol{\beta})^T \boldsymbol{q} + \frac{1}{2}v^2 \boldsymbol{d}(\boldsymbol{\beta})^T \boldsymbol{Q}\boldsymbol{d}(\boldsymbol{\beta})$$
$$= f(\boldsymbol{\beta}) - v\boldsymbol{d}(\boldsymbol{\beta})^T \boldsymbol{d}(\boldsymbol{\beta}) + \frac{1}{2}v^2 \boldsymbol{d}(\boldsymbol{\beta})^T \boldsymbol{Q}\boldsymbol{d}(\boldsymbol{\beta})$$

is minimized in v for given $\boldsymbol{\beta}$. Therefore,

$$v^* = \frac{\boldsymbol{d}(\boldsymbol{\beta})^T \boldsymbol{d}(\boldsymbol{\beta})}{\boldsymbol{d}(\boldsymbol{\beta})^T \boldsymbol{Q}\boldsymbol{d}(\boldsymbol{\beta})}. \tag{4.10}$$

In each iteration of the gradient descent method we obtain for given $\boldsymbol{\beta}^{(k)}$ the following new values:

$$\beta^{(k+1)} = \beta^{(k)} + \nu^* d(\beta^{(k)}) = \beta^{(k)} + \frac{d(\beta^{(k)})^T d(\beta^{(k)})}{d(\beta^{(k)})^T Q d(\beta^{(k)})} d(\beta^{(k)}),$$

$$f(\beta^{(k+1)}) = f(\beta^{(k)}) - \frac{1}{2} \frac{(d(\beta^{(k)})^T d(\beta^{(k)}))^2}{d(\beta^{(k)})^T Q d(\beta^{(k)})}. \tag{4.11}$$

Therefore,

$$\begin{aligned}
\frac{f(\beta^{(k+1)}) - f(\beta^*)}{f(\beta^{(k)}) - f(\beta^*)} &= \frac{f(\beta^{(k)}) - \frac{1}{2} \frac{(d(\beta^{(k)})^T d(\beta^{(k)}))^2}{d(\beta^{(k)})^T Q d(\beta^{(k)})} - f(\beta^*)}{f(\beta^{(k)}) - f(\beta^*)} \\
&= 1 - \frac{\frac{1}{2} \frac{(d(\beta^{(k)})^T d(\beta^{(k)}))^2}{d(\beta^{(k)})^T Q d(\beta^{(k)})}}{f(\beta^{(k)}) - f(\beta^*)} \\
&= 1 - \frac{\frac{1}{2} \frac{(d(\beta^{(k)})^T d(\beta^{(k)}))^2}{d(\beta^{(k)})^T Q d(\beta^{(k)})}}{\frac{1}{2}(Q\beta^{(k)} + q)^T Q^{-1} (Q\beta^{(k)} + q)} \\
&= 1 - \frac{\frac{1}{2} \frac{(d(\beta^{(k)})^T d(\beta^{(k)}))^2}{d(\beta^{(k)})^T Q d(\beta^{(k)})}}{\frac{1}{2} d(\beta^{(k)})^T Q^{-1} d(\beta^{(k)})} \\
&= 1 - \frac{(d(\beta^{(k)})^T d(\beta^{(k)}))^2}{d(\beta^{(k)})^T Q d(\beta^{(k)}) d(\beta^{(k)})^T Q^{-1} d(\beta^{(k)})} \\
&= 1 - \frac{1}{\gamma_k},
\end{aligned}$$

where

$$\gamma_k := \frac{d(\beta^{(k)})^T Q d(\beta^{(k)}) d(\beta^{(k)})^T Q^{-1} d(\beta^{(k)})}{(d(\beta^{(k)})^T d(\beta^{(k)}))^2}.$$

For the method to converge quickly, γ_k should be small. What this means can be illustrated by the following inequality (Kantorovich inequality; see, e.g., Newman (1959)):

$$\gamma_k \leq \frac{(\lambda_1 + \lambda_n)^2}{4\lambda_1 \lambda_n}.$$

Therefore, for the convergence rate ρ we have

$$\rho \leq 1 - \frac{4\lambda_1 \lambda_n}{(\lambda_1 + \lambda_n)^2} = \left(\frac{\lambda_1/\lambda_n - 1}{\lambda_1/\lambda_n + 1} \right)^2,$$

This proves the theorem. □

This theorem can even be generalized as follows:

Theorem 4.3: Convergence Rate of Steepest Descent
If f is twice differentiable, then the gradient descent method linearly converges for f with a convergence rate

$$\rho \leq \left(\frac{\lambda_1/\lambda_n - 1}{\lambda_1/\lambda_n + 1}\right)^2.$$

Here, λ_1 designates the largest and λ_n the smallest eigenvalue of the Hessian $\nabla^2 f(\beta)$.

Proof. see Luenberger (1973, pp. 148 – 155). □

Approximation by Linear Least Squares Problem
There is also a connection between the minimization of f using the gradient descent method and the least squares problem from Chapter 3. Linearizing the function f in the optimum β^* helps to further illustrate the convergence of the method. For the linear model we know that the following holds:

$$\begin{aligned} f(\beta) &= \|\boldsymbol{y} - \boldsymbol{X}\beta\|_2^2 \\ &= \beta^T \boldsymbol{X}^T \boldsymbol{X} \beta - 2\beta^T \boldsymbol{X}^T \boldsymbol{y} + \boldsymbol{y}^T \boldsymbol{y}. \end{aligned}$$

Then $f(\beta)$ is a quadratic function in β and large λ_1/λ_n indicates elongated contours of the objective function near the optimum, since the contours of a quadratic function f are n-dimensional ellipsoids with axes in the directions of the n orthogonal eigenvectors of the positive definite hessian $\nabla^2 f(\beta^*)$. Also, the length of the axis corresponding to the ith eigenvector is given by the ith eigenvalue λ_i. Moreover, the square of the spectral condition number κ^* of $\boldsymbol{X}$ is the ratio of the largest and the smallest eigenvalue of $\nabla^2 f(\beta^*) = 2\boldsymbol{X}^T\boldsymbol{X}$, since

$$\begin{aligned} {\kappa^*}^2 &= \|\boldsymbol{X}\|^2 \|\boldsymbol{X}^+\|^2 \\ &= \frac{(s_1(\boldsymbol{X}))^2}{(s_n(\boldsymbol{X}))^2} \qquad (s_i = i\text{th singular value of } \boldsymbol{X}(\beta^*)) \\ &= \frac{\lambda_1(2\boldsymbol{X}^T\boldsymbol{X})}{\lambda_n(2\boldsymbol{X}^T\boldsymbol{X})}. \end{aligned}$$

Therefore, the spectral condition number κ^* is proportional to λ_1/λ_n and the larger the quotient is, the narrower the ellipsoid, i.e. the narrower the valley that leads to the optimum. This fits well with the characterization of convergence speed by means of the convergence rate ρ in the previous Theorems. Indeed, the larger λ_1/λ_n is, the larger ρ is, and the slower the algorithm converges. Examples of well and badly conditioned quadratic objective functions f are given in Figure 4.7.

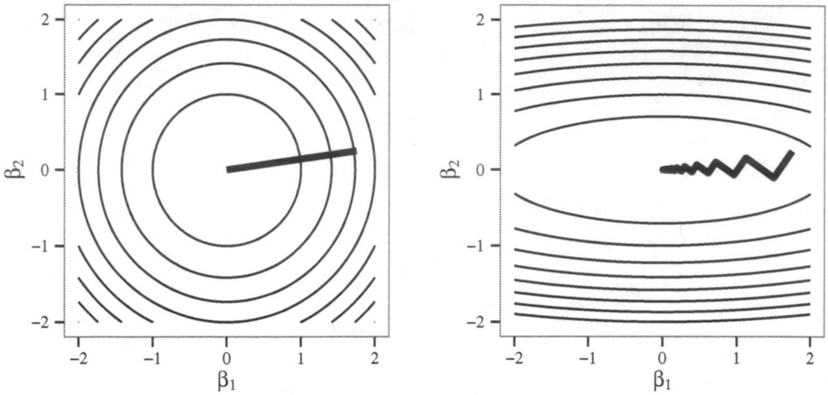

Figure 4.7: Examples: well-conditioned problem (left), badly conditioned problem (right) for gradient descent.

4.4.1.4 Newton Algorithm

If f is twice continuously differentiable, then we can use an even more powerful method in analogy to the Newton method for univariate functions. In each iteration, we approximate f by a quadratic function determined by the Taylor expansion of f in $\boldsymbol{\beta}^{(k)}$:

$$f(\boldsymbol{\beta}^{(k)} + \Delta\boldsymbol{\beta}) \approx f(\boldsymbol{\beta}^{(k)}) + \nabla f(\boldsymbol{\beta}^{(k)})^T \Delta\boldsymbol{\beta} + \frac{1}{2}\Delta\boldsymbol{\beta}^T \nabla^2 f(\boldsymbol{\beta}^{(k)})\Delta\boldsymbol{\beta}. \quad (4.12)$$

The gradient corresponding to $\Delta\boldsymbol{\beta}$ in Equation 4.12 is

$$\nabla f(\boldsymbol{\beta}^{(k)} + \Delta\boldsymbol{\beta}) = \nabla f(\boldsymbol{\beta}^{(k)}) + \nabla^2 f(\boldsymbol{\beta}^{(k)})\Delta\boldsymbol{\beta}. \quad (4.13)$$

Thus, in the (global) minimum $\Delta\boldsymbol{\beta}$ should fulfill

$$\nabla f(\boldsymbol{\beta}^{(k)}) + \nabla^2 f(\boldsymbol{\beta}^{(k)})\Delta\boldsymbol{\beta} = 0.$$

Therefore, the new approximation is chosen as

$$\boldsymbol{\beta}^{(k+1)} = \boldsymbol{\beta}^{(k)} - \nabla^2 f(\boldsymbol{\beta}^{(k)})^{-1} \nabla f(\boldsymbol{\beta}^{(k)}). \quad (4.14)$$

If Equation 4.14 is used and no line search is performed, we call the step taken a **full Newton step**. For this step size one can show, analogously to the univariate case, that there is an ε ball around the global minimum $\boldsymbol{\beta}^*$, in which the method converges quadratically.

Unfortunately, it is also true that for this method termination is not even guaranteed outside of this ball. Thus, the method is in every respect a generalization of the Newton method for the minimization of univariate functions, both in its fast convergence as well as its drawbacks of being not as robust as we might like it to be. To alleviate some of the drawbacks, equation (4.14) is often supplemented by a parameter ν such that

$$\beta^{(k+1)} = \beta^{(k)} - \nu \nabla^2 f(\beta^{(k)})^{-1} \nabla f(\beta^{(k)}).$$

Now the Newton method is a special case of our general iterative minimization procedure that we used for coordinate and gradient descent. In each step we then solve the following minimization problem:

$$\underset{\nu \in \mathbb{R}^+}{\text{minimize}} \quad \tilde{f}^{(k)}(\nu) := f(\beta^{(k)} - \nu \nabla^2 f(\beta^{(k)})^{-1} \nabla f(\beta^{(k)})).$$

Often, the restriction $\nu \leq 1$ is used. This ensures that the largest step we take is a full Newton step. Note that the larger the step size is, the more trust we must place in our quadratic approximation of the function because the distance we extrapolate is larger.

The pseudocode of the Newton method is given in Algorithm 4.7.

Algorithm 4.7 Multivariate Newton Method

Require: $f\colon \mathbb{R}^n \to \mathbb{R}$
gradient $\nabla f\colon \mathbb{R}^n \to \mathbb{R}^n$
Hessian $\nabla^2 f\colon \mathbb{R}^n \to \mathbb{R}^{n \times n}$
Initial solution β

while not done **do**
  $\Delta\beta \leftarrow \nabla^2 f(\beta)^{-1} \nabla f(\beta)$
  $g(\nu) := f(\beta - \nu \Delta\beta)$
  $\nu \leftarrow \text{linesearch}(g)$
  $\beta \leftarrow \beta - \nu \Delta\beta$
end while

4.4.1.5 Quasi-Newton Methods

A disadvantage of the Newton method is that we need the Hessian in analytical form. This can be a hindrance in two respects. On the one hand it could be quite expensive to compute, especially if we have many parameters to optimize, and for some problems computing the analytical form can be impractical. Even if it is possible to numerically approximate the Hessian, in practice this is seldom done because of the large number of function evaluations required. Instead we replace $\nabla^2 f$ by an approximation based solely on the known gradient ∇f. Instead of Equation 4.12 we then have

$$f(\boldsymbol{\beta}^{(k)} + \Delta\boldsymbol{\beta}) \approx f(\boldsymbol{\beta}^{(k)}) + \nabla f(\boldsymbol{\beta}^{(k)})^T \Delta\boldsymbol{\beta} + \frac{1}{2}\Delta\boldsymbol{\beta}^T \boldsymbol{Q}^{(k)} \Delta\boldsymbol{\beta}.$$

Here, $\boldsymbol{Q}^{(k)} \in L(n,n)$ is required to be positive definite so that we can be sure that f is convex. If the matrix were not positive definite, we could not be sure that

$$-(\boldsymbol{Q}^{(k)})^{-1} \nabla f(\boldsymbol{\beta}^{(k)})$$

is a descent direction (see below).

There are obviously many possible ways to choose $\boldsymbol{Q}^{(k)}$. The simplest choice is the identity matrix, which leads to the gradient descent method. At the other end of the spectrum, we could plug in a numerical approximation of ∇^2 and would arrive at an approximate Newton method. However, what we are really interested in are methods that lead to better descent directions than the naive gradient descent method but do not suffer the high computational cost of the approximate Newton method.

To derive such descent directions, let $\boldsymbol{\beta}^{(k)}$ be a point in the coefficient space. We are looking for a direction $\Delta\boldsymbol{\beta}$, in which the objective function f decreases. When we proceed in such a direction we have to ensure that we do not go too far so that we do not overshoot and miss the minimum—possibly arriving at a solution worse than our current solution. On the other hand, if we take a step that is too small, our algorithm will need more iterations to cover the same distance and eventually converge. In other words, we need to find a balance such that the chosen step size ν and direction $\Delta\boldsymbol{\beta}$ lead to

$$f(\boldsymbol{\beta}^{(k)} + \nu\Delta\boldsymbol{\beta}) < f(\boldsymbol{\beta}^{(k)}). \tag{4.15}$$

If $\Delta\boldsymbol{\beta}$ points downhill, a small step in this direction should definitely lead to a reduction of the objective function. Thus, we look for a $\Delta\boldsymbol{\beta}$ so that $f(\boldsymbol{\beta}^{(k)} + \nu\Delta\boldsymbol{\beta})$ is a decreasing function of ν for ν sufficiently small. This leads to

$$\begin{aligned} \left.\frac{\mathrm{d}f(\boldsymbol{\beta}^{(k)} + \nu\Delta\boldsymbol{\beta})}{\mathrm{d}\nu}\right|_{\nu=0} &= \left(\left.\frac{\partial f}{\partial \boldsymbol{\beta}}\right|_{\boldsymbol{\beta}^{(k)}}\right)\left(\left.\frac{\mathrm{d}(\boldsymbol{\beta}^{(k)} + \nu\Delta\boldsymbol{\beta})}{\mathrm{d}\nu}\right|_{\nu=0}\right) \\ &= (\nabla f(\boldsymbol{\beta}^{(k)}))^T \Delta\boldsymbol{\beta} < 0. \end{aligned} \tag{4.16}$$

Then, we choose

$$\Delta\boldsymbol{\beta} = -\boldsymbol{P}\nabla f(\boldsymbol{\beta}^{(k)}), \tag{4.17}$$

where $\boldsymbol{P}$ is any positive definite matrix, i.e. $(\nabla f)^T \boldsymbol{P} \nabla f > 0$ for all vectors $\nabla f \neq 0$. This implies that $(\nabla f(\boldsymbol{\beta}^{(k)}))^T \Delta\boldsymbol{\beta} = -(\nabla f(\boldsymbol{\beta}^{(k)}))^T \boldsymbol{P}(\nabla f(\boldsymbol{\beta}^{(k)})) < 0$, if $\nabla f(\boldsymbol{\beta}^{(k)}) \neq 0$, which is what we want from a descent direction.

In order to simplify notation let $\boldsymbol{P}^{(k)} := (\boldsymbol{Q}^{(k)})^{-1}$. Then, $-\boldsymbol{P}^{(k)}\nabla f(\boldsymbol{\beta}^{(k)})$ is a descent direction if

$$-\nabla f(\boldsymbol{\beta}^{(k)})^T \boldsymbol{P}^{(k)} \nabla f(\boldsymbol{\beta}^{(k)}) < 0.$$

Obviously, this is the case for a deliberate gradient iff $\boldsymbol{Q}^{(k)}$ is positive definite and therefore also $\boldsymbol{P}^{(k)}$.

If the search space Θ is not one-dimensional, then $\boldsymbol{Q}$ or $\boldsymbol{P}$ is not uniquely determined. In the literature there are different suggestions for the choice of $\boldsymbol{Q}$ or $\boldsymbol{P}$, respectively. All of these methods have in common that in each iteration a new matrix $\boldsymbol{P}^{(k+1)}$ or $\boldsymbol{Q}^{(k+1)}$ is generated from the current matrix $\boldsymbol{P}^{(k)}$ or $\boldsymbol{Q}^{(k)}$ and the current gradient $\nabla f(\boldsymbol{\beta}^{(k)})$.

As noted above, the gradient descent ($\boldsymbol{Q}^{(k)} = I_n$) as well as the Newton method ($\boldsymbol{Q}^{(k)} = \nabla^2 f(\boldsymbol{\beta}^{(k)})$) are special cases of the quasi-Newton method.

The iterative approach becomes clear by inserting $\boldsymbol{Q}^{(k+1)}$ for the Hessian in Equation 4.13 and rewriting it as

$$\boldsymbol{P}^{(k+1)}(\nabla f(\boldsymbol{\beta}^{(k+1)}) - \nabla f(\boldsymbol{\beta}^{(k)})) = \boldsymbol{\beta}^{(k+1)} - \boldsymbol{\beta}^{(k)}.$$

If $\boldsymbol{P}^{(k+1)}$ is decomposed into the sum of $\boldsymbol{P}^{(k)}$ and an **update** $\boldsymbol{U}^{(k)}$, then we get:

$$\boldsymbol{U}^{(k)}(\nabla f(\boldsymbol{\beta}^{(k+1)}) - \nabla f(\boldsymbol{\beta}^{(k)})) = \boldsymbol{\beta}^{(k+1)} - \boldsymbol{\beta}^{(k)} - \boldsymbol{P}^{(k)}(\nabla f(\boldsymbol{\beta}^{(k+1)}) - \nabla f(\boldsymbol{\beta}^{(k)})).$$

With the notation

$$\begin{aligned}
\Delta\boldsymbol{\beta}^{(k)} &:= \boldsymbol{\beta}^{(k+1)} - \boldsymbol{\beta}^{(k)} = -\nu^* \boldsymbol{P}^{(k)} \nabla f(\boldsymbol{\beta}^{(k)}), \\
\boldsymbol{g}^{(k)} &:= \nabla f(\boldsymbol{\beta}^{(k+1)}) - \nabla f(\boldsymbol{\beta}^{(k)}), \\
\boldsymbol{s}^{(k)} &:= \boldsymbol{P}^{(k)} \boldsymbol{g}^{(k)}, \\
\boldsymbol{e}^{(k)} &:= \Delta\boldsymbol{\beta}^{(k)} - \boldsymbol{s}^{(k)}
\end{aligned}$$

this equation simplifies to

$$\boldsymbol{U}^{(k)} \boldsymbol{g}^{(k)} = \Delta\boldsymbol{\beta}^{(k)} - \boldsymbol{s}^{(k)} = \boldsymbol{e}^{(k)}. \tag{4.18}$$

$\boldsymbol{U}^{(k)}$ has to be symmetrical and positive definite so that $\boldsymbol{P}^{(k+1)}$ is a positive definite matrix given that $\boldsymbol{P}^{(k)}$ is positive definite. Note that $\boldsymbol{U}$ is not updating the approximation of the Hessian $\boldsymbol{Q}$, but directly its inverse $\boldsymbol{P}$. By updating the inverse directly we do not have to store the Hessian and perform the costly and potentially numerically unstable inversion to find $\boldsymbol{P}$.

In what follows, we will introduce some possible update formulas. They are structured by the rank of the update. There is only one possible update $\boldsymbol{U}$ that has rank 1. This update is called SR1 for symmetrical rank 1. The update matrix of Davidon, Fletcher, and Powell (DFP) has rank 2, and the update proposed by Broyden, Fletcher, Goldfarb, and Shanno (BFGS) rank 3.

Quasi-Newton Methods: SR1 (Broyden, 1967)

The symmetrical rank 1 correction matrix $\boldsymbol{U}^{(k)}$ has the form

$$\boldsymbol{U}_{SR1}^{(k)} := \frac{\boldsymbol{e}^{(k)}(\boldsymbol{e}^{(k)})^T}{(\boldsymbol{e}^{(k)})^T \boldsymbol{g}^{(k)}}.$$

This update matrix obviously fulfills Equation (4.18). The update formula thus has the form

$$\boldsymbol{P}^{(k+1)} = \boldsymbol{P}^{(k)} + \frac{\boldsymbol{e}^{(k)}(\boldsymbol{e}^{(k)})^T}{(\boldsymbol{e}^{(k)})^T \boldsymbol{g}^{(k)}}.$$

The disadvantage of this update rule is that $\boldsymbol{P}^{(k+1)}$ is not always positive definite. Additionally, in practice it might occur that the denominator of the quotient is close to zero leading to numerical problems.

Quasi-Newton Methods: DFP (Fletcher and Powell, 1963)

The update matrix proposed by Davidon, Fletcher, and Powell has the form

$$\boldsymbol{U}_{DFP}^{(k)} := \frac{\Delta\boldsymbol{\beta}^{(k)}(\Delta\boldsymbol{\beta}^{(k)})^T}{(\boldsymbol{g}^{(k)})^T \Delta\boldsymbol{\beta}^{(k)}} - \frac{\boldsymbol{P}^{(k)}\boldsymbol{g}^{(k)}(\boldsymbol{g}^{(k)})^T\boldsymbol{P}^{(k)}}{(\boldsymbol{g}^{(k)})^T\boldsymbol{P}^{(k)}\boldsymbol{g}^{(k)}}.$$

$\boldsymbol{U}_{DFP}^{(k)}$ is a rank 2 matrix and we can easily show that it fulfills Equation 4.18:

$$\begin{aligned}\boldsymbol{U}_{DFP}^{(k)}\boldsymbol{g}^{(k)} &= \frac{\Delta\boldsymbol{\beta}^{(k)}(\Delta\boldsymbol{\beta}^{(k)})^T}{(\boldsymbol{g}^{(k)})^T \Delta\boldsymbol{\beta}^{(k)}}\boldsymbol{g}^{(k)} - \frac{\boldsymbol{P}^{(k)}\boldsymbol{g}^{(k)}(\boldsymbol{g}^{(k)})^T\boldsymbol{P}^{(k)}}{(\boldsymbol{g}^{(k)})^T\boldsymbol{P}^{(k)}\boldsymbol{g}^{(k)}}\boldsymbol{g}^{(k)} \\ &= \Delta\boldsymbol{\beta}^{(k)} - \boldsymbol{P}^{(k)}\boldsymbol{g}^{(k)}.\end{aligned}$$

For this update matrix, Fletcher and Powell (1963) have shown that the resulting direction matrix $\boldsymbol{P}^{(k+1)}$ is always positive definite if $\boldsymbol{P}^{(k)}$ is positive definite and $\Delta\boldsymbol{\beta}^{(k)}$ is chosen optimally. Therefore, this choice of the direction matrix $\boldsymbol{P}^{(k)}$ in iteration k theoretically leads to an acceptable step. Here, the identity matrix $\boldsymbol{I}_n$ is a reasonable value for $\boldsymbol{P}^{(1)}$, but any prior knowledge about the structure of the Hessian and the starting point could be used to construct a different initial approximation of the Hessian as long as it is positive definite.

In practice, however, we can expect numerical errors to accumulate over time. This will eventually lead to a $\boldsymbol{P}^{(k+1)}$ that is not a positive definite matrix (Bard, 1968). For this reason, the DFP update is seldom used in modern implementations of a quasi-Newton type algorithm. Instead, a dual update is used called the BFGS update according to its discoverers.

Quasi-Newton Methods: BFGS (Fletcher, 1970)
Broyden, Fletcher, Goldfarb, and Shanno did not propose a rank 2 update for the direction matrix $\boldsymbol{P}^{(k)}$ but a dual update for the approximation of the Hessian $\boldsymbol{Q}^{(k)}$:

$$\frac{\boldsymbol{g}^{(k)}(\boldsymbol{g}^{(k)})^T}{(\boldsymbol{g}^{(k)})^T\Delta\boldsymbol{\beta}^{(k)}} - \frac{\boldsymbol{Q}^{(k)}\Delta\boldsymbol{\beta}^{(k)}(\Delta\boldsymbol{\beta}^{(k)})^T\boldsymbol{Q}^{(k)}}{(\Delta\boldsymbol{\beta}^{(k)})^T\boldsymbol{Q}^{(k)}\Delta\boldsymbol{\beta}^{(k)}}.$$

Using the following transformation due to Sherman and Morrison (1950)

$$(\boldsymbol{A}+\boldsymbol{u}\boldsymbol{v}^T)^{-1} = \boldsymbol{A}^{-1} - \frac{\boldsymbol{A}^{-1}\boldsymbol{u}\boldsymbol{v}^T\boldsymbol{A}^{-1}}{1+\boldsymbol{v}^T\boldsymbol{A}^{-1}\boldsymbol{u}}$$

the above rank 2 update of $\boldsymbol{Q}^{(k)}$ can be transformed into a rank 3 update for $\boldsymbol{P}^{(k)}$:

$$\boldsymbol{U}_{BFGS}^{(k)} := \frac{(\boldsymbol{e}^{(k)})^T\boldsymbol{g}^{(k)}\Delta\boldsymbol{\beta}^{(k)}(\Delta\boldsymbol{\beta}^{(k)})^T}{((\Delta\boldsymbol{\beta}^{(k)})^T\boldsymbol{g}^{(k)})^2} - \frac{\boldsymbol{e}^{(k)}(\Delta\boldsymbol{\beta}^{(k)})^T + \Delta\boldsymbol{\beta}^{(k)}(\boldsymbol{e}^{(k)})^T}{(\Delta\boldsymbol{\beta}^{(k)})^T\boldsymbol{g}^{(k)}}.$$

$\boldsymbol{U}_{BFGS}^{(k)}$ fulfills Equation 4.18 and the corresponding direction matrix $\boldsymbol{P}^{(k+1)}$ is positive definite if $\boldsymbol{P}^{(k)}$ is and the gradient is reduced (see Exercise 4.4.3):

$$\nabla f(\boldsymbol{\beta}^{(k+1)})^T\boldsymbol{P}^{(k)}\nabla f(\boldsymbol{\beta}^{(k)}) - \nabla f(\boldsymbol{\beta}^{(k)})^T\boldsymbol{P}^{(k)}\nabla f(\boldsymbol{\beta}^{(k)}) < 0$$
$$\Leftrightarrow (\Delta\boldsymbol{\beta}^{(k)})^T\boldsymbol{g}^{(k)} > 0.$$

The BFGS update is the most popular quasi-Newton update rule. We can, however, still arrive at a nonpositive definite $\boldsymbol{P}^{(k+1)}$ due to rounding errors that accumulate over time. Any implementation should therefore add safeguards for such a situation. A simplistic solution is to fix such a nonpositive definite $\boldsymbol{P}^{(k+1)}$ by adding a small $\varepsilon > 0$ to all elements on the diagonal (see the Levenberg-Marquardt method in Section 4.4.2.1).

The pseudocode for the BFGS method is given in Algorithm 4.8.

Quasi-Newton Methods: Broyden's Method (Fletcher, 1970)
Since there really is no clear-cut criterion for choosing either the BFGS or the DFP update, we can construct a combination of the two. This leads us to a whole family of update formulas:

$$\boldsymbol{U}_{\gamma}^{(k)} := \gamma\boldsymbol{U}_{DFP}^{(k)} + (1-\gamma)\boldsymbol{U}_{BFGS}^{(k)},$$

where $0 < \gamma < 1$ is a fixed constant. Again, this update rule leads to a positive definite $\boldsymbol{P}^{(k+1)}$ if $\boldsymbol{P}^{(k)}$ is positive definite and if $(\Delta\boldsymbol{\beta}^{(k)})^T\boldsymbol{g}^{(k)} > 0$. While this class of update formulas is not in widespread use, it is interesting from a theoretical standpoint because it allows a smooth transition from the BFGS to the DFP strategy.

Algorithm 4.8 BFGS Quasi-Newton Method

Require: $f\colon \mathbb{R}^n \to \mathbb{R}$
gradient $\nabla f\colon \mathbb{R}^n \to \mathbb{R}^n$
Initial solution $\boldsymbol{\beta}$

```
P ← I_n
while not done do
    Δβ ← −P∇f(β)
    fn(ν) := f(β + νΔβ)
    ν ← linesearch(fn)
    Δβ ← νΔβ
    g ← ∇f(β + Δβ) − ∇f(β)
    e ← Δβ − Pg
    P ← P + (e^T g Δβ(Δβ)^T)/(g^T Δβ)^2 − (eΔβ^T + Δβe^T)/(g^T Δβ)
    β ← β + Δβ
end while
```

Quasi-Newton Methods: Inexact Line Search
In the previous discussion we have not focused on the line search component of the procedure and only stated that we could use one of the methods from the univariate optimization section. In almost all cases the search direction $-\boldsymbol{P}^{(k)}\nabla f(\boldsymbol{\beta}^{(k)})$ in iteration k will not point in the direction of the global minimum. It is therefore often not wise to find the exact minimum in the line search direction. That is, we do not need to minimize

$$\varphi(\nu) := f(\boldsymbol{\beta}^{(k)} - \nu \boldsymbol{P}^{(k)}\nabla f(\boldsymbol{\beta}^{(k)})).$$

Instead, it is adequate to only **sufficiently** reduce φ. Thus, we may choose ν such that

$$\varphi(\nu) \le f(\boldsymbol{\beta}^{(k)}) - c_1 \nu \nabla f(\boldsymbol{\beta}^{(k)})^T \boldsymbol{P}^{(k)} \nabla f(\boldsymbol{\beta}^{(k)}) \tag{4.19}$$

with $0 < c_1 < 1$. This condition, the so-called **Armijo condition**, is always satisfied for ν close to 0 since $f(\boldsymbol{\beta}^{(k)} + \nu \boldsymbol{d})$ can be approximated by $f(\boldsymbol{\beta}^{(k)}) + \nu \nabla f(\boldsymbol{\beta}^{(k)})^T \boldsymbol{d}$ with arbitrary accuracy in an ε-ball around $\boldsymbol{\beta}^{(k)}$. $\nabla f(\boldsymbol{\beta}^{(k)})^T \boldsymbol{d}$ is also called the **directional derivative** of f in $\boldsymbol{\beta}^{(k)}$ in the direction $\boldsymbol{d}$. Here, $\boldsymbol{d} = -\boldsymbol{P}^{(k)}\nabla f(\boldsymbol{\beta}^{(k)})$.

Equation 4.19 suggests the following approach:

1. Fix c_1 (e.g. $c_1 = 10^{-4}$).
2. Start with a sufficiently large ν (e.g. $\nu = 1$).
3. If ν fulfills condition (4.19), take ν as the step size.

4. Else let $\nu = \nu \cdot \gamma$ for a fixed $\gamma \in (0,1)$.
5. Go to step 3.

The resulting step size is used for the calculation of the quasi-Newton step:

$$\Delta\beta^{(k)} = -\nu \boldsymbol{P}^{(k)} \nabla f(\beta^{(k)}).$$

Alternatively, we could also start with a sufficiently small ν and then increase it until the Armijo condition is not satisfied anymore, taking the last valid value of ν as the step size.

The above so-called **inexact line search** is an essential part of all efficient quasi-Newton optimization algorithms. In practice, exact minimization of φ is generally not worth while and wastes precious resources in the optimization. In fact, it can lead to a situation where the algorithm zigzags through the parameter space (cf. Figure 4.7).

The ν chosen this way may, however, not lead to a positive definite direction matrix $\boldsymbol{P}^{(k+1)}$. Therefore, we add a second condition that leads to a positive definite $\boldsymbol{P}^{(k+1)}$ for DFP and BFGS. This condition is called the **curvature condition** and is given by

$$\nabla f(\beta^{(k+1)})^T \boldsymbol{P}^{(k)} \nabla f(\beta^{(k)}) < \nabla f(\beta^{(k)})^T \boldsymbol{P}^{(k)} \nabla f(\beta^{(k)}). \tag{4.20}$$

If a ν fulfills both the Armijo and the curvature condition, then we advance by a sufficient amount in each iteration. Therefore, we do not need to find the global minimum in the line search step of each iteration thereby saving time and function evaluations.

Since there is more than one ν that satisfies the previous conditions, we introduce one more set of conditions, the so-called **Wolfe conditions**:

$$f(\beta^{(k+1)}) \le f(\beta^{(k)}) - c_1 \nu \nabla f(\beta^{(k)})^T \boldsymbol{P}^{(k)} \nabla f(\beta^{(k)}),$$
$$\nabla f(\beta^{(k+1)})^T \boldsymbol{P}^{(k)} \nabla f(\beta^{(k)}) \le c_2 \nabla f(\beta^{(k)})^T \boldsymbol{P}^{(k)} \nabla f(\beta^{(k)}),$$

where $0 < c_1 < c_2 < 1$. In the literature values of $c_1 = 10^{-4}$ and $c_2 = 0.9$ have proven to be effective (see, e.g., Nocedal and Wright, 1999).

4.4.1.6 Convex Functions: Summary

We have introduced four different iterative methods for the global optimization of convex functions of multiple variables. These methods are consecutive improvements of each other and culminate in the class of quasi-Newton methods, which are even by today's standards excellent general purpose unrestricted convex optimization algorithms.

In order to really understand the strengths and weaknesses of these methods, it is sensible to implement them once. For real optimization problems, however, one should rely on well-established program libraries. These implementations, which have been honed for years, often contain additional implementation-specific heuristics and stop criteria that are indispensable for a robust algorithm. We have omitted many of these gritty details in the discussion to aid in understanding the key aspects of these algorithms.

As just one example of the types of optimizations used by modern implementations of quasi-Newton methods, let us mention that they do not store the approximation of the Hessian nor the inverse of the Hessian but a Cholesky decomposition of $\boldsymbol{Q} = \boldsymbol{G}\boldsymbol{G}^T$. This provides superior numerical accuracy and it is much easier to calculate the inverse ($\boldsymbol{P}$) given the decomposition than just using the Hessian. The decomposition has the additional benefit that the matrix $\boldsymbol{G}$ is an invertible lower triangle matrix. This implies that the Cholesky decomposition cannot be singular even in the presence of rounding errors[3] and that it always leads to a positive definite matrix $\boldsymbol{Q}$.

4.4.1.7 Example: Logistic Regression

To illustrate the strengths and weaknesses of these methods, we will now consider a simple example from classical statistics. We want to consider a two-class **classification problem**. Given m observations $(\boldsymbol{x}_i, y_i)$, $i = 1, \ldots, m$, where $\boldsymbol{x}_i \in \mathbb{R}^n$ and $y_i \in \{0,1\}$, we use the following notation:

$$\boldsymbol{X} := \begin{bmatrix} \boldsymbol{x}_1^T \\ \vdots \\ \boldsymbol{x}_m^T \end{bmatrix} \in L(m,n) \quad \text{and} \quad \boldsymbol{y} := \begin{bmatrix} y_1 \\ \vdots \\ y_m \end{bmatrix} \in \{0,1\}^m.$$

In addition, one of the columns of $\boldsymbol{X}$ is assumed to be $\mathbf{1}$. This simplifies the following notation because we can drop the intercept term from all formulas.

The so-called **decision border** of a classification problem is built of the values $\boldsymbol{x}$, for which

$$P(Y = 1 \mid X = \boldsymbol{x}) = P(Y = 0 \mid X = \boldsymbol{x}).$$

For **logistic regression**, which we will consider here, we assume that

$$\log\left(\frac{P(Y = 1 \mid X = \boldsymbol{x})}{P(Y = 0 \mid X = \boldsymbol{x})}\right) = \boldsymbol{x}^T \boldsymbol{\beta}. \tag{4.21}$$

---

[3]Do note that numerical instabilities may still lead to a numerically singular matrix.

Since the right-hand side can take arbitrary values from $\mathbb{R}$, but the ratio of the two probabilities has to be positive, a logarithmic transformation is applied.

The ratio on the left-hand side of Equation 4.21 is called the **odds ratio** and the decision boundary is then given by

$$\log\left(\frac{P(Y=1 \mid X=\boldsymbol{x})}{P(Y=0 \mid X=\boldsymbol{x})}\right) = \mu = \boldsymbol{x}^T\boldsymbol{\beta}.$$

So for logistic regression we arrive at a decision boundary that is **linear** in $\boldsymbol{\beta}$. A standard value for μ is zero, which is based on the assumption, that we are interested in representing $P(Y=1 \mid X=\boldsymbol{x}) = P(Y=0 \mid X=\boldsymbol{x})$.

Since we only consider two-class problems, Equation 4.21 simplifies to

$$\log\left(\frac{P(Y=1 \mid X=\boldsymbol{x})}{1-P(Y=1 \mid X=\boldsymbol{x})}\right) = \boldsymbol{x}^T\boldsymbol{\beta}$$

and via transformation to

$$P(Y=1 \mid X=\boldsymbol{x}) = \frac{\exp(\boldsymbol{x}^T\boldsymbol{\beta})}{1+\exp(\boldsymbol{x}^T\boldsymbol{\beta})} =: p(\boldsymbol{x},\boldsymbol{\beta})$$

as well as

$$P(Y=0 \mid X=\boldsymbol{x}) = 1-P(Y=1 \mid X=\boldsymbol{x}) = 1-p(\boldsymbol{x},\boldsymbol{\beta}) = \frac{1}{1+\exp(\boldsymbol{x}^T\boldsymbol{\beta})}.$$

Since $y_i \in \{0,1\}$ is Bernoulli distributed, this leads to the likelihood

$$\begin{aligned} L(\boldsymbol{\beta}) &= \prod_{i=1}^{m}\left(p(\boldsymbol{x}_i,\boldsymbol{\beta})^{y_i}(1-p(\boldsymbol{x}_i,\boldsymbol{\beta}))^{(1-y_i)}\right) \\ &= \prod_{i=1}^{m}\left(\left(\frac{\exp(\boldsymbol{x}_i^T\boldsymbol{\beta})}{1+\exp(\boldsymbol{x}_i^T\boldsymbol{\beta})}\right)^{y_i}\left(\frac{1}{1+\exp(\boldsymbol{x}_i^T\boldsymbol{\beta})}\right)^{(1-y_i)}\right) \end{aligned}$$

and to the log-likelihood

$$\begin{aligned} \log L(\boldsymbol{\beta}) &= \sum_{i=1}^{m}\left(y_i\log(p(\boldsymbol{x}_i,\boldsymbol{\beta})) + (1-y_i)\log((1-p(\boldsymbol{x}_i,\boldsymbol{\beta})))\right) \\ &= \sum_{i=1}^{m}\left(y_i\boldsymbol{x}_i^T\boldsymbol{\beta} - \log(1+\exp(\boldsymbol{x}_i^T\boldsymbol{\beta}))\right). \end{aligned}$$

For this we can also calculate the gradient and Hessian, which are given by

$$\begin{aligned} \nabla \log L(\boldsymbol{\beta}) &= \sum_{i=1}^{m}\boldsymbol{x}_i(y_i - p(\boldsymbol{x}_i,\boldsymbol{\beta})), \\ \nabla^2 \log L(\boldsymbol{\beta}) &= -\sum_{i=1}^{m}\boldsymbol{x}_i\boldsymbol{x}_i^T p(\boldsymbol{x}_i,\boldsymbol{\beta})(1-p(\boldsymbol{x}_i,\boldsymbol{\beta})) \end{aligned}$$

or in matrix notation:

$$\begin{aligned}\nabla \log L(\boldsymbol{\beta}) &= \boldsymbol{X}^T(\boldsymbol{y}-\boldsymbol{p}(\boldsymbol{X},\boldsymbol{\beta})),\\ \nabla^2 \log L(\boldsymbol{\beta}) &= -\boldsymbol{X}^T \mathrm{diag}(\boldsymbol{p}(\boldsymbol{X},\boldsymbol{\beta})(1-\boldsymbol{p}(\boldsymbol{X},\boldsymbol{\beta})))\boldsymbol{X}\\ &=: -\boldsymbol{X}^T \boldsymbol{D}\boldsymbol{X},\end{aligned}$$

where $\boldsymbol{p}(\boldsymbol{X},\boldsymbol{\beta}) := [p(\boldsymbol{x}_1,\boldsymbol{\beta})\ldots p(\boldsymbol{x}_m,\boldsymbol{\beta})]^T$.

Moreover, from $\nabla^2 \log L(\boldsymbol{\beta}) = -\boldsymbol{X}^T\boldsymbol{D}\boldsymbol{X}$ it follows that the Hessian of $-\log L$ is positive semidefinite, so we may conclude that $-\log L$ is convex, but not strictly convex! Indeed, there are data sets $(\boldsymbol{X},\, \boldsymbol{y})$, for which $-\log L$ is not bounded below. It is especially unfortunate that this is true iff (at least part of) the data are linear separable, meaning that at least one of the $p(\boldsymbol{x}_i,\boldsymbol{\beta}) = 0$ or 1, i.e. in a rather **trivial** case. So logistic regression does not find a classification rule in one of the simplest cases. In practice, it is therefore indispensable to check the convergence of the optimization process in the case of logistic regression which is often neglected!

If the data are indeed linearly separable, then other classification algorithms should be used. One simple alternative is the **Linear Discriminant Analysis (LDA)** (see Section 7.3.1). Alternatively, one can apply **restricted** logistic regression with the added constraint that $||\boldsymbol{\beta}|| < c$ but this leads to a constrained optimization problem that can be tedious to solve.

Logistic Regression: Optimization Problem

The optimization problem induced by logistic regression is given by

$$\underset{\boldsymbol{\beta}\in\mathbb{R}^n}{\text{minimize}} \quad f(\boldsymbol{\beta}) := -\log L(\boldsymbol{\beta}) = -\sum_{i=1}^{m}\left(y_i\boldsymbol{x}_i^T\boldsymbol{\beta} - \log(1+\exp(\boldsymbol{x}_i^T\boldsymbol{\beta}))\right).$$

The objective function is twice continuously differentiable, convex, and if the data are not linearly separable, even strictly convex. Therefore, all introduced optimization methods can be used to minimize the negative log-likelihood function.

Simulation: Comparison of Gradient Methods for Logistic Regression

To compare the gradient descent method (G-D), the Newton method, and the BFGS quasi-Newton method, we will use the following **simulation design**:

1. Given an observation number m and a dimension n, generate m n-dimensional vectors whose first entry is 1 and whose other $(n-1)$ elements are standard normally distributed random numbers. From these vectors form the matrix $\tilde{\boldsymbol{X}}$.
2. Generate a uniform random m-dimensional vector $\boldsymbol{y} \in \{0,1\}^m$.

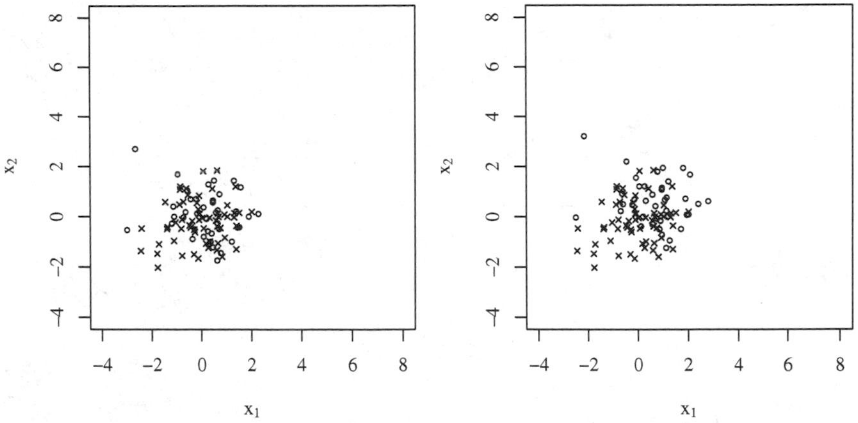

Figure 4.8: Point clouds for $\kappa = 0$ (left) and $\kappa = 0.5$ (right).

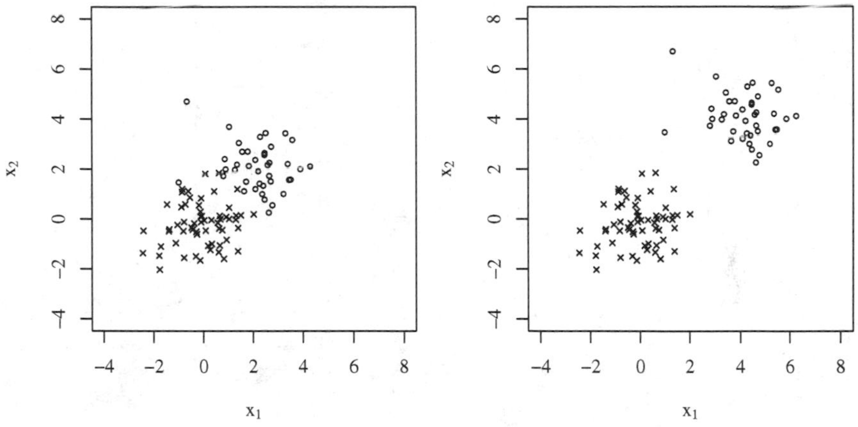

Figure 4.9: Point clouds for $\kappa = 2$ (left) and $\kappa = 4$ (right).

3. Calculate $\boldsymbol{X} := \tilde{\boldsymbol{X}} + \kappa \boldsymbol{y} \boldsymbol{1}^T$, where $\kappa \geq 0$.
4. $(\boldsymbol{X}, \boldsymbol{y})$ then represents a logistic regression problem.

The difficulty of the problem is controlled by the parameter κ. The larger κ is, the further apart the two distributions are, and the more probable it is that the two classes are linearly separable (see Figures 4.8 and 4.9). Moreover, the contour plots in Figures 4.10 and 4.11 show that the contours contract for increasing κ. Instead of nearly spherical contours for $\kappa = 0$, we get strongly elliptical contours for $\kappa = 2$. This implies that the optimization problem is paradoxically less well-conditioned when the corresponding classification problem gets simpler. Notice that linearly separable classes are to be

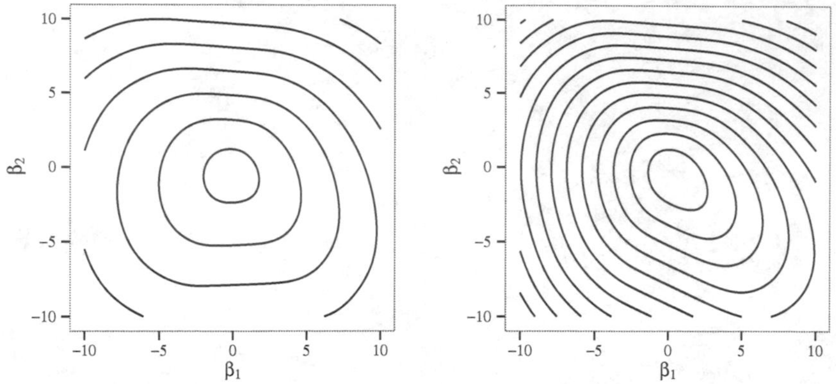

Figure 4.10: Contours for $\kappa = 0$ (left) and $\kappa = 0.5$ (right).

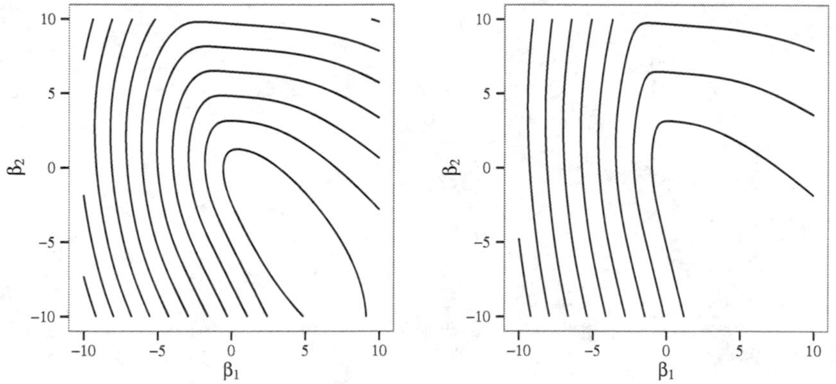

Figure 4.11: Contours for $\kappa = 2$ (left) and $\kappa = 4$ (right).

expected if there are many factors in the data with many different levels each. Also, for real-world problems we usually cannot visualize the contours of the log-likelihood because we want to estimate more than two parameters.

For the simulation study we generate $\tilde{\boldsymbol{X}}$ and $\boldsymbol{y}$ for every possible combination of $m = \{100, 200, 500, 1000\}$ and $n = \{5, 10, 15, 20\}$. From this we generate $(\boldsymbol{X}, \boldsymbol{y})$ for $\kappa = \{0, 0.25, 0.5, 1\}$. This leads to a total of $4 \times 4 \times 4 = 64$ different classification problems. For each problem we have to choose a starting point $\boldsymbol{\beta}^{(1)}$ for the algorithms. In practice, one often uses $\boldsymbol{\beta}^{(1)} = \mathbf{0}$. In our experiment we instead vary the starting values systematically. For each of the 64 problems we first numerically determine the minimum $\boldsymbol{\beta}^*$ using multiple

BFGS runs with extremely tight stop criteria, and subsequently choose a random direction $\beta_u \in \mathbb{R}^n$ ($||\beta_u||_2 = 1$). Then, for $\varepsilon = \{1,2,3,4\}$ we determine a ν such that

$$(1+\varepsilon)(-\log L(\beta^*)) = -\log L(\beta^* + \nu\beta_u).$$

As a starting value we then choose $\beta^{(1)} = \beta^* + \nu\beta_u$. This procedure is repeated for different β_u. Therefore, we start our optimization at different distances from the optimum so that we can see if the convergence speed is related in any way to the distance to the optimum. This is what we expect from the theoretical analysis.

We, thus, vary the following influential factors:

- Method (G-D, Newton, BFGS),
- Number of observations ($m = \{100, 200, 500, 1000\}$),
- Number of parameters ($n = \{5, 10, 15, 20\}$),
- "Condition" ($\kappa = \{0, 0.25, 0.5, 1\}$),
- Goodness of starting point ($\varepsilon = \{1,2,3,4\}$),
- Direction of starting point (β_{u1} till β_{u4}).

Overall, we then have 3072 different optimization runs. To evaluate the quality of the result we measure the

- Number of function evaluations,
- Distance to the minimum ($||\hat{\beta}^* - \beta^*||_2$),
- Suboptimality of the returned solution ($f(\hat{\beta}^*) - f(\beta^*)$).

Simulation Results

When we analyze the results, we first need to make sure we understand some of the oddities in the data set. Since we do not know the exact global minimum, we might observe better solutions than our found minimum. In this study, we got lucky and such a situation did not occur. However, we do have 72 runs, all using the gradient descent strategy, which do not converge at all. Lastly, some runs hit the global minimum up to the numerical precision.

In our analysis we will focus on the second and third quality criteria from above. The analysis of the number of function evaluations required for convergence is left as an exercise for the reader (see Exercise 4.4.5). When analyzing the distance to the optimum and the suboptimality of the final solution, the dependence on parameter ε is not relevant to be analyzed. Any algorithm that converges from an $\varepsilon = 4$ starting point obviously passed through the region of $\varepsilon = 2$ starting points, and therefore is only at a disadvantage when analyzing the number of function evaluations until convergence is reached.

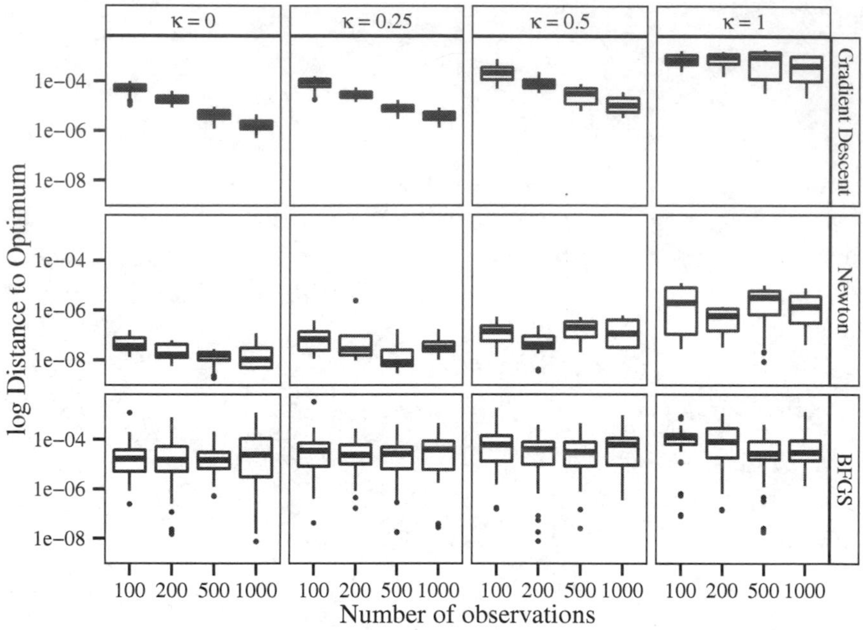

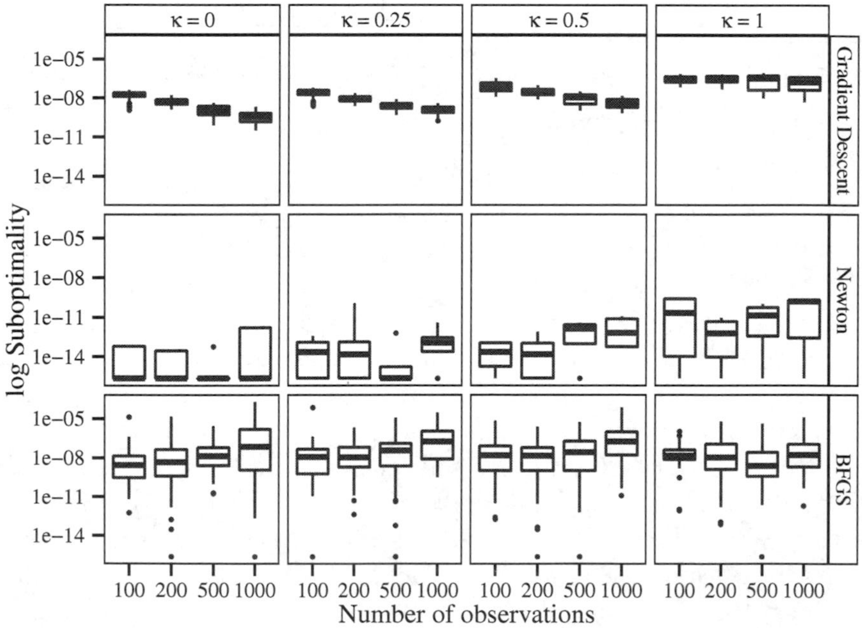

Figure 4.12: Logistic regression: Boxplot of the distribution of the distance to the minimum (top) and suboptimality (bottom).

The distribution of the distance to the minimum and the suboptimality is shown in Figure 4.12. Both plots show the distribution by κ (along the top), algorithm (along the side) and the number of observations m (along the x-axis). The two figures look quite similar. The most surprising result is probably that BFGS is not influenced by any of the parameters. Neither the training set size nor the "condition" of the problem (κ) appears to have any influence on the quality of the result. We also see that BFGS has the lowest average solution quality, but this may well be due to the chosen convergence criteria. For the gradient descent strategy we observe the exact opposite behavior. We see that the solution quality increases with the size of the training set and decreases, as we might expect, with increasing κ. Newton's method lies somewhere in between the two other algorithms; only κ has a small influence on the solution quality, and again, with increasing κ, the quality of the solution decreases.

4.4.2 Nonconvex Problems

In the previous section we exclusively studied convex problems. While these are not as seldom as one might think in practice, it is still often the case that we encounter nonconvex problems. So how do we deal with these? One solution is to try and find a suitable transformation so that the transformed problem is convex. This is often possible for quasi-convex problems, for example. For all other problems that are at least once continuously differentiable, we can only find **local optima** efficiently. For this optimization, we can again employ the methods we already know but have to take care that they now might not terminate because of the ill-posed (nonconvex) problem. Sometimes a third case arises; that is, we have to tackle a nonconvex problem that has a known structure which we can exploit. One special class of problems that fall under this category will be studied in the next section.

4.4.2.1 Nonlinear Least Squares (NLS) Problems

Nonlinear Least Squares (NLS) problems often arise in real-world regression problems. While they are not convex in general, they are usually well behaved and do not have many local optima. Because of their importance to statistics, there are several specialized algorithms for these types of problems. Before we get to them, let us fix some notation.

We assume that the **statistical model** for which we want to estimate the coefficients has the following form:

$$y = h(X, \beta) + e = h(\beta) + e,$$

where $\boldsymbol{y} = [y_1 \dots y_m]^T$, $\boldsymbol{h}(\boldsymbol{\beta}) = [h(\boldsymbol{x}_1, \boldsymbol{\beta}) \dots h(\boldsymbol{x}_m, \boldsymbol{\beta})]^T$, $\boldsymbol{\beta} = [\beta_1 \dots \beta_n]^T$ and $\boldsymbol{e} = [e_1 \dots e_m]^T$, and that the **objective function** is the sum of squared model errors. We will use the shorthand $h_i(\boldsymbol{\beta})$ for $h(\boldsymbol{x}_i, \boldsymbol{\beta})$.

$$S(\boldsymbol{\beta}) = (\boldsymbol{y} - \boldsymbol{h}(\boldsymbol{\beta}))^T (\boldsymbol{y} - \boldsymbol{h}(\boldsymbol{\beta})) = \boldsymbol{e}(\boldsymbol{\beta})^T \boldsymbol{e}(\boldsymbol{\beta}).$$

Then, the corresponding **normal equations** have the form

$$\nabla S(\boldsymbol{\beta}) = -2\boldsymbol{Z}(\boldsymbol{\beta})^T (\boldsymbol{y} - \boldsymbol{h}(\boldsymbol{\beta})) = \mathbf{0},$$

where $\boldsymbol{Z}(\boldsymbol{\beta}) := \boldsymbol{J_h}(\boldsymbol{\beta})$ is the $m \times n$ **Jacobian matrix** of the first derivatives of $\boldsymbol{h}(\boldsymbol{\beta})$.

For the $n \times n$ **Hessian matrix** of the second derivatives of $S(\boldsymbol{\beta})$ we have

$$\boldsymbol{H}(\boldsymbol{\beta}) = 2\boldsymbol{Z}(\boldsymbol{\beta})^T \boldsymbol{Z}(\boldsymbol{\beta}) - 2 \sum_{t,t'=1}^{m} (y_t - h_t(\boldsymbol{\beta})) \nabla^2 h_{t'}(\boldsymbol{\beta}).$$

Please note that in the special case of a **linear model** we obtain

$$\begin{aligned} \boldsymbol{h}(\boldsymbol{\beta}^*) &= \boldsymbol{X}\boldsymbol{\beta}^*, \\ \nabla S(\boldsymbol{\beta}^*) &= -2\boldsymbol{X}^T(\boldsymbol{y} - \boldsymbol{X}\boldsymbol{\beta}^*) = 0 \qquad \Leftrightarrow \qquad \boldsymbol{X}^T\boldsymbol{X}\boldsymbol{\beta}^* = \boldsymbol{X}^T\boldsymbol{y}, \\ \boldsymbol{H}(\boldsymbol{\beta}^*) &= 2\boldsymbol{X}^T\boldsymbol{X}. \end{aligned}$$

If the Jacobian matrix $\boldsymbol{Z}(\boldsymbol{\beta})$ and the Hessian matrix $\boldsymbol{H}(\boldsymbol{\beta})$ can be calculated analytically, then any of the (quasi-)Newton type methods can be used to find a **local** minimum of $S(\boldsymbol{\beta})$. Since this problem is nonconvex in general, it is important to start the algorithms from multiple points and to check the gradient and possibly Hessian after termination to ensure that indeed a minimum and not a saddle point or even a maximum was found.

Nonlinear Least Squares: Gauss-Newton Method

Instead of using the full Hessian, we can use its special structure to find approximations to it that are easier to calculate. One such approximation leads directly to the so-called **Gauss-Newton method** or **Gauss method**. If the expected value of $e_t = y_t - h_t(\boldsymbol{\beta}^*)$ is assumed to be zero, then the second term of the Hessian can, on average, be ignored and the first term can be used as an approximation of the Hessian of $S(\boldsymbol{\beta})$. This leads to

$$\boldsymbol{P}^{(k)} = (2\boldsymbol{Z}(\boldsymbol{\beta}^{(k)})^T \boldsymbol{Z}(\boldsymbol{\beta}^{(k)}))^{-1}$$

and, since

$$\nabla S(\boldsymbol{\beta}^{(k)}) = -2\boldsymbol{Z}(\boldsymbol{\beta}^{(k)})^T (\boldsymbol{y} - \boldsymbol{h}(\boldsymbol{\beta}^{(k)})),$$

we get with step length 1:

$$\begin{aligned}\beta^{(k+1)} &= \beta^{(k)} + (Z(\beta^{(k)})^T Z(\beta^{(k)}))^{-1} Z(\beta^{(k)})^T (y - h(\beta^{(k)})) \\ &= (Z(\beta^{(k)})^T Z(\beta^{(k)}))^{-1} Z(\beta^{(k)})^T (y - h(\beta^{(k)}) + Z(\beta^{(k)})\beta^{(k)}).\end{aligned} \tag{4.22}$$

This is the **least squares estimator** for the model

$$\bar{y}(\beta^{(k)}) = Z(\beta^{(k)})\beta + e, \tag{4.23}$$

where

$$\bar{y}(\beta^{(k)}) = y - h(\beta^{(k)}) + Z(\beta^{(k)})\beta^{(k)}. \tag{4.24}$$

The Gauss method can therefore be interpreted as a sequence of linear regressions. In every step we calculate the least squares estimator for a linear approximation of the nonlinear model.

The simplicity of this method and its good local convergence properties (Dennis, 1973) make this method attractive, though $P^{(k)}$ is often singular and not positive definite if $\beta^{(k)}$ is far away from β^* minimizing $S(\beta)$. Moreover, if the linear approximation of the nonlinear model is bad or the model residuals are large, then convergence might also be slow for this method.

Like Broyden's method, we can combine the Gauss method with a quasi-Newton type algorithm. Since the performance of the Gauss method depends on the size of the model residuals, more exactly on

$$\sum_{t,t'=1}^{m} (y_t - h_t(\beta))\nabla^2 h_{t'}(\beta), \tag{4.25}$$

Brown and Dennis (1971) suggested the following combination of the Gauss method and a quasi-Newton method.

Instead of approximating the inverse of the Hessian of the objective function, now the Hessian of the function $h_t(\beta)$ is iteratively approximated; i.e. we choose:

$$M_t^{(k+1)} = M_t^{(k)} + K_t^{(k)},$$

where $M_t^{(k)}$ is an approximation of $\nabla^2 h_t(\beta^{(k)})$. Since

$$\nabla h_t(\beta^{(k)}) \cong \nabla h_t(\beta^{(k+1)}) + \nabla^2 h_t(\beta^{(k+1)})(\beta^{(k)} - \beta^{(k+1)}),$$

by inserting $M_t^{(k+1)} = M_t^{(k)} + K_t^{(k)}$ for $\nabla^2 h_t(\beta^{(k+1)})$ we get

$$K_t^{(k)}(\beta^{(k+1)} - \beta^{(k)}) = \mu_t^{(k)} \tag{4.26}$$

with

$$\mu_t^{(k)} = \nabla h_t(\beta^{(k+1)}) - \nabla h_t(\beta^{(k)}) - M_t^{(k)}(\beta^{(k+1)} - \beta^{(k)}).$$

Therefore, we use

$$K_t^{(k)} = \frac{\mu_t^{(k)}(\beta^{(k+1)} - \beta^{(k)})^T}{(\beta^{(k+1)} - \beta^{(k)})^T(\beta^{(k+1)} - \beta^{(k)})}$$

as a rank 1 correction matrix because then (4.26) is obviously valid.

The direction matrix for this method is

$$P^{(k)} = \left(Z(\beta^{(k)})^T Z(\beta^{(k)}) - \sum_{t,t'=1}^{m} (y_t - h_t(\beta^{(k)}))M_{t'}^{(k)} \right)^{-1}.$$

A possible choice for $M_t^{(1)}$ is the identity matrix for all t.

Nonlinear Least Squares: Levenberg-Marquardt Method

The Marquardt method (Marquardt, 1963), sometimes also called **Levenberg-Marquardt method**, can be used to modify methods not guaranteeing positive definite direction matrices $P^{(k)}$. This method utilizes the fact that

$$P^{(k)} + \lambda^{(k)} \bar{P}^{(k)} \tag{4.27}$$

is always positive definite if $\bar{P}^{(k)}$ is positive definite and the scalar $\lambda^{(k)}$ is sufficiently large.

A possible choice for $\bar{P}^{(k)}$ is the identity matrix. Typically, this method is used in combination with the Gauss method and $Z(\beta^{(k)})^T Z(\beta^{(k)})$ is modified and not its inverse. The new direction matrix is then given by:

$$P^{(k)} = (Z(\beta^{(k)})^T Z(\beta^{(k)}) + \lambda^{(k)} \bar{P}^{(k)})^{-1}, \tag{4.28}$$

where I_n is used as $\bar{P}^{(k)}$.

For $\lambda^{(k)}$ near zero this method is very similar to the Gauss method, but as we increase $\lambda^{(k)}$, the method transitions into a steepest descent type algorithm. Since the good performance of the Gauss method is particularly high near the minimum, we start with a small $\lambda^{(1)}$ and decrease $\lambda^{(k)} > 0$ in every iteration, unless this leads to an unacceptable step. Notice that the step length and the step direction are jointly determined. Typically, the step length is set to 1 for the Marquardt method.

Analogously, we could modify the Hessian of the objective function and use

$$P^{(k)} = (H^{(k)} + \lambda^{(k)} I_n)^{-1} \tag{4.29}$$

as direction matrix. This method is generally called **quadratic hill climbing method**, since it was introduced in a maximization context (Goldfeld et al., 1966).

The performance of these methods is not invariant to transformations of the coefficient space. Therefore, the choice of $\bar{\boldsymbol{P}}^{(k)}$ has an influence on the convergence and different choices might be appropriate for different problems (Marquardt (1963), Goldfeld and Quandt (1972, chapter 1)).

The Levenberg-Marquardt method has very good convergence properties in practice, even if the starting vector $\boldsymbol{\beta}^{(1)}$ is far away from the minimum of the objective function. To illustrate this, let us look at a simple nonlinear model.

4.4.2.2 Example: Comparing the Algorithms

We will study the following artificial nonlinear statistical model:

$$\begin{aligned} y_t &= \beta_1^* + \beta_2^* x_{t,2} + \beta_2^{*2} x_{t,3} + e_t, \qquad t = 1,\ldots,20, \text{ or} \\ \boldsymbol{y} &= \boldsymbol{h}(\boldsymbol{\beta}^*) + \boldsymbol{e}, \end{aligned}$$

where $\boldsymbol{y} = [y_1\ y_2 \ldots y_{20}]^T$, $\boldsymbol{\beta}^* = [\beta_1^*\ \beta_2^*]^T$ is the true coefficient vector, $\boldsymbol{e} = [e_1\ e_2 \ldots e_{20}]^T$, and

$$\boldsymbol{h}(\boldsymbol{\beta}) = \begin{bmatrix} \beta_1 + \beta_2 x_{1,2} + \beta_2^2 x_{1,3} \\ \beta_1 + \beta_2 x_{2,2} + \beta_2^2 x_{2,3} \\ \vdots \\ \beta_1 + \beta_2 x_{20,2} + \beta_2^2 x_{20,3} \end{bmatrix}.$$

The values of $x_{t,2}$ and $x_{t,3}$ are taken to be **pseudo-random numbers** from a uniform distribution on $[0,1)$ (cp. Chapter 6). The y_t are calculated by summation of $\beta_1^* + \beta_2^* x_{t,2} + \beta_2^{*2} x_{t,3}$ and a standard normally distributed random number, where the true coefficients are set to $\beta_1^* = \beta_2^* = 1$. That means we added the random error to

$$1 + x_{t,2} + x_{t,3}.$$

In order to determine the least squares estimator of $\boldsymbol{\beta}^*$, we have to minimize

$$S(\boldsymbol{\beta}) = (\boldsymbol{y} - \boldsymbol{h}(\boldsymbol{\beta}))^T (\boldsymbol{y} - \boldsymbol{h}(\boldsymbol{\beta})). \tag{4.30}$$

Figure 4.13 shows a contour plot of $S(\boldsymbol{\beta})$. In the figure we see that $S(\boldsymbol{\beta})$ has two local minima. Which optimum is found largely depends on the starting vector $\boldsymbol{\beta}^{(1)}$ (see Figure 4.14). This influence will be studied systematically in what follows.

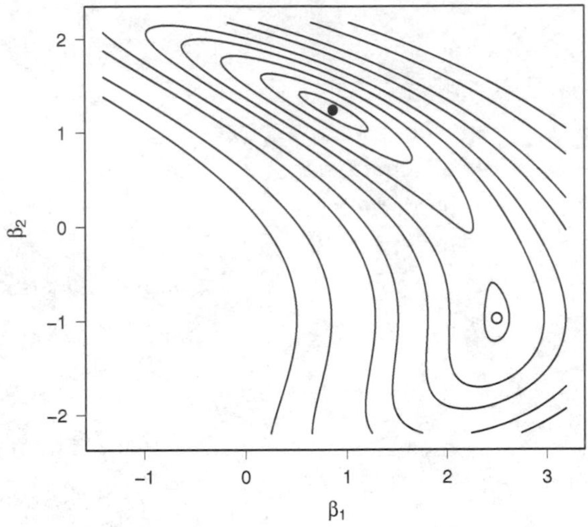

Figure 4.13: Contour plot of $S(\boldsymbol{\beta})$ in the example. The two circles indicate the minima of the function, where the filled one is the global minimum.

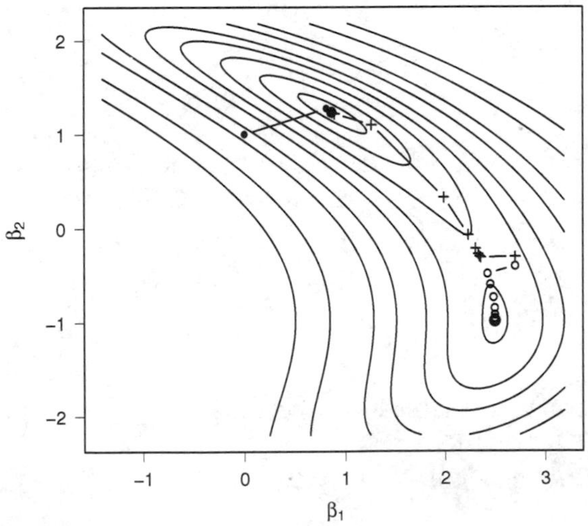

Figure 4.14: Example: Levenberg-Marquardt paths; nearby starting points (+ and o) may lead to different solutions.

Table 4.4: Runtimes of the Different Algorithms on Problem (4.30)

		Quantiles of FE Distribution				
Algorithm	Function	Min.	0.25	Med.	0.75	Max.
Gradient-Descent	S	28	278	557	2457	7024
	∇S	4	27	51	228	637
Newton	S	5	15	73	104	105
	∇S	5	15	73	104	105
	$\nabla^2 S$	5	15	73	104	105
BFGS	S	11	24	30	38	60
	∇S	6	11	11	14	25
Gauss-Newton	S	26	32	38	38	65
	J_h	16	20	23	24	37
Levenberg-Marquardt	S	4	6	7	7	12
	J_h	3	5	6	6	8
Nelder-Mead	S	45	57	65	75	143

Note: Depicted are the minimum, lower quantile, median, upper quantile and maximum number of function, gradient, Hessian and Jacobian evaluations (FE) used by the algorithms.

All the following figures (see, e.g., Figure 4.15) used to show the convergence of the methods use the same display. At the location of each initial parameter value an arrow is shown whose size is proportional to the number of function evaluations and which points toward the local minimum that was found by the algorithm. The scaling of the arrow size is not comparable between figures because some algorithms require many more function evaluations than others. In principle, it is hard to compare, e.g., the Newton method with the method of steepest descent because the Newton method additionally calculates the (expensive) Hessian in each iteration. Therefore, it has much more information about the local shape of the function landscape than most of the other methods. The number of function, gradient, Hessian, and Jacobian evaluations used by the algorithms are summarized in Table 4.4. The Nelder-Mead method mentioned in the table will be described in Section 4.4.3.

Gradient Descent

In most cases the method of steepest descent converges to the minimum that

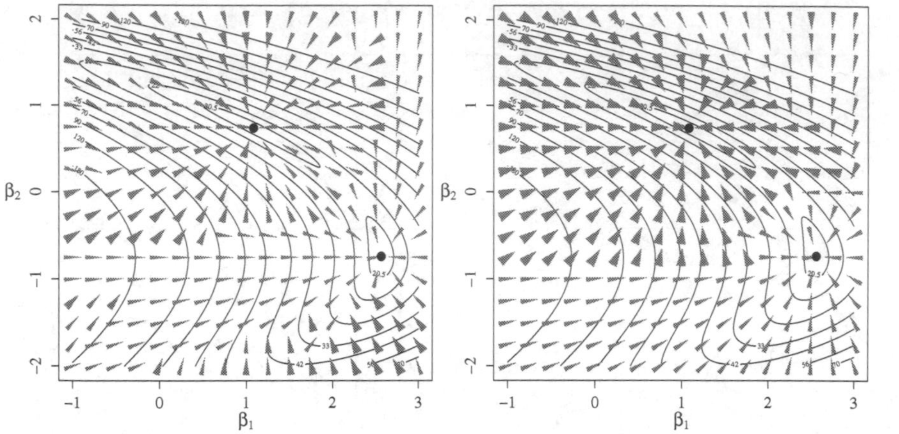

Figure 4.15: Example: Steepest descent attraction (left), Newton attraction (right).

is nearest to the starting vector, but frequently only after a tremendous number of function evaluations, in particular in the case of the global minimum. The reason for this is the long-stretched valley in which the global minimum is located (see Figure 4.15). Here, the line search requires many iterations to find a good new intermediate solution. This can be inferred from the ratio of function evaluations to gradient evaluations. In each iteration the algorithm will only perform one gradient evaluation but many function evaluations.

Newton Method
The Newton-Raphson method converges against both local minima very quickly (see Figure 4.15). Note that there is an area in the top right of the parameter space from which the method does not converge toward the closer (upper) minimum, but instead ends up in the minimum farther away (lower right). Also note that the Newton method converges extremely fast if the initial parameter vector lies in the flat lower left region of the parameter space.

BFGS
For the BFGS method, the assignment of a starting vector to a local minimum is not so clear. Nevertheless, the method converges to the nearest minimum relatively fast in an acceptable surrounding (see Figure 4.16). In Table 4.4 we can see that although the BFGS method does not use second-order information, it is on average faster than the Newton method and is noticeably faster in the worst case. This is all the more true if we factor in the fact that evaluating the Hessian $\nabla^2 S$ is much more expensive than a function or gradient evaluation.

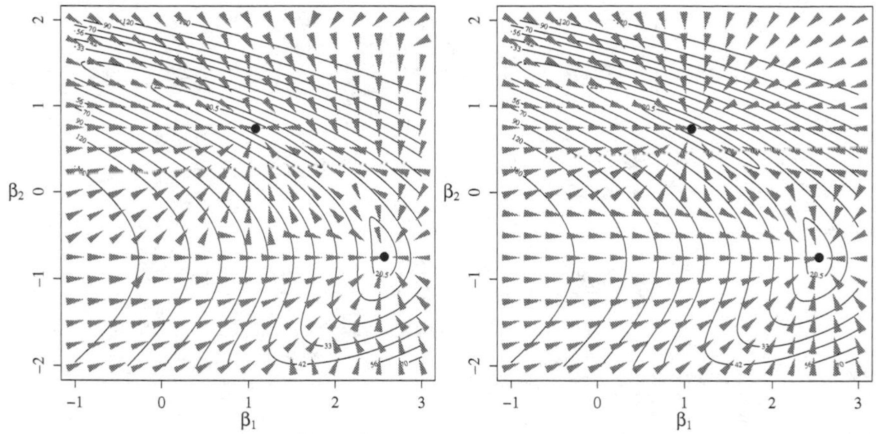

Figure 4.16: Example: BFGS attraction (left), Gauss-Newton attraction (right).

Gauss-Newton

The convergence landscape of the Gauss-Newton method shows a very interesting structure (see Figure 4.16). Apparently, only the location of the starting value of β_2 decides to which minimum the algorithm converges. The median number of necessary function evaluations is similar to the BFGS method, the worst case behavior is also similar, but in the best case the method performs poorly when compared to BFGS. This is surprising, given that it is one of two methods in the comparison specifically tailored to nonlinear least squares problems.

Levenberg-Marquardt

The Levenberg-Marquardt method works similar to the Gauss-Newton method (see Figure 4.17). It can clearly be seen that the initial value of β_2 is decisive for the found local minimum. In terms of number of function or Jacobian evaluations the algorithm is by far the best candidate in the comparison. Clearly, the specialization is worth the extra implementation effort. In practice, Levenberg-Marquardt type algorithms are the de facto standard when it comes to solving large scale nonlinear least squares problems. We will see an example of this in Section 4.5.

Let us try to explain the somewhat strange behavior of the Gauss-Newton method and the Levenberg-Marquardt method that the initial value of β_2 in a way determines the found local minimum. In order to better understand this,

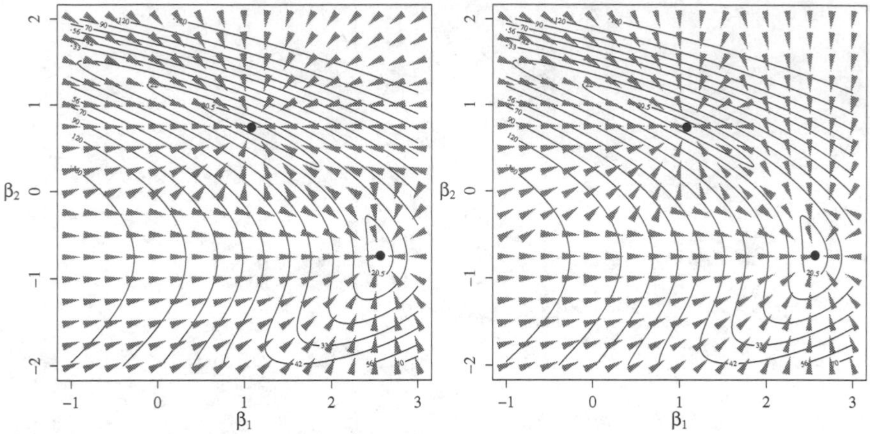

Figure 4.17: Example: Levenberg-Marquardt attraction (left), Nelder-Mead attraction (right).

let us analyze the structure of Gauss-Newton steps:

$$\begin{aligned}
&-(\boldsymbol{Z}(\boldsymbol{\beta}^{(k)})^T\boldsymbol{Z}(\boldsymbol{\beta}^{(k)}))^{-1}\boldsymbol{Z}(\boldsymbol{\beta}^{(k)})^T(\boldsymbol{y}-\boldsymbol{h}(\boldsymbol{\beta}^{(k)}))\\
=&-(\boldsymbol{Z}(\boldsymbol{\beta}^{(k)})^T\boldsymbol{Z}(\boldsymbol{\beta}^{(k)}))^{-1}\boldsymbol{Z}(\boldsymbol{\beta}^{(k)})^T\boldsymbol{e}(\boldsymbol{\beta}^{(k)})).
\end{aligned}$$

Here, $2\boldsymbol{Z}(\boldsymbol{\beta}^{(k)})^T\boldsymbol{Z}(\boldsymbol{\beta}^{(k)})$ replaces the Hessian

$$\boldsymbol{H}(\boldsymbol{\beta}^{(k)}) = 2\boldsymbol{Z}(\boldsymbol{\beta}^{(k)})^T\boldsymbol{Z}(\boldsymbol{\beta}^{(k)}) - 2\sum_{t,t'=1}^{m}(y_t - h_t(\boldsymbol{\beta}^{(k)}))\nabla^2 h_{t'}(\boldsymbol{\beta}^{(k)}).$$

So, let us have a look at $\boldsymbol{Z}$:

$$\boldsymbol{Z}(\boldsymbol{\beta}^{(k)}) = J_{\boldsymbol{h}}(\boldsymbol{\beta}^{(k)}) = \begin{bmatrix} 1 & x_{1,2}+2\beta_2^{(k)}x_{1,3} \\ 1 & x_{2,2}+2\beta_2^{(k)}x_{2,3} \\ \vdots & \vdots \\ 1 & x_{20,2}+2\beta_2^{(k)}x_{20,3} \end{bmatrix}.$$

The approximation $2\boldsymbol{Z}(\boldsymbol{\beta}^{(k)})^T\boldsymbol{Z}(\boldsymbol{\beta}^{(k)})$ does not depend on $\beta_1^{(k)}$ in contrast to the Hessian $\boldsymbol{H}(\boldsymbol{\beta}^{(k)})$ via the terms $(y_t - h_t(\boldsymbol{\beta}^{(k)}))$. This might be one reason for the behavior of the Gauss-Newton method.

There is one method in Table 4.4 that we have not discussed, the Nelder-Mead method. It only performs function evaluations, avoiding all costly gradient or Hessian computations and performs quite well considering this. In fact, except for the Levenberg-Marquardt algorithm and maybe BFGS,

Nelder-Mead dominates all other algorithms. How optimization can be performed without using first- or second-order information will be the topic of the next section.

4.4.3 *Derivative-Free Methods* [4]

A major drawback of all of the methods we have seen so far is that they need an analytical form of the first partial derivatives in order to determine a descent direction. If these are not available, some of these methods use a numerical approximation scheme as a substitute:

$$\frac{\partial S}{\partial \beta_i}\bigg|_{\beta} \cong \left(\frac{S(\beta_1,\ldots,\beta_{i-1},\beta_i+\Delta\beta_i,\beta_{i+1},\ldots,\beta_n)}{2\Delta\beta_i}\right. \\ \left. -\frac{S(\beta_1,\ldots,\beta_{i-1},\beta_i-\Delta\beta_i,\beta_{i+1},\ldots,\beta_n)}{2\Delta\beta_i}\right)\bigg|_{\beta}$$

for sufficiently small $\Delta\beta_i$. Similar one-sided approximations are also sometimes used. It should be clear that using such a crutch not only increases the number of function evaluations but also degrades the quality of solution we can expect. If desired, higher-order derivatives can be approximated using the same idea. Details for such methods can be found in Bard (1974, Section 5-18), and Quandt (1983).

Another possibility to avoid partial derivatives is the application of so-called **direct search methods**. These procedures are especially useful if the first derivatives of the objective function are nonexistent or complicated to calculate.

Simplex Method

One such direct search method is the so-called **simplex method** proposed by Spendley et al. (1962) in connection with statistical design of experiments. Notice that this method has nothing to do with the simplex method in linear programming (see Section 4.6.3).

A **simplex** is set up by $n+1$ vectors $\beta_1,\beta_2,\ldots,\beta_{n+1}$ in a n-dimensional space. These vectors are the corners of the simplex. In 2D, simplexes of maximal dimension are triangles in the plane. In 3D, they form tetrahedrons (cp. Figure 4.18) and generally in $\mathbb{R}^n$, simplexes are regular polyhedrons.

From analytical geometry we know that the coordinates of the corners of a regular simplex can be represented by the following matrix $\boldsymbol{D}$, where each column stands for one corner, numbered $i=1,\ldots,(n+1)$, and the rows

[4]Partly based on Himmelblau (1972, pp. 148 – 157).

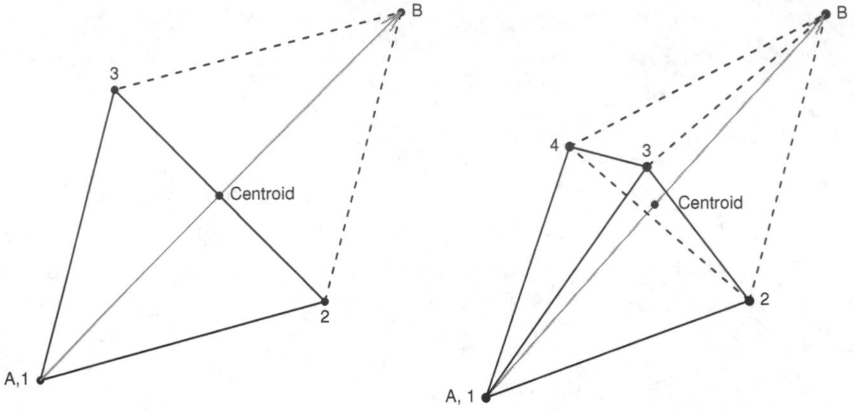

Figure 4.18: Regular simplexes for two and three independent variables. The highest function value $f(x)$ is in (1). The arrow points into the direction of largest improvement.

represent the coordinates $j = 1, \dots, n$:

$$\boldsymbol{D} = \begin{bmatrix} 0 & d_1 & d_2 & \dots & d_2 \\ 0 & d_2 & d_1 & \dots & d_2 \\ 0 & d_2 & d_2 & \dots & d_2 \\ \vdots & \vdots & \vdots & \dots & \vdots \\ 0 & d_2 & d_2 & \dots & d_1 \end{bmatrix} \qquad \text{is a } n \times (n+1) \text{ matrix with}$$

$$d_1 = \frac{t}{n\sqrt{2}}(\sqrt{n+1} + n - 1),$$

$$d_2 = \frac{t}{n\sqrt{2}}(\sqrt{n+1} - 1), \qquad t = \text{distance between two corners.}$$

So the triangle in 2D with $n = 2$ and $t = 1$ in Figure 4.18 has the following coordinates:

Corner	$\beta_{i,1}$	$\beta_{i,2}$
1	0	0
2	0.965	0.259
3	0.259	0.965

The objective function is evaluated in each corner. Then, the point A with the highest function value is reflected at the centroid of the remaining points (see Figure 4.18). Point A is then eliminated and a new simplex (the reflection) is generated by the remaining points of the old simplex and the new reflected

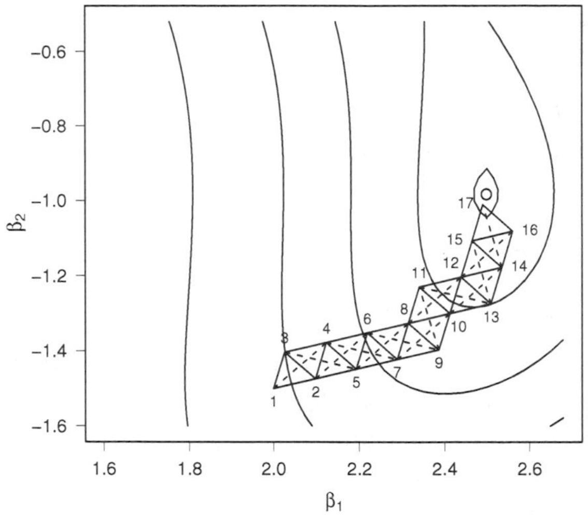

Figure 4.19: Sequence of regular simplexes for the minimization of $f(\boldsymbol{\beta})$. Dashed lines represent reflection directions.

point B in the direction of the reflection so that the new simplex has the same volume as before.

The search is then continued in the same way by always replacing the point with the largest function value by its reflection. Additionally, rules for simplex reduction and to prevent loops near the optimum are applied. This leads to a derivative-free search with, in principle, fixed step length in each iteration, but with changing search direction. Figure 4.19 illustrates the sequence of simplexes in 2D for a well-conditioned objective function.

Nelder-Mead Method

Nelder and Mead (1965) proposed a search method which generalizes the simplex method and is more effective but nevertheless easy to implement. Some practical difficulties of the simplex method, in particular that it is not possible to speedup the search for problems with curvy valleys, lead to different improvements of the method, especially to the method of Nelder and Mead, in which the simplex is allowed to change its shape, becoming a "flexible polyhedron".

The method of Nelder and Mead minimizes the function $f(\boldsymbol{\beta})$ of n independent variables based on the $(n+1)$ corners of a flexible polyhedron in $\mathbb{R}^n$. Each corner is defined by a vector $\boldsymbol{\beta}$. The corner with the largest function value $f(\boldsymbol{\beta})$ is projected through the centroid of the remaining corners. Better (meaning smaller) values of the objective function are found by itera-

Algorithm 4.9 Nelder-Mead Method

Require: $f\colon \mathbb{R}^n \to \mathbb{R}$

    Coefficients $\alpha > 0,\ 0 < \beta < 1,\ \gamma > 1,\ \varepsilon > 0.$

    Initial simplex defined by $\beta_1, \ldots, \beta_n, \beta_{n+1}.$

1: **while** Size of polyhedron $> \varepsilon$ **do**

2:      $l \leftarrow \arg\min_{i \in \{1,\ldots,n+1\}} f(\beta_i)$

3:      $h \leftarrow \arg\max_{i \in \{1,\ldots,n+1\}} f(\beta_i)$

4:      $\beta_{\text{centroid}} \leftarrow$ centroid of $\{\beta_1, \ldots, \beta_{n+1} \text{ without } \beta_h\}$

5:      $\beta_{\text{candidate}} \leftarrow \beta_{\text{centroid}} + \alpha(\beta_{\text{centroid}} - \beta_h)$ {Reflection}

6:      **if** $f(\beta_{\text{candidate}}) \leq f(\beta_l)$ **then**

7:          $\beta'_{\text{candidate}} \leftarrow \beta_{\text{centroid}} + \gamma(\beta_{\text{candidate}} - \beta_{\text{centroid}})$ {Expansion}

8:          **if** $f(\beta'_{\text{candidate}}) < f(\beta_l)$ **then**

9:              $\beta_h \leftarrow \beta'_{\text{candidate}}$

10:         **else**

11:             $\beta_h \leftarrow \beta_{\text{candidate}}$

12:         **end if**

13:      **else if** $\forall i \neq h\colon f(\beta_{\text{candidate}}) > f(\beta_i)$ **then**

14:         $\beta_{\text{candidate}} \leftarrow \beta_{\text{centroid}} + \beta(\beta_h - \beta_{\text{centroid}})$

15:         **if** $f(\beta_{\text{candidate}}) \leq f(\beta_h)$ **then**

16:             $\beta_h \leftarrow \beta_{\text{candidate}}$ {Contraction}

17:         **else**

18:             **for** $i = 1$ **to** $n+1$ **do**

19:                 $\beta_i \leftarrow \beta_l + 0.5(\beta_i - \beta_l)$ {Reduction}

20:             **end for**

21:         **end if**

22:      **else**

23:         $\beta_h \leftarrow \beta_{\text{candidate}}$

24:      **end if**

25: **end while**

for the starting vector $\beta^{(1)} = [-0.5\ 0.5]^T$ is the first corner of the following simplex:

Corner	$\beta_{i,1}$	$\beta_{i,2}$
1	-0.500	0.500
2	-0.017	0.629
3	-0.371	0.983

It is clear that the flexible polyhedron, in contrast to a rigid simplex, can adapt to the topography of the objective function. It will expand along long

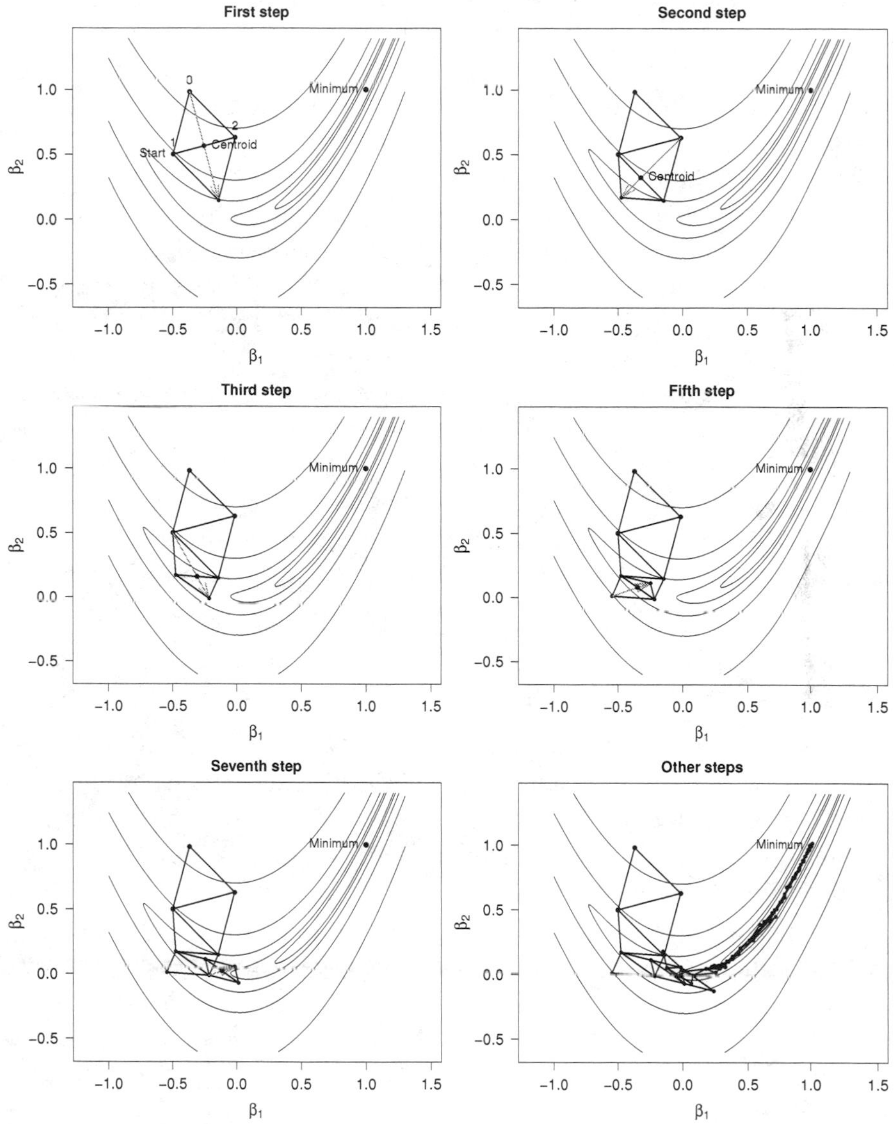

Figure 4.20: Construction of Nelder-Mead procedure in several steps.

inclined planes, change direction in curvy valleys, and can contract when it is close to a minimum (see contour plots in Figure 4.20).

What we have neglected so far is the choice of parameters α, β and γ. Once the size of the flexible polyhedron has adapted to the local landscape it should not change until the topography of the problem has changed. This is only possible if we set $\alpha = 1$. Moreover, Nelder and Mead demonstrated that $\alpha = 1$ requires fewer function evaluations than $\alpha < 1$ in their example cases. On the other hand, α should not be much greater than 1, since smaller values of α lead to a faster adaptation of the flexible polyhedron to the local problem topology. This is especially true if it is necessary to change direction in a curvy valley. Finally, for large values of α, we would need many steps to reduce the size of the polyhedron in situations close to a minimum. Overall, $\alpha = 1$ appears to be the right compromise.

In order to study the influence of β and γ on the progress of the search, Nelder and Mead solved test problems with a large number of different combinations of values for β and γ. Nelder and Mead then recommended the values $\alpha = 1$, $\beta = 0.5$, and $\gamma = 2$ for a general unrestricted problem.

Nelder and Mead also found that the size and orientation of the starting polyhedron influence the number of function evaluations. However, the adequate choice of α, β, and γ appeared to be more important.

4.4.4 Summary

In this section we introduced a general framework for nonlinear multivariate convex and nonconvex optimization. We discussed methods for differentiable and non-differentiable objective functions. The most prominent representatives are quasi-Newton methods in the convex differentiable case, the Levenberg-Marquardt method in the (nonconvex) nonlinear least-squares case, and the Nelder-Mead method in the non-differentiable case. We compared the behavior of the methods on a nonlinear least-squares example showing that specialization may, on the one hand, help in that the Levenberg-Marquardt method is fastest. It may, however, on the other hand, also lead to unexpected behavior, again illustrated by the Levenberg-Marquardt method, which shows nearly no dependence on one of the coefficient values.

4.5 Example: Neural Nets

Let us continue with a famous but somewhat involved example of a nonlinear model where nonlinear least squares estimation is typically used in a special way, namely, the Artificial Neural Net (ANN).

4.5.1 Concept[5]

Definition 4.7: Artificial Neural Net (ANN)
An **Artificial Neural Net (ANN)** consists of a set of processing units, the so-called **nodes** simulating neurons, which are linked by a set of **weights** analogous to the synaptic connections in the nervous system. The nodes represent very simple calculation components based on the observation that a neuron behaves like a switch: if sufficient neurotransmitters have been accumulated in the cell body, an action potential is generated. This potential is mathematically modeled as a weighted sum of all signals reaching the node, and is compared to a given limit. Only if this limit is exceeded, is the node "firing".

That a neural net can model complex interrelations is, similar to a computer, not enabled by the complexity of a single processing unit, but by the density and complexity of the connections. Concerning **computability**, neural nets are an equally powerful model as the Turing machine (see `http://www.math.rutgers.edu/~sontag/FTP_DIR/aml-turing.ps.gz`).

In contrast to **conventional computers**, the storage of a neural net is distributed over the whole structure and is modified by experience, i.e. by so-called learning from new observations. In comparison, conventional computers work with fixed programs on very complex central processing units. Structurally, an ANN is more comparable to a **natural (biological) neural net** as, e.g., the human brain.

Definition 4.8: Multilayer Networks
The most well-known neural net is the so-called **Multilayer Neural Network** or **Multilayer Perceptron**. Such a network can be hierarchically organized into **layers** of neurons, namely the input layer, the interior layers, and the output layer. A **feed-forward net** only allows for signals in one direction, namely, from the input nodes toward the output nodes. As in any neural network, in a multilayer network a weight is assigned to every connection between two nodes. These weights represent the influence of the input node on the output node.

In what follows, we will only consider special multilayer neural networks, namely, networks with only one interior (or hidden) layer (see Figure 4.21). Please note that sometimes the weighted connections rather than the neurons are called layers. Then, the network in the figure would be called a two layer feed-forward network instead of a three layer feed-forward network.

[5]Partly based on Cross et al. (1995) and Hwang and Ding (1997).

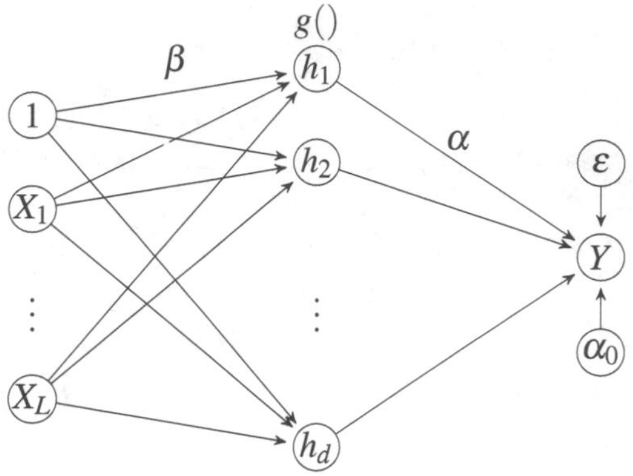

Figure 4.21: Model of a multi-layer neural network. There are L input neurons and d hidden neurons in one hidden layer.

In an artificial neural network (ANN), linear combinations of the input signals $X_1, \ldots, X_L$ with individual weights β_l are used as input for each node of the hidden layer. Each node then transforms this input signal using an **activation function** g to derive the output signal. These output signals are then again linearly combined with weights α_i to determine the value y of the only node in the output layer to arrive at the output signal Y. In addition to the input signals $X_1, \ldots, X_L$, a constant term α_0, the so-called **bias**, is added to the output, analogous to the intercept term of the linear model. As usual, we assume that the model has an additive stochastic term ε that complements the deterministic part of the model. We assume that this error term has zero mean and finite variance.

Definition 4.9: Activation Function
The **activation function** is generally not chosen as a jump function "firing" only beyond a fixed activation potential, as originally proposed, but as a symmetrical sigmoid function with the following properties:

$$\lim_{x \to -\infty} g(x) = 0, \tag{4.38}$$

$$\lim_{x \to \infty} g(x) = 1, \tag{4.39}$$

$$g(x) + g(-x) = 1. \tag{4.40}$$

A popular choice for the activation function is the **logistic function** (see Figure 4.22):

$$g(x) = \frac{1}{1 + e^{-x}}. \tag{4.41}$$

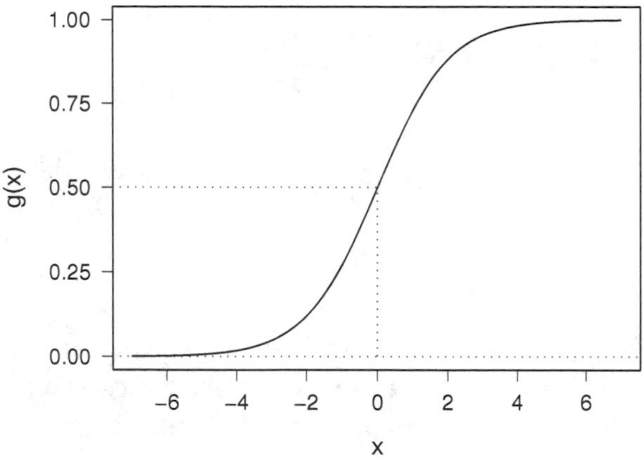

Figure 4.22: Logistic activation function.

Another obvious choice for the activation function is the cumulative distribution function of any symmetrical distribution.

Definition 4.10: Statistical Model for Neural Nets
The model corresponding to the multilayer network with one hidden layer has the form:

$$Y = \alpha_0 + \sum_{i=1}^{d} \alpha_i g(\beta_i^T \boldsymbol{X} + \beta_{i0}) + \varepsilon =: f(\boldsymbol{X}; \boldsymbol{\Theta}) + \varepsilon, \qquad (4.42)$$

where $\boldsymbol{X} = [X_1 \ \dots \ X_L]^T$ is the vector of input signals, $\beta_i = [\beta_{i1} \ \dots \ \beta_{iL}]^T$ is the vector of the weights of the input signals for the ith node of the hidden layer, β_{i0} is the input weight of the constant, $\alpha = [\alpha_1 \ \dots \ \alpha_d]^T$ is the vector of the weights of the output signals of the nodes of the hidden layer, α_0 is the bias, and ε is a random variable with expected value 0.

The unknown **model coefficients** of this model are:

$$\boldsymbol{\Theta} = [\alpha_0 \ \dots \ \alpha_d \ \beta_{10} \ \dots \ \beta_{d0} \ \beta_1^T \ \dots \ \beta_d^T]^T.$$

They have to be estimated (as statisticians would say) or learned (in the language of neural networks and machine learning). The process of learning the coefficients of a neural network is also referred to as the **training** of the network, whereas predictions are then used to **test** the network.

4.5.2 *Coefficient Estimation with Neural Nets* [6]

The model coefficients are estimated by means of the **nonlinear least squares method** (cp. Section 4.4.2.1).

We have to minimize the following criterion function corresponding to Θ:

$$\sum_{j=1}^{m} (f(x_j;\Theta) - y_j)^2. \tag{4.43}$$

The established method used to estimate the parameters is called **backpropagation**. At least two variants are in widespread use. The so-called **batch backpropagation** is nothing more than the well-known steepest descent method (cp. Section 4.4.1.3) using all observations for the calculation of the criterion and its gradient in the inner minimization step. In contrast, in **online backpropagation** the observations are individually presented to the neural network and the weights of the nodes are continuously updated as new observations pass through the network. The gradient is therefore only calculated using the jth error term $(f(x_j;\Theta) - y_j)^2$. Then, the change in the coefficients after the presentation of the complete set of observations only approximately corresponds to one iteration step of the batch backpropagation, since during the step Θ is changed after every observation. It is clear that the observations have to be presented several times to fully reflect the steepest descent method. We may want to permute the observations periodically so that they are presented to the network in a different order in each iteration. In other variants, each iteration is based on a different random subset of the observations. This variant is called the **mini-batches** method, representing a compromise between the online and the batch optimization.

The derivation of the online backpropagation rule is left as an exercise to the reader. It is based on a clever application of the chain rule of differential calculus. Instead, we will concentrate on the properties of neural networks.

The **online backpropagation rule** has the form:

$$\Theta_{pq}^{(j;k)} = \Theta_{pq}^{(j-1;k)} + t^{(j;k)} o_p^{(j;k)} \delta_q^{(j;k)}, \tag{4.44}$$

where:

- The lower indices of the coefficients $\Theta_{pq}^{(j;k)}$ indicate that the link between the nodes p and q is represented for any numbering of the nodes, where node q is in a layer that directly follows the layer of node p. The upper indexes indicate that the jth observation is presented the kth time to the net (i.e. we are in the kth iteration).

---

[6] Partly based on Zell (1995, pp. 105–114).

– $t^{(j;k)}$ is a step size, called **learning rate**, which should be chosen as adequately as possible.
– $o_p^{(j;k)}$ is the output of the pth node for the jth observation in the kth iteration.
– $\delta_q^{(j;k)}$ is the negative error derivative, indicating how strong the model error in the kth iteration reacts to changes of the output of node q when presenting observation j. One can show that

$$\delta_q^{(j;k)} = \left\{ \begin{array}{ll} o_q^{(j;k)}(1-o_q^{(j;k)})(y_{jq}-o_q^{(j;k)}), & \text{if } q \text{ is an output node,} \\ o_q^{(j;k)}(1-o_q^{(j;k)})\sum_N \delta_N^{(j;k)} \Theta_{qN}^{(j;k)}, & \text{if } q \text{ is a hidden node} \end{array} \right\},$$

where:

- y_{jq} is the true value of the output of the output node q for observation j.
- The index N runs through all nodes succeeding the hidden node q.

Please note that this formulation of the backpropagation rule allows more than one hidden layer and more than one output node. For our net with only one hidden layer and one output node the index N in the previous formula only represents the one output node, and the true value y_{jq} of the output of the one output node for observation j is just referred to as y_j.

So far, nothing was said about the order in which the weights are updated. This is, however, specific for the algorithm, and even giving the method its name.

Indeed, in the first step in iteration k, observation j is "shifted forward" through the net with the actual weights $\Theta_{pq}^{(j-1;k)}$. This leads to the outputs $o_q^{(j;k)}$ of all nodes. With these, we can calculate the error derivatives of the output nodes. Then, we step backwards and calculate the error derivatives of the hidden nodes, especially from the error derivatives of the output nodes. This leads to the term **backward propagation** or **backpropagation**, since the error derivatives propagate backwards through the net, and the updating of the weights also runs backwards, starting with the edges leading to output nodes.

Shape of the Criterion Function

The error function is highly nonlinear and nonconvex in Θ because of the activation function. This implies that every descent method will probably only find a **local minimum** of the criterion function.

In practice, there are some tricks to handle the problem of local minima. E.g., we could start the estimation method repeatedly with different starting vectors. However, nothing can guarantee the efficient (or even the successful)

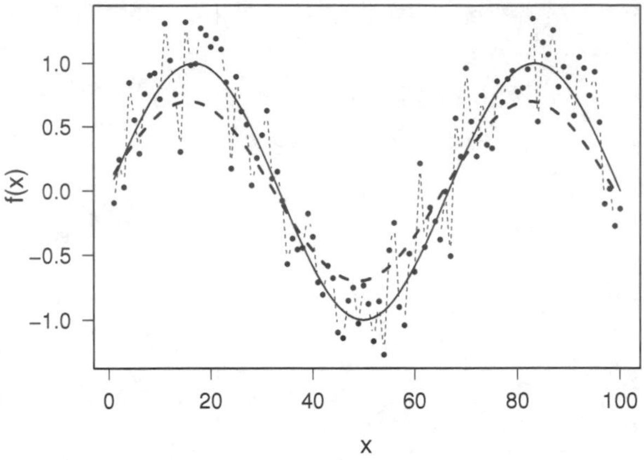

Figure 4.23: Noisy sine function (solid line). Overfitting: Interpolating function with zero errors (thin dashed line). Underfitting: Badly approximating function (broken line). We look for something "adequate inbetween".

finding of the global optimum. Most of the time one even has no information about the realized distance of the found local optimum to the global optimum.

Interpolation or Approximation?
Neural nets are a very flexible nonparametric regression method. Indeed, neural nets are **universal approximators** (see below) and approximate any unknown relationship between x and y. The question is: What should be optimized? In general, we are not really looking for a function which reduces the criterion $\frac{1}{m}\sum_{j=1}^{m}(f(x_j;\Theta)-y_j)^2$ (nearly) to 0! The solving of the nonlinear least squares problem is only a surrogate for the minimization of the expected value $E_{X,Y}[(f(X,\Theta)-Y)^2]$ for new (!) data, where X and Y are random variables with realizations x_i and y_i.

Universal Approximation
For neural nets we can show that they can approximate any smooth function arbitrarily well if the number of nodes in the hidden layer is not limited. This is called universal approximation property of neural nets.

Positively said, we can represent every interesting function by an ANN. Negatively said, if too many neurons are chosen, there is the chance to nearly interpolate, i.e. to also explain, the noise and not only the underlying function (cp. Figure 4.23).

Hold-Out Assessment
Neural nets tend to overfit because of their already mentioned complexity. One way out is the error assessment on a so-called **hold-out** set, not used

during training (see Section 7.4). If this error is increasing after a while, then training is stopped.

Increase of Speed of Convergence?
From the standpoint of a numerical mathematician, backpropagation should be replaced by a faster converging method like, e.g., the Levenberg-Marquardt algorithm in Section 4.4.2.1. Moreover, online backpropagation has the same convergence problems as the standard batch steepest descent (cp. Section 4.4.1.3). Therefore, more advanced optimization methods are often used in practice in order to reduce training time.

Efficiency on Large Data Sets
However, especially for large data sets, (online) backpropagation can be much more efficient than an advanced (second-order) optimization algorithm, because the calculation of a complete update step scales only linearly with the number of parameters. For second-order methods scaling is quadratic or even cubic, because of the inversion of the Jacobian matrix term, e.g. for the Levenberg-Marquardt method.

4.5.3 Identifiability of Neural Nets

Unfortunately, the model function of the neural nets yields an important problem, since model (4.42) is **not identifiable** in the sense that there exist several coefficient vectors $\mathbf{\Theta}$ with the same value of the **model function** $f(\mathbf{X};\mathbf{\Theta})$. Identifiability is a general problem of nonlinear models. Therefore, we will discuss it here in more detail using neural nets to illustrate the concept.

Unfortunately, **neural nets with the logistic activation function (4.41) are never identifiable**, and therefore interpretation of the estimated coefficients of a model (4.42) should be avoided in any case.

In the following, two concepts of identifiability will be discussed that were particularly introduced for neural nets: namely redundancy and reducibility.

Definition 4.11: Redundancy and Reducibility
A neural net is called **redundant** if another neural net exists with less nodes and identical values of the model function $f(X;\mathbf{\Theta})$. A coefficient vector $\mathbf{\Theta} = [\alpha_0 \ldots \alpha_d\ \beta_{10} \ldots \beta_{d0}\ \beta_1^T \ldots \beta_d^T]^T$ is called **reducible**, if one of the following three conditions is valid for $i \neq 0$ and $j \neq 0$:

1. $\alpha_i = 0$ for any $i = 1, \ldots, d$.
2. $\beta_i = \mathbf{0}$ for any $i = 1, \ldots, d$, $\mathbf{0}$ = null vector of length L.
3. $(\beta_{i0}, \beta_i^T) = \pm(\beta_{j0}, \beta_j^T)$ for any $i \neq j$.

These two notions are related if we are using the logistic activation function.

Theorem 4.4: Reducibility with Logistic Activation Function
If Θ is reducible and g is the logistic activation function (4.41), then the corresponding neural network is redundant.

Proof. We handle the three conditions of reducibility:

1. The ith neuron of the hidden layer can be eliminated without changing the value of f.
2. If the ith neuron of the hidden layer is eliminated and α_0 is replaced by $\alpha_0 + \alpha_i g(\beta_{i0})$, then the value of f is not changed.
3. If $(\beta_{i0}, \beta_i^T) = (\beta_{j0}, \beta_j^T)$, then α_j is replaced by $\alpha_j + \alpha_i$. If $(\beta_{i0}, \beta_i^T) = -(\beta_{j0}, \beta_j^T)$, then α_j is replaced by $\alpha_j - \alpha_i$ and α_0 by $\alpha_0 + \alpha_i$. In both cases the value of f is not changed. Indeed, in the second case the use of the logistic activation function g implies

$$\begin{aligned}\alpha_i g(\beta_i^T x + \beta_{i0}) &= \frac{\alpha_i}{1 + e^{-(\beta_i^T x + \beta_{i0})}} \\ &= \frac{\alpha_i e^{(\beta_i^T x + \beta_{i0})}}{e^{(\beta_i^T x + \beta_{i0})} + 1} = \frac{\alpha_i (1 + e^{(\beta_i^T x + \beta_{i0})}) - \alpha_i}{1 + e^{(\beta_i^T x + \beta_{i0})}} \\ &= \alpha_i - \alpha_i g(-\beta_i^T x - \beta_{i0}) = \alpha_i - \alpha_i g(\beta_j^T x + \beta_{j0}).\end{aligned}$$

□

On the other hand, a not reducible Θ does not necessarily lead to a non-redundant neural network. A sufficient condition for this will be given later.

Unfortunately, redundancy is not synonymous with nonidentifiability, since several neural nets with the same number of neurons can lead to the same values of the model function $f(X;\Theta)$:

Theorem 4.5: Nonidentifiability of Neural Nets
Neural networks with a logistic activation function (4.41) are never identifiable, since there are two kinds of transformations of the model coefficients Θ that leave the model function invariant:

(i) permutations of the neurons, respectively of the coefficients $\mu_i = [\alpha_i\ \beta_{i0}\ \beta_i^T]^T$ of the neurons and

(ii) the transformations

$$(\alpha_0, \mu_1, \ldots, \mu_i, \ldots, \mu_d) \to (\alpha_0 + \alpha_i, \mu_1, \ldots, -\mu_i, \ldots, \mu_d).$$

Proof.

1. Is clear.
2. Follows from $\alpha_i g(\beta_i^T x + \beta_{i0}) = \alpha_i - \alpha_i g(-\beta_i^T x - \beta_{i0})$ as in 3. of the previous proof. □

Serial connection of these transformations results in $2^d d!$ transformations τ with $f(X;\Theta) = f(X;\tau(\Theta))$.

However, for certain activation functions one can show that 1. and 2. are the only transformations leaving the model function invariant. Decisive for this is that the contribution of any neuron to the model function cannot be represented by the contributions of the other neurons and of the model constant α_0. That means any neuron delivers an innovative contribution to the model function. Condition 1 formulates the corresponding requirements concerning the activation function g.

Condition 1: The class of functions $\{g(bx + b_0), b > 0\} \cup \{g \equiv 1\}$ is linearly independent. In more detail, for every $d \in \mathbb{N}$ and scalars $a_0, a_i, b_{i0} \in \mathbb{R}$ and $b_i > 0$, $i = 1, \ldots, d$, with $(b_i, b_{i0}) \neq (b_j, b_{j0})$ for every $i \neq j$:

$$a_0 + \sum_{i=1}^{d} a_i g(b_i x + b_{i0}) = 0 \qquad \forall x \in \mathbb{R} \qquad \Rightarrow a_0 = a_1 = \cdots = a_d = 0.$$

Notice that the condition $b > 0$ does not cause any restriction since $\alpha_i g(\beta_i^T x + \beta_{i0}) = \alpha_i - \alpha_i g(-\beta_i^T x - \beta_{i0})$.

A further reaching condition on the activation function additionally guarantees the invertibility of the covariance matrix of the model coefficients and thus, at least asymptotically, of an adequate estimator of this covariance matrix. An estimator for the covariance matrix is, e.g., required for the determination of prediction intervals (cp. Section 7.4).

Condition 2: Let g be differentiable and g' its first derivative. The class of functions

$$\{g(bx + b_0), g'(bx + b_0), xg'(bx + b_0), b > 0\} \cup \{g \equiv 1\}$$

is linearly independent, i.e. for every $d \in \mathbb{N}$ and scalars $a_0, a_i, e_i, f_i, b_{i0} \in \mathbb{R}$ and $b_i > 0$, $i = 1, \ldots, d$, with $(b_i, b_{i0}) \neq (b_j, b_{j0})$ for every $i \neq j$:

$$a_0 + \sum_{i=1}^{d} [a_i g(b_i x + b_{i0}) + e_i g'(b_i x + b_{i0}) + f_i x g'(b_i x + b_{i0})] = 0 \; \forall x \in \mathbb{R}$$

$$\Rightarrow a_0 = a_1 = \cdots a_d = e_1 = \cdots = e_d = f_1 = \cdots = f_d = 0.$$

If the coefficients vector Θ is not reducible, then Condition 1 guarantees identifiability except for the above transformations, and Condition 2 even guarantees invertibility of the covariance matrix of the coefficients.

Theorem 4.6: Identifiability of the Model and Invertibility of the Covariance Matrix of the Coefficients
For model (4.42) of neural nets with symmetrical sigmoid activation function g as in (4.38–4.40) let condition 1 be valid. Let the vector of model coefficients Θ be not reducible. Moreover, let X be a vector of random variables with range $\mathbb{R}^L$. Then:

1. Θ is identifiable, except for the above inevitable transformations τ, i.e. if there is a Θ^* with $f(X;\Theta) = f(X;\Theta^*)$, then there is an inevitable transformation τ so that $\Theta^* = \tau(\Theta)$.
2. If g is additionally differentiable and condition 2 is valid, then matrix $B_E := \mathrm{E}\left[\frac{\partial f}{\partial \Theta}(X;\Theta)^T \frac{\partial f}{\partial \Theta}(X;\Theta)\right]$ is invertible, where $\frac{\partial f}{\partial \Theta}(X;\Theta)^T$ is the gradient vector dependent on the random vector X.

Proof. See Hwang and Ding (1997). □

The first part of this result implies the following corollary, giving a sufficient condition for non-redundancy of the model.

Corollary 4.1: Non-Redundancy of the Model
Under condition 1 the non-reducibility of the model coefficients guarantees the non-redundancy of the model.

Therefore, reducibility and redundancy are equivalent if the logistic activation function satisfies condition 1. Indeed, it is even satisfying condition 2, as well as some important distribution functions of symmetrical distributions, e.g. the standard normal distribution.

Theorem 4.7: Activation Functions Satisfying Condition 2
The distribution function of the standard normal distribution $\mathcal{N}(0,1)$ and the logistic function (4.41) satisfy condition 2.

Proof. See Hwang and Ding (1997, pp. 754 – 756). □

Here, this concludes our discussion of neural nets. In Section 7.4, we will come back to neural nets considering the following topics:

- We will show that neural nets are totally unproblematic for prediction, if their activation function satisfies condition 2, though the coefficients of the neural nets are not interpretable.

– We will give a procedure for identifying the number of nodes in the hidden layer of a neural net.

4.5.4 *Summary*

In this section we discussed (artificial) neural networks as a prominent example for a nonlinear least-squares problem. We discussed specialized estimation methods for this model and the nonidentifiability of the model coefficients. Notice that nonidentifiability is a general problem, at least in nonlinear models.

4.6 Constrained Optimization

In this section we will show how the idea of iteration can be applied to the problem of solving constrained optimization problems. For this, we assume the reader to have a rudimentary understanding of Lagrangian duality, which is crucial when dealing with a constrained optimization problem. Recall that for a solution β^* to a constrained optimization problem

$$\begin{aligned} \underset{\beta \in \mathbb{R}^n}{\text{minimize}} \quad & f(\beta) \\ \text{subject to} \quad & \boldsymbol{g}(\beta) \leq \mathbf{0}, \\ & \boldsymbol{h}(\beta) = \mathbf{0} \end{aligned}$$

to be optimal, the Karush-Kuhn-Tucker (KKT) conditions and some suitable regularity[7] conditions must hold. These can be formulated as follows.

Suppose that the objective function $f\colon\ : \mathbb{R}^n \to \mathbb{R}$ and the constraint functions $g_i\colon \mathbb{R}^n \to \mathbb{R}$, $i = 1, \ldots, m$, and $h_j\colon \mathbb{R}^n \to \mathbb{R}$, $j = 1, \ldots, l$, are continuously differentiable at a point β^*. Let $\boldsymbol{g} := [g_1 \; \ldots \; g_m]^T$, $\boldsymbol{h} := [h_1 \; \ldots \; h_l]^T$ and $J_{\boldsymbol{g}} \in L(m,n)$, $J_{\boldsymbol{h}} \in L(l,n)$ the corresponding Jacobians . If β^* is a local minimum that satisfies some regularity conditions, then there exist constants $\boldsymbol{\mu} := [\mu_1 \; \ldots \; \mu_m]^T$ and $\boldsymbol{\lambda} := [\lambda_1 \; \ldots \; \lambda_l]^T$, called **Karush-Kuhn-Tucker (KKT) multipliers**, such that:

Stationarity:

$$\nabla f(\beta) + J_{\boldsymbol{g}}(\beta)^T \boldsymbol{\mu} + J_{\boldsymbol{h}}(\beta)^T \boldsymbol{\lambda} = 0,$$

Primal feasibility:

$$\boldsymbol{g}(\beta) \leq \mathbf{0} \quad \text{and} \quad \boldsymbol{h}(\beta) = \mathbf{0},$$

[7]For convex problems one set of regularity conditions is the Slater conditions. They require that there is a strictly feasible point β.

Dual feasibility:

$$\mu \geq 0,$$

Complementary slackness:

$$\mu^T g(\beta) = 0.$$

If we only have equality constraints, these reduce to the well-known optimality conditions obtained using Lagrange multipliers.

4.6.1 Equality-Constrained Optimization

Let us begin by restricting ourselves to equality-constrained problems.

Definition 4.12: Equality-Constrained Optimization Problem
A convex equality-constrained optimization problem with linear constraints is given by

$$\begin{aligned} \underset{\beta \in \mathbb{R}^n}{\text{minimize}} \quad & f(\beta) \\ \text{subject to} \quad & A\beta = b, \end{aligned} \tag{4.45}$$

where $f\colon \mathbb{R}^n \to \mathbb{R}$ is convex, $A \in \mathbb{R}^{l \times n}$ and $b \in \mathbb{R}^l$.

In the following, we will additionally assume f to be differentiable. We then know that for a solution β^* to (4.45) the following must hold for some vector λ^*:

$$\nabla f(\beta^*) + A^T \lambda^* = 0 \quad \text{and} \quad A\beta^* = b. \tag{4.46}$$

This requires solving a system of $n + l$ equations of which n are nonlinear in general. For the special case that f is quadratic and convex, that is, of the form

$$f(\beta) = \frac{1}{2}\beta^T P \beta + q^T \beta + r$$

with $P \in \mathbb{R}^{n \times n}$ positive definite (p.d.), $q \in \mathbb{R}^n$, and $r \in \mathbb{R}$, the optimality conditions from (4.46) are given by

$$P\beta^* + q + A^T \lambda^* = 0, \quad \text{and} \quad A\beta^* = b.$$

We can obtain $(\beta\ \lambda)^T$ by solving

$$\begin{bmatrix} P & A^T \\ A & 0 \end{bmatrix} \begin{bmatrix} \beta \\ \lambda \end{bmatrix} = \begin{bmatrix} -q \\ b \end{bmatrix} \tag{4.47}$$

in $\boldsymbol{\beta}$ and $\boldsymbol{\lambda}$. The matrix

$$\boldsymbol{K} = \begin{bmatrix} \boldsymbol{P} & \boldsymbol{A}^T \\ \boldsymbol{A} & \boldsymbol{0} \end{bmatrix}$$

is often referred to as the **KKT** matrix in reference to the Karush-Kuhn-Tucker conditions that an optimal solution must satisfy. Solving this system can be done both analytically as well as by numerical methods, depending on the size and structure of $\boldsymbol{K}$.

To arrive at an iterative method for the general case where f is convex but not quadratic, we use the same trick we used for the multivariate optimization problem (see Equation 4.12). We again arrive at

$$\nabla f(\boldsymbol{\beta}^{(k+1)}) \cong \nabla f(\boldsymbol{\beta}^{(k)}) + \nabla^2 f(\boldsymbol{\beta}^{(k)})(\boldsymbol{\beta}^{(k+1)} - \boldsymbol{\beta}^{(k)}). \tag{4.48}$$

By combining (4.48) and (4.47) we obtain as our Newton step for a feasible $\boldsymbol{\beta}^{(k)}$

$$\begin{bmatrix} \Delta\boldsymbol{\beta} \\ \boldsymbol{\lambda} \end{bmatrix} = \begin{bmatrix} \nabla^2 f(\boldsymbol{\beta}^{(k)}) & \boldsymbol{A}^T \\ \boldsymbol{A} & 0 \end{bmatrix}^{-1} \begin{bmatrix} -\nabla f(\boldsymbol{\beta}^{(k)}) \\ 0 \end{bmatrix}, \tag{4.49}$$

which we can plug into Algorithm 4.7 to solve any convex equality-constrained minimization problem with linear constraints and twice differentiable objective function f if we are given an initial feasible point $\boldsymbol{\beta}^{(1)}$ from which to start the iteration.

The second set of equations in (4.49) may not be immediately intuitive, but if we rewrite it as

$$\boldsymbol{A}\Delta\boldsymbol{\beta} = 0$$

it is clear that its purpose is to ensure that our new intermediate solution $\boldsymbol{\beta}^{(k)} + \Delta\boldsymbol{\beta}$ is still feasible.

The convergence properties of this algorithm are the same as those for the Newton method for unconstrained minimization. In particular, its convergence for $\boldsymbol{\beta}^{(k)}$ close to $\boldsymbol{\beta}^*$ will be quadratic.

Finding an initial solution can be thought of as an LLS type problem and can be solved using any of the methods discussed in the previous chapter. Another possibility is to use the so-called infeasible start Newton method. There, the Newton step is calculated as follows:

$$\begin{bmatrix} \Delta\boldsymbol{\beta} \\ \boldsymbol{\lambda} \end{bmatrix} = \begin{bmatrix} \nabla^2 f(\boldsymbol{\beta}^{(k)}) & \boldsymbol{A}^T \\ \boldsymbol{A} & 0 \end{bmatrix}^{-1} \begin{bmatrix} -\nabla f(\boldsymbol{\beta}^{(k)}) \\ -\boldsymbol{A}\boldsymbol{\beta}^{(k)} + \boldsymbol{b} \end{bmatrix}. \tag{4.50}$$

Here, we want to minimize both the quadratic approximation of f and the gap between $\boldsymbol{A}\boldsymbol{\beta}$ and $\boldsymbol{b}$. Once we hit a feasible solution $\boldsymbol{\beta}^{(k)}$ for some k, all further iterations reduce to the regular Newton method for equality-constrained problems. Since we can only approximately find a feasible solution, this will seldom happen in practice. This leads to more complex termination criteria for this method, and care should therefore be taken when implementing it.

For those familiar with duality theory, it will be obvious that the infeasible Newton method is a primal-dual method because it simultaneously improves the primal ($\boldsymbol{\beta}$) and the dual ($\boldsymbol{\lambda}$) solution. For a much more in-depth discussion of the methods presented here as well as the duality theory that motivates them and gives further insight into their convergence guarantees, see the excellent book by Boyd and Vandenberghe (2004, chap. 5, 10, and 11), on which parts of this and the following subsection are based.

4.6.2 Inequality-Constrained Optimization

By broadening Definition 4.12 to include convex inequality constraints we arrive at inequality constraint optimization. Such problems are defined as follows.

Definition 4.13: Inequality-constrained Optimization Problem
A convex inequality-constrained optimization problem with additional linear equality constraints is given by

$$\begin{aligned} \underset{\boldsymbol{\beta}\in\mathbb{R}^n}{\text{minimize}} \quad & f(\boldsymbol{\beta}) \\ \text{subject to} \quad & \boldsymbol{A}\boldsymbol{\beta} = \boldsymbol{b}, \\ & \boldsymbol{g}(\boldsymbol{\beta}) \leq 0, \end{aligned} \tag{4.51}$$

where $f\colon \mathbb{R}^n \to \mathbb{R}$ and $\boldsymbol{g}\colon \mathbb{R}^n \to \mathbb{R}^m$ are convex, $\boldsymbol{A} \in \mathbb{R}^{l\times n}$, $\boldsymbol{b} \in \mathbb{R}^l$.

There is a wealth of literature, both theoretical and practical, which covers the characterization and solution of problems that fall under Definition 4.13. Solving (4.51) is what is referred to as convex optimization. Depending on the structure of f and $\boldsymbol{g}$ there may be specialized algorithms that are more efficient than what follows. It is therefore wise to check if such an algorithm exists before attempting to solve a general convex optimization problem with the following method. On the other hand, many of the specialized methods are not very different from what follows, so the loss in performance does not have to be as dramatic as, say, not using the (known) derivatives of f and $\boldsymbol{g}$ to guide the optimization process.

The class of iterative solvers we will briefly explore in the following are called **interior-point methods**. This name comes from the fact that they move along a path inside the feasible region defined by the constraints towards the optimal solution $\boldsymbol{\beta}^*$. The theory behind these methods is quite involved and requires a good grasp of duality theory, which we will not cover here. For a gentle introduction to duality theory and the theory behind interior point methods we refer the reader to Boyd and Vandenberghe (2004, chap. 11). Skipping over a few technical details, the main idea of interior point methods is to augment the objective function f with a **barrier** defined by the constraints to hinder the optimization algorithm from moving outside of the feasible region. We then solve this relaxed problem, ignoring the constraints using a Newton or quasi-Newton method for equality-constrained optimization.

An obvious choice for the barrier function would be

$$\varphi_I(\boldsymbol{\beta}) := \sum_{i=1}^{m} \begin{cases} 0 & g_i(\boldsymbol{\beta}) \leq 0 \\ 1 & \text{else} \end{cases},$$

i.e. the indicator function for the constraint is 1 iff the constraint is violated. Here we denote with $g_i(\boldsymbol{\beta})$ the ith entry in the vector $\boldsymbol{g}(\boldsymbol{\beta})$. We would then solve

$$\begin{aligned} \underset{\boldsymbol{\beta}\in\mathbb{R}^n}{\text{minimize}} \quad & f(\boldsymbol{\beta}) + \varphi_I(\boldsymbol{\beta}) \\ \text{subject to} \quad & \boldsymbol{A}\boldsymbol{\beta} = \boldsymbol{b} \end{aligned}$$

which is similar to Problem 4.51 except that the inequality constraints have been replaced by the barrier function φ_I.

The problem with this barrier function is that it leads to an objective function that is neither continuous nor differentiable. It is therefore customary to approximate the indicator function in the feasible region with the so-called **logarithmic barrier**:

$$\varphi(\boldsymbol{\beta}, t) := -\frac{1}{t}\sum_{i=1}^{m} \log(-g_i(\boldsymbol{\beta})).$$

Here t is a tuning constant. The larger t, the better is the approximation of φ_I. An example with different values of t is shown in Figure 4.24. Another interpretation of t is the relative weight of the barrier function compared to the objective function.

$$\begin{aligned} \underset{\boldsymbol{\beta}\in\mathbb{R}^n}{\text{minimize}} \quad & f(\boldsymbol{\beta}) + \varphi(\boldsymbol{\beta}, t) \\ \text{subject to} \quad & \boldsymbol{A}\boldsymbol{\beta} = \boldsymbol{b} \end{aligned} \tag{4.52}$$

This optimization problem is not only equality-constrained, but also still

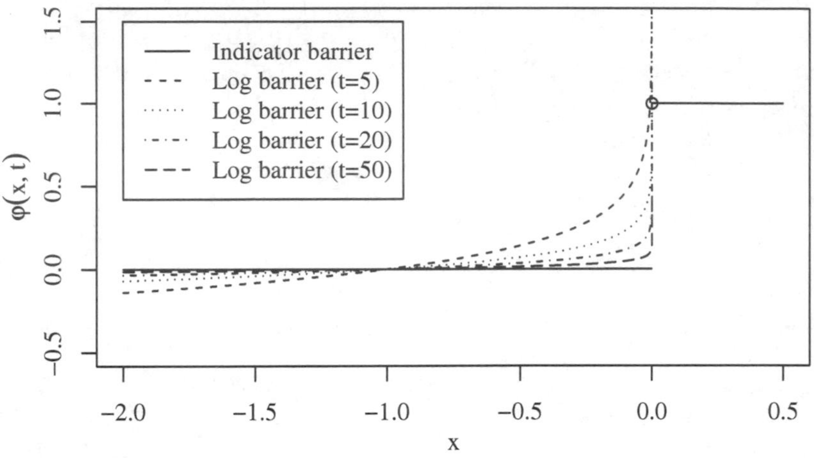

Figure 4.24: Plot showing the approximation quality of the log barrier function to the step barrier for different values of t. Here we have only one constraint.

twice differentiable if f and $\boldsymbol{g}$ are twice differentiable. We can therefore solve it efficiently using either of the Newton methods we discussed in the previous section. If we have no equality constraints, we can even use one of the quasi-Newton methods for unconstrained minimization.

If we denote with $\boldsymbol{\beta}^*(t)$ the optimal solution to (4.52) and view it as a function of t, we arrive at the notion of a **central path**. As mentioned, t controls the relative weight of the objective function and the barrier during the optimization. If we start our unconstrained minimization from a feasible point $\boldsymbol{\beta}$, we expect the barrier function to trap the optimizer in the feasible region. One can even prove that as $t \to \infty$, $\boldsymbol{\beta}^*(t)$ converges toward the solution of (4.51) since the weight of the barrier decreases compared to the objective function, while at the same time still bounding the algorithm to the feasible region. This motivates the sequential unconstrained minimization Algorithm 4.10.

4.6.2.1 Applications

So, why is constrained optimization of interest to statisticians? We will try to illustrate this using two problems. First, we will revisit the problem of quantile estimation, and then look at one of the most popular classification methods of the last 10 years—the support vector machine (SVM).

Algorithm 4.10 Sequential Unconstrained Minimization

Require: $f\colon \mathbb{R}^n \to \mathbb{R}$ and $g\colon \mathbb{R}^n \to \mathbb{R}^m$ convex, $A \in \mathbb{R}^{l\times n}$, $b \in \mathbb{R}^l$, initial weight $t > 0$, and strictly feasible $\beta \in \mathbb{R}^n$

$k \leftarrow 0$
$\beta^{(k)} \leftarrow \beta$
repeat
  $k \leftarrow k+1$
  $\beta^{(k)} \leftarrow \beta^*(t)$ {solve using (quasi-)Newton method starting from $\beta^{(k-1)}$}
  $t \leftarrow 50t$
until $k/t < \varepsilon$ {Stop criterion}
return $\beta^{(k)}$

4.6.3 Linear Programming

In the introductory quantile estimation problem we saw that when we replaced the regular L2 loss function used for linear regression with the L1 loss function, we arrived at a problem that we could not solve. In this section, we will derive methods to solve problems like the one given in Equation 4.3 efficiently. Recall, given m observations $(x_i, y_i) \in \mathbb{R}^n \times \mathbb{R}$ we seek a solution to the problem

$$\underset{\beta\in\mathbb{R}^n}{\text{minimize}} \sum_{i=1}^{m} |x_i^T\beta - y_i|. \tag{4.53}$$

This problem is a convex optimization problem, but it is not differentiable. While the lack of a derivative means we cannot apply any of the powerful gradient based search methods we know, it does not prevent us from using a Nelder-Mead type algorithm to find an optimal β. Due to the convex nature of the problem, we can even be fairly certain that any solution returned by such an algorithm will be close to the true optimal solution.

In the next few sections we will see that we can rewrite the above problem into one that superficially looks more complicated but will turn out to be easily solvable. To simplify notation let

$$X := [x_1 \ \dots \ x_m]^T \quad \text{and} \quad y := [y_1 \ \dots \ y_m]^T.$$

Then we can rewrite (4.53) as

$$\begin{aligned} &\underset{(\beta,r)\in\mathbb{R}^{n+m}}{\text{minimize}} && \|r\|_1 \\ &\text{subject to} && X\beta - y = r. \end{aligned} \tag{4.54}$$

This problem does not look simpler. In fact, we now have m equality constraints we must satisfy. Before we can see why this transformation is useful, we have to perform another substitution. We will eliminate our residual vector $\boldsymbol{r}$ and replace it by the sum of two vectors that fulfill the constraints $\boldsymbol{r}^+ + \boldsymbol{r}^- = \boldsymbol{r}$, $\boldsymbol{r}^+ \geq 0$ and $\boldsymbol{r}^- \leq 0$. This leads to the even more complicated problem

$$\begin{aligned} \underset{(\boldsymbol{\beta},\boldsymbol{r}^+,\boldsymbol{r}^-)\in\mathbb{R}^{n+2m}}{\text{minimize}} \quad & \|\boldsymbol{r}^+ + \boldsymbol{r}^-\|_1 \\ \text{subject to} \quad & \boldsymbol{X\beta} - \boldsymbol{y} = \boldsymbol{r}^+ + \boldsymbol{r}^-, \\ & \boldsymbol{r}^+ \geq 0, \\ & \boldsymbol{r}^- \leq 0. \end{aligned} \tag{4.55}$$

Not only did we increase the number of parameters by another m, but we also added inequality constraints to our problem. Why go through all this trouble? Not only do we now have one equality and two inequality constraints, but we have also increased the number of variables over which we have to optimize by $2m$. The only bright side to all of this is that we can eliminate the norm in our objective function, which gives us

$$\begin{aligned} \underset{(\boldsymbol{\beta},\boldsymbol{r}^+,\boldsymbol{r}^-)\in\mathbb{R}^{n+2m}}{\text{minimize}} \quad & \mathbf{1}^T\boldsymbol{r}^+ - \mathbf{1}^T\boldsymbol{r}^- \\ \text{subject to} \quad & \boldsymbol{X\beta} - \boldsymbol{r}^+ - \boldsymbol{r}^- = \boldsymbol{y}, \\ & \boldsymbol{r}^+ \geq 0, \\ & \boldsymbol{r}^- \leq 0. \end{aligned} \tag{4.56}$$

This problem is equivalent to (4.55) since $|r_i^+ + r_i^-| \leq r_i^+ - r_i^-$ and one of the two entries r_i^+ or r_i^- can always be set to zero.

Why is this desirable? The objective function has suddenly become a simple linear function. In fact, if we did not have the restrictions, this problem would be unbounded below and we would be done. What we can infer from this is that the solution to this problem must lie somewhere on the boundary of the **feasible region**. We also observe that all of the constraints in problem (4.56) are linear in our parameter vector $[\boldsymbol{\beta} \ \ \boldsymbol{r}^+ \ \ \boldsymbol{r}^-]^T$. These types of problems where both the objective function as well as the constraints are linear are called **linear programming problems (LP problems)**. They arise frequently in many fields, including, as we just saw, statistics. There is a wealth of literature on how to solve them more or less efficiently. It is interesting to know that no polynomial time algorithm was known for the solution of LPs until 1979. In the LP literature many different representations and standardizations of LPs can be found. The most widely used form is the so-called **standard form**.

Definition 4.14: Standard Form Linear Program
A **linear program in standard form** is given by

$$\begin{aligned} \underset{\boldsymbol{x}\in\mathbb{R}^n}{\text{minimize}} \quad & \boldsymbol{c}^T\boldsymbol{x} \\ \text{subject to} \quad & \boldsymbol{A}\boldsymbol{x} \leq \boldsymbol{b}, \\ & \boldsymbol{x} \geq 0, \end{aligned} \tag{4.57}$$

where $\boldsymbol{A} \in \mathbb{R}^{m\times n}$, $\boldsymbol{b} \in \mathbb{R}^m$ and $\boldsymbol{c} \in \mathbb{R}^n$.

All linear programs can be transformed into standard form linear programs using a few simple transformations: negation of inequalities to change their direction, substitution of equality constraints by two inequalities and the trick used above to split a real-valued parameter into two parameters that only take positive values. Notice that the latter two transformations increase the size of the problem by adding more constraints or variables.

In practice, almost all modern software packages either do not require an LP in standard form or can do these transformations for you, but for the rest of this chapter we will assume that our LP is in standard form. We therefore only need to deal with inequality constraints.

We, nevertheless, will not completely transform problem (4.56) into the standard form. However, we will replace the equality with inequality constraints to arrive at

$$\begin{aligned} \underset{(\boldsymbol{\beta},\boldsymbol{r}^+,\boldsymbol{r}^-)\in\mathbb{R}^{n+2m}}{\text{minimize}} \quad & \boldsymbol{1}^T\boldsymbol{r}^+ - \boldsymbol{1}^T\boldsymbol{r}^- \\ \text{subject to} \quad & \boldsymbol{X}\boldsymbol{\beta} - \boldsymbol{r}^+ - \boldsymbol{r}^- \leq \boldsymbol{y}, \\ & \boldsymbol{X}\boldsymbol{\beta} - \boldsymbol{r}^+ - \boldsymbol{r}^- \geq \boldsymbol{y}, \\ & \boldsymbol{r}^+ \geq 0, \\ & \boldsymbol{r}^- \leq 0. \end{aligned} \tag{4.58}$$

Each of our inequality constraints partitions the search space $\mathbb{R}^{n+2m}$ into two half-spaces, the **feasible** region where the constraint is not violated and the **infeasible** region where we violate the constraint condition. If we denote the half-space that is feasible w.r.t. the ith constraint with F_i, then our solution must lie in the intersection of all of these half-spaces for the $2m$ constraints:

$$F := \bigcap_{i=1}^{2m} F_i.$$

Note that we ignore the additional $2m$ constraints on $\boldsymbol{r}^+$ and $\boldsymbol{r}^-$ here because it is customary to view them as direct constraints on the search space and not as generic linear constraints.

F is a, possibly unbounded, convex polytope. Since we know that our objective function is linear and we know that we are optimizing over a convex polytope, we can conclude that if there is a global minimum, i.e. the polytope is bounded in the direction of the negative gradient of our objective function, then it must be one of the vertexes of the polytope.

This insight gave rise to one of the popular methods for solving these types of problems. The key idea is that we can start at any vertex of the polytope and jump to one of the neighboring vertexes if it improves our objective function. If no neighboring vertex has a lower function value, we have found a global minimum for our problem under the condition that the polytope is bounded. This idea and the resulting class of algorithms go back to Dantzig (1951) and are confusingly also called **simplex algorithms**, but have nothing in common with the algorithms of Section 4.4.3.

In practice, there are many subtle issues that must be resolved before we arrive at a robust implementation of this algorithm. The first problem we would encounter is that we need a feasible vertex of the convex polytope defined by our constraints. Finding such a vertex amounts to solving another LP! You might ask yourself how we can solve this auxiliary LP—the trick is that by construction this LP has a trivial feasible solution from which to start. There are quite a few high quality simplex type LP solvers available but all suffer from a key weakness. There are pathological LP instances for which the runtime of all known simplex type algorithms is exponential.

However, there is another class of algorithms that do not suffer from this drawback and we already know about it. These are the **interior-point methods** like in Algorithm 4.10, and instead of walking along the outside edges of the polytope, they move along the inside of the polytope along a path that ends in the global minimum. For real-world problems there are more advanced and specialized interior point methods for LPs available. Even using the simple algorithm given here and a high quality Newton type algorithm implementation, we can reliably and quickly solve LPs in a few hundred variables with a few hundred constraints.

After quite some work we have finally arrived at a solution for our motivating example. Instead of an algebraic solution that we can find for the linear least squares regression problem, we have an algorithmic solution to the linear L1 regression problem. Along the way, we saw that using clever transformations we turned our problem of minimizing an unconstrained non-differentiable function into a linear programming problem which we can solve reliably and efficiently. This is not a coincidence. Frequently, seemingly hard problems can be transformed into linear or at least convex optimization problems for which there are established solvers. The hard part is therefore not solving the problem, but finding a mapping to an easily solvable problem. In fact, the L1 regression problem is often solved using another trick we did not discuss. Instead of solving the problem as stated above, its Lagrange dual, a maximization problem in m variables, is solved. Additionally, the sparse structure of the constraint matrix is often exploited by the algorithm.

4.6.3.1 Simulation

Let us now illustrate why we would want to reformulate an L1 regression problem as an LP instead of trying to solve it directly with one of the methods from Section 4.4.3. Specifically, we will use a state-of-the-art LP solver to determine the true minimum of the objective function from Equation 4.53 and then compare this solution to the solution obtained by solving the optimization problem directly using a Nelder-Mead type algorithm.

Note that numerical stability is generally not as much of an issue for LP solvers as it is for LLS solvers because they know that the solution must be one of the vertices of the polytope defined by the constraints. We therefore focus not on verifying the LP solver as was done in Chapter 3, but instead study when the solution obtained via a direct minimization of Equation 4.53 is inferior to the solution obtained via the LP formulation. We will study the influence of the following factors:

m: The number of observations / rows of $\boldsymbol{X}$.

n: The number of parameters in the model / columns of $\boldsymbol{X}$.

v: The variance of the noise term in our model.

For $m = 20, 50, 100, 200, 500$, $n = 2, 3, 4, 5, 7, 9, 13, 17, 20$, and $v = 0, 0.1, 0.5, 1$ we will generate an $m \times n$ matrix $\boldsymbol{X}$ whose entries are uniform random numbers in the range $[-2, 2]$ and a random parameter setting $\boldsymbol{\beta}$ that is drawn from an n-dimensional multivariate standard Gaussian distribution. We then generate a noise vector $\boldsymbol{\epsilon}$ with $\varepsilon_i \sim \text{Laplace}(0, \sqrt{v/2}), i = 1, \ldots, m$ (see Exercise 4.6.1). Finally, we calculate the response to be

$$\boldsymbol{y} = \boldsymbol{X}\boldsymbol{\beta} + \boldsymbol{\epsilon}.$$

We can now solve for $\hat{\boldsymbol{\beta}}^*$, the optimal solution to Equation 4.53, using an LP solver[8] and also compute $\hat{\boldsymbol{\beta}}_{\text{NM}}$, the solution obtained using the Nelder-Mead algorithm. From this we can derive the lack of fit of the Nelder-Mead solution

[8]Here the solver from the GNU Linear Programming Kit was used, but any other LP solver would have worked as well.

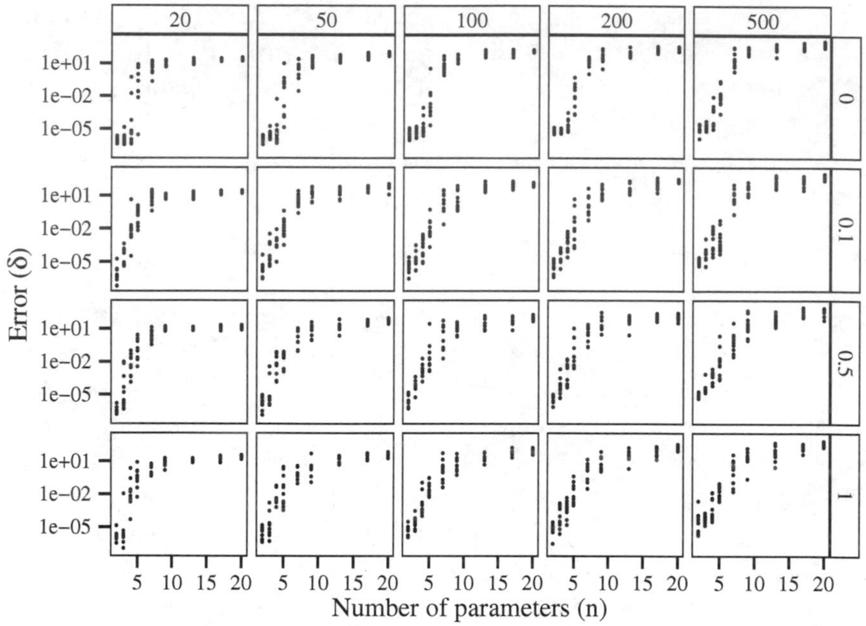

Figure 4.25: Error (δ) of the Nelder-Mead algorithm for different numbers of parameters. Each plot shows one combination of number of observations (m) and variance (v). The different numbers of observations are shown in the columns and the different variance levels in the rows of the plot.

as

$$\delta = \|\boldsymbol{y} - \boldsymbol{X}\hat{\boldsymbol{\beta}}_{\text{NM}}\|_1 - \|\boldsymbol{y} - \boldsymbol{X}\hat{\boldsymbol{\beta}}^*\|_1$$

Because the solution obtained by the Nelder-Mead algorithm can depend on the initial solution used to construct the simplex, the error δ is computed for 10 different initial solutions. To summarize, we generate $5 \times 9 \times 4 = 180$ L1 regression problems and solve each of these problems 10 times using the Nelder-Mead algorithm. For each run we record the difference between the loss of the Nelder-Mead solution and the loss of the optimal solution.

The results of this simulation are shown in Figure 4.25. We can clearly see that, at least for a moderate to high number of parameters, we would always prefer the LP solution to the estimate obtained with the Nelder-Mead algorithm. Furthermore, it is interesting to see that as the variance rises, the error of the Nelder-Mead algorithm decreases ever so slightly. This is surprising since in the case of zero variance it is trivial to construct the optimal solution. All one has to do is pick n points at random and calculate the line going through these points.

4.6.4 *Quadratic Programming*

Another prominent class of problems is those where the constraints are all linear, just like they would be in a linear program, but the objective function is now quadratic. One prominent example of such a problem is the support vector machine (SVM). The SVM is a two-class classifier just like the logistic regression. We again assume that we are given some training examples $\boldsymbol{X} := [\boldsymbol{x}_1 \ \dots \ \boldsymbol{x}_m]^T$ and their classes $\boldsymbol{y} := [y_1 \ \dots \ y_m]^T$. But we now want the class labels to be $\{-1,1\}$ instead of $\{0,1\}$ which will simplify notation.

The idea of the SVM is that we seek to find a linear hyperplane that separates the positive ($y_i = 1$) and negative ($y_i = -1$) examples. That is, we seek $\boldsymbol{\beta}$ and β_0 such that

$$\forall i \in \{1,\dots,m\}: y_i(\boldsymbol{\beta}^T \boldsymbol{x}_i - \beta_0) - 1 \geq 0$$

which can be rewritten as

$$\forall i \in \{1,\dots,m\}: y_i(\boldsymbol{\beta}^T \boldsymbol{x}_i - \beta_0) \geq 1.$$

Clearly, if such a hyperplane exists, it is not necessarily well defined. So instead of finding just any separating hyperplane, we wish to find the one to which the points have maximal distance. That is, we seek the solution to

$$\begin{aligned} \underset{(\boldsymbol{\beta},\beta_0,t)}{\text{minimize}} \quad & -t \\ \text{subject to} \quad & y_i(\boldsymbol{\beta}^T \boldsymbol{x}_i - \beta_0) \geq t, \\ & \|\boldsymbol{\beta}\|_2 \leq 1. \end{aligned}$$

The second constraint is necessary because if we do not constrain $\boldsymbol{\beta}$, the problem is unbounded. If the optimal t^* is positive, then the points can be separated using a linear hyperplane, but often we will not have the luxury that our two classes are linearly separable. In such a case it makes sense to relax the constraints on a case-by-case basis and then minimize the amount of relaxation we perform. This leads to the following optimization problem:

$$\begin{aligned} \underset{(\boldsymbol{\beta},\beta_0,\boldsymbol{\xi})}{\text{minimize}} \quad & \sum_{i=1}^{m} \xi_i \\ \text{subject to} \quad & y_i(\boldsymbol{\beta}^T \boldsymbol{x}_i - \beta_0) \geq 1 - \xi_i, \\ & \xi_i \geq 0. \end{aligned}$$

Here, $\boldsymbol{\xi} = [\xi_1 \ \ldots \ \xi_m]^T$ measures how much each one of the original constraints is violated. Note that this is a linear programming problem that we can solve using any LP solver. We were able to drop the constraint on $\boldsymbol{\beta}$ here because there is now a trade-off between $\boldsymbol{\beta}$ and $\boldsymbol{\xi}$ in the optimization and the problem is therefore bounded in $\boldsymbol{\beta}$.

More generally we may want to find a trade-off between the amount of constraint violation and the width of the margin of the separating hyperplane. We can encode this in the following quadratic programming problem:

$$\begin{aligned}
\underset{(\boldsymbol{\beta},\beta_0,\boldsymbol{\xi})}{\text{minimize}} \quad & \frac{1}{2}\|\boldsymbol{\beta}\|_2 + C\sum_{i=1}^{m}\xi_i \\
\text{subject to} \quad & y_i(\boldsymbol{\beta}^T\boldsymbol{x}_i - \beta_0) \geq 1 - \xi_i, \\
& \xi_i \geq 0.
\end{aligned}$$

Instead of solving the above problem directly, the Lagrangian dual problem is often optimized. Recall that the dual problem is obtained by forming the Lagrangian

$$\Lambda(\boldsymbol{\beta},\beta_0,\boldsymbol{\xi},\boldsymbol{\lambda},\boldsymbol{\eta}) = \frac{1}{2}\|\boldsymbol{\beta}\|_2 + C\sum_{i=1}^{m}\xi_i + \sum_{i=1}^{m}\lambda_i(1-\xi_i - y_i(\boldsymbol{\beta}^T\boldsymbol{x}_i - \beta_0)) - \sum_{i=1}^{m}\eta_i\xi_i$$

and then maximizing

$$\inf_{(\boldsymbol{\beta},\beta_0,\boldsymbol{\xi})} \Lambda(\boldsymbol{\beta},\beta_0,\boldsymbol{\xi},\boldsymbol{\lambda},\boldsymbol{\eta})$$

w.r.t. the Lagrange multipliers $\boldsymbol{\lambda}$ and $\boldsymbol{\eta}$. Luckily, there exists an analytic form for the above infimum. Using it, we can write down the dual problem as

$$\begin{aligned}
\underset{\boldsymbol{\lambda}}{\text{maximize}} \quad & \sum_{i=1}^{m}\lambda_i - \frac{1}{2}\sum_{i=1}^{m}\sum_{j=1}^{m}\lambda_i\lambda_j y_i y_j \boldsymbol{x}_i^T\boldsymbol{x}_j \\
\text{subject to} \quad & \boldsymbol{y}^T\boldsymbol{\lambda} = 0 \\
& 0 \geq \boldsymbol{\lambda} \geq C
\end{aligned}$$

Note that the new problem is still convex but only has m parameters over which we must optimize. Even more surprising, the problem does not depend on $\boldsymbol{\eta}$!

Again, this problem can be solved using any quadratic programming solver, but in practice, one will want to use specialized solvers for larger data sets since there is one constraint for each observation. We also need to fine-tune the parameter C to explore different levels of slack. Methods to do this will be discussed in Chapter 7.

Table 4.5: Misclassification Rate of a Linear SVM for Different Values of C

C	Misclass. Error
10^{-2}	0.067
10^{-1}	0.068
10^{+0}	0.067
10^{+1}	0.066
10^{+2}	0.071

Once we have the optimal $\boldsymbol{\lambda}^*$, we can recover the desired parameters $\boldsymbol{\beta}^*$ and β_0^* using the following identities:

$$\boldsymbol{\beta}^* = \sum_{i=1}^{m} \lambda_i \alpha_i \boldsymbol{x}_i$$

$$\beta_0^* = y_j - \sum_{i=1}^{m} \lambda_i \alpha_i \boldsymbol{x}_i^T \boldsymbol{x}_j$$

with j such that $0 < \lambda_j < C$. These can be derived from the KKT conditions for the original (primal) problem.

Classification Example

We will illustrate solving the SVM problem using a music retrieval problem taken from Section 7.3.2. The aim is to determine if a given piece of music is played by the piano or a guitar. We are given 5654 observations of 423 different features that characterize the spectrum of the music being played. For a detailed description of these please refer to Section 7.3.2.

From the 5654 observations, we draw 600 at random to form our training set $[\boldsymbol{x}_1 \cdots \boldsymbol{x}_{600}]^T$ and corresponding $[y_1 \cdots y_{600}]^T$. The remaining 5054 observations will be used to assess the quality of our decision rule

$$\text{sign}\left(\boldsymbol{\beta}^{*T} \boldsymbol{x} + \beta_0^*\right).$$

This process is repeated 10 times and the average misclassification rate is recorded. The results of this comparison are summarized in Table 4.5. We see that for all chosen values of C the error rate of the decision rule is rather similar. This is explained by the fact that the data are almost linearly separable which is not surprising given that we have only 600 observations and 423 features.

Kernel Support Vector Machine
Notice that the observed data, the $\boldsymbol{x}_i$, in the above optimization problem only appear as an inner product, i.e. a distance calculation between the ith and jth observation. We may therefore be tempted to replace the term $\boldsymbol{x}_i^T \boldsymbol{x}_j$ with a function $k\colon \mathbb{R}^n \times \mathbb{R}^n \to \mathbb{R}$. Indeed, by imposing some regularity conditions on k, the above problem will still be a convex quadratic program and therefore easily solvable. We call the function k a **kernel**.

The use of kernels is not limited to support vector machines and there is quite a bit of literature which explains what advantages and disadvantages the use of kernels in machine learning has. See, for example, Steinwart and Christmann (2008). A common interpretation of the kernel function is that it is the distance between the ith and jth observation in some transformed feature space. That is, there exists a function $\psi\colon \mathbb{R}^n \to \mathbb{A}$ such that $k(\boldsymbol{x}, \boldsymbol{x}') = \psi(\boldsymbol{x})^T \psi(\boldsymbol{x}')$.

Our discussion of SVMs is of course far from complete. It only aims to illustrate that constrained optimization problems appear in many branches of statistics. For a thorough discussion of support vector machines, see, for example, Steinwart and Christmann (2008). Other examples of constrained optimization in statistics include lasso regression (L1 constrained regression), nonparametric estimation and problems from experimental design theory. All of these topics are explored in Boyd and Vandenberghe (2004).

4.6.5 Summary

In this section we have taken a brief look at the vast field of constrained optimization. We showed how we can easily solve equality-constrained convex optimization problems by adapting the Newton method from the previous section. When additional linear inequality constraints were added, we learned how to reduce the solution of such a problem to a series of equality-constrained optimization problems. This once more illustrated the power of **iteration** as a general problem solving strategy. Finally, we studied two special cases of constrained optimization problems, the linear programming problem (LP) and the quadratic programming problem (QP).

4.7 Evolutionary Computing

So far we have discussed methods that work well for finding the global optimum of convex or quasi-convex functions. But what if our optimization problem does not meet the strict requirements of the algorithms described in the previous sections? What if our function is multi-modal or is not smooth?

There are a plethora of methods that have been developed for this case. They all belong to the family of black-box optimization techniques. Black-box in this context means that the function we want to optimize is viewed as if it was a black-box. You input a parameter combination on one side, and a function value is returned on the other side. For now, we will make the additional assumption, that the parameter space of the function is not discrete. Moreover, since there are so many different approaches to black-box optimization we cannot present every single strategy in detail in this section. Instead, we will describe the general concept behind most of them and then focus on several popular algorithms.

The first idea you might have when tackling a nonconvex problem is to use one of the aforementioned algorithms for convex optimization and execute several runs of the algorithm, starting from different locations in the parameter space. Such a strategy is called a **restart strategy** and can be applied to any optimization algorithm. It is just another outer iteration layer that wraps the actual optimization procedure. This outer iteration will uncover several different local optima, of which you then choose the best one. If the function has only a few stationary points, this is certainly a viable option. If, on the other hand, the function has hundreds, maybe even thousands, of local optima, this strategy can fail miserably. What we really want is a procedure that, once it finds a local optimum, gets "unstuck" and tries to find an even better local minimum.

What is especially obvious with this procedure is that, unless we know a priori how many local optima there are, we cannot be sure that we have found the global optimum unless we test every point in the parameter space. For most real-world problems this means that we will not be able to solve them to **global** optimality and prove this. However, for most of these types of problems, we will be happy with a good enough solution in practice.

The big challenge for traditional convex optimization algorithms when used to solve a black-box optimization problem is that they can get stuck in local optima. We just saw how we can restart the search from a different location to unwedge the search, but this seems like a crutch. What we really want is a search procedure that will explore the whole parameter space and, once it has found an interesting region, exploit that knowledge to produce a succession of better solutions. We also want the algorithm to be able to unwedge itself if it gets stuck in one area of the search space while in exploitation mode.

Many different classes of algorithms have been developed independently that try to implement such a strategy. They are all roughly similar and either explicitly or implicitly follow the general design of Algorithm 4.11. These

---

Algorithm 4.11 General Search Distribution Based Optimization Scheme

---

Require: $f \colon \mathbb{R}^n \to \mathbb{R}$
    Initial search distribution $\mathscr{D}^{(1)}$
    Initial solution β, population size p, iteration number limit K

1: **for** $i = 1 \to K$ **do**
2:     $\mathbb{P} \leftarrow \emptyset$
3:     **for** $j = 1 \to p$ **do**
4:         $\beta_{\text{candidate}} \leftarrow$ sample observation from $\mathscr{D}^{(i)}$
5:         **if** $f(\beta_{\text{candidate}}) < f(\beta)$ **then**
6:             $\beta \leftarrow \beta_{\text{candidate}}$
7:         **end if**
8:         $\mathbb{P} \leftarrow \mathbb{P} \cup \beta_{\text{candidate}}$
9:     **end for**
10:    $\mathbb{S} \leftarrow \emptyset$
11:    **for all** $\beta' \in \mathbb{P}$ **do**
12:       **if** some condition **then**
13:          $\mathbb{S} \leftarrow \mathbb{S} \cup \beta'$
14:       **end if**
15:    **end for**
16:    $\mathscr{D}^{(i+1)} \leftarrow$ adapt $\mathscr{D}^{(i)}$ based on $\mathbb{S}$
17: **end for**
18: **return** β

---

include **simulated annealing, genetic algorithms, evolutionary strategies, particle swarm optimization, ant colony optimization, bee colony optimization,** and **estimate of distribution algorithms.** All of these algorithms have a search distribution $\mathscr{D}$ over the parameter space, and draw one or several parameter combinations from this distribution (data set $\mathbb{P}$) which are then evaluated. Based on a subset $\mathbb{S}$, usually selected based on the fitness of the individuals, the search distribution is finally adapted for the next iteration. This scheme is depicted in Algorithm 4.11 and even though many things are not specified in the algorithm description, it is still not general enough to truly overview every variant of every algorithm class mentioned above.

Historically, most of these algorithms can be traced back to two schools of thought. On the one hand, there is the idea of mimicking evolution by creating an artificial genome and then applying the same mutation and crossover operations to them as those observed in biological systems. This train of thought, which resulted in the development of genetic algorithms (GAs), can be traced back to Holland, who developed it at the University of Michigan in the late

1960s and early 1970s (Holland, 1975). The critical part of these algorithms is a good mapping of problem representation to genetic representation. Because the genes (in their most basic form) in a GA can only encode discrete values, they are best suited to combinatorial or other discrete optimization problems that we have not covered in this chapter.

The competing idea is due to Rechenberg and Schwefel, who, while at the TU Berlin, started to use what they called evolutionary strategies (ES) to solve complex engineering problems (Rechenberg, 1973; Schwefel, 1975). Their idea does not rely on a gene to encode the problem. Instead, they use a natural representation of the problem domain (for continuous optimization problems this would be a vector), and apply mutation to this representation. This idea has historically been the more fruitful approach bringing forth some of the most powerful general purpose optimization algorithms that we know of. These include the class of covariance matrix adapting algorithms (CMA-ES) due to Ostermeier et al. (1994).

A third class of algorithms is those that try to mimic physical processes. A prominent example of this class of algorithms is simulated annealing, whose idea comes from the observation of how metals are strengthened by annealing, that is, heating and then slowly cooling them. Its heritage can be traced back to the early 1960s when Monte-Carlo sampling techniques were developed that play a key role in Section 6.4. Because it is probably the simplest of all the methods mentioned here, we will start with a discussion of this method and then present some more advanced evolutionary strategies that generalize some of the ideas of simulated annealing.

4.7.1 Simulated Annealing

Simulated Annealing (SA) is based on the observation that when metals are heated and then slowly cooled back down, the configuration of the metal atoms can reach a state that has lower total energy. The key observation here is that we need to pass a state of higher energy before we can reach the final state of lower energy. So how can we apply this observation to our problem of optimizing continuous multi-modal functions? To escape from a local minimum, we also have to pass through a region of higher fitness values to finally reach the global minimum.

Based on these observations, the simulated annealing procedure can be described as follows. In each iteration, pick a new vector β' in the neighborhood of the current solution β, and then calculate the reduction in energy, i.e. the improvement we make if we choose this solution, $\Delta = f(\beta') - f(\beta)$. Now, we want to unconditionally accept β' as the new current solution if it is an improvement ($\Delta < 0$). If it is not an improvement ($\Delta \geq 0$), then we want to accept the solution with some probability that depends on the loss we incur and the current temperature of the system. Finally, we decrease the temperature of the system; that is, we decrease the probability of accepting a worse solution.

The critical parts of the algorithm are the schedule by which the temperature is lowered in the system and how new solutions are sampled from the neighborhood of the current solution. In continuous optimization it is customary to use Gaussian random numbers with mean β and standard deviation proportional to the current temperature. Choosing the standard deviation proportional to the temperature implies that as the system cools down, our search becomes more and more local—it starts to exploit the region it has found instead of exploring a large part of the parameter space. What is missing is a schedule used to lower the temperature and a corresponding probability of accepting a worse solution. A simple but effective approach is to lower the temperature logarithmically and to decide whether to accept a solution based on

$$\exp\left(-\frac{t}{\Delta}\right) > u,$$

where t is the current temperature and u is a uniform random number from the interval $[0,1]$. The beauty of this scheme is that since t is always positive and $\Delta = f(\beta') - f(\beta)$ is negative, if β' is an improvement, $\exp(-\frac{t}{\Delta})$ will be greater than 1 in all such cases. This leads us to Algorithm 4.12, which implements the ideas of this section. It also includes some technical improvements. For example, it has a scaling vector σ that allows us to control the standard deviation for each coordinate direction separately. What the algorithm does not have is a natural stop criterion. Here, we only rely on an iteration limit; in practice, we might augment this with a criterion that terminates the optimization run if little or no improvement has been made during the past k iterations for some suitable value of k. We would also like to terminate if the standard deviation gets too small, this is, when we stop searching at all, but only sample from a very small region. Numerically, we may even end up sampling only one point!

The choice of t_0 and σ in Algorithm 4.12 can have a large impact on the performance of the algorithm. It is wise to use different values and compare the results. If we suspect that the problem is highly multi-modal, we should also employ a restart strategy because SA will quickly focus its search on a relatively small area of the parameter space and have a hard time moving out of this region because the size of the steps the algorithm can take decreases in each iteration. A good starting point for t_0 is 10. For σ, we choose a starting

Algorithm 4.12 Simulated Annealing

Require: $f\colon \mathbb{R}^n \to \mathbb{R}$
Initial solution $\boldsymbol{\beta}$, iteration limit K, scaling vector $\boldsymbol{\sigma}$, initial temperature t_0.

for $i = 1 \to K$ **do**
  $t \leftarrow \frac{t_0}{\log(i+1)}$
  $\boldsymbol{\beta}' \leftarrow$ sample observation from $\mathcal{N}(\boldsymbol{\beta}, t \cdot \mathrm{diag}(\boldsymbol{\sigma}))$
  $\Delta \leftarrow f(\boldsymbol{\beta}') - f(\boldsymbol{\beta})$
  $u \leftarrow$ uniform random number from $[0, 1]$
  if $\exp(-\frac{t}{\Delta}) > u$ **then**
    $\boldsymbol{\beta} \leftarrow \boldsymbol{\beta}'$
  end if
end for
return $\boldsymbol{\beta}$

point such that the initial search distribution covers about 50% of the parameter space with a probability of about 95%.

There is a wealth of theory and quite a few extensions to the basic idea presented here. Many of them focus on changes to the cooling schedule and the acceptance criterion. Particularly notice the combination of simulated annealing and the Nelder-Mead method as introduced in Press et al. (1995, pp. 451 – 455). What remains remarkable to the authors is that such a simple algorithm, less than 10 lines of code in many modern programming languages, is already capable of performing much better than pure random search.

4.7.2 Evolutionary Strategies

One of the weaknesses of simulated annealing is that at any given time we only have one candidate solution. What if we had a whole **population** of **offspring** from which to **select** the new best solution or solutions from which we then update our search distribution? You can tell by the terminology that the methods we will discuss now are not motivated by physical processes but by biology. If we take a simulated annealing strategy and instead of generating one candidate in each iteration, generate p candidates, these p candidate solutions are, in analogy to biological evolution, often referred to as offspring.

Instead of having to decide if we accept the new solution, we have to select one or more of the p offspring as our new best solution. Let us focus on the case of selecting exactly one offspring for a moment. Then it would be natural to select the best offspring w.r.t. the fitness function. If we accept this candidate even if it is not better than the current parent solution, then we arrive at what is called a comma selection scheme. Such an ES is often denoted as a $(1,\lambda)$-ES, which means that we have one parent solution from which λ offspring are generated and the best offspring becomes the parent of the next generation. If we allow the parent to survive, that is, we only select the best offspring if it is indeed better than the current parent, then we have what is called a plus strategy: $(1+\lambda)$-ES. Both of these strategies fit well within the framework described by Algorithm 4.11.

Before we consider the case of multiple parents, we have to revisit the generation of offspring. In SA we added normally distributed random noise with a standard deviation proportional to the current temperature. Since we do not have a temperature here, we have to use some other mechanism to control the variance of the distribution. In the original ES a $\frac{1}{5}$ success rule is used. That is, the standard deviation is increased or decreased such that on average $\frac{1}{5}$ of all offspring are better than their respective parent. There are other methods of self-adaption that can be used, but for simplicity we will only mention this rule.

Now the question arises, what do we do if we select multiple offspring as the new parents? Imagine how offspring are produced in nature. Accordingly, we select two parents at random, recombine their vectors to form a new offspring, and then mutate this offspring. Recombination can be done in many different ways. Again, the simplest approach is to choose one of the parents at random for each element of the parameter vector. Mutation is performed by adding Gaussian noise whose standard deviation is usually controlled by self-adaption.

These ideas give rise to what are then denoted as $(\mu+\lambda)$- and (μ,λ)-ES. These evolutionary strategies have μ parents and λ offspring in each iteration of the algorithm. The general evolutionary strategy is given in Algorithm 4.13. A much more thorough introduction into the history and current state-of-the-art evolutionary strategies can be found in Beyer and Schwefel (2002).

In practice evolutionary strategies perform remarkably well when you consider that they do not fully exploit the best candidate in each iteration and do not have an explicit mechanism to decrease their search radius as SA does. The one weakness of standard evolutionary strategies is that they are not invariant under rotation. It is easy for an ES to solve a problem that is separable—that is, we can optimize the problem one variable at a time. But as soon as we rotate the parameter space, it becomes much harder for the ES to find good new solutions. Part of the problem here is that the mutation does not adapt to include any notion of correlation of the parameters. The so-called covariance matrix adapting evolutionary strategy (CMA-ES) tries

Algorithm 4.13 Evolutionary Strategy

Require: Fitness function $f\colon \mathbb{R}^n \to \mathbb{R}$, population $\mathscr{P} = \{\beta_1, \ldots, \beta_\mu\}$, fixed number $\lambda > \mu$ of offspring, fixed number of parents ρ, initial mutation strength σ, iteration limit K.

for $i = 1 \to K$ **do**
  if "+" strategy **then**
    $\mathscr{C} \leftarrow \mathscr{P}$
  else
    $\mathscr{C} \leftarrow \emptyset$
  end if
  for $j = 1 \to \lambda$ **do**
    $\beta^c \leftarrow$ combination of ρ random elements from $\mathscr{P}$
    $\beta^c \leftarrow \beta^c + \mathscr{N}(0, \sigma \boldsymbol{I}_n)$ {mutation}
    $\mathscr{C} \leftarrow \mathscr{C} \cup \{\beta^c\}$
  end for
  Update σ based on success probability {e.g. target is $\frac{1}{5}$}
  $\mathscr{P} \leftarrow \mu$ best individuals in $\mathscr{C}$.
end for
return best solution in $\mathscr{P}$

to solve this problem by continually updating not just the step size parameter, but also the complete covariance matrix used in the Gaussian distribution of the mutation. This idea was introduced by Ostermeier et al. (1994) and continually refined. Its current incarnations can be considered the state-of-the-art in continuous black-box optimization.

4.7.2.1 Desirabilities Example

In this section we will study a real life problem in the context of process optimization. More specifically, we will optimize a ball mill used in the production of chocolate. The data used for this example are taken from (Alamprese et al., 2007). Let us begin by defining the process parameters we can vary and the objectives we want to optimize. There are only two process parameters, the runtime of the ball mill ($r \in [20, 45]$) and the speed of the shaft that agitates the balls in the mill ($s \in [5, 90]$). Instead of optimizing just one objective, we are interested in optimizing several at the same time. These are the runtime of the mill (r), the total energy consumption of the mill (E), the particle size (d_{90}), the iron content of the chocolate after milling (Fe) and two parameters that characterize the melting properties of the chocolate (η_{Ca} and τ_{Ca}). In the cited article, the authors use a straightforward designed experiment to capture the relationship between the parameter settings of the mill and the objectives. They use a linear regression to model this relationship, and we will follow their lead here. The fully quadratic models we obtain are summarized in the following set of equations:

$$\begin{bmatrix} r \\ E \\ d_{90} \\ Fe \\ \eta_{\mathrm{Ca}} \\ \tau_{\mathrm{Ca}} \end{bmatrix} = \begin{bmatrix} 0.00000 & 1.00000 & 0.00000 & 0.00000 & 0.00000 & 0.00000 \\ -7.15454 & 0.10496 & 0.15219 & -0.00049 & -0.00097 & 0.00004 \\ 42.63923 & -0.81017 & -0.03155 & 0.01089 & 0.00418 & -0.00842 \\ 29.28979 & -0.26587 & -0.01191 & 0.00334 & 0.00142 & -0.00175 \\ 3.14204 & -0.02727 & -0.01515 & 0.00015 & 0.00005 & 0.00010 \\ -6.37205 & 0.19362 & 0.15108 & -0.00003 & -0.00038 & -0.00145 \end{bmatrix} \begin{bmatrix} 1 \\ r \\ s \\ r^2 \\ s^2 \\ rs \end{bmatrix}$$

Notice that r, the runtime of the mill, is both a parameter and an objective. It is a parameter because we can change the runtime of the mill at will, and at the same time, we want to minimize the runtime so that we can maximize production. The shorter a batch of chocolate is milled, the more chocolate we can mill in a single day.

Next, we need some way to combine our objectives into a single objective that we can minimize. To accomplish this, we use so called **desirability functions** that map each objective into the interval from 0 to 1, where 0 denotes a completely undesirable result and 1 a totally desirable one. Different methods have been proposed to construct these mapping functions. One such method, which we will use here, are the so called Derringer-Suich desirability functions. For each objective, their definition follows.

First we give the desirability function for the runtime r. Here we say that a runtime of 30 minutes or less is completely desirable and a runtime of 45 minutes or more is not tolerable. All runtimes between 30 and 45 minutes are given a desirability between 0 and 1 based on a linear scale.

$$d_r(r) := \begin{cases} 1 & \text{if } r < 30 \\ \frac{r-45}{30-45} & \text{if } 30 \le r < 45 \\ 0 & \text{else} \end{cases}$$

The energy consumption E should also be minimized. We set an energy consumption of less than 3 kWh to be perfect and a consumption of more than 4 kWh as completely undesirable. Again, all values in between are assigned desirabilities based on a linear interpolation.

$$d_E(E) := \begin{cases} 1 & \text{if } E < 3 \\ \frac{E-4}{3-4} & \text{if } 3 \le E < 4 \\ 0 & \text{else} \end{cases}$$

The optimal particle size d_{90} for our chocolate is 21. Values lower than 20 or higher than 23 are completely undesirable. All values in-between are again linearly interpolated.

$$d_{d_{90}}(d_{90}) := \begin{cases} 0 & \text{if } d_{90} < 20 \\ \frac{d_{90}-20}{21-20} & \text{if } 20 \leq d_{90} < 21 \\ \frac{d_{90}-23}{21-23} & \text{if } 21 \leq d_{90} < 23 \\ 0 & \text{else} \end{cases}$$

During the milling process some iron shavings contaminated the chocolate. As long as there are less than 20 ppm, the chocolate is completely OK. If, on the other hand there, are more than 30 ppm, the chocolate is not usable. For values in-between, we assign desirabilities based on linear interpolation.

$$d_{Fe}(Fe) := \begin{cases} 1 & \text{if } Fe < 20 \\ \frac{Fe-30}{20-30} & \text{if } 20 \leq Fe < 30 \\ 0 & \text{else} \end{cases}$$

The next two properties (η_{Ca} and τ_{Ca}) characterize the melting properties of the chocolate. For both properties we again have an optimal value ($\eta_{\mathrm{Ca}} = 1.5$, $\tau_{\mathrm{Ca}} = 8$) as well as lower and upper bounds outside of which the chocolate is not usable. Values inside the bounds are, as in the previous cases, calculated using linear interpolation.

$$d_{\eta_{\mathrm{Ca}}}(\eta_{\mathrm{Ca}}) := \begin{cases} 0 & \text{if } \eta_{\mathrm{Ca}} < 1 \\ \frac{\eta_{\mathrm{Ca}}-1}{1.5-1} & \text{if } 1 \leq \eta_{\mathrm{Ca}} < 1.5 \\ \frac{\eta_{\mathrm{Ca}}-2}{1.5-2} & \text{if } 1.5 \leq \eta_{\mathrm{Ca}} < 2 \\ 0 & \text{else} \end{cases}$$

$$d_{\tau_{\mathrm{Ca}}}(\tau_{\mathrm{Ca}}) := \begin{cases} 0 & \text{if } \tau_{\mathrm{Ca}} < 5 \\ \frac{\tau_{\mathrm{Ca}}-5}{8-5} & \text{if } 5 \leq \tau_{\mathrm{Ca}} < 8 \\ \frac{\tau_{\mathrm{Ca}}-10}{8-10} & \text{if } 8 \leq \tau_{\mathrm{Ca}} < 10 \\ 0 & \text{else} \end{cases}$$

Since all the desirabilities of our objectives are of the same magnitude, we can use some form of average as the objective for our optimization problem. Here it is customary to use the (weighted) geometric average instead of the usual arithmetic average. This has the advantage that as soon as one of the desirabilities is zero, the geometric mean is also zero.

Combining all the above, we arrive at the following optimization problem:

$$\underset{r,s}{\text{minimize}} \quad -\Big(d_r(r)^{10} \cdot d_E(E(r,s))^{\frac{1}{10}} \cdot d_{d_{90}}(d_{90}(r,s))^5 \cdot d_{Fe}(Fe(r,s)) \cdot d_{\eta_{\text{Ca}}}(\eta_{\text{Ca}}(r,s)) \cdot d_{\tau_{\text{Ca}}}(\tau_{\text{Ca}}(r,s))\Big)^{\frac{10}{181}}.$$

Here we minimize the negative geometric mean of all the desirability functions. The additional exponents on the desirability functions act as weights that encode the relative importance of the objectives. We use the linear models we derived from the experimental data to feed the desirability functions with predictions of the objectives.

What makes this a **hard** optimization problem? First, we can expect the objective function to be zero for most of the parameter settings. Second, since the desirability functions are not piecewise linear and not convex, the problem is neither convex nor differentiable. It is therefore quite a challenge to find any point with a desirability greater than zero and from there it is even harder to maximize the geometric mean of all the desirabilities. This is illustrated by Exercise 4.7.5.

4.7.2.2 Maximum-Likelihood Example

Most introductory textbooks on statistical estimation theory cover maximum likelihood estimation in one of their chapters. There, it is usually required, and henceforth assumed, that the negative log-likelihood will be a convex function. If this is not the case, maximizing the likelihood, which amounts to minimizing the negative log-likelihood, is a formidable problem. In this section we will look at one simple example were the negative log-likelihood is not convex. We will use some of the algorithms covered in this section to nevertheless find the most probable parameters given some observed data.

Consider the random variable

$$X \sim p\mathcal{N}(\mu_1, \sigma_1) + (1-p)\mathcal{N}(\mu_2, \sigma_2)$$

and m observations $x_1, \ldots, x_m$ from X. The random variable X is a simple Gaussian mixture. If p, μ_1, σ_1, μ_2 and σ_2 are unknown, we can try to estimate them using the maximum likelihood approach from our observations $x_1, \ldots, x_m$.

Let $\beta := [p \;\; \mu_1 \;\; \sigma_1 \;\; \mu_2 \;\; \sigma_2]^T$; then

$$p(x_i, \beta) := \frac{p}{\sqrt{2\pi\sigma_1^2}} \exp\left(\frac{(x_i-\mu_1)^2}{2\pi\sigma_1^2}\right) + \frac{1-p}{\sqrt{2\pi\sigma_2^2}} \exp\left(\frac{(x_i-\mu_2)^2}{2\pi\sigma_2^2}\right)$$

and our likelihood and corresponding negative log-likelihood functions are given by

$$
\begin{aligned}
L(\beta) &:= \prod_{i=1}^{m} p(x_i, \beta) \\
&= \prod_{i=1}^{m} pd(x_i, \mu_1, \sigma_1) + (1-p)d(x_i, \mu_2, \sigma_2) && (4.59) \\
-\log L(\beta) &:= -\sum_{i=1}^{m} \log\left(pd(x_i, \mu_1, \sigma_1) + (1-p)d(x_i, \mu_2, \sigma_2)\right). && (4.60)
\end{aligned}
$$

Here we denote by

$$
d(x, \mu, \sigma) := \frac{1}{\sqrt{2\pi\sigma^2}} \exp\left(\frac{(x-\mu)^2}{2\pi\sigma^2}\right)
$$

the value of the probability density function of the Normal distribution with parameters μ and σ in x.

It is easy to show that $-\log L(\beta)$ is not convex (see Exercise 4.7.6). We can nevertheless try to estimate β by applying one of the optimization procedures introduced in this section. The resulting optimization problem is given by

$$
\begin{aligned}
\underset{(p,\mu_1,\sigma_1,\mu_2,\sigma_2)}{\text{minimize}} \quad & -\sum_{i=1}^{m} \log\left(pd(x_i, \mu_1, \sigma_1) + (1-p)d(x_i, \mu_2, \sigma_2)\right) \\
\text{subject to} \quad & 0 < p \leq 0.5 \\
& \sigma_1 \geq 0 \\
& \sigma_2 \geq 0
\end{aligned}
$$

Note the additional constraints required to obtain a valid parameter vector. The estimated standard deviations should be positive, and the probability should lie between 0 and $\frac{1}{2}$. To see why p is not constrained to $[0,1]$, consider the log-likelihood of the parameter vector $\beta = [p \ \ \mu_1 \ \ \sigma_1 \ \ \mu_2 \ \ \sigma_2]^T$ and of the vector $\tilde{\beta} = [1-p, \mu_2, \sigma_2, \mu_1, \sigma_1]^T$. They are identical! So to ensure that there is a unique global minimum, we restrict p to the range $[0, 0.5]$. In Exercise 4.7.7 a simple example mixture is analyzed and the corresponding maximum likelihood problem solved. There are other approaches to deal with such a mixture of distributions. One of these approaches is covered in Section 5.2.3.

4.7.3 Summary

We concluded this chapter with a section describing modern methods that are inspired by natural processes, be they biological evolution or physical cooling processes. All of these optimization procedures can be used for non-convex and non-differentiable optimization problems. While they are slower at finding an optimal solution and usually give no guarantees that the returned solution is even optimal, sometimes we must resort to these methods because none of the specialized methods discussed in previous sections are applicable. We illustrated this with two examples, one resulting from a process optimization task and the other being a classical maximum-likelihood estimation problem.

4.8 Implementation in R

In R itself and many contributed R packages there are various implementations of the aforementioned and even more sophisticated methods.

In base R, we will find functions for root finding (`uniroot`, `polyroot`) and univariate optimization (`optimize`). The function `optim` provides "general-purpose optimization based on Nelder-Mead, quasi-Newton and conjugate-gradient algorithms" and also "includes an option for box-constrained optimization and simulated annealing". The latter is easy to use. Consider that you want to minimize the bivariate function $f(x) := 2x_1^2 + 3(x_2 - 4)^2 - 5$. Then the corresponding R code is:

```
f <- function(x) 2 * x[1]^2 + 3 * (x[2] - 4)^2 - 5
## define starting values for the vector of parameters:
start <- c(0, 0)
optim(par = start, fn = f, method = "Nelder-Mead")
```

which will use Nelder-Mead optimization to find the minimum at $x_1 = 0$ and $x_2 = 4$ minus numerical inaccuracies.

An excellent overview of the contributed R packages is given at the CRAN Task View called "Optimization and Mathematical Programming" (Theussl, 2013). Among the mentioned packages, optimx (Nash and Varadhan, 2011) is meant as "a replacement and extension of the `optim()` function" that provides infrastructure for accessing methods in other packages; also, some that specialize in problems with different kind of constrains, quadratic approximation and large-scale problems.

A highly competitive implementation of simulated annealing is provided by the GenSA package (Xiang et al., 2012). A pure R implementation of a simple covariance matrix adapting evolutionary strategy is contained in the cmaes package (Trautmann et al., 2011), and finally a large and growing collection of test problems for continuous optimization algorithms can be found in the soobench package (Mersmann and Bischl, 2012). The ROI package (Hornik et al., 2011), which is still under development, provides a concise interface to many mathematical programming problem solvers in R. It is especially helpful if one wants to try out different LP or QP solvers for a given problem.

4.9 Conclusion

Superficially, this chapter has been a review of almost all optimization techniques a statistician might encounter. The deeper insight, however, is that all of these techniques hinge on the powerful idea of **iteration**. Instead of solving the often difficult task using a single, complex and expensive operation, we broke the process down into what might be called baby steps. Instead of a globally best solution, each **iteration** merely provided a better solution. This powerful idea can also be applied in many other areas of computing.

Iteration, as we have seen, is not to be confused with the divide-and-conquer technique we learned about while studying the Quick sort algorithm (Algorithm 2.3 on page 28). Divide and conquer has a fixed recursion depth because, given finite data, at some point there is nothing left to divide and we start with the conquer phase. In contrast, the iterative methods we studied in this chapter usually do not have an a-priory known number of iterations they will perform. Instead, a stop criterion is required that, after each iteration, judges the quality of the current solution and terminates if it is deemed "good enough". This alludes to one of the drawbacks of iterative techniques, they always return an approximation of the true optimal solution. Often times we can live with this since we already know from Chapter 3 that all computations we perform are inexact.

At the heart of many classical and modern statistical inference techniques lies an optimization problem, be it explicitly written out or implicitly formulated. And almost all non-trivial instances of such problems are solved using iterative techniques. At the same time, classical statistical procedures such as sequential design of experiments or quality control can be recast as iterative methods. As you can see, iteration is present in every branch of statistics and at every scale. It is likely one of the two most powerful problem-solving strategies in our arsenal, the other strategy being divide and conquer.

4.10 Exercises

Exercise 4.2.1: Show that f_{1D-LAS} in equation (4.3) is convex w.r.t. β.

Exercise 4.3.1: Implement Algorithm 4.2. Use your implementation to minimize the following functions:

- $f(x) := x^2, -2 \leq x \leq 1$
- $f(x) := |x|, -2 \leq x \leq 1$
- $f(x) := |x| + x^2, -2 \leq x \leq 1$

What do you observe?

Exercise 4.3.2: What happens if you use the **absolute error** instead of the **relative error** as the stop criterion for your Golden-Section search from Exercise 4.3.1?

Exercise 4.3.3: Implement Algorithm 4.3. Use the functions from Exercise 4.3.1 to test your implementation. What do you observe?

Exercise 4.3.4: Why do all convex twice differentiable univariate functions start to look like a quadratic function in an ε environment around their minimum (hint: look at the Taylor expansion of f)?

Exercise 4.4.1: Write a simple gradient descent optimization procedure. Test your algorithm using a quadratic function $f(\beta) = \frac{1}{2}\beta^T \boldsymbol{Q}\beta$. Vary the condition number of the defining matrix $\boldsymbol{Q}$ of the quadratic problem. What do you observe? What happens if $\boldsymbol{Q}$ is not positive definite?

Exercise 4.4.2: Extend your gradient descent implementation from Exercise 4.4.1 to the full Newton method. If no Hessian function is provided, your implementation should gracefully degrade to the BFGS method.

Exercise 4.4.3: Show that $\boldsymbol{U}_{BFGS}^{(k)}$ fulfills Equation 4.18 and the corresponding direction matrix $\boldsymbol{P}^{(k+1)}$ is positive definite if $\boldsymbol{P}^{(k)}$ is and the gradient is reduced:

$$(\Delta\beta^{(k)})^T \boldsymbol{g}^{(k)} > 0 \Leftrightarrow \nabla f(\beta^{(k+1)})^T \boldsymbol{P}^{(k)} \nabla f(\beta^{(k)}) - \nabla f(\beta^{(k)})^T \boldsymbol{P}^{(k)} \nabla f(\beta^{(k)}) < 0.$$

Exercise 4.4.4: Write a small program to implement the logistic regression simulation study. What difficulties do you encounter?

Exercise 4.4.5: Use the results from the previous exercise and analyze the number of function evaluations until convergence is reached. Use plots similar to those used to visualize the suboptimality and the distance to the optimum. What do you observe? Also consider the number of gradient and Hessian evaluations required by each algorithm.

Exercise 4.5.1: Show that the Equation (4.44) is equivalent to gradient descent for error term $(f(x_j;\Theta) - y_j)^2$.

Exercise 4.6.1: Look at the simulation study in Section 4.6.3.1. Explain why the Laplace distribution was chosen as the error distribution instead of the standard Normal distribution.

Exercise 4.6.2: Design a simulation study, similar to the logistic regression study, to compare the strengths and weaknesses of different LP solvers that are readily available in R when applied to L1 regression problems. Do you observe any differences in solution quality or runtime of the algorithms? Note that there are exact solvers for LPs that will calculate the optimal solution to arbitrary precision.

Exercise 4.6.3: Analyze the results of your L1 regression simulation study.

Exercise 4.6.4: Use the Nelder-Mead algorithm to minimize the error function of the L1 Regression problem directly. How does the returned solution compare to the solution returned by an LP solver? Include a comparison of the runtimes of the algorithms.

Exercise 4.7.1: Implement Algorithm 4.12. Use it to optimize the sphere function $f(\beta) = \beta^T \beta$, $\beta \in \mathbb{R}^3$. How well does the algorithm perform with 1 000, 10 000 and 100 000 iterations? Try different starting temperatures t_0 and scaling vectors σ. Can you think of a simple trick to improve the algorithm?

Exercise 4.7.2: Compare your simulated annealing algorithm from the previous exercise with pure random search on many different test functions. What do you observe? Note that you can find a large selection of test functions in the R package soobench.

Exercise 4.7.3: Implement both a simple $(1+\lambda)$ and a $(1,\lambda)$-ES (see Algorithm 4.13) and compare its performance, measured in number of function evaluations, on several functions to the performance of simulated annealing.

Exercise 4.7.4: Implement the objective function from the desirability example (Section 4.7.2.1) and plot the desirability as a function of the runtime ($r \in [20, 45]$) and the shaft speed ($s \in [5, 90]$). What do you observe?

Exercise 4.7.5: Optimize the objective function you implemented in the previous exercise using a suitable algorithm. Try different starting values. What do you observe? Do the returned optimal values for r and s make sense? Calculate the individual desirabilities for each objective. Which objectives are close to optimal and which, if any, are almost undesirable?

Exercise 4.7.6: Show that the negative log-likelihood function (Equation 4.60) in the Gaussian mixture example is not convex in β.

Hint

Pick values for $x_1, \ldots, x_m$ (for suitably small m) and then find parameter vectors β_1, β_2 and β_3 for which the convexity property does not hold.

Exercise 4.7.7: Estimate p, μ_1, σ_1, μ_2, σ_2 in the setting described in Section 4.7.2.2 using the following 100 observations

```
 2.81  5.10  2.60  5.38  7.60  2.93  3.52  0.09  2.96  4.23
 6.75  6.94  5.77  1.30  4.89  7.13  6.63  1.04 -0.40  5.12
 6.15 -0.09  5.31  5.86 -0.54  7.62  5.94  2.51  7.76  7.41
 6.65  1.67  3.86  6.27 -1.19  5.70 -0.22 -0.18  0.77  5.55
 3.62  5.89  0.65  9.42  4.75  4.05  4.67  6.73  5.19  1.75
 4.99  6.52  5.08  6.47 -0.67  4.88  5.96 -0.12  2.51  4.93
 4.86  3.48  2.93  3.74  6.17 -0.38  3.43  5.33  2.53  0.14
-1.49 -1.47  0.12  5.00  0.00  3.35 -0.61  5.06  3.33  0.18
 6.49 -0.49  3.46  5.31  6.98  4.85 -0.12  1.20 -0.47  3.95
 4.96 -0.89 -0.44  2.77  6.21  5.55  7.31  1.64  5.17  7.71
```

Optimize the negative log-likelihood (Equation 4.60) using a suitable optimization algorithm that is included in R or available as a package. To choose the initial parameter settings, look at a histogram of the data and try to guess the parameters. The data were generated using the parameters $p = 0.25$, $\mu_1 = 0$, $\sigma_1 = 1$, $\mu_2 = 5$ and $\sigma_2 = 1$. How close is your estimate to these values? Does your chosen algorithm consistently converge to this solution?

Chapter 5

Deduction of Theoretical Properties

Often, intuition comes first when building an algorithm. "Couldn't we make this in that way?" This leads to many so-called **heuristics**, which do not stand the test of time, when their theoretical properties are undesirable. Only if there are favorable theoretical properties for an algorithm, is there a very good chance that it will be used in practice for a longer period of time. At times the algorithms first have to stand the practice test, since deduction of theoretical properties generally needs time. This typically leads to two different situations. Some algorithms are not as much used as they should be as long as, e.g., their optimality properties are unclear - like the Partial Least Squares (PLS) method below. Other algorithms are used frequently since practice has uncovered the not yet proven theory - like in the case of the Expectation Maximization (EM) algorithm below. We include this chapter in the book as a motivation to take up the challenge to deduce theoretical properties, since only through them the practical properties of the heuristics become clear. In the following, we will study two meanwhile established algorithms regarding their theoretical properties:

- the PLS algorithm regarding its optimality properties and
- the EM algorithm regarding its convergence.

We will start with the PLS algorithm.

5.1 PLS – from Algorithm to Optimality[1]

5.1.1 Motivation and History

If we can split the observed features into two groups, where the so-called **responses** should be predicted by the other group, the **influential factors**, we

[1]Partly based on Geladi and Kowalski (1986, pp. 1 – 17) and Weihs and Jessenberger (1998, pp. 170 – 172).

are interested in methods for **optimal prediction** with, e.g., **linear multivariate multiple models**.

Definition 5.1: Multivariate Multiple Linear Model
Let $\boldsymbol{Y}$ be the data matrix of the M responses and $\boldsymbol{X}$ be the data matrix of the L influential factors, each with m observations for the same objects or subjects. Then, a **multivariate multiple linear model** has the form

$$\boldsymbol{Y} = \boldsymbol{X}\boldsymbol{B} + \boldsymbol{E},$$

where $\boldsymbol{B}$ is a matrix of unknown coefficients and $\boldsymbol{E}$ a matrix of model errors.

If, as in this definition, the structure of the matrix $\boldsymbol{B}$ of the coefficients is completely left open, then the **linear least squares estimator** has the property that every response can be considered individually. Thus, we can restrict ourselves to multiple linear models for one response. For other estimation or prediction methods this might not be true.

5.1.2 NIPALS Algorithm

Modern prediction methods often first construct so-called latent variables, to which regression is then applied.

Idea of the Algorithm
Instead of directly using the generally highly correlated original influential factors in the design matrix $\boldsymbol{X}$ as regressors, first adequate linear combinations of these factors are built, so-called **latent variables**, which have preferable properties as regressors.

Formally, this relates to the following transformation:

$$\boldsymbol{Y} = \boldsymbol{X}\boldsymbol{B} + \boldsymbol{E} = (\boldsymbol{X}\boldsymbol{G})\boldsymbol{A} + \boldsymbol{E} =: \boldsymbol{Z}\boldsymbol{A} + \boldsymbol{E},$$

where the matrix $\boldsymbol{G}$ determines the linear combinations of the original factors and $\boldsymbol{B} = \boldsymbol{G}\boldsymbol{A}$.

The art is to define useful properties of $\boldsymbol{G}$ and methods for its construction. Let us consider here only two **desirable properties** of G:

- Orthogonality of the columns of $\boldsymbol{X}\boldsymbol{G}$, since then the coefficients in $\boldsymbol{A}$ can be determined independently of each other, and
- utilization of as much information in the data and in the model as possible, here in particular of the assumed relationship of $\boldsymbol{Y}$ and $\boldsymbol{X}$.

A modern prediction method that takes into account the relationship between responses and influential factors is the **Partial Least Squares method**

(PLS). In the literature this method was first introduced as an iterative method without any optimality property. Moreover, at first the proposed iteration was neither really understood by most users nor often applied. For a long time the econometrician Hermann Wold and later his son, the chemist Svante Wold, in Sweden as well as the group of Bruce Kowalski in the U.S. were more or less the only ones who "raised the flag of PLS".

In our opinion this only changed fundamentally after Höskuldsson (1988) proved an optimality property, which we will discuss in Section 5.1.3. We will start, however, introducing the historical PLS method using the NIPALS iteration.

The basic idea of the **NIPALS (Nonlinear Iterative Partial Least Squares) algorithm** is described in Geladi and Kowalski (1986) as follows:

- The design matrix $\boldsymbol{X}$ as well as the response matrix $\boldsymbol{Y}$ are represented as a sum of rank-1 matrices.
- These rank-1 matrices are outer products of vectors called **scores** and **loadings**.
- These scores and loadings are calculated step by step for $\boldsymbol{X}$ and $\boldsymbol{Y}$.

More formally, the following matrix decompositions are studied:

$$\boldsymbol{X} = \boldsymbol{T}\boldsymbol{P}^T + \boldsymbol{F} = \sum_h \boldsymbol{t}_h \boldsymbol{p}_h^T + \boldsymbol{F} \quad \text{and}$$

$$\boldsymbol{Y} = \boldsymbol{U}\boldsymbol{Q}^T + \boldsymbol{H} = \sum_h \boldsymbol{u}_h \boldsymbol{q}_h^T + \boldsymbol{H}$$

with vectors $\boldsymbol{t}_h$, $\boldsymbol{p}_h$, $\boldsymbol{u}_h$, and $\boldsymbol{q}_h$, and matrices

$$\begin{aligned}
\boldsymbol{X} &= [\boldsymbol{x}_1 \ \ldots \ \boldsymbol{x}_L] \in L(m,L),\\
\boldsymbol{Y} &= [\boldsymbol{y}_1 \ \ldots \ \boldsymbol{y}_M] \in L(m,M),\\
\boldsymbol{T} &= [\boldsymbol{t}_1 \ \ldots \ \boldsymbol{t}_l] \in L(m,l),\\
\boldsymbol{P} &= [\boldsymbol{p}_1 \ \ldots \ \boldsymbol{p}_l] \in L(L,l),\\
\boldsymbol{U} &= [\boldsymbol{u}_1 \ \ldots \ \boldsymbol{u}_l] \in L(m,l),\\
\boldsymbol{Q} &= [\boldsymbol{q}_1 \ \ldots \ \boldsymbol{q}_l] \in L(M,l),
\end{aligned}$$

where $l \leq \max(L,M)$ is prefixed, as well as error matrices $\boldsymbol{F}$ and $\boldsymbol{H}$.

Oversimplified, the decompositions of $\boldsymbol{X}, \boldsymbol{Y}$ are calculated independently of each other:

$\boldsymbol{X}$-chunk:

1. Set $\boldsymbol{t} := \boldsymbol{x}_j$ for any j.
2. Calculate the (transposed) loadings $\boldsymbol{p}^T = \boldsymbol{t}^T\boldsymbol{X}/\boldsymbol{t}^T\boldsymbol{t}$ (regression of $\boldsymbol{X}$ on $\boldsymbol{t}$).

3. Normalize: $p_{new}^T = \frac{p_{old}^T}{\|p_{old}\|}$.
4. Calculate the scores for the loadings p: $t = Xp$.
5. Compare the t in steps (2) and (4). If they are (approximately) equal, then STOP. Else got to (2).

Y-chunk:

1. Set $u := y_j$ for any j.
2. Calculate the (transposed) loadings $q^T = u^T Y / u^T u$ (regression of Y on u).
3. Normalize: $q_{new}^T = \frac{q_{old}^T}{\|q_{old}\|}$.
4. Calculate the scores for the loadings q: $u = Yq$.
5. Compare the u in steps (2) and (4). If they are (approximately) equal, then STOP. Else go to (2).

As a motivation for the iteration in the X-chunk, notice that in the limit

$$p^T = \frac{\frac{t^T X}{t^T t}}{\|\frac{t^T X}{t^T t}\|} = \frac{t^T X}{\|t^T X\|} = \frac{p^T X^T X}{\|p^T X^T X\|},$$

and thus

$$\|p^T X^T X\| p = X^T X p,$$

i.e. p is an eigenvector of $X^T X$ for the singular value $\|p^T X^T X\|$.

Analogously:

$$t = Xp = \frac{\frac{XX^T t}{t^T t}}{\|\frac{X^T t}{t^T t}\|} = \frac{XX^T t}{\|X^T t\|} = \frac{XX^T t}{\|X^T X p\|},$$

i.e. t is an eigenvector of XX^T for the singular value $\|X^T X p\| = \|p^T X^T X\|$.

Notice that the singular value decomposition of a matrix delivers the **best approximation of the matrix with prefixed rank.**

Naturally, it is not sensible to determine the decompositions in the X- and the Y-chunk independently of each other, since then the model link between the matrices would be ignored. The, at first sight somewhat abstruse, idea of the NIPALS algorithm is to construct the link by interchanging the roles of t and u in the steps (2) of the two chunks. Overall, this leads to the NIPALS (Nonlinear Iterative Partial Least Squares) Algorithm 5.1.

This algorithm generally converges very fast.

Algorithm 5.1 NIPALS (Nonlinear Iterative Partial Least Squares) Algorithm

Require: Matrix of responses $\boldsymbol{Y} = [\boldsymbol{y}_1 \ \ldots \ \boldsymbol{y}_M]$ and matrix of influential variables $\boldsymbol{X}$.

1: Set $\boldsymbol{u}_{start} := \boldsymbol{y}_j$ for any j.
2: Calculate the $\boldsymbol{X}$-loadings: $\boldsymbol{p}^T = \frac{\boldsymbol{u}^T\boldsymbol{X}}{\boldsymbol{u}^T\boldsymbol{u}}$ (regression of $\boldsymbol{X}$ on $\boldsymbol{u}$).
3: Normalize: $\boldsymbol{p}_{new}^T = \frac{\boldsymbol{p}_{old}^T}{\|\boldsymbol{p}_{old}\|}$.
4: Calculate the $\boldsymbol{X}$-scores: $\boldsymbol{t} = \boldsymbol{X}\boldsymbol{p}$.
5: Calculate the $\boldsymbol{Y}$-loadings: $\boldsymbol{q}^T = \frac{\boldsymbol{t}^T\boldsymbol{Y}}{\boldsymbol{t}^T\boldsymbol{t}}$ (regression of $\boldsymbol{Y}$ on $\boldsymbol{t}$).
6: Normalize: $\boldsymbol{q}_{new}^T = \frac{\boldsymbol{q}_{old}^T}{\|\boldsymbol{q}_{old}\|}$.
7: Calculate the $\boldsymbol{Y}$-scores: $\boldsymbol{u} = \boldsymbol{Y}\boldsymbol{q}$.
8: **if** $\boldsymbol{t}$ in step (4) $\approx \boldsymbol{t}$ in the previous iteration step **then**
9: Calculate the $\boldsymbol{X}$-loadings: $\boldsymbol{p}^T = \frac{\boldsymbol{t}^T\boldsymbol{X}}{\boldsymbol{t}^T\boldsymbol{t}}$ (regression of $\boldsymbol{X}$ on $\boldsymbol{t}$)
10: Normalize: $\boldsymbol{p}_{new}^T = \frac{\boldsymbol{p}_{old}^T}{\|\boldsymbol{p}_{old}\|}$.
11: Calculate the $\boldsymbol{X}$-scores: $\boldsymbol{t} = \boldsymbol{X}\boldsymbol{p}$.
12: STOP
13: **end if**
14: go to (2)

A motivation for this algorithm can be:

$$\boldsymbol{p}^T = \frac{\frac{\boldsymbol{u}^T\boldsymbol{X}}{\boldsymbol{u}^T\boldsymbol{u}}}{\|\frac{\boldsymbol{u}^T\boldsymbol{X}}{\boldsymbol{u}^T\boldsymbol{u}}\|} = \frac{\boldsymbol{u}^T\boldsymbol{X}}{\|\boldsymbol{u}^T\boldsymbol{X}\|} = \frac{\boldsymbol{q}^T\boldsymbol{Y}^T\boldsymbol{X}}{\|\boldsymbol{q}^T\boldsymbol{Y}^T\boldsymbol{X}\|}$$
$$= \frac{\boldsymbol{t}^T\boldsymbol{Y}\boldsymbol{Y}^T\boldsymbol{X}}{\|\boldsymbol{t}^T\boldsymbol{Y}\boldsymbol{Y}^T\boldsymbol{X}\|} = \frac{\boldsymbol{p}^T\boldsymbol{X}^T\boldsymbol{Y}\boldsymbol{Y}^T\boldsymbol{X}}{\|\boldsymbol{p}^T\boldsymbol{X}^T\boldsymbol{Y}\boldsymbol{Y}^T\boldsymbol{X}\|}.$$

Therefore:

$$\|\boldsymbol{p}^T\boldsymbol{X}^T\boldsymbol{Y}\boldsymbol{Y}^T\boldsymbol{X}\|\boldsymbol{p} = \boldsymbol{X}^T\boldsymbol{Y}\boldsymbol{Y}^T\boldsymbol{X}\boldsymbol{p}.$$

So rather than using the singular values of $\boldsymbol{X}$ or $\boldsymbol{Y}$, we focus on those of $\boldsymbol{X}^T\boldsymbol{Y}$.

If $\boldsymbol{Y}$ only consists of one column, then steps (5)-(7) can be left out setting $q = 1$. Steps (9)-(11) are carried out in order to construct **orthogonal $\boldsymbol{X}$-scores**.

After the first iteration, $\boldsymbol{X}$ and $\boldsymbol{Y}$ are replaced by the corresponding **residuals** of the rank-1 decomposition. For example, we have to calculate:

$$\boldsymbol{X}_R = \boldsymbol{X} - \boldsymbol{t}\boldsymbol{p}^T = \boldsymbol{X} - \frac{\boldsymbol{t}\boldsymbol{t}^T\boldsymbol{X}}{\boldsymbol{t}^T\boldsymbol{t}}.$$

Then, we continue with another iteration of the algorithm. For a proof of orthogonality, see Section 5.1.3.

The columns of the matrix $\boldsymbol{G}$ are the $\boldsymbol{X}$-loadings $\boldsymbol{p}$ from the different iterations. The scores $\boldsymbol{t}$ build the columns of the matrix $\boldsymbol{XG}$.

In the 1980s many people were discouraged by the opacity of this algorithm. However, the results of the next section were convincing to use the PLS method.

5.1.3 Covariance Optimality

Höskuldsson (1988) has shown that the **1st PLS component** represents the weighted sum of the influential factors with the greatest covariance with a weighted sum of the responses.

For illustration, we first restrict ourselves to one (mean-adjusted) response $\boldsymbol{y}_1$. Let $\boldsymbol{X}$ be the (mean-adjusted) data matrix of the L possible influential factors with m observations. Then, the scores $\boldsymbol{z}_1 = \boldsymbol{X}\,\boldsymbol{g}_1$ of the 1st PLS component with loading vector $\boldsymbol{g}_1$ should have maximum empirical covariance with the vector $\boldsymbol{y}_1$ of the observations of the response for all vectors $\boldsymbol{g}$ of length 1, i.e. with $\boldsymbol{g}^T\boldsymbol{g} = \sum_{i=1}^{L} g_i^2 = 1$. More formally,

$$|\widehat{\text{cov}}(\boldsymbol{X}\boldsymbol{g}_1, \boldsymbol{y}_1)| = \max_{\|\boldsymbol{g}\|=1} |\widehat{\text{cov}}(\boldsymbol{X}\boldsymbol{g}, \boldsymbol{y}_1)|.$$

The restriction to loading vectors of length 1 is necessary, since the covariance is increasing with the length of the observation vectors without limit.

Theorem 5.1: Covariance Optimality
For one response $\boldsymbol{y}_1$, the covariance optimal scores $\boldsymbol{z}_1$ and loadings $\boldsymbol{g}_1$ have the form

$$\boldsymbol{z}_1 = \boldsymbol{X}\,\boldsymbol{g}_1 = \frac{\boldsymbol{X}\boldsymbol{X}^T\boldsymbol{y}_1}{\sqrt{\boldsymbol{y}_1^T\boldsymbol{X}\boldsymbol{X}^T\boldsymbol{y}_1}}.$$

Proof. Following the Lagrange multiplier method we are looking for a vector $\boldsymbol{g}$ such that

$$(\boldsymbol{y}_1^T\boldsymbol{X}\boldsymbol{g})^2 - \mu(\boldsymbol{g}^T\boldsymbol{g} - 1) = \boldsymbol{g}^T\boldsymbol{X}^T\boldsymbol{y}_1\boldsymbol{y}_1^T\boldsymbol{X}\boldsymbol{g} - \mu(\boldsymbol{g}^T\boldsymbol{g} - 1) = \text{max!}$$

for a certain $\mu \in \mathbb{R}$. Differentiating with respect to $\boldsymbol{g}$ leads to:

$$\boldsymbol{X}^T\boldsymbol{y}_1\boldsymbol{y}_1^T\boldsymbol{X}\boldsymbol{g} - \mu\boldsymbol{g} = 0,$$

i.e. $\boldsymbol{g}$ is the only non-vanishing eigenvector (of length 1) of the rank-1 matrix

$X^T y_1 y_1^T X$.

$$g_1 = \frac{X^T y_1}{\sqrt{y_1^T X X^T y_1}}$$

has this property, since

$$X^T y_1 y_1^T X g_1 = \frac{X^T y_1 y_1^T X X^T y_1}{\sqrt{y_1^T X X^T y_1}} = \sqrt{y_1^T X X^T y_1} X^T y_1$$
$$= (y_1^T X X^T y_1) g_1.$$

□

Obviously, the found vector g_1 corresponds to the X-loadings p_1 of the NIPALS algorithm for one column y_1, and z_1 corresponds to t_1 (see Section 5.1.2), since in the NIPALS algorithm

$$p = \frac{\frac{X^T u}{u^T u}}{\|\frac{X^T u}{u^T u}\|} = \frac{X^T y_1}{\|X^T y_1\|} = \frac{X^T y_1}{\sqrt{y_1^T X X^T y_1}} = g_1.$$

In the more general case of more than one response the 1st PLS component has to satisfy:

$$|\widehat{\text{cov}}(Xg_1, Yc_1)| = \max_{\|g\|=1, \|c\|=1} |\widehat{\text{cov}}(Xg, Yc)|.$$

Unfortunately, in this case no explicit representation of the 1st PLS component is possible. Instead, the 1st PLS component results from a decomposition of the matrix $X^T Y$ (cp. the motivation of the NIPALS algorithm in Section 5.1.2).

Theorem 5.2: Calculation of the 1st PLS Component
Let $S := X^T Y$ and $S = V \Phi U^T$ be the singular value decomposition of S, where $V^T V = I$, $U^T U = I$, and Φ is a diagonal matrix with elements ≥ 0. Then, $g_1 := v_1$ is the 1st PLS component, where v_1 is the first column of V.

After the determination of the 1st PLS component the fraction $\frac{z_1 z_1^T X}{z_1^T z_1}$ corresponding to this component is subtracted from the influential factors X and the effect $\frac{z_1 z_1^T Y}{z_1^T z_1}$ of the component is subtracted from the responses Y. That means we calculate the **residuals** X_R of the influential factors and Y_R

of the responses, respectively, corresponding to the regression of each $\boldsymbol{X}$ and $\boldsymbol{Y}$ on the 1st PLS component $\boldsymbol{z}_1$:

$$\boldsymbol{X}_R := \boldsymbol{X} - \frac{\boldsymbol{z}_1 \boldsymbol{z}_1^T \boldsymbol{X}}{\boldsymbol{z}_1^T \boldsymbol{z}_1}, \quad \boldsymbol{Y}_R := \boldsymbol{Y} - \frac{\boldsymbol{z}_1 \boldsymbol{z}_1^T \boldsymbol{Y}}{\boldsymbol{z}_1^T \boldsymbol{z}_1}.$$

Please note the remark on the residualization of $\boldsymbol{X}$ in Section 5.1.2. The residualization of $\boldsymbol{Y}$ is carried out analogously.

The 2nd PLS component $\boldsymbol{z}_2$ is chosen as the 1st, however using $\boldsymbol{X}_R$ and $\boldsymbol{Y}_R$ instead of $\boldsymbol{X}$ and $\boldsymbol{Y}$. The further PLS components are derived analogously.

The PLS components are pairwise **uncorrelated**, since, e.g.,

$$\boldsymbol{z}_1^T \boldsymbol{X}_R = \boldsymbol{z}_1^T \boldsymbol{X} - \frac{\boldsymbol{z}_1^T \boldsymbol{z}_1 \boldsymbol{z}_1^T \boldsymbol{X}}{\boldsymbol{z}_1^T \boldsymbol{z}_1} = \boldsymbol{z}_1^T \boldsymbol{X} - \boldsymbol{z}_1^T \boldsymbol{X} = 0.$$

An important problem of the PLS method is the fact that it is **not scale-invariant** to both influential factors and responses. Therefore, one of the following scales is typically chosen. Either one is looking for "natural" units of the involved variables, or all variables are standardized to variance 1. In the first case, we talk of **PLS based on covariances**, and in the second case of **PLS based on correlations**.

5.1.4 PLS Method: Examples

Example 5.1: Exact Linear Dependency
Let us demonstrate the work of PLS for a simple case, where the one response is an exact linear combination of the columns of the matrix $\boldsymbol{X}$. Then, this linear combination is identified by the loadings of the 1st PLS component. We will demonstrate this by means of a simple matrix of influential factors typically used in **design of experiments**. Let

$$\boldsymbol{X} = \begin{bmatrix} 1 & 1 & 1 \\ 1 & 1 & -1 \\ 1 & -1 & 1 \\ 1 & -1 & -1 \end{bmatrix},$$

and $\boldsymbol{y} = [y_1\ y_2\ y_3\ y_4]^T$ be mean-centered. The mean-centered $\boldsymbol{X}$ has the form

$$\boldsymbol{X} = \begin{bmatrix} 0 & 1 & 1 \\ 0 & 1 & -1 \\ 0 & -1 & 1 \\ 0 & -1 & -1 \end{bmatrix}.$$

Table 5.1: Goodness of Fit for PLS Components

		HUEREM		HUEREMAL	
		PLS1	PLS1+2	PLS1	PLS1+2
individual PLS	R^2	0.47	0.62	0.74	0.81
combined PLS	R^2	0.45	0.62	0.74	0.80

Then, the loadings have the form

$$g_1 = \begin{bmatrix} 0 & y_1 + y_2 - y_3 - y_4 & y_1 - y_2 + y_3 - y_4 \end{bmatrix}^T / \|.\|,$$

where $/\|.\|$ indicates normalization. Therefore, if $\boldsymbol{y}$ is a linear combination of the columns $\boldsymbol{x}_2, \boldsymbol{x}_3$ of $\boldsymbol{X}$, namely $\boldsymbol{y} = \alpha_2 \boldsymbol{x}_2 + \alpha_3 \boldsymbol{x}_3$, then

$$g_1 = \begin{bmatrix} 0 & 4\alpha_2 & 4\alpha_3 \end{bmatrix}^T / \|.\|.$$

Thus, the loadings reproduce the weighting vector $\boldsymbol{\alpha}$ of the linear combination except for the normalization.

Example 5.2: Application to Dyestuff Production
We are interested in the prediction of the hue of a dyestuff under daylight (HUEREM) and Artificial Light (HUEREMAL). Measuring was realized by means of REMission spectra. For the construction of prediction models 93 observations (= lots of dyestuff) of 18 (chemical-)analytical properties together with the realized hues were available. We calculated the PLS components on the basis of correlations and considered the following models for the two response variables:

$$HUEREM(AL) - \overline{HUEREM(AL)} = \beta_1 PLS1 + \varepsilon \quad \text{and}$$
$$HUEREM(AL) - \overline{HUEREM(AL)} = \beta_1 PLS1 + \beta_2 PLS2 + \varepsilon,$$

where PLS1 and PLS2 are the first and second PLS component, respectively. Table 5.1 shows a comparison of the goodness of fit of the two models for the two response variables, once for models only based on one response (individual PLS), and once for models based on the two responses together (combined PLS). Obviously, the two PLS methods appear to produce similar results for this data situation.

What is also left open for applications is the determination of the **number of PLS components** to be used in models to optimize predictive power. This is typically identified by so-called resampling methods. Please refer to Section 7.4.

5.1.5 PLS Method: Summary and Outlook

Summary
In this section we introduced and discussed the PLS method in its historical development from an often used (NIPALS) algorithm to a theoretically well-based optimization method.

Outlook
The PLS method is also intensively studied today, even in special workshops and conferences. For an overview over the state of the art in PLS research, you might want to take a look into the *Handbook of Partial Least Squares* by Vincenzo Esposito Vinzi, Wynne W. Chin, Joerg Henseler, and Huiwen Wang (Vinzi et al., 2010).

5.2 EM Algorithm[2]

5.2.1 Motivation and Introduction

We will now switch to the EM algorithm.

The term **EM algorithm** stands for **E**xpectation **M**aximization algorithm and originates from Dempster et al. (1977). However, the algorithm is based on earlier ideas, e.g. of Sundberg (1974). The oldest paper cited by Dempster et al. (1977) as a reference for the EM algorithm stems from McKendrick (1926), where missing data were replaced by estimates from already observed data.

The EM algorithm solves the **maximum likelihood problem for incomplete data**. For this estimation problem it is assumed that information is missing for the random variables in the model: Either some random variables are not observed, or new, unobserved random variables would simplify the problem considerably.

The idea of the EM algorithm is to replace the missing information by expected values given certain estimates of the unknown coefficients. Using these expected values of the missing data, we reestimate the coefficients, and so on iteratively.

The **principle of the EM algorithm** is given in Algorithm 5.2, which shows one iteration step of the iterative estimation of unknown variables and coefficients.

[2]Partly based on Ng et al. (2004) and Hastie et al. (2001).

Algorithm 5.2 Iteration Step of the EM Algorithm (General Case)

1: **[E-step]** Estimate the unknown variables by their expected values, given actual estimations of the unknown coefficients.
2: **[M-step]** Estimate the unknown coefficients by the Maximum-Likelihood principle, given actual estimations of the unknown (latent) variables.

Although this procedure immediately appears natural, its properties are not obvious at all, in particular convergence is unclear.

5.2.2 *Definition and Convergence*

A very important merit of Dempster et al. (1977) was that they first proved convergence of the EM algorithm for the case of the exponential family.

Let $\boldsymbol{X} = [\boldsymbol{Y}^T \quad \boldsymbol{Z}^T]^T$ be a random vector of augmented data combined of $\boldsymbol{Y}$, the random vector of observed, and $\boldsymbol{Z}$, the vector of missing data. On the one hand, the **augmented data** $\boldsymbol{X}$ are derived from the observed data $\boldsymbol{Y}$ by inserting the **conditional expectation** for the missing data, given the actual model coefficient estimates $\boldsymbol{\beta}$. On the other hand, if $g_e(\boldsymbol{x} \mid \boldsymbol{\beta})$ is the density function of $\boldsymbol{X}$ given the model coefficients $\boldsymbol{\beta}$, then the log-likelihood function $l_e(\boldsymbol{\beta}) := \log(g_e(\boldsymbol{x} \mid \boldsymbol{\beta}))$ should be optimized for the determination of the optimal $\boldsymbol{\beta}$ given the augmented data $\boldsymbol{x}$. Notice, however, that the explicit knowledge of the augmented data is not even necessary. What we only have to determine in order to be able to optimize the log-likelihood with respect to $\boldsymbol{\beta}$ is the expected value of the log-likelihood function with respect to the conditional distribution of $\boldsymbol{Z}$ given the observed data and the actual $\boldsymbol{\beta}^{(k)}$.

This leads to the formulation of the EM algorithm in Algorithm 5.3.

Algorithm 5.3 $(k+1)$th Iteration Step of the EM Algorithm (General Case)

1: **[E-step]** Determine the expected value of the log-likelihood function with respect to the conditional distribution of $\boldsymbol{Z}$:

$$\varphi(\boldsymbol{\beta} \mid \boldsymbol{\beta}^{(k)}) := \mathrm{E}_{\boldsymbol{Z}|\boldsymbol{Y},\boldsymbol{\beta}^{(k)}}(l_e(\boldsymbol{\beta})).$$

2: **[M-step]** Choose $\boldsymbol{\beta}^{(k+1)}$ such that $\varphi(\boldsymbol{\beta} \mid \boldsymbol{\beta}^{(k)})$ is maximal with respect to $\boldsymbol{\beta}$.

We can start the algorithm with random values for $\boldsymbol{\beta}^{(1)}$, and iterate until, e.g., $|l(\boldsymbol{\beta}^{(k+1)}) - l(\boldsymbol{\beta}^{(k)})| < \varepsilon$ for an adequate (log-)likelihood function l of the observable data $\boldsymbol{Y}$ (see the theorem below).

Now consider the exponential family.

Definition 5.2: Exponential Family
Distributions with densities of the form

$$g_e(\boldsymbol{x} \mid \boldsymbol{\beta}) = a(\boldsymbol{x}) \exp\left(\boldsymbol{b}(\boldsymbol{\beta})^T \boldsymbol{c}(\boldsymbol{x})\right) / d(\boldsymbol{\beta}),$$

where a and d are scalar-valued and $\boldsymbol{b}$ and $\boldsymbol{c}$ vector-valued functions, belong to the **exponential family**. $\boldsymbol{c} = \boldsymbol{c}(x)$ is the vector of **sufficient statistics** for $\boldsymbol{x}$.

The exponential family contains the most important distribution functions like, e.g., the multivariate normal distribution, the Poisson distribution, and the multinomial distribution.

For the exponential family, it is true that

$$\begin{aligned}\varphi(\boldsymbol{\beta} \mid \boldsymbol{\beta}^{(k)}) :=& \mathrm{E}_{\boldsymbol{Z}|\boldsymbol{Y},\boldsymbol{\beta}^{(k)}}(l_e(\boldsymbol{\beta})) \\ =& \mathrm{E}_{\boldsymbol{Z}|\boldsymbol{Y},\boldsymbol{\beta}^{(k)}}(\log(a(\boldsymbol{x}))) + \boldsymbol{b}(\boldsymbol{\beta})^T \boldsymbol{c}^{(k)} - \log(d(\boldsymbol{\beta})),\end{aligned}$$

where

$$\boldsymbol{c}^{(k)} = \mathrm{E}_{\boldsymbol{Z}|\boldsymbol{Y},\boldsymbol{\beta}^{(k)}}(\boldsymbol{c}(x)).$$

Obviously, the first term of the log-likelihood function does not depend on $\boldsymbol{\beta}$.

This leads to the formulation of the EM algorithm for exponential families in Algorithm 5.4.

Algorithm 5.4 $(k+1)$h Iteration Step of the EM Algorithm (Exponential Family)

1: **[E-step]** $\boldsymbol{c}^{(k)} = \mathrm{E}_{\boldsymbol{Z}|\boldsymbol{Y},\boldsymbol{\beta}^{(k)}}(\boldsymbol{c}(x))$
2: **[M-step]** $\boldsymbol{\beta}^{(k+1)} = \arg\max_{\boldsymbol{\beta}}(\boldsymbol{b}(\boldsymbol{\beta})^T \boldsymbol{c}^{(k)} - \log(d(\boldsymbol{\beta})))$

Note that the sufficient statistics of the missing variables $\boldsymbol{Z}$ will typically be dependent on the observable variables $\boldsymbol{Y}$ and the actual model coefficients $\boldsymbol{\beta}^{(k)}$, see the example below.

Theorem 5.3: Monotonicity and Convergence
For the above exponential family, the EM algorithm leads to non-decreasing values of $\log(g(\boldsymbol{y} \mid \boldsymbol{\beta}^{(k)}))$ and to convergence if the log-likelihood of the density function

$$\log(g(\boldsymbol{y} \mid \boldsymbol{\beta}^{(k)}))$$

of the observable variables is bounded.

Proof. We transform the representation to the observable data:

$$\varphi(\boldsymbol{\beta} \mid \boldsymbol{\beta}^{(k)}) := \mathrm{E}_{\boldsymbol{Z}|\boldsymbol{Y},\boldsymbol{\beta}^{(k)}}(l_e(\boldsymbol{\beta})) = \mathrm{E}_{\boldsymbol{Z}|\boldsymbol{Y},\boldsymbol{\beta}^{(k)}}(\log(g(\boldsymbol{y} \mid \boldsymbol{\beta})) + \log(f_e(\boldsymbol{x} \mid \boldsymbol{y}, \boldsymbol{\beta}))),$$

where

$$f_e(\boldsymbol{x} \mid \boldsymbol{y}, \boldsymbol{\beta}) = \frac{g_e(\boldsymbol{x} \mid \boldsymbol{\beta})}{g(\boldsymbol{y} \mid \boldsymbol{\beta})}$$

is the conditional density of $\boldsymbol{X}$ given $\boldsymbol{Y}$. Then

$$\begin{aligned}\varphi(\boldsymbol{\beta} \mid \boldsymbol{\beta}^{(k)}) &= \log(g(\boldsymbol{y} \mid \boldsymbol{\beta})) + \mathrm{E}_{\boldsymbol{Z}|\boldsymbol{Y},\boldsymbol{\beta}^{(k)}}(\log(f_e(\boldsymbol{x} \mid \boldsymbol{y}, \boldsymbol{\beta}))) \\ &=: \log(g(\boldsymbol{y} \mid \boldsymbol{\beta})) + h(\boldsymbol{\beta}, \boldsymbol{\beta}^{(k)})\end{aligned}$$

and

$$\begin{aligned}\log(g(\boldsymbol{y} \mid \boldsymbol{\beta}^{(k+1)})) - \log(g(\boldsymbol{y} \mid \boldsymbol{\beta}^{(k)})) =&(\varphi(\boldsymbol{\beta}^{(k+1)} \mid \boldsymbol{\beta}^{(k)}) - \varphi(\boldsymbol{\beta}^{(k)} \mid \boldsymbol{\beta}^{(k)})) \\ &- (h(\boldsymbol{\beta}^{(k+1)}, \boldsymbol{\beta}^{(k)}) - h(\boldsymbol{\beta}^{(k)}, \boldsymbol{\beta}^{(k)})).\end{aligned}$$

With the Jensen inequality for logarithms, $-\log(\mathrm{E}(X)) \le \mathrm{E}(-\log(X))$ (since log is concave), one can show that

$$h(\boldsymbol{\beta}^{(k+1)}, \boldsymbol{\beta}^{(k)}) \le h(\boldsymbol{\beta}^{(k)}, \boldsymbol{\beta}^{(k)}) \tag{5.1}$$

as follows:

$$\begin{aligned}&h(\boldsymbol{\beta}^{(k)}, \boldsymbol{\beta}^{(k)}) - h(\boldsymbol{\beta}^{(k+1)}, \boldsymbol{\beta}^{(k)}) \\ &= \int \log(f_e(\boldsymbol{x} \mid \boldsymbol{y}, \boldsymbol{\beta}^{(k)}))gz(z \mid \boldsymbol{y}, \boldsymbol{\beta}^{(k)})dz - \int \log(f_e(\boldsymbol{x} \mid \boldsymbol{y}, \boldsymbol{\beta}^{(k+1)}))gz(z \mid \boldsymbol{y}, \boldsymbol{\beta}^{(k)})dz \\ &= \int \log(\frac{g_e(\boldsymbol{x} \mid \boldsymbol{\beta}^{(k)})}{g(\boldsymbol{y} \mid \boldsymbol{\beta}^{(k)})})gz(z \mid \boldsymbol{y}, \boldsymbol{\beta}^{(k)})dz - \int \log(\frac{g_e(\boldsymbol{x} \mid \boldsymbol{\beta}^{(k+1)})}{g(\boldsymbol{y} \mid \boldsymbol{\beta}^{(k+1)})})gz(z \mid \boldsymbol{y}, \boldsymbol{\beta}^{(k)})dz \\ &= \int \log(\frac{g_e(\boldsymbol{x} \mid \boldsymbol{\beta}^{(k)})}{g_e(\boldsymbol{x} \mid \boldsymbol{\beta}^{(k+1)})})gz(z \mid \boldsymbol{y}, \boldsymbol{\beta}^{(k)})dz + \int \log(\frac{g(\boldsymbol{y} \mid \boldsymbol{\beta}^{(k+1)})}{g(\boldsymbol{y} \mid \boldsymbol{\beta}^{(k)})})gz(z \mid \boldsymbol{y}, \boldsymbol{\beta}^{(k)})dz \\ &= \int -\log(\frac{g_e(\boldsymbol{x} \mid \boldsymbol{\beta}^{(k+1)})}{g_e(\boldsymbol{x} \mid \boldsymbol{\beta}^{(k)})})gz(z \mid \boldsymbol{y}, \boldsymbol{\beta}^{(k)})dz + \log(\frac{g(\boldsymbol{y} \mid \boldsymbol{\beta}^{(k+1)})}{g(\boldsymbol{y} \mid \boldsymbol{\beta}^{(k)})}) \\ &\ge -\log(\frac{g(\boldsymbol{y} \mid \boldsymbol{\beta}^{(k+1)})}{g(\boldsymbol{y} \mid \boldsymbol{\beta}^{(k)})}) + \log(\frac{g(\boldsymbol{y} \mid \boldsymbol{\beta}^{(k+1)})}{g(\boldsymbol{y} \mid \boldsymbol{\beta}^{(k)})})) \\ &=0.\end{aligned}$$

Moreover, by construction:

$$\varphi(\boldsymbol{\beta}^{(k+1)} \mid \boldsymbol{\beta}^{(k)}) \ge \varphi(\boldsymbol{\beta}^{(k)} \mid \boldsymbol{\beta}^{(k)}).$$

Therefore:

$$\log(g(\boldsymbol{y} \mid \boldsymbol{\beta}^{(k+1)})) \ge \log(g(\boldsymbol{y} \mid \boldsymbol{\beta}^{(k)})),$$

i.e. $\log(g(\boldsymbol{y} \mid \boldsymbol{\beta}^{(k)}))$ is monotonically non-decreasing and converges to $\log(L^*)$ for bounded likelihood functions. □

Unfortunately, the limit of the EM algorithm might not maximize the log-likelihood. Indeed, there are examples for convergence to a minimum or a saddle point (McLachlan and Krishnan, 1997). Moreover, the algorithm may converge to a local maximum. This shows that the intuitively very appealing idea of the EM algorithm has its limits, since convergence to the global maximum of the log-likelihood cannot be guaranteed in practice. Therefore, a plausibility check of EM results is urgently needed. Note that Wu (1983) discussed the convergence of the EM algorithm outside the exponential family.

Let us conclude with a theorem on convergence speed.

Theorem 5.4: Convergence Speed
The convergence speed of the EM algorithm is linear, and the convergence rate depends on the share of information in the observed data. That is, if an important part of data is missing, then convergence can be very slow.

Proof. If β^* is the limit of $\beta^{(k)}$, then $\beta^{(k+1)} - \beta^* \approx \boldsymbol{J_M}(\beta^*)(\beta^{(k)} - \beta^*)$, where
$\boldsymbol{J_M}(\beta)$ is the Jacobi matrix of the function $\boldsymbol{M}(\beta)$ with $\beta^{(k+1)} = \boldsymbol{M}(\beta^{(k)})$.
Meng and van Dyk (1997) showed that:
$\boldsymbol{J_M}(\beta) = \boldsymbol{I} - (-\boldsymbol{H}(\boldsymbol{y} \mid \beta))(\mathrm{E}(-\boldsymbol{H}_e(\boldsymbol{x} \mid \boldsymbol{y}, \beta)))^{-1}$, where
$-\boldsymbol{H}(\boldsymbol{y} \mid \beta) = -\mathrm{Hessian}(\log(g(\boldsymbol{y} \mid \beta))) = \text{information matrix}(\boldsymbol{y})$,
$-\boldsymbol{H}_e(\boldsymbol{x} \mid \boldsymbol{y}, \beta) = -\mathrm{Hessian}(\log(g_e(\boldsymbol{x} \mid \beta))) = \text{information matrix}(\boldsymbol{x})$.

Therefore, the global **convergence rate** (cp. Chapter 4)
$r = \lim_{k\to\infty} \|\beta^{(k+1)} - \beta^*\| / \|\beta^{(k)} - \beta^*\| = \text{largest eigenvalue of } \boldsymbol{J_M}(\beta^*)$
is small (and therefore convergence fast!), if the share of information in the observed data $\boldsymbol{y}$ is large. □

5.2.3 *Example: Mixture of Gaussians*

Consider the following Gaussian mixture model with two normal distributions:

$$
\begin{aligned}
&Y_1 \sim \mathcal{N}(\mu_1, \sigma_1^2) \\
&Y_2 \sim \mathcal{N}(\mu_2, \sigma_2^2) \\
&Y = (1-R)Y_1 + RY_2 \\
&R \in \{0,1\} \text{ (missing class information)} \\
&P(R=1) = \pi
\end{aligned}
$$

Let $\varphi_{\beta_i}(x) = \mathcal{N}(\mu_i, {\sigma_i}^2)$, $\beta_i = (\mu_i, {\sigma_i}^2)$, then the density of Y is given by

$$g_Y(y) = (1-\pi)\varphi_{\beta_1}(y) + \pi\varphi_{\beta_2}(y).$$

The following coefficients should be estimated by means of maximum likelihood:

$$\theta = (\pi, \beta_1, \beta_2) = (\pi, \mu_1, \sigma_1^2, \mu_2, \sigma_2^2).$$

The log-likelihood function based on the m sample data has the form

$$\begin{aligned} l(\theta; y) &= \log(\prod_{i=1}^{m}(1-\pi)\varphi_{\beta_1}(y_i) + \pi\varphi_{\beta_2}(y_i)) \\ &= \sum_{i=1}^{m} \log((1-\pi)\varphi_{\beta_1}(y_i) + \pi\varphi_{\beta_2}(y_i)). \end{aligned}$$

Direct optimization is difficult because of the sum of terms inside the logarithm (also for numerical optimization). It would be much easier if we would know the class R_i for all training data. The joint density, the marginal density of Y, and the conditional density of R given Y have the form

$$\begin{aligned} \varphi(R, y) &= ((1-\pi)\varphi_{\beta_1}(y))^{1-R}(\pi\varphi_{\beta_2}(y))^R, \\ \varphi(y) &= (1-\pi)\varphi_{\beta_1}(y) + \pi\varphi_{\beta_2}(y), \\ \varphi(R \mid y) &= \frac{((1-\pi)\varphi_{\beta_1}(y))^{1-R}(\pi\varphi_{\beta_2}(y))^R}{(1-\pi)\varphi_{\beta_1}(y) + \pi\varphi_{\beta_2}(y)}. \end{aligned}$$

This leads to the conditional expectation

$$\mathrm{E}(R \mid y, \theta) = \frac{\pi\varphi_{\beta_2}(y)}{(1-\pi)\varphi_{\beta_1}(y) + \pi\varphi_{\beta_2}(y)},$$

since in a Bernoulli experiment the expected value is equal to the probability of $R = 1$. Now, we augment each observation y_i by the missing data R_i. Then, the log-likelihood function for the augmented data $[R \;\; y]^T$ has the form

$$\begin{aligned} l_e(\theta; R, y) = &\sum_{i=1}^{m}((1-R_i)\log\varphi_{\beta_1}(y_i) + R_i \log\varphi_{\beta_2}(y_i)) \\ &+ \sum_{i=1}^{m}((1-R_i)\log(1-\pi) + R_i \log\pi). \end{aligned}$$

The maximum likelihood estimators are now the sample mean and the sample variance for the subsets corresponding to data with $R_i = 0$ and $R_i = 1$. However, in a real situation the values of R_i are unknown so that we proceed iteratively as in the variant of the EM algorithm in Algorithm 5.5.

Example 5.3: Application of the EM Algorithm
We generated 50 observations each of $Y_1 \sim \mathcal{N}(-2, 2.25)$ and $Y_2 \sim$

Algorithm 5.5 $(k+1)$th Iteration Step of the EM Algorithm (Mixture of Two Gaussians Y_1 and Y_2)

1: **[E-step]** Substitute each R_i by its expected value, i.e. by the **posterior probability** of component Y_2 for observation i, corresponding to the actual parametric model:
$\gamma_i(\boldsymbol{\theta}) = \mathrm{E}(R_i \mid y, \boldsymbol{\theta}) = P(R_i = 1 \mid y, \boldsymbol{\theta})$.

2: **[M-step]** Calculate new Maximum-Likelihood estimations of the coefficients $\hat{\theta}$.

$\mathcal{N}(2, 2.25)$. These data we analyzed by means of the EM algorithm ignoring the knowledge of the two classes, i.e. the two different distributions. Thus, we would like to fit the above Gaussian mixture model with two normal distributions. The EM algorithm now has the form as in Algorithm 5.6.

Algorithm 5.6 EM Algorithm for Mixtures of two Gaussians Y_1 and Y_2

1: **[Initialization]**: Choose starting values for $\hat{\pi}$, $\hat{\mu}_1$, $\hat{\sigma}_1^2$, $\hat{\mu}_2$, and $\hat{\sigma}_2^2$.

2: **[E-step]**: Calculate the **posterior probabilities** of component Y_2:

$$\hat{\gamma}_i = \frac{\hat{\pi}\varphi_{\hat{\beta}_2}(y_i)}{(1-\hat{\pi})\varphi_{\hat{\beta}_1}(y_i) + \hat{\pi}\varphi_{\hat{\beta}_2}(y_i)}, \qquad i = 1, \ldots, m.$$

3: **[M-step]**: Calculate weighted means and variances:

$$\hat{\mu}_1 = \frac{\sum_{i=1}^m (1-\hat{\gamma}_i) y_i}{\sum_{i=1}^m (1-\hat{\gamma}_i)}, \qquad \hat{\mu}_2 = \frac{\sum_{i=1}^m \hat{\gamma}_i y_i}{\sum_{i=1}^m \hat{\gamma}_i},$$

$$\hat{\sigma}_1^2 = \frac{\sum_{i=1}^m (1-\hat{\gamma}_i)(y_i - \hat{\mu}_1)^2}{\sum_{i=1}^m (1-\hat{\gamma}_i)}, \qquad \hat{\sigma}_2^2 = \frac{\sum_{i=1}^m \hat{\gamma}_i (y_i - \hat{\mu}_2)^2}{\sum_{i=1}^m \hat{\gamma}_i},$$

$$\hat{\pi} = \frac{\sum_{i=1}^m \hat{\gamma}_i}{m}.$$

4: **[Iteration]**: Repeat the E- and M-steps until convergence.

Problem formulation and choice of good starting values
Notice that the global maximum of this log-likelihood function is degenerated. For example, the solution $\hat{\mu}_1 = y_i$ for an arbitrary $i \in \{1, \ldots, m\}$ and $\hat{\sigma}_1^2 = 0$ leads to an infinite value for the likelihood function. Since this is no sensible maximum, we restrict the problem by $\hat{\sigma}_1, \hat{\sigma}_2 > 0$.

Unfortunately, then the maximum is not unique. Therefore, we might have to start with many random starting values. For $\hat{\mu}_1$ and $\hat{\mu}_2$ we may randomly

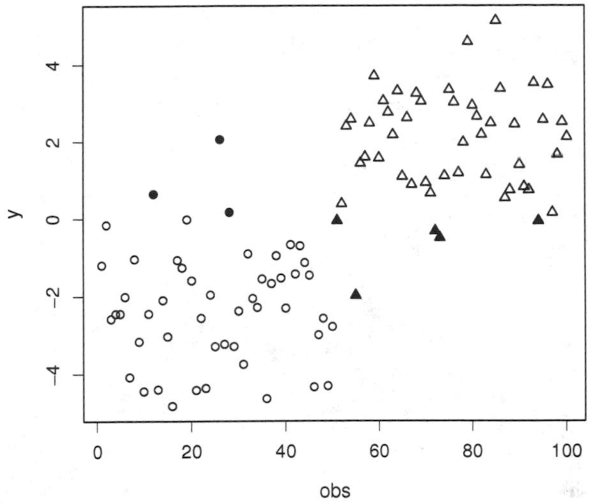

Figure 5.1: Plot of example data; symbol shapes indicate original classes, filled symbols indicate incorrect EM estimates.

choose any two different y_i. For $\hat{\sigma}_1$ and $\hat{\sigma}_2$ we may choose any starting values with $\hat{\sigma}_1, \hat{\sigma}_2 > 0.5$. For $\hat{\pi}$ we may start randomly in $(0,1)$. Considering the results for the different starting values, we choose the result with the largest likelihood value.

Consider Figure 5.1 for the data situation and the outcome of the EM algorithm. The estimated class is the one with highest responsibility. Only 8 observations were "misclassified". The convergence of the EM algorithm is very fast in the beginning (3 digits of log-likelihood correct after 2 iterations), but very slow at the end (4 digits correct after 5 iterations, 5 digits correct after 55 iterations, and the $6th$ digit only after 103 iterations). The slow convergence is illustrated in Figure 5.2. Note that there is some overlap of the two original classes, i.e. the information about the classes is not clear from the data. This, apparently, influences convergence negatively. See Figure 5.3 for the posterior probability of class 1 at convergence. Obviously, for some observations the class is not very clear. The final estimates of the unknown coefficients are:

$$\hat{\mu}_1 = -2.13, \quad \hat{\mu}_2 = 2.05, \quad \hat{\sigma}_1^2 = 2.24, \quad \hat{\sigma}_2^2 = 1.59, \quad \hat{\pi} = 0.472.$$

Obviously, the distribution of the first class is not well estimated, also leading to a too high estimated probability of the second class.

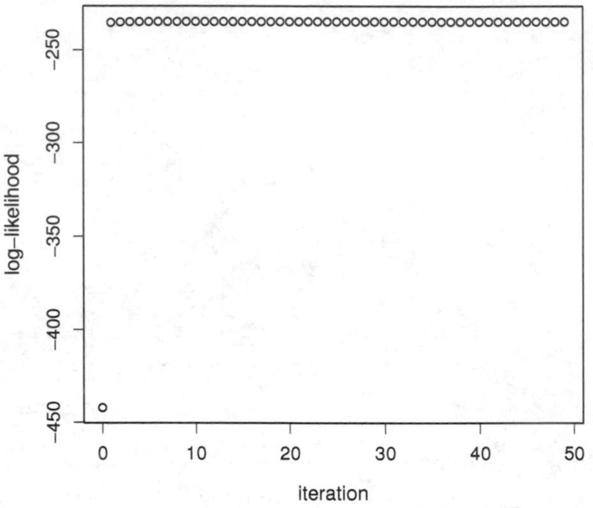

Figure 5.2: Convergence of log-likelihood.

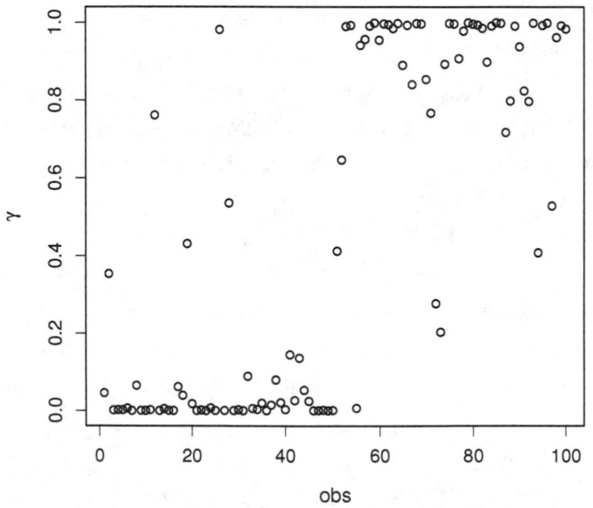

Figure 5.3: Posterior probability of class 1 at convergence.

5.2.4 *Example: k-means Clustering*

In this example we will look at an intuitive clustering algorithm that will turn out to be an application of the **EM principle**. Say we are given m observations $\boldsymbol{x}_1$ to $\boldsymbol{x}_m$ from $\mathbb{R}^p$ and we want to cluster them into k clusters. If one thinks about this for a while, one might get the idea to characterize the k clusters by k points from $\mathbb{R}^p$. Let us call these points $\boldsymbol{\mu}_1$ to $\boldsymbol{\mu}_k$. Now if we ask ourselves to which of the k clusters an observation $\boldsymbol{x}_i$ belongs, one intuitive approach is to assign it so that a distance measure

$$\delta(\boldsymbol{x}_i, \boldsymbol{\mu}_j)$$

becomes minimal w.r.t. j. For the sake of simplicity, we will not go into any details of how to choose the distance measure δ, but one usual choice for normalized data is the L2 norm. Other options include the Mahalanobis distance and if the $\boldsymbol{x}_i$ were discrete an appropriate distance measure for the set. With

$$c_i := \arg\min_j \delta(\boldsymbol{x}_i, \boldsymbol{\mu}_j)$$

we denote the cluster the ith observation belongs to. That is, we **expect** the observation to come from the cluster it is closest to. This idea is formalized in Algorithm 5.7.

Algorithm 5.7 Expectation Step of k-means ($kmeans\_e\_step$)

Require: $\boldsymbol{X} = (\boldsymbol{x}_1 \dots \boldsymbol{x}_m)$ with $\boldsymbol{x}_i \in \mathbb{R}^p$, cluster centers $\boldsymbol{\mu} = (\boldsymbol{\mu}_1 \cdots \boldsymbol{\mu}_k)$ with $\boldsymbol{\mu}_j \in \mathbb{R}^p$

for $i = 1 \to m$ **do**
  $c_i \leftarrow 1$
  for $j = 2 \to k$ **do**
    if $\delta(\boldsymbol{x}_i, \boldsymbol{\mu}_j) < \delta(\boldsymbol{x}_i, \boldsymbol{\mu}_{c_i})$ **then**
      $c_i \leftarrow j$
    end if
  end for
end for
return $\boldsymbol{c} = (c_1 \cdots c_m)$

So now that we know how to assign observations to clusters, how do we determine the locations of the cluster centroids $\boldsymbol{\mu}_j$? Let us assume that we already know to which cluster each observation belongs, i.e. we have a c_i for every $\boldsymbol{x}_i$ that tells us to which of the k clusters the ith observation belongs. Let us, again for the sake of simplicity, assume that we are using the L2 norm as our distance measure. In that case, wouldn't it be natural to choose the μ_i as the mean of all the x_i belonging to cluster j, or more formally,

$$\mu_j = \frac{1}{\sum_{i=1}^{m} I(c_i = j)} \sum_{i=1}^{m} I(c_i = j)\boldsymbol{x}_i,$$

where $I()$ is the indicator function. This procedure would **maximize** the fit of our cluster centroids to the data or, put another way, **minimize** the mean distance from the cluster centroid to the observations belonging to the centroid. Again, we can turn this idea into an algorithm, which is given in Algorithm 5.8

Algorithm 5.8 Maximization Step of k-means (*kmeans_m_step*)

Require: $\boldsymbol{X} = (\boldsymbol{x}_1 \ldots \boldsymbol{x}_m)$ with $\boldsymbol{x}_i \in \mathbb{R}^p$, cluster assignment $\boldsymbol{c} = (c_1 \ldots c_m)$, $c_i \in \{1, \ldots, k\}$

$\boldsymbol{s} = (\boldsymbol{s}_1 \ldots \boldsymbol{s}_k) = \boldsymbol{0}$
$\boldsymbol{m} = (m_1 \ldots m_k) = \boldsymbol{0}$
for $i = 1 \to m$ **do**
    $\boldsymbol{s}_{c_i} \leftarrow \boldsymbol{s}_{c_i} + x_i$
    $m_{c_i} \leftarrow m_{c_i} + 1$
end for
for $j = 1 \to k$ **do**
    $\boldsymbol{\mu}_j = \frac{\boldsymbol{s}_j}{n_j}$
end for
return $\boldsymbol{\mu} = (\boldsymbol{\mu}_1 \cdots \boldsymbol{\mu}_k)$

So what we now have is a way to compute cluster assignments c_i given that we know the cluster centroids $\boldsymbol{\mu}_j$ and another method to compute the cluster centroids $\boldsymbol{\mu}_j$ given cluster assignments c_i. But these two parts are exactly the ingredients that are necessary for an EM type algorithm. Combining the two previous algorithms results in Algorithm 5.9. In this particular implementation, we have chosen to perform the M step before the E step in each loop. That is, we start with a (random) cluster assignment and from that calculate the initial cluster centroids. It is also possible to initialize the $\boldsymbol{\mu}_j$ by assigning them each a different random observation. It is, however, important to note that one should refrain from assigning completely random values to the initial cluster centroids. If this is done, one can end up in a situation where no observation is assigned to a cluster and the following M step fails to compute a new cluster centroid for that particular cluster.

To illustrate the operation of the algorithm, the first five iterations on an

Algorithm 5.9 k-means Clustering

Require: $\boldsymbol{X} = (\boldsymbol{x}_1 \ldots \boldsymbol{x}_m)$ with $\boldsymbol{x}_i \in \mathbb{R}^p$, number of clusters k

for $i = 1 \rightarrow m$ **do**
    $c_i \leftarrow$ random integer between 1 and k {Initialize cluster assignment}
end for
$\boldsymbol{c} \leftarrow (c_1 \ldots c_m)$
for $j = 1 \rightarrow k$ **do**
    $\boldsymbol{\mu}_j \leftarrow$ random vector from $\mathbb{R}^p$ {Initialize cluster centers}
end for
$\boldsymbol{\mu} \leftarrow (\boldsymbol{\mu}_1 \cdots \boldsymbol{\mu}_k)$
repeat
    $\mu' \leftarrow \boldsymbol{\mu}$
    $\boldsymbol{\mu} \leftarrow$ kmeans_m_step$(\boldsymbol{X}, \boldsymbol{c})$
    $\boldsymbol{c} \leftarrow$ kmeans_e_step$(\boldsymbol{X}, \boldsymbol{\mu})$
until $\mu \approx \mu'$

artificial data set are visualized in Figure 5.4. Initially all the cluster centroids appear bunched up in the center. As the algorithm progresses they slowly move outward toward the five clusters. We can see that after five iterations the cluster centroids appear to reflect the clusters in the data almost perfectly.

If one looks into the literature there are many variants and extensions of the basic k-means algorithm presented here. Most of them try to deal with data that are not normalized and for which we do not wish to burn ourselves with the explicit specification of a suitable distance measure. Others extend the approach to different domains, such as time series or add constraints.

Finally, we must note that, like almost all EM type algorithms, the solution returned is not guaranteed to be globally optimal. Therefore, it is advisable to run the algorithm with several different starting configurations and choose the clustering that best fits the data.

5.2.5 *EM Algorithm: Summary and Outlook*

5.2.5.1 *Summary*

In this section we introduced and discussed the EM algorithm in its general form as well as in special forms for the exponential family, for mixture of Gaussians, and for k-means clustering. We showed that the convergence properties of the EM algorithm should be well-known, since convergence to the global maximum of the log-likelihood cannot be guaranteed in practice. Therefore, a plausibility check of EM results is always required.

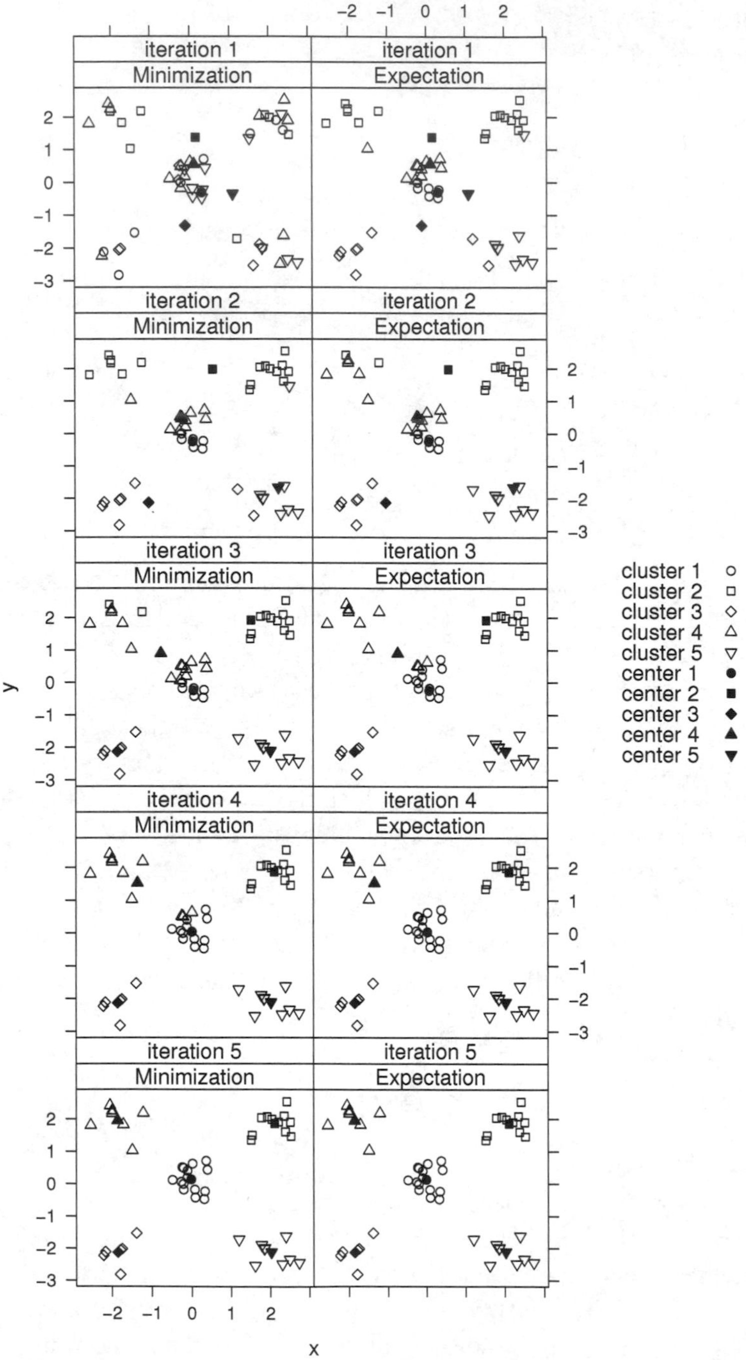

Figure 5.4: Example k-means run. In each row the state of Algorithm 5.9 after one iteration is shown after the M step (first column) and the E step (second column).

Outlook

Notice that the EM algorithm is always especially suited to the problem at hand. Different problems need different EM algorithms. Often, there are special variants of the EM algorithm for different problems, also in order to improve convergence speed. In particular, several have tried to optimize the E step (cp. Meng and van Dyk (1997)).

Notice that because of the unknown latent variables, there is the danger of nonidentifiability, i.e. that there are different β^* with $g(y \mid \beta^*) = L^*$ (cp. Wu (1983)). Finally, the maximum property of the result of the EM algorithm should be plausibility checked.

5.3 Implementation in R

5.3.1 PLS

An R implementation of several PLS methods is available in the contributed CRAN package pls by Mevik and Wehrens (2007). It provides the traditional NIPALS algorithm (function `oscorespls.fit`) as well as many other PLS algorithms that have not been discussed here. Among these methods are kernel PLS (function `kernelpls.fit`) and wide kernel PLS (function `widekernelpls.fit`). The package supports multi-response models as well as partial least squares and principal component regression in function `mvr`. For the correct syntax, please see the help pages for the functions. For example, `oscorespls.fit(X, Y)` applies PLS via the NIPALS algorithm on a matrix of observations X and a vector or matrix of responses Y.

5.3.2 EM Algorithm

Since the EM algorithm is a concept that can be applied rather more widely than a specific method for solving a particular task, there can't be a very general implementation of it. The most complicated part for implementing the EM algorithm is that the functions to be optimized must be known in advance, e.g., the likelihood to be optimized must be known. That means we find many implementations in places where specific tasks are to be solved by means of the EM algorithm.

There are several R packages that implement various EM type algorithms available in the CRAN Task View called "Cluster Analysis & Finite Mixture Models" (Leisch and Grün, 2013). This is not a big surprise given that the EM works very well for the estimation of parameters of mixtures. Among the many mentioned packages are the following widely used packages with EM implementations:

- mclust (Fraley and Raftery, 2002): "Fits mixtures of Gaussians using the EM algorithm and provides comprehensive strategies using hierarchical clustering, EM and the Bayesian Information Criterion (BIC) for clustering, density estimation, and discriminant analysis."
- flexmix (Grün and Leisch, 2008): "Implements a user-extensible framework for EM estimation of mixtures of regression models, including mixtures of (generalized) linear models."
- mixPHM (Mair and Hudec, 2008): "Fits mixtures of proportional hazard models with the EM algorithm."

Since the likelihood that is to be optimized is known for the specific applications of these packages, they can implement their own versions of the EM algorithm.

5.4 Conclusion

In this chapter we showed that the deduction of theoretical properties of a heuristic algorithm might, on the one hand, motivate the usage of a method, like in the case of the optimality property of PLS. On the other hand, the knowledge of theoretical shortcomings might lead to caution applying a method, like in the case of the convergence properties of the EM algorithm. Overall, this chapter should motivate the reader to take up the challenge to prove theoretical properties, since only through them the practical properties of the heuristics become clear.

5.5 Exercises

Exercise 5.1.1: Implement the two PLS algorithms described in Section 5.1.3 (based on covariances and on correlations), both for one response only (individual PLS) and for more than one response (combined PLS).

Exercise 5.1.2: Using the PLS implementation of the PLS algorithms of Section 5.1.3 in Exercise 5.1.1 generate PLS1 and PLS2 for the two response variables HUEREM and HUEREMAL individually and combined, based on the 93 observations of the 18 physico-chemical properties in the data available from the exercises section under `http://www.statistik.tu-dortmund.de/fostal.html`. Calculate the goodness-of-fit R^2 as in Table 5.1, and compare the results.

Exercise 5.1.3:

a. As in Example 5.1, choose $y = a_2x_2 + a_3x_3$ with a_2, a_3 fixed individually,

Table 5.2: Multinomial Example

Blood Type	Genotype	Freq.	Blood Type	Genotype	Freq.
A	AA	p^2	A	A0	$2pr$
B	BB	q^2	B	B0	$2qr$
AB	AB	$2pq$	0	00	r^2

and x_2, x_3 chosen from different distributions. Compare the outcome of PLS with the theoretical result in Example 5.1.

b. Choose $y = a_2x_2 + a_3x_3 + \varepsilon$ with a_2, a_3 fixed individually, x_2, x_3 chosen from different distributions with the same variance V, and ε chosen from $\mathcal{N}(0, \sigma \cdot V)$. Vary $\sigma = 0.1, 1, 10, 100$. Observe the effect on the structure of the first PLS component compared with the result in Example 5.1.

Exercise 5.2.1: One might use population data to validate the hypothesis that human AB0 blood types are determined by 3 alleles, A, B, and 0 at a single genetic locus, rather than being 2 independent factors A/not-A, B/not-B. Suppose that the population frequencies of A, B, and 0 are p, q, and r $(p+q+r=1)$. We want to estimate $[p \quad q \quad r]^T$. We assume that the types of the two alleles carried by an individual are independent (Hardy-Weinberg equilibrium (1908)), and that individuals are independent (unrelated). AB0 blood types are determined as shown in Table 5.2.

Obviously, the easily observable blood type Y and the much harder observable (unknown) genotype X are multinomially distributed, namely:

$$Y \sim M_4(n, (p^2 + 2pr, q^2 + 2qr, 2pq, r^2)),$$
$$X \sim M_6(n, (p^2, 2pr, q^2, 2qr, 2pq, r^2)).$$

Explain that the log-likelihood of Y is hard to maximize in contrast to the log-likelihood of X. Derive an EM algorithm to estimate $[p \quad q \quad r]^T$. This method was already known to geneticists in the 1950s as "genecounting", whereas the EM algorithm was invented much later.

Exercise 5.2.2: Implement Algorithm 5.9 and reproduce Figure 5.4 for a self-chosen artificial data set.

Exercise 5.2.3: Modify your implementation from Exercise 5.2.2 of Algorithm 5.9 to initialize the cluster centroids instead of the cluster assignments. Compare the two algorithms for different artificial data sets. Which one converges faster?

Exercise 5.2.4: How would you choose k for the k-means procedure if it is not known a priori?

Exercise 5.2.5: Look at the R implementation of `kmeans` in the `stats` package. How does the algorithm used differ from the basic k-means algorithm described here?

Chapter 6

Randomization

6.1 Motivation and History

On the one hand, randomness is the basis for statistics. If you do not accept the concept of randomness, then you cannot practice statistics. On the other hand, statistical algorithms on the computer are, for the most part, deterministic. Even though they might produce so-called random numbers, they are usually designed to generate reproducible numbers. Indeed, reproducibility is even one of the intrinsic requirements for any scientific study. This kind of pseudorandomness will be discussed in the chapter. This means that we will introduce so-called generators of pseudorandom numbers, i.e. we will discuss the **randomization** of numbers.

Note that there were already examples in previous chapters where random entries were assumed to be available. For example, the test matrices used for the verification of the LLS-Solvers in Chapter 3 have free elements that can be fixed deliberately. Moreover, there were many stochastic optimizers in Chapter 4 like, e.g., simulated annealing, which utilize random numbers. For example, in order to be equipped with a multiplicity of test matrices it is favorable to fix the free elements with numbers that are "randomly" drawn. Such "random" numbers should possibly be uniformly distributed. This is also true for the random numbers needed for stochastic optimizers. However, in what follows, we will also introduce methods for the generation of realizations of other important discrete and continuous distributions.

6.1.1 Preliminaries

In this subsection, we will give notation, definitions and results about function norms that we assume to be known. In the following sections we will not explicitly refer to this subsection when using these facts. All other theories needed will be cited or proved.

Definition 6.1: L1 Norm
The **L1 norm** of a function g: $\mathbb{R}^d \to \mathbb{R}$ is defined by

$$\|g\|_1 := \int_{\mathbb{R}^d} |g(x)|dx \tag{6.1}$$

Definition 6.2: Variation Norm
To measure the distance between two distributions the **variation norm**

$$\|\mu - \pi\|_{Var} := \sup_{A \subseteq E} |\mu(A) - \pi(A)|, \tag{6.2}$$

can be used, where μ and π are density functions over E and A is an arbitrary Borel-measurable subset. As a special case, pointwise distances for $A = \{x\}$, $x \in E$, are covered.
One can show:

$$\|\mu - \pi\|_{Var} = \frac{1}{2} \sum_{y \in E} |\mu(y) - \pi(y)|, \tag{6.3}$$

when E is finite.
Example: Let $E = \{0, 1\}$ and π be the density of the Bernoulli distribution with $\pi(1) = 0.75$ and $\mu(0) = \mu(1) = 0.5$. Then:

$$\|\mu - \pi\|_{Var} := \max_{A \subseteq E} |\mu(A) - \pi(A)| = 0.25 = \frac{1}{2} \sum_{y \in E} |\mu(y) - \pi(y)|$$

For continuous $\pi(x)$ one can correspondingly show:

$$\|\mu - \pi\|_{Var} = \frac{1}{2} \int_E |\mu(x) - \pi(x)|dx. \tag{6.4}$$

6.2 Theory: Univariate Randomization

6.2.1 *Introduction*

In what follows we first introduce a method for the generation of sequences that one might be willing to call (uniformly distributed) random. The "randomness" of the elements of such a sequence very much depends on the choice of certain parameters.

We will discuss criteria for "randomness" of such sequences, and we will demonstrate how to choose such parameters adequately. Then, we will introduce methods with which uniformly or otherwise distributed random sequences can be generated.

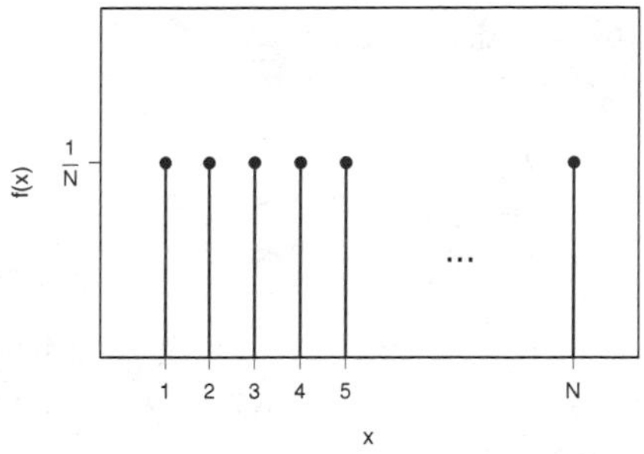

Figure 6.1: A discrete uniform distribution.

Definition 6.3: Random Sequence, According to Lehmer (1951) (Knuth, 1969, p. 127)
The term **random sequence (RS)** conveys the vague notion of a sequence in which each element appears to be unpredictable for the uninitiated and whose elements pass certain statistical tests dependent on its designated use.

Obviously, this definition is not exact. What is needed for its clarification is a preferably short list of mathematical properties, all of which fulfill our intuitive idea of random sequences and which is that complete that we are ready to name **every** sequence with these properties random (Knuth, 1969, p. 128).

Further clarification of this term turns out to be fairly complicated and will not be discussed here (see instead Knuth, 1969, pp. 127 – 151).

6.2.2 Uniform Distribution

An obviously very important property of an RS is its distribution. In general, one first tries to find discrete uniformly distributed RSs since transformations to other distributions are often easy to realize. Therefore, we first define:

Definition 6.4: Uniformly Distributed RS
A **uniformly distributed RS (URS)** is a sequence that satisfies tests on discrete uniform distributions (see Figure 6.1) and other tests on randomness described in the following.

The general discrete uniform distribution is defined as follows:

Definition 6.5: Discrete Uniform Distribution
Each function of the type

$$f(x) = f(x,N) = \begin{cases} \frac{1}{N} & x = x_1, x_2, \ldots, x_N \\ 0 & \text{else,} \end{cases}$$

where $N \in \mathbb{N}$, is called (discrete) density (or probability function) of a **discrete uniform distribution**. A random variable with such a density is called **discretely uniformly distributed**.

For the generation of RSs so-called random number generators are used:

Definition 6.6: Random Number Generator
A method for the generation of RSs is called a **random number generator** (RNG).

Possible RNGs

1. One type of RNGs generates the RS so that no one can predict any element of the sequence. (**Example**: Measurement of the emission of a radioactive source by means of a Geiger counter.)
2. Another type of generator, which is preferred in computers, generates the $(j+1)$th element of the sequence from the first j elements, as exemplified in the following.

Basic RNG
On a computer, a basic random number generator typically generates a discrete uniform distribution on the set $\{0,1,2,\ldots,2^{31}-1\}$ or $\{0,1,2,\ldots,2^{32}-1\}$, depending on whether signed or unsigned integers are required. This choice reflects the fact that most modern machines have a word size of 32 bits.

Such a random number generator is the basis for the generation of random numbers of many distributions, as exemplified in what follows.

Generation of Uniformly Distributed Random Numbers
In order to generate a discrete uniform distribution on the numbers $x_1, \ldots, x_N$, e.g., the interval $\left[0, 2^{31}-1\right]$ is divided into N equal-sized parts that are matched with one x_i each.

Certain types of RNGs for which it is known that they generate sequences with desirable properties are often used in practice, though they do not correspond to the intuitive notion of a random sequence since they are deterministic. In the literature such sequences are often called **pseudo- or quasi-random numbers**. The basic type of such generators is introduced now.

Definition 6.7: Linear Congruential Method
A **Linear Congruential Sequence (LCS)** $\{x_j\}_{j\in\mathbb{N}}$ is recursively defined by:

$$x_{j+1} := \text{mod}_m(ax_j + c), \tag{6.5}$$

where
$x_0 \in \mathbb{N}$ is the **starting value**,
$a \in \mathbb{N}$ is the **multiplier**,
$c \in \mathbb{N}$ is the **increment**,
$m \in \mathbb{N}$ is the **modulus** with $m > x_0, a, c$.

The modulo-operation mod_m is defined as follows:

Definition 6.8: Modulo-Operation
Let $a, b \in \mathbb{Z}$ and $m \in \mathbb{N}$. Then the equivalence relation

$$a \equiv b :\Leftrightarrow a - b \text{ divisible by m}$$

partitions the integer numbers $\mathbb{Z}$ into so-called **residue classes** modulo m. We use the notation $a = \text{mod}_m(b)$ and denote the set of representatives of the residue classes modulo m as $\mathbb{Z}_m = \{0, 1, 2, \ldots, m-1\}$.

Example 6.1: Examples of LCSs

1. Let $a := 5, \quad c := 7, \quad x_0 := 4, \quad m := 13$. Then
$$x_1 = \text{mod}_{13}(27) = 1, \quad x_2 = 12, \quad x_3 = 2, \quad x_4 = 4 = x_0,$$
$$x_5 = x_1 \ldots$$
2. Let $a := 5, \quad c := 7, \quad x_0 := 4, \quad m := 8$. Then
$$x_1 = 3, \quad x_2 = 6, \quad x_3 = 5, \quad x_4 = 0,$$
$$x_5 = 7, \quad x_6 = 2, \quad x_7 = 1, \quad x_8 = 4 = x_0,$$
$$x_9 = x_1 \quad \ldots$$

The fact that in both examples the sequences are cyclical motivates the following definition:

Definition 6.9: Period of an RS
A sequence $\{x_j\}_{j\in\mathbb{N}}$ has **period** μ iff there are $\lambda \in \mathbb{N}_0$ and $\mu \in \mathbb{N}$ such that $x_0, \ldots, x_\lambda, \ldots, x_{\lambda+\mu-1}$ are different and $x_{j+\mu} = x_j$ for $j \geq \lambda$.

In the above examples the periods are 4 and 8, respectively. One can show:

Theorem 6.1: Bound for LCS Period
Every LCS has a period $\leq m$.

Proof. $x_{j+1} := \text{mod}_m(ax_j + c) < m$ for all $j \in \mathbb{N}_0$. Thus, among the first $m+1$ elements of the sequence, at least two elements are equal, e.g. x_λ and $x_{\lambda+\mu}$ with $1 \leq \mu \leq m$.
Since (6.5) uniquely defines the successor x_{j+1} of x_j, it is clear that $x_{j+\mu} = x_j \forall j \geq \lambda$. Thus, the period is $\leq \mu \leq m$. □

Now, we will discuss the choice of $m, a, c,$ and x_0.

Choice of Modulus m
In order that the number of different random numbers is not too small, m should not be too small because of Theorem 6.1.

On computers, m is often taken to be the **word length** since this speeds up the modulo-operation considerably. This typically leads to $m = 2^{31}$ or $m = 2^{32}$, depending on the use of signed or unsigned integers.

Machine Realization of a Generator with $m = 2^{31}$
Each division of an integer by a power of 2 can be realized by a right shift in dual representation. Solving, e.g., $6/4$ by shifting two places to the right in the dual representation $(110)_2$ leads to the result $(1.1)_2$, which is rounded to $(1)_2$, i.e. 1, ignoring the residue 2. The residue is the part of the original dual representation lost by shifting, i.e. $(10)_2$. Thus, the modulo-operation for an LCS can be realized as follows.

Algorithm 6.1 Modulo-Operation for an LCS with $m = 2^{31}$

Require: Integer number representation: (1 sign bit, 31 dual bits)
1: Calculate $ax_j + c$ in a "double-exact accumulator" using two words: (1 sign bit, 63 dual bits).
2: Realize the division by $m = 2^{31}$ by eliminating the 1*st* word, i.e. sign bit and first 31 dual digits.
3: In the remaining second word, add the lost sign bit. {The sign bit always represents "+".}

This obviously makes the modulo-operation very fast for $m = 2^{31}$.

Choice of Multiplier a
Given the modulus m, the choice of a should guarantee a long period. Therefore, we choose a by the following theorem to guarantee maximum period $(= m)$:

Theorem 6.2: Maximal LCS Period
An LCS has period m iff

1. c and m are coprime,

2. $a-1$ is a multiple of every prime dividing m,
3. $a-1$ is a multiple of 4 if m is a multiple of 4.

Proof. see (Knuth, 1969, pp. 15 – 18). □

If $m = 2^{31}$, Theorem 6.2 gives the following conditions for maximum period:

1. c odd,
2. $a \equiv 1 \pmod 4$, i.e. $a \equiv 1 \pmod 8$ or $a \equiv 5 \pmod 8$.

Further, a number of theoretic considerations restrict the possible multipliers a by (Knuth, 1969, pp. 21 – 24, 78, 155):

1. $a \equiv 5 \pmod 8$,
2. $\sqrt{m} < a < m - \sqrt{m}$,
3. the dual representation should not follow a simple scheme.

Choice of Increment c
For fixed a, m the generation is the fastest for $c = 0$. Unfortunately, by Theorem 6.2 a maximum period cannot be reached for $c = 0$. The maximum period for $m = 2^{31}$ is 2^{29} (Knuth, 1969, pp. 18 – 19). Though this might even be long enough, normally another c is used. (Knuth, 1969, p. 78) recommends to choose c so that

$$\frac{c}{m} \approx \frac{1}{2} - \frac{\sqrt{3}}{6}. \tag{6.6}$$

Choosing an odd c so that this is true guarantees, by Theorem 6.2, a maximum period for a corresponding a. Thus, for $m = 2^{31}$ we might want to use

$$c = 453816693.$$

Choice of Starting Value x_0
Since the recommended m, a, c guarantee maximum period, i.e. the LCS randomly runs through all numbers between 0 and $m-1$ until the first repetition, the choice of x_0 is unimportant. Nevertheless, x_0 should be fixed for reproducibility.

Let us now discuss alternatives to the LCS somewhat more briefly.

Inversive Generators (Niederreiter, 1995; Eichenauer and Lehn, 1986, pp. 315 – 326)
Alternatives to LCSs are the so-called inversive generators. There, in the modulo-operation, instead of x_j the corresponding inverse x_j^{-1} in $\mathbb{Z}_m$ is used. Naturally, one first has to clarify how such an inverse is defined.

The transfer of addition and multiplication from $\mathbb{Z}$ to $\mathbb{Z}_m$ in a canonical way leads to a commutative ring $\mathbb{Z}_m$ with a multiplicative identity element. Therefore, the ring $\mathbb{Z}_m$ is called **residue class ring modulo** m. A residue class ring modulo a prime number m is a **field** (not valid for general m). **In what follows, we assume that** m **is prime.**

Definition 6.10: Inverse in a Residue Class Field
If it exists, the **inverse element** $x^{-1} \in \mathbb{Z}_m$ for $x \in \mathbb{Z}_m$ is defined by:

$$mod_m(x \cdot x^{-1}) = 1.$$

Theorem 6.3: Multiplicative Inverse
An element $x \in \mathbb{Z}_m$ has a multiplicative inverse x^{-1} iff $gcd(x,m) = 1$, where gcd stands for "greatest common divisor".

Proof. The proof of this theorem is constructive:
The extended Euclidean algorithm (cp. Example 2.1) delivers the proposition that the gcd can be written as an integer linear combination of x and m:

$$u \cdot x + v \cdot m = gcd(x,m).$$

If $gcd(x,m) = 1$, then the coefficient u of x in this linear combination is the multiplicative inverse of x in the sense of the above definition. □

Example 6.2: Extended Euclidean Algorithm: GCD of the Numbers 17 and 5

$$\begin{aligned} \underline{17} &= 3 \cdot \underline{5} + \underline{2} \quad \text{and} \\ \underline{5} &= 2 \cdot \underline{2} + \underline{1} \end{aligned}$$

Thus,

$$\begin{aligned} \underline{1} &= \underline{5} - 2 \cdot \underline{2} \\ &= \underline{5} - 2 \cdot (\underline{17} - 3 \cdot \underline{5}) = 7 \cdot \underline{5} - 2 \cdot \underline{17} \end{aligned}$$

Thus, $mod_{17}(5 \cdot 7) = 1$, meaning $7 = 5^{-1}$ in $\mathbb{Z}_{17}$.

If m is prime, obviously for all $0 \leq x < m$ it is true that $gcd(x,m) = 1$. Notice that one can even show that $0^{-1} = 0$ and $x^{-1} = mod_m(x^{m-2})$ for all other $x \in \mathbb{Z}_m$ if m is prime (see, e.g., Hellekalek, 1995).

Definition 6.11: Inversive Generator
A random sequence is generated by an **inversive generator** iff

$$x_{j+1} := mod_m(a \cdot x_j^{-1} + c),$$

where x_j^{-1} is the inverse of x_j in the residue class field $\mathbb{Z}_m$ and m is prime.

As the modulus, the prime number $m = 2^{31} - 1$ could be used. For sensible choices of the parameters a, c and for further considerations about inversive generators, cp., e.g., (Hellekalek, 1995).

6.2.2.1 *Multiply-with-Carry Generators*

In the previous sections we have looked at several historic RNG concepts. These are not considered state of the art, and their use is usually frowned upon. In this section we will learn about a structurally simple extension of the LCS idea that can create competitive random number generators that are both fast on modern hardware and easy to implement.

The class of **multiply-with-carry** (MWC) generators was first proposed by Marsaglia (1996). It can be thought of as a generalization of the LCS generator by replacing the fixed increment c with a varying increment. For the simplest MWC generator, a lag-1 MWC generator, the recursive definition is given by

$$x_j := \text{mod}_m(ax_{j-1} + c_{j-1}) \quad \text{and} \quad c_j := \left\lfloor \frac{ax_{j-1} + c_{j-1}}{m} \right\rfloor. \tag{6.7}$$

Notice how similar the recursion for x is to the one defined in Equation 6.5. The only difference is that for the MWC the increment or **carry** changes for every x. To see why this is called a multiply-with-carry generator, let us look at an example.

Example 6.3: MWC Sequence
Let $a = 6$ and $m = 10$, then the first 12 elements of the MWC sequence for $x_0 = 4$ and $c_0 = 4$ are given by

j	0	1	2	3	4	5	6	7	8	9	10	11	...
c_j	4	2	5	0	3	0	1	4	5	5	2	2	...
x_j	4	8	0	5	0	3	8	9	8	3	3	0	...
$ax_j + c_j$	28	50	5	30	3	18	49	58	53	23	20	2	...

Notice that x_j is the last digit of $ax_{j-1} + c_{j-1}$, and that c_j is the leading digit, or carry, of $ax_{j-1} + c_{j-1}$.

Before we continue, let us ask ourselves why we would want to use this generator instead of the LCS generator? The construction looks very similar, but we have to do more work per random number we wish to calculate. But do we really need to do more work? If we choose $m = 2^{32}$, ignoring for a minute if this leads to a decent generator, then x_{j-1} and c_{j-1} can be stored in word-sized, i.e. 32-bit, integers and $t = ax_{j-1} + c_{j-1}$ is a 64-bit integer whose lower 32 bits are x_j and whose upper 32 bits are c_j. Most modern CPUs have instructions to calculate the product of two 32-bit integers in full 64-bit precision with little or no extra overhead. The rest is then just a matter of bit shifting and masking the result to extract the new x and c by implementing the modulo-operation in analogy to Algorithm 6.1. Therefore, there really isn't much more work involved in this generator compared to an LCS type generator. It does have one major advantage, though: it has a much longer period.

If you pick, e.g., $a = 698,769,069$, $m = 2^{32}$ and pick your seed so that $0 \leq c_0 < a$ and $0 \leq x_0 < m$, $[c_0 \;\; x_0] \neq [0 \;\; 0]$ and $[c_0 \;\; x_0] \neq [a-1 \;\; m-1]$, then the resulting MWC sequence will have a period of $am-1$, i.e. approximately $2^{50.4} \approx 10^{18.2}$, quite a bit longer than the best possible period length that can be achieved with an LCS type generator. But we have to pay a price for this: We need two seeds and have a higher, although in most cases negligibly higher, computational burden per generated random number.

So how do we pick a and m so that we may achieve a long period for the generator for almost all seeds? Without going into all the details, all sensible values of a and m are such that $p = am - 1$ is prime. The reasoning behind this is that for each seed pair x and c we may obtain a different period length of the associated MWC. The length of the period is of the order of m in the group of residues relatively prime to some d, where d is one of the divisors of p. If p is prime, then there are only two trivial divisors, 1 and d. The former leads to two trivial seeds that need to be avoided (see above example) and the latter to a period length of

$$\arg\min_i \operatorname{mod}_p m^i = 1. \tag{6.8}$$

This motivates the period given in the previous paragraphs. For a full description of the theory behind the calculation of the period, see Marsaglia (1996).

Algorithm 6.2 gives an efficient algorithm for most modern CPUs to calculate consecutive elements of the MWC sequence. In practice, the achievable period length of these generators is still not satisfactory. It can be extended almost arbitrarily by using a lag-r MWC generator. The lag r MWC uses the rth from last generated value rather than the last one. So the defining equation is given by

$$x_j = \operatorname{mod}_m(ax_{j-r} + c_{j-r}) \quad \text{and} \quad c_j = \left\lfloor \frac{ax_{j-r} + c_{j-r}}{m} \right\rfloor. \tag{6.9}$$

Choosing a so that the period length is maximal for these types of generators is nontrivial, for details see Marsaglia (2003).

Algorithm 6.2 Calculation of Next Element in a Multiply-with-Carry Sequence

Require: x and c are 32 bit unsigned integers
Require: $c \neq 0$ or $x \neq 0$
Require: $c \neq a-1$ or $x \neq 2^{32}-1$
Require: $p = 2^{32}a - 1$ is prime and $(p-1)/2$ is prime. I.e. p is a safeprime.
1: $t \leftarrow ax + c$, where t is a 64 bit integer
2: $x_o \leftarrow$ lower 32 bit of t
3: $c_o \leftarrow$ upper 32 bit of t
4: **return** (x_o, c_o)

Table 6.1: List of Other Common Random Number Generators

RNG Class	Comment
lagged Fibonacci	Sensitive to seeds
Mersenne Twister	Widely deployed and popular because of the long period length
WELL	Improvement on the Mersenne Twister

6.2.2.2 Overview of Other Generators

A plethora of other generators have been proposed in the literature, but many of them have been proven to be inadequate for modern usage. A summary of the most common algorithms not mentioned so far found in use today is given in Table 6.1. The lagged Fibonacci sequence type generators are popular because they are discussed at length in volume II of Knuth (1998). In 2002 the generator recommended by him had to be revised because it showed some weaknesses if the initial seed was chosen unfortunately. Both the Mersenne Twister (Matsumoto and Nishimura, 1998) as well as the WELL generator (Panneton et al., 2006) are structured somewhat differently in that they do not operate on integers directly, but on 1-bit vectors of length 32, that is, they view a 32-bit integer as a vector of 32 1-bit integers. These generators generate a true stream of random bits. Their structure does not lend itself to easy implementation, however, and if possible, one should use the code provided by the original authors.

Let us now discuss some of the tests proposed to check the randomness of a number sequence. Other tests can be found in the exercises.

6.2.2.3 Empirical Tests on Randomness

There are lots of different tests candidate random sequences should pass before their randomness is accepted. We will only discuss some simple tests here to show the principle, give some more examples in the exercises, and refer to so-called test suites for the interested reader.

The following tests check a given section from the period of a number sequence on "randomness". The tests will use the following sequences:

$\{x_i\}_{i\in\mathbb{N}_0}$, an RS with elements in $\mathbb{N}_0$,

$\{u_i\}_{i\in\mathbb{N}_0}$, the corresponding sequence with elements $u_i := \frac{x_i}{m} \in [0,1)$,

$\{y_i^{(D)}\}_{i\in\mathbb{N}_0}$, the corresponding sequence with elements $y_i^{(D)} := \lfloor Du_i \rfloor \in [0,D)$, $D \in \mathbb{N}$.

Most of the following tests utilize the auxiliary chi-square test.

Chi-Square Test

Let M be the number of independent observations falling in K different categories, then the following hypothesis is tested:

H_0**:** The probability that an observation falls into category s is equal to p_s, $1 \le s \le K$.

M is to be chosen so that

$$Mp_s \ge 5, \qquad 1 \le s \le K. \tag{6.10}$$

Now let M_s be the number of observations in category s with $\sum_{s=1}^{K} M_s = M$. To assess the hypothesis on the basis of the observations, the realization v of the statistic

$$V := \sum_{s=1}^{K} \frac{(M_s - Mp_s)^2}{Mp_s} = \frac{1}{M}\sum_{s=1}^{K} \frac{M_s^2}{p_s} - \sum_{s=1}^{K} (2M_s - Mp_s) = \frac{1}{M}\sum_{s=1}^{K} \frac{M_s^2}{p_s} - M,$$

is calculated. Assuming that the null-hypothesis is correct, V is approximately chi-square distributed with $K-1$ degrees of freedom if (6.10) is satisfied for M.

Notice: For a chi-square distribution: variance $= 2\cdot$ (expected value).

We say that the hypothesis is

$$\begin{array}{ll} \textbf{rejected} & \text{iff } P(V < v) < 0.01 \text{ or } P(V \ge v) < 0.01, \\ \textbf{suspicious} & \text{iff } 0.01 \le P(V < v) < 0.05 \text{ or } 0.01 \le P(V \ge v) < 0.05, \\ \textbf{nearly suspicious} & \text{iff } 0.05 \le P(V < v) < 0.1 \text{ or } 0.05 \le P(V \ge v) < 0.1, \\ \textbf{fair} & \text{iff } 0.1 \le P(V \ge v) \le 0.9. \end{array} \tag{6.11}$$

The parameters of the following tests using the chi-square test are always chosen so that:

1. (6.10) is satisfied and
2. M is not too big (because of computer time), e.g. $M = 10,000$.

Now, we will introduce some tests on different aspects of uniformity and independence:

- a basic "Uniform distribution test" as a simple test on uniformity and independence,
- the "Gap test", representative for tests on whether number sizes do not have special structure,
- the "Permutation test", testing the size order in tuples of numbers,
- the "Maximum-of-$\mathscr{T}$ test on the distribution of maxima of tuples, and
- the "Correlation test" on the correlation in a sequence.

Moreover, we will indicate that tests on structured subsequences might be sensible.

Definition 6.12: Uniform Distribution Test
k-tuples of succeeding elements of a sequence $\{x_i\}_{i \in \mathbb{N}_0}$ are tested on independence and uniform distribution by means of a chi-square test on $\{y_i^{(D)}\}_{i \in \mathbb{N}_0}$, $D \in \mathbb{N}$.

For LCSs $K := D$ is used in the chi-square test and probabilities $\frac{1}{D^k}$ in the cases:

$$k = 1 \quad \text{and} \quad (y_j^{(D)}) \qquad , j = 1, 2, \ldots$$

$$k = 2 \quad \text{and} \quad \left.\begin{matrix} (y_{2j-1}^{(D)}, y_{2j}^{(D)}) \\ (y_{2j}^{(D)}, y_{2j+1}^{(D)}) \end{matrix}\right\} \qquad , j = 1, 2, \ldots$$

$$k = 3 \quad \text{and} \quad \left.\begin{matrix} (y_{3j-2}^{(D)}, y_{3j-1}^{(D)}, y_{3j}^{(D)}) \\ (y_{3j-1}^{(D)}, y_{3j}^{(D)}, y_{3j+1}^{(D)}) \\ (y_{3j}^{(D)}, y_{3j+1}^{(D)}, y_{3j+2}^{(D)}) \end{matrix}\right\} \qquad , j = 1, 2, \ldots$$

For example, we could check the following cases: $\begin{cases} k = 1 & \text{with } D = 128, \\ k = 2 & \text{with } D = 32, \\ k = 3 & \text{with } D = 8. \end{cases}$

Definition 6.13: Gap Test
For $0 \leq \alpha < \beta \leq 1$ the length of the subsequences $\{u_j, u_{j+1}, \ldots, u_{j+r}\}$ is analyzed for which $u_{j+r} \in (\alpha, \beta)$, but the other elements not. Such a subsequence of $r + 1$ numbers represents a gap of length r. For a given $z \in \mathbb{N}$ the number of

gaps of length $0,1,\dots,z-1$ are counted as well as the gaps of lengths greater than or equal to z.

These counts are analyzed by means of a chi-square test with $K := z+1$ categories and probabilities $p_r := q(1-q)^r$, $r = 0,\dots,z-1$, and $p_z = (1-q)^z$, where $q := \beta - \alpha$.

For example, the case $\alpha = \frac{1}{4}$, $\beta = \frac{3}{4}$, $z = 5$ could be analyzed.

Definition 6.14: Permutation Test
Divide the sequence $\{u_i\}_{i\in\mathbb{N}_0}$ into N disjoint groups

$$(u_{\mathcal{T}\cdot j}, u_{\mathcal{T}\cdot j+1}, \dots, u_{\mathcal{T}\cdot j+\mathcal{T}-1}),$$

$0 \le j < N$, with $\mathcal{T}$ elements each. The elements in such a group can be size ordered in $\mathcal{T}!$ different ways.

The realized number of each of these orderings is determined, and then a chi-square test is applied with $K := \mathcal{T}!$ categories and probability $\frac{1}{\mathcal{T}!}$ for each category.

For example, for $\mathcal{T} = 3$, a sequence $\{u_i\}$ of length 100 would be partitioned into $N = \lfloor 100/3 \rfloor = 33$ disjoint groups

$$(u_0, u_1, u_2),\ (u_3, u_4, u_5), \dots, (u_{96}, u_{97}, u_{98}),$$

and the size ordering inside each of these groups is determined. The $6 = 3!$ possible orderings are: $u_i < u_{i+1} < u_{i+2}$, $u_i < u_{i+2} < u_{i+1}$, $u_{i+1} < u_i < u_{i+2}$, $u_{i+1} < u_{i+2} < u_i$, $u_{i+2} < u_i < u_{i+1}$, and $u_{i+2} < u_{i+1} < u_i$. Note that all elements of $\{x_i\}$, and thus all elements of $\{u_i\}$, are different.

Definition 6.15: Maximum-of-$\mathcal{T}$ Test
For $0 \le j < N$ let

$$z_j := \max(u_{\mathcal{T}\cdot j}, u_{\mathcal{T}\cdot j+1}, \dots, u_{\mathcal{T}\cdot j+\mathcal{T}-1}),$$

where $\mathbb{N} \ni \mathcal{T} < M$. The distribution of Z is of the form:

$$\begin{aligned}
P(Z \le x) &= P(\max(U_1, U_2, \dots, U_{\mathcal{T}}) \le x) \\
&= P(U_1 \le x, U_2 \le x, \dots, U_{\mathcal{T}} \le x) \\
&= P(U_1 \le x)P(U_2 \le x)\dots P(U_{\mathcal{T}} \le x) \\
&= x \cdot x \cdot \dots \cdot x \\
&= x^{\mathcal{T}}
\end{aligned}$$

if the U_t are all independently uniformly distributed on the interval $[0,1)$. Therefore:

$$P(Z^{\mathcal{T}} \leq x^{\mathcal{T}}) = x^{\mathcal{T}}.$$

Thus, $Z^{\mathcal{T}}$ is uniformly distributed on the interval $[0,1)$ and $\lfloor DZ^{\mathcal{T}} \rfloor$ is uniformly distributed on the integers of the interval $[0,D)$.

Therefore, a uniform distribution test is applied to the section $\lfloor Dz_0^{\mathcal{T}} \rfloor, \ldots, \lfloor Dz_{N-1}^{\mathcal{T}} \rfloor$ of the sequence with $k=1$.

For example, the cases $\mathcal{T} = 2,3,4,5$ with $D = 32$ could be analyzed.

Definition 6.16: Correlation Test
This test calculates the empirical (cyclical) **autocorrelation coefficient**

$$C := \frac{N(u_0u_1 + u_1u_2 + \cdots + u_{N-2}u_{N-1} + u_{N-1}u_0) - (u_0 + u_1 + \cdots + u_{N-1})^2}{N(u_0^2 + u_1^2 + \cdots + u_{N-1}^2) - (u_0 + u_1 + \cdots + u_{N-1})^2}$$

of a section of length N of the sequence $\{u_i\}_{i \in \mathbb{N}_0}$, which is a measure for the dependency of consecutive elements.

The coefficient is called cyclical because of the term $u_{N-1}u_0$ in the numerator. One can show that $-1 \leq C \leq 1$, $C = \pm 1$ indicates complete dependence and $C = 0$ independence.

In general, it cannot be expected that the autocorrelation coefficient is exactly zero for random sequences. For example, the above autocorrelation coefficient is **always** equal to -1 for $N = 2$, even when the realizations of U_i are independent, unless the numerator is zero:

$$C = \frac{4u_0u_1 - (u_0 + u_1)^2}{2(u_0^2 + u_1^2) - (u_0 + u_1)^2} = -\frac{(u_0 - u_1)^2}{(u_0 - u_1)^2} = -1.$$

The general exact distribution of this autocorrelation coefficient for uniform U is unfortunately unknown. However, this distribution is, at least for normal U, for sufficiently large N fairly well approximated by a normal distribution with expected value and variance:

$$\mu_N := -\frac{1}{N-1} \quad \text{and} \quad \sigma_N^2 := \frac{\frac{N(N-3)}{N+1}}{(N-1)^2}, \quad N > 3$$

(cp. Dixon, 1944, pp. 119 – 144).

For a suitable sequence $\{u_i\}_{i \in \mathbb{N}_0}$

$$\mu_N - 2\sigma_N \leq C \leq \mu_N + 2\sigma_N. \tag{6.12}$$

should thus be true in 95% of the cases. Indeed, (6.12) is also essentially true in 95% of the cases for uniform $\{u_i\}_{i\in\mathbb{N}_0}$ (see (Knuth, 1969, pp. 64 – 65)).

The correlation test can be rated as **passed** (corresponding to "fair" in (6.11)) iff (6.12) is true, and as **not passed** otherwise (corresponding to "rejected" in (6.11)).

Definition 6.17: Tests for Subsequences
Often, groups of, say, q random numbers are needed in applications. Therefore, the above tests are also applied to subsequences of the types:

$$\{x_0, x_q, x_{2q}, \ldots\}, \{x_1, x_{q+1}, x_{2q+1}, \ldots\}, \{x_{q-1}, x_{2q-1}, x_{3q-1}, \ldots\}.$$

For example, subsequences of the first type with $q = 2, 3, 4$ could be analyzed.

For LCSs the following formula can be used for simplification (Knuth, 1969, p. 10):

$$x_{i+k} = (a^k x_i + \frac{(a^k - 1)c}{a - 1}) \bmod (m), \qquad k \geq 0, i \geq 0.$$

For more examples of empirical tests on independence and randomness see the exercises.

6.2.3 *Test Suites for Random Number Generators*

There are two well-established test suites that combine all of the tests described above and many more. These are the DieHard and its descendant DieHarder (Brown et al., 2010) test suites, as well as the newer TestU01 suite (L'Ecuyer and Simard, 2007), which improves on the DieHard tests in several ways. Instead of simply running each of the statistical tests once, they test multiple random sequences and collect the p-values. These p-values are then tested for uniformity, a strong indicator that the null hypothesis holds, using a Kolmogorov-Smirnov test. This is repeated for all tests in the battery, and the number of times the KS test rejected the null hypothesis is counted, and this is then reported as the result of the test battery.

6.2.3.1 *Unrecommended Generator*

It is well-known that the (at least in the 1970s) often used IBM generator **RANDU**:

$$x_{j+1} := 65539 x_j \bmod (2^{31})$$

cannot be recommended. Marsaglia (1968) has stated that for every LCS "random numbers fall mainly in the planes". For RANDU one can show that the

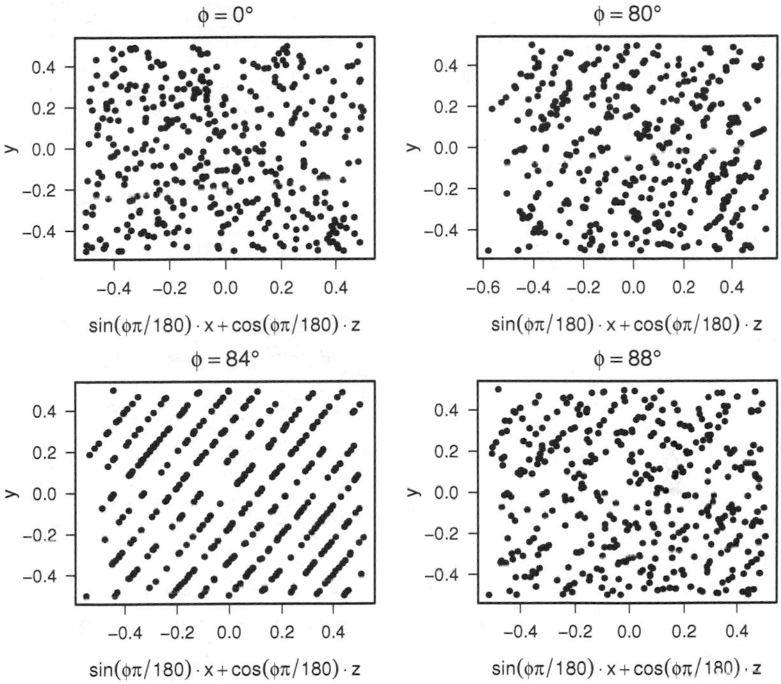

Figure 6.2: Rotation of RANDU outcomes.

relation $x_j = 6x_{j-1} - 9x_{j-2}$ is valid, i.e. the random numbers fall into planes in 3D. Let $(x_j, y_j, z_j) = (x_j, x_{j-1}, x_{j-2})/2^{31} - (.5, .5, .5)$. In Figure 6.2 the (x,y)–coordinates of 400 such points of the RANDU generator are presented after rotation around the y-axis with rotation angles $0^o, 80^o, 84^o, 88^o$. Obviously, the planes are the clearest for a rotation of 84°, but a small deterioration already blurs the picture. Moreover, the number of planes the RANDU realizations lie on is very small, i.e. 15.

6.2.3.2 Recommended Generators

A good starting point when choosing a generator is L'Ecuyer and Simard (2007). The authors compare about 60 different generators using their TestU01 suite. Of the generators presented so far, only the lagged MWC, the WELL, and the Mersenne Twister are deemed fit. Whether a period length of much larger than 2^{60} is required depends on the use case.

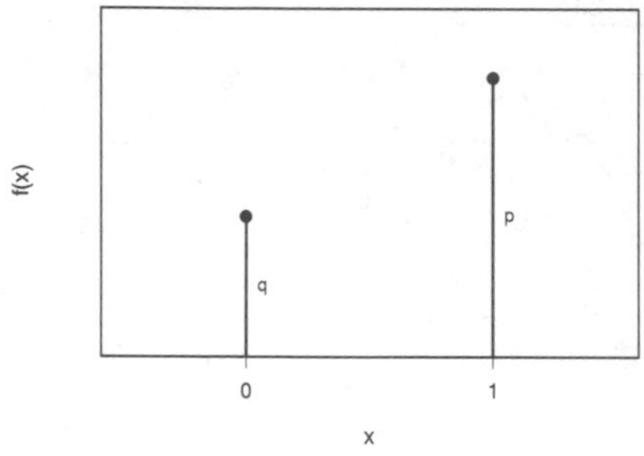

Figure 6.3: Bernoulli distribution.

6.2.4 *Other Distributions*

Random numbers from other distributions than the discrete uniform distribution can often be easily derived from random realizations of this basic distribution. This will be demonstrated in what follows by means of the following basic discrete distributions: Bernoulli distribution, Binomial distribution, hypergeometrical distribution, Poisson distribution, and negative binomial distributions, including the Pascal distribution. Along the way, we will also introduce a general method to the generation of random numbers for discrete distributions.

6.2.4.1 *Bernoulli Distribution*

Definition 6.18: Bernoulli Distribution
Each discrete density function of the type

$$f(x) = f(x;p) = \begin{cases} p^x(1-p)^{1-x} & \text{for } x = 0,1 \\ 0 & \text{else,} \end{cases}$$

where $0 \leq p \leq 1$, defines a density of a **Bernoulli distribution (with parameter p)**. For an example see Figure 6.3. **Notation**: $q := 1-p$

Bernoulli Experiment
A Bernoulli experiment has the two possible outcomes 0 or 1, or “failure” and “success”, respectively.

Bernoulli distributed random realizations are generated by the partition of the interval $[0, 2^{31}-1]$ into two subintervals, $I_0 := [0, q(2^{31}-1))$ and $I_1 := [q(2^{31}-1), 2^{31}-1]$, and by the allocation of 0 and 1, respectively, to these subintervals.

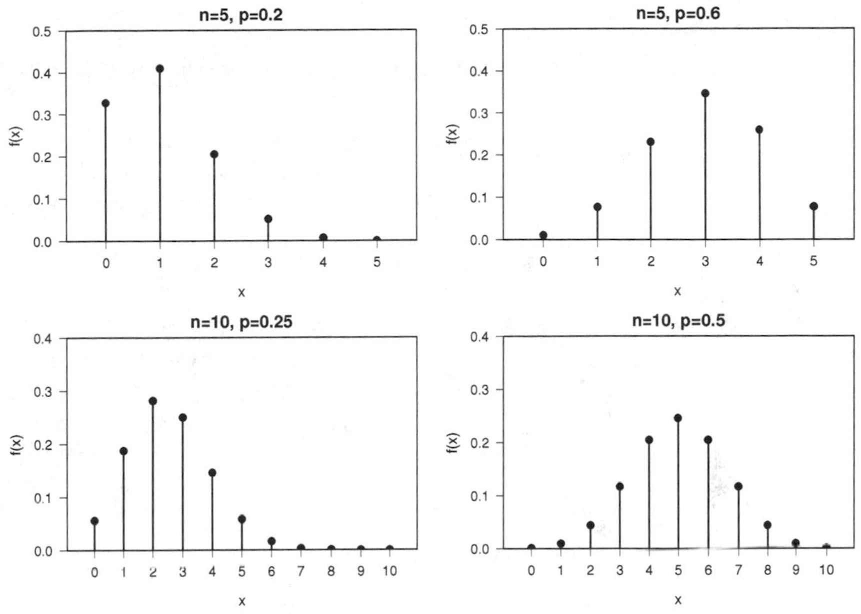

Figure 6.4: Densities of binomial distributions with different parameters.

6.2.4.2 Binomial Distribution

Definition 6.19: Binomial Distribution
Each discrete density function of the type

$$f(x) = f(x; n, p) = \begin{cases} \binom{n}{x} p^x q^{n-x} & \text{for } x = 0, 1, \ldots, n \\ 0 & \text{else,} \end{cases}$$

where $n \in \mathbb{N}$, $0 \leq p \leq 1$ and $q := 1 - p$, defines a density of a **binomial distribution (with parameters n, p)**. A random variable with such a density is called **binomially distributed** and is abbreviated $Bin(n, p)$.

A binomial distribution can take very different forms. For examples, see Figure 6.4.

Binomial Experiment
Consider the random experiment consisting of n "independent" repetitions of the same Bernoulli experiment. Then, the sample space has the form:

$$\Omega = \{\omega = (\omega_1, \omega_2, \ldots, \omega_n) \mid \omega_i = \text{success or } \omega_i = \text{failure}\}.$$

Since the single experiments are independent of each other, the probability of a result of the total experiment is given by multiplication of the probability of the results of the single experiments.

Let the random variable
$X :=$ number of successes in n independent Bernoulli experiments. Then:

$$P(X=x) = \binom{n}{x} p^x q^{n-x} \qquad \text{for } x = 0, 1, \ldots, n,$$

since the probability of x successes and $(n-x)$ failures is equal to $p^x q^{n-x}$, and since $\binom{n}{x}$ different combinations of Bernoulli experiments can have x successes.

Therefore, a **random realization of a binomial distribution with parameters n, p** is generated as the sum of n random realizations of a Bernoulli distribution with parameter p.

6.2.4.3 Hypergeometrical Distribution

Definition 6.20: Hypergeometrical Distribution
Each discrete density function of the type

$$f_X(x; M, K, n) = \begin{cases} \frac{\binom{K}{x}\binom{M-K}{n-x}}{\binom{M}{n}} & \text{for } x = 0, 1, \ldots, n \\ 0 & \text{else,} \end{cases}$$

where $M \in \mathbb{N}$, $M \geq K \in \mathbb{N}_0$ and $M \geq n \in \mathbb{N}$, defines a density of a **hypergeometrical distribution (with parameters M, K, n)** (for examples see Figure 6.5).

Example 6.4: Hypergeometrical Distribution
Consider an urn containing K red and $(M-K)$ white balls: How big is the chance to get exactly k red balls when drawing n balls **without replacement**? Let the corresponding event A be defined by $A :=$ "k red balls and $(n-k)$ white balls drawn", then: $P(A) = \frac{\binom{K}{k}\binom{M-K}{n-k}}{\binom{M}{n}}$.

Then a **random realization of a hypergeometrical distribution with parameters M, K, n** is generated by shuffling the numbers 1 to M. Draw n times with replacement and count the numbers $\leq K$. This count is the requested realization.

6.2.4.4 Poisson Distribution

Definition 6.21: Poisson Distribution
Each discrete density function of the type

$$f_X(x; \lambda) = \begin{cases} \frac{e^{-\lambda}\lambda^x}{x!} & \text{for } x = 0, 1, 2 \ldots \\ 0 & \text{else,} \end{cases}$$

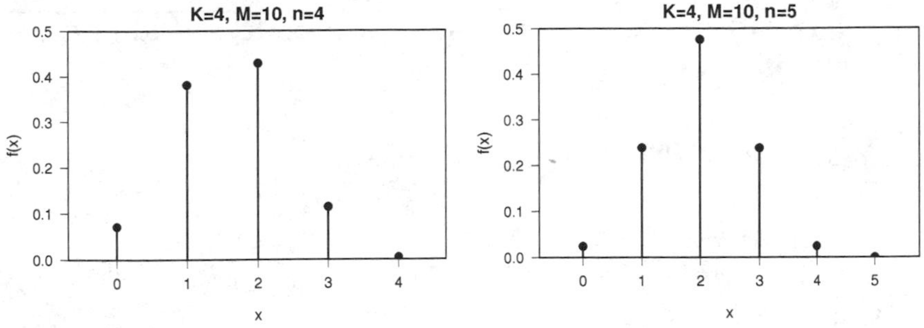

Figure 6.5: Densities of hypergeometrical distributions with different parameters.

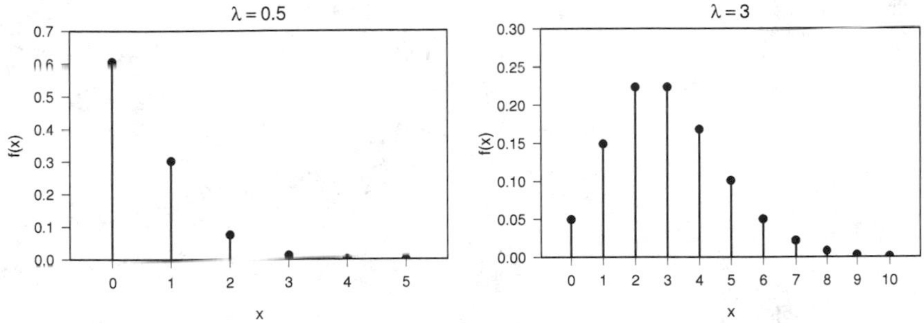

Figure 6.6: Densities of Poisson distributions with different parameters.

where $\lambda > 0$, defines a density of a **Poisson distribution (with parameter λ)** (for examples see Figure 6.6).

Poisson Distributed Random Realizations
Poisson distributed random realizations are generated similar to uniformly distributed random numbers by a **general method for the generation of random realizations of discrete distributions** (see Algorithm 6.3). In the case of the Poisson distribution set, $p_i := f_X(i; \lambda)$.

6.2.4.5 *Waiting Time Distributions*

Definition 6.22: Waiting Time Distributions
Another type of distribution for counting problems is the so-called **waiting time distributions**. Such distributions specify the probability of the waiting time for one or more successes in, e.g., Bernoulli experiments. In other words, we look for the probability of the rth success in the yth experiment.

Algorithm 6.3 General Method for the Generation of Random Realizations x of Discrete Distributions on $x = 0, 1, 2, \ldots$

Require: Basic random number generator RNG in $[0, 2^{31} - 1]$
1: Divide the interval $[0, 2^{31} - 1]$ into subintervals
$I_0 := [0, p_0(2^{31} - 1))$,
$I_1 := [p_0(2^{31} - 1), (p_0 + p_1)(2^{31} - 1)), \ldots$,
$I_i := [(p_0 + \ldots + p_{i-1})(2^{31} - 1), (p_0 + \ldots + p_i)(2^{31} - 1)), \ldots$,
where $p_i :=$ probability of realization i.
2: Generate a random realization z from $0, 1, \ldots, 2^{31} - 1$ by RNG.
3: Sum up the p_j, $j = 0, 1, 2, \ldots$, until for the first time $\sum_{j=0}^{i} p_j(2^{31} - 1) > z$.
4: Set $x := i$ {since $z \in I_i$}

Let **X := number of failures before the** r**th success = waiting time for the** r**th success.** Then, the latest experiment was a success, having probability p. Among the first $x + r - 1$ experiments there were $r - 1$ successes and x failures. The probability of such an event is obviously:

$$p\binom{x+r-1}{r-1}p^{r-1}q^x = \binom{r+x-1}{x}p^r q^x$$

The corresponding distribution is called negative binomial distribution.

Definition 6.23: Negative Binomial Distribution
Each discrete density function of the type

$$f(x) = f(x; r, p) = \begin{cases} \binom{r+x-1}{x}p^r q^x & \text{for } x = 0, 1, 2 \ldots \\ 0 & \text{else,} \end{cases}$$

where $r \in \mathbb{N}$, $0 < p \leq 1$ and $q := 1 - p$, defines a density of a **negative binomial distribution (with parameters r, p)**.

Random realizations of a negative binomial distribution with parameters r, p are generated by counting the number of failures in Bernoulli experiments with parameter p before the rth success.

An especially important negative binomial distribution is the geometrical distribution.

Definition 6.24: Geometrical Distribution
Each discrete density function of the type

$$f(x) = f(x; p) = \begin{cases} pq^x & \text{for } x = 0, 1, 2 \ldots \\ 0 & \text{else,} \end{cases}$$

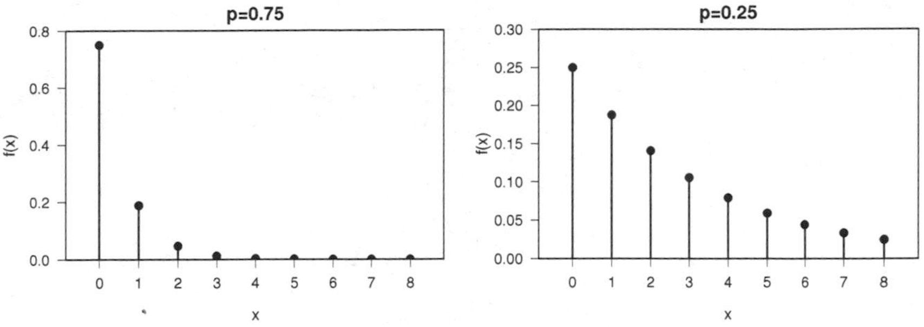

Figure 6.7: Densities of geometrical distributions with different parameters.

where $0 < p \leq 1$ and $q := 1 - p$, defines a density of a **geometrical** (or **Pascal**) **distribution** (for examples see Figure 6.7).

A Pascal distribution is a negative binomial distribution with $r = 1$. A geometrical distribution characterizes the waiting for the first success.

Realizations of a geometrical distribution can be generated by counting the number of Bernoulli experiments before the first success.

6.2.5 *Continuous Distributions*

Let us now switch to continuous distributions. Again, we will concentrate on the most important distributions of this type. We will comment on the generation of random realizations of the continuous uniform distribution, the triangular distribution, the normal distribution, the exponential distribution, and the lognormal distribution. Along the way, we will also introduce a general method to the generation of random numbers for continuous distributions with strictly monotonic distribution functions, the so-called **inversion method**.

6.2.5.1 *Continuous Uniform Distribution*

Definition 6.25: Continuous Uniform Distribution
A continuous density function of the type

$$f(x) = f(x; a, b) = \begin{cases} \frac{1}{b-a} & \text{for } x \in [a, b] \\ 0 & \text{else,} \end{cases}$$

where $a, b \in \mathbb{R}$ and $a < b$, defines the density of the **continuous uniform distribution on the interval** $[a, b]$. A random variable with such a density is called **(continuously) uniformly distributed**.

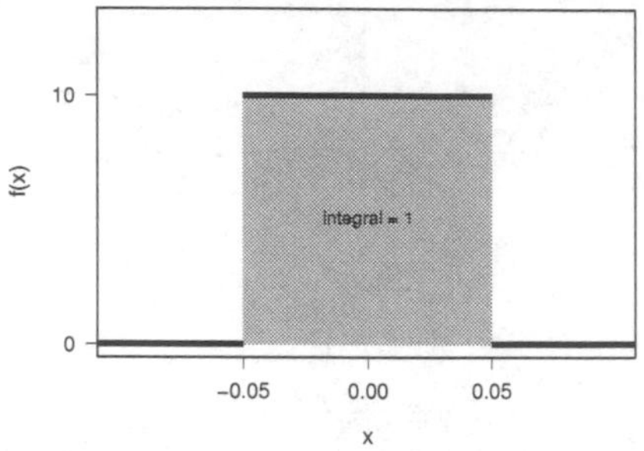

Figure 6.8: Density of a rectangular distribution of rounding errors.

Because of the shape of the density function, this distribution is also sometimes referred to as the **rectangular distribution** $R[a,b]$.

Example 6.5: Rounding Errors
Let $X =$ "rounding error of measurements to one decimal place" ranging from -0.05 to $+0.05$ (see Figure 6.8). Obviously, rounding errors are uniformly distributed on $[-0.05, 0.05]$.

A random number from the interval $[0,1)$ is a **realization of a uniformly distributed random variable in** $[0,1)$. Such a random number u can be approximated by

$$u := \frac{x}{2^{31}},$$

where x was generated by a discrete uniform random number generator on the numbers $0,1,2,\ldots,2^{31}-1$. Similarly, if x were distributed according to the discrete uniform distribution on the set $\{0,1,2,\ldots,2^{32}-1\}$, then

$$u := \frac{x}{2^{32}}$$

approximates a continuous uniform random variate.

Since computers generally use finite floating-point approximations to represent numbers from the real axis, even u follows a discrete distribution in practice. We should therefore take a special look at the case of converting a random 32-bit integer x into a random double floating-point value u that approximates the $R[0,1)$ distribution. The naive way of going about this, and the one chosen most of the time, is to convert x into a double floating-point value[1] and then multiply it by $2.3283064365386962890 6 \times 10^{-10}$, which is

[1]Remember that a double has a mantissa of 52 bits, so any 32-bit integer can be represented exactly by a double floating-point value.

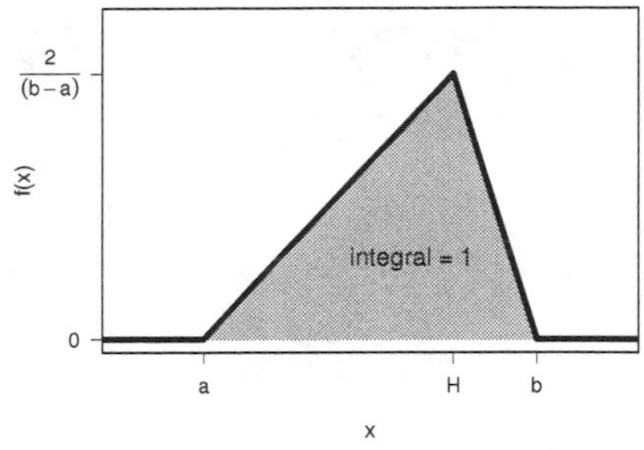

Figure 6.9: Density of the triangular distribution.

the closest approximation of 2^{-32} that can be stored in a double-precision variable. The problem with this approach is that while we potentially have 52 bits in double-precision, only 32 bits are used to fill them. So we are only using about, in decimal terms, 9 to 10 digits of the available 15 to 16 digits of the mantissa. For a detailed description of the problem as well as an elegant solution that uses two 32-bit integers to produce one double-precision random variate, see Doornik (2007). However, we are not aware of any major numerical package that uses Doornik's approaches, and they do come at the cost of slowing down the continuous uniform random number generator by approximately half because instead of one, two 32-bit integers need to be produced. Finally, it should be mentioned that the lagged Fibonacci generator mentioned earlier can be modified to produce values in the range $[0,1)$ directly, thereby eliminating the need for any conversion.

6.2.5.2 Triangular Distribution

Definition 6.26: Triangular Distribution
A continuous density function of the type

$$f(x) = f(x;a,b,H) = \begin{cases} \frac{2}{b-a} \cdot \frac{x-a}{H-a} & \text{for } x \in [a,H] \\ \frac{2}{b-a} \cdot \frac{b-x}{b-H} & \text{for } x \in [H,b] \\ 0 & \text{else,} \end{cases}$$

where a,b,H are real numbers, defines the density of the **triangular distribution with mode H on the interval [a, b]**. A random variable with such a density is called **triangularly distributed** (see Figure 6.9).

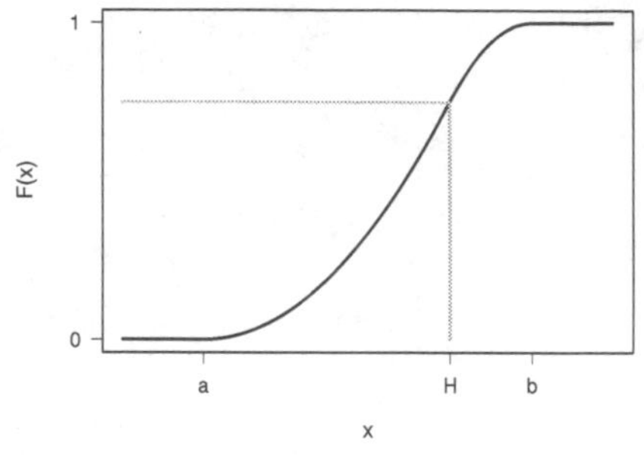

Figure 6.10: Distribution function of a triangular distribution.

The **distribution function of the triangular distribution** has the form:

$$F(x) = \begin{cases} 0 & x < a \\ \int_a^x \frac{2(y-a)}{(b-a)(H-a)} dy = \frac{(x-a)^2}{(b-a)(H-a)} & x \in [a,H] \\ \frac{H-a}{b-a} + \frac{(b-H)^2-(b-x)^2}{(b-a)(b-H)} = 1 - \frac{(b-x)^2}{(b-a)(b-H)} & x \in [H,b] \\ 1 & x > b. \end{cases}$$

For $H = \frac{a+b}{2}$ we have an **equilateral triangular distribution**. If $H = a$ or $H = b$, we talk of a **one-sided (left- or right-sided) triangular distribution**.

Example 6.6: Sketch of the Distribution Function
Let $a := 0$, $b := 1$, $H := 0.8$, then:

$$F(x) = \begin{cases} \frac{5}{4}x^2 & x \in [0, 0.8] \\ 1 - 5(1-x)^2 & x \in [0.8, 1] \end{cases}$$

See Figure 6.10.

In order to generate random **realizations of the triangular distribution** we introduce a general method for the generation of random realizations of continuous strictly monotonic distribution functions.

Inversion Method: Random Realizations of a Continuous, Strictly Monotonic Distribution

Lemma 6.1: Distribution of a Distribution Function
Let the distribution function of X be continuous and strictly monotonically increasing. Then, $F(X)$ is uniformly distributed on the interval $[0, 1]$.

Proof. For the uniform distribution on the interval $[0, 1]$ it is true that $P(Y \leq y) = y$. For each strictly monotonic distribution function F:

$$P(F(X) \leq F(x)) = P(X \leq x) = F(x),$$

since F is continuous and strictly monotonic.
Thus, the random variable $Y := F(X)$ is uniformly distributed on the interval $[0,1]$. □

This leads to Algorithm 6.4. This algorithm is easily applied to triangular distributions as follows:

Algorithm 6.4 Inversion Method: Realizations of a Continuous, Strictly Monotonic Distribution

Require: Basic random number generator RNG in $[0, 2^{31}-1]$.
1: Based on RNG generate a realization $u \in [0,1)$ of a uniformly distributed random variable.
2: $F^{-1}(u)$ is a realization of a random variable with distribution function F.

Random Realizations from a Triangular Distribution (With Fixed A,B,H)
Solve one of the following relations for x depending on where u lies:

$$F(x) = \frac{(x-a)^2}{(b-a)(H-a)} = u \in [0, \tfrac{H-a}{b-a}], \quad F(x) = 1 - \frac{(b-x)^2}{(b-a)(b-H)} = u \in [\tfrac{H-a}{b-a}, 1].$$

6.2.5.3 Normal Distribution

Definition 6.27: Normal Distribution
A continuous density function (see Figure 6.11) of the type

$$f(x) = f(x;\mu,\sigma^2) = \frac{1}{\sqrt{2\pi}\sigma} e^{-\frac{1}{2}(\frac{x-\mu}{\sigma})^2},$$

where $\sigma > 0$ and $\mu \in \mathbb{R}$, defines the density of a **normal distribution** with expected value μ and variance σ^2. A random variable X with such a density is called **normally distributed**. **Notation:** $X \sim \mathcal{N}(\mu,\sigma^2)$. The $\mathcal{N}(0,1)$ distribution is called **standard normal distribution**.

The **distribution function of a normal distribution** has the form

$$\begin{aligned} \Phi_{\mu,\sigma^2}(x) &:= \int_{-\infty}^{x} f(z)dz \\ &= \int_{-\infty}^{x} \frac{1}{\sqrt{2\pi}\sigma} e^{-\frac{1}{2}(\frac{z-\mu}{\sigma})^2} dz \\ &= \frac{1}{\sqrt{2\pi}} \int_{-\infty}^{\frac{x-\mu}{\sigma}} e^{-\frac{1}{2}y^2} dy \\ &= \Phi_{0,1}(\tfrac{x-\mu}{\sigma}) \end{aligned}$$

using the substitution rule with $z = \sigma y + \mu$.

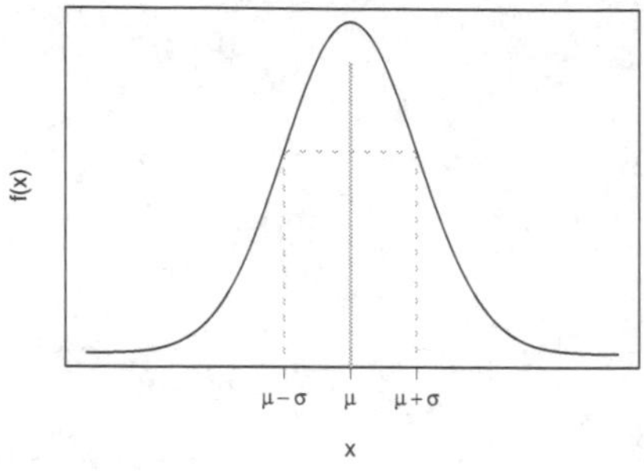

Figure 6.11: Density of normal distribution.

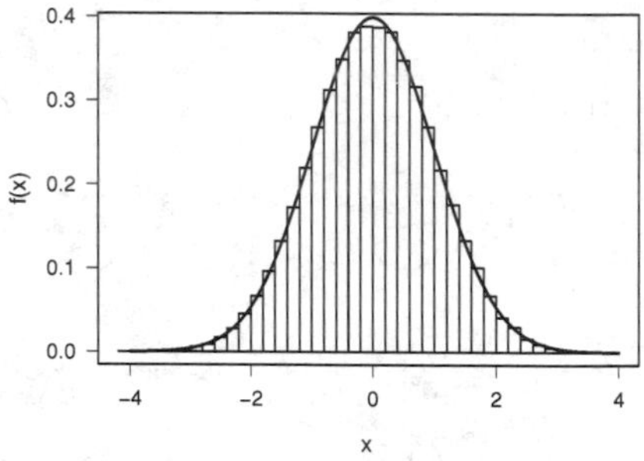

Figure 6.12: Density of standard normal distribution with approximation by empirical distribution of sums.

Normally Distributed Random Numbers: Approximation
Unfortunately, the inversion method from above is not applicable for generating normally distributed random numbers, since the antiderivative of e^{-x^2} does not exist. However, one can approximate the quantile function and then apply this method (see 6.13 below). Nevertheless, there are other **generators for (approximately) normally distributed random numbers**. The simplest example is $x := \sum_{i=1}^{12} u_i - 6$, where the u_i are independent uniformly distributed random numbers in $[0,1)$. The corresponding random variable X is approximately $\mathcal{N}(0,1)$ distributed, since $\mathrm{E}(X) = 0$, $\mathrm{var}(X) = 1$ due to $\mathrm{var}(U) = 1/12$, and since by the central limit theorem the sum of arbitrarily distributed random variables is approximately normally distributed. One might want to check the goodness of such an approximation. For this, one might evaluate the distance between the approximation and the true $\mathcal{N}(0,1)$ distribution function. Here, we show a histogram of 100000 random numbers generated by this approximation and the actual density of the $\mathcal{N}(0,1)$ distribution (see Figure 6.12).

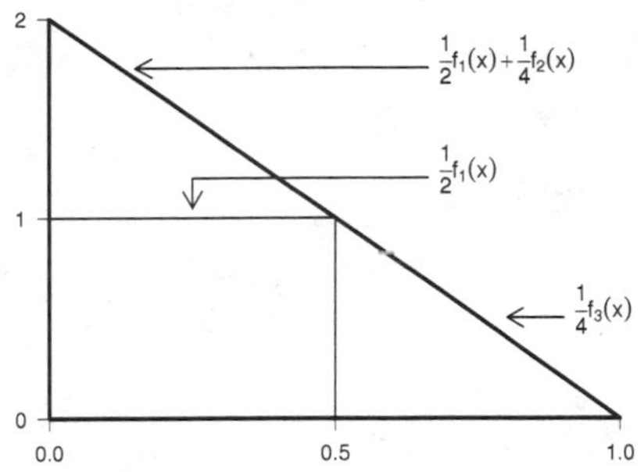

Figure 6.13: Partition of density.

6.2.5.4 Mixture of Distributions

Another class of methods to generate normal realizations is motivated by mixtures of distributions.

Definition 6.28: Mixture of Distributions
Let $F, F_i, i = 1, \ldots, k$, be distribution functions. Then, the distribution function

$$F(x) = p_1 F_1(x) + p_2 F_2(x) + \ldots + p_k F_k(x), \qquad p_i > 0, i = 1, \ldots, k, \sum_{i=1}^{k} p_i = 1$$

characterizes a **mixture of distributions**.

To generate a realization of a random variable X with such a distribution function F the general method for the generation of random realizations of discrete distributions in Section 6.2.4 can be combined with the above inverse method for the generation of realizations of continuous distributions.

Example 6.7: Mixture of Distributions (Kennedy and Gentle, 1980, p. 179 – 180)
Let

$$f(x) := \begin{cases} 2 - 2x, & 0 \le x \le 1 \\ 0, & \text{else} \end{cases} \qquad f_1(x) := \begin{cases} 2, & 0 \le x \le \frac{1}{2} \\ 0, & \text{else} \end{cases}$$

$$f_2(x) := \begin{cases} 4 - 8x, & 0 \le x \le \frac{1}{2} \\ 0, & \text{else} \end{cases} \qquad f_3(x) := \begin{cases} 8 - 8x, & \frac{1}{2} \le x \le 1 \\ 0, & \text{else} \end{cases}$$

Then $f(x) = \frac{1}{2} f_1(x) + \frac{1}{4} f_2(x) + \frac{1}{4} f_3(x)$ (cp. Figure 6.13).

f_1 is the density of a $R[0, \frac{1}{2})$-distributed random variable, and the densities

f_2, f_3 represent triangular distributions with distribution functions

$$F_2(x) = 4x - 4x^2 \text{ on } [0, \frac{1}{2}], \qquad F_3(x) = -4x^2 + 8x - 3 \text{ on } [\frac{1}{2}, 1].$$

To construct a realization of a random variable with distribution function F_2 or F_3, the inverse method is applied, i.e. we solve:

$$g = 4x - 4x^2 \qquad \text{or} \qquad g = -4x^2 + 8x - 3, \qquad g \in (0,1).$$

If g_1, g_2 are independent realizations of an R[0,1)-distributed random variable, then, as a realization of a random variable with density f, choose:

$$x = \begin{cases} \frac{1}{2}g_2 & 0 \leq g_1 < \frac{1}{2} \\ \frac{1}{2} - \sqrt{\frac{1}{4} - \frac{g_2}{4}} & \frac{1}{2} \leq g_1 < \frac{3}{4} \\ 1 - \sqrt{1 - \frac{3+g_2}{4}} & \frac{3}{4} \leq g_1 < 1 \end{cases}$$

i.e. apply the general method for the generation of discrete random realizations to the case differentiation in g_1 representing the weights of the single densities, and choose those solutions of the equations of the triangular distributions so that x lies in the interval corresponding to the density.

This method is applied to generate normal random realizations.

Definition 6.29: $\mathcal{N}(0,1)$ Distributed Random Numbers: Rectangle-Wedge–Tail Method (Marsaglia et al., 1964, pp. 4 – 10)
To generate normally distributed random numbers partition the area under the density function of the normal distribution (similar to Figure 6.13) into rectangles, wedges, and a tail (cp. Figure 6.14).

This partition represents (similar to the above example) a mixture of relatively simple distributions. In the version of this method originally proposed in the literature in 88% of the cases only a simple transformation of the type $y^{(D)}$ of an $R[0,1)$ realization is needed to generate a normal random realization.

Normally Distributed Random Numbers by Means of the Inversion Method
We can also generate normally distributed random numbers by means of the inversion method using the quantile function of the normal distribution on a linear combination of two $R[0,1)$ random numbers u_1 and u_2. As realization of a random variate with a general normal density we obtain

$$x = \mu + \sigma \cdot \Phi_{0,1}^{-1}\left(\frac{\lfloor 2^{27}u_1 \rfloor + u_2}{2^{27}}\right) \tag{6.13}$$

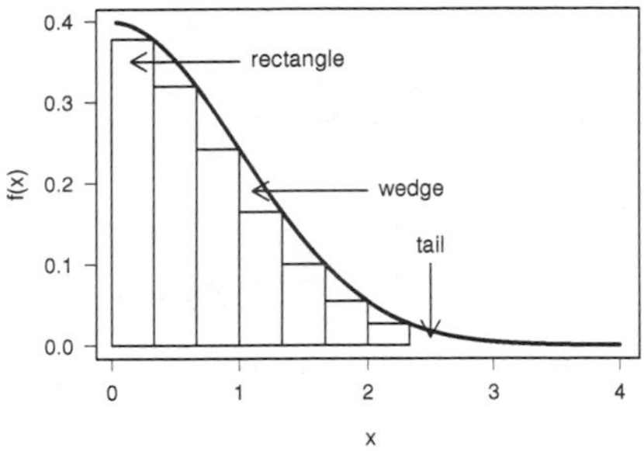

Figure 6.14: Approximation of the density of a normal distribution.

with the standard normal quantile function $\Phi_{0,1}^{-1}(p)$. This is the default method used in R, for example. The peculiar use of two uniform deviates is a safety precaution because most uniform random number generators, as previously detailed, are not random enough. Assuming u_1 and u_2 were generated from 32-bit discrete uniform random variables, we can combine them to generate another approximately uniform random number of higher quality. To obtain this random number, we sum the two lower-quality deviates after "shifting" one of them so that their range does not overlap and normalize afterwards. The major difficulty with this method is that its accuracy highly depends on the accuracy of the quantile function, particularly in the tails. Wichura (1988) describes an accurate algorithm (used in R) to calculate the quantile function of a normal distribution.

6.2.5.5 Exponential Distribution

In the continuous case the waiting time distributions are often called **lifetime distributions**. The lifetime of an object is understood as the time between start of operation (birth) and failure (death) of the object. For the start of operation generally $x = 0$ is used.

Let X := **lifetime of an object of a certain kind**. The following distribution fits, e.g., only for the modeling of the lifetime of objects that do not age.

Definition 6.30: Exponential Distribution
Each continuous density function of the type

$$f_X(x) = f_X(x;\lambda) = \begin{cases} \lambda e^{-\lambda x} & x \geq 0, \text{ where } \lambda > 0 \\ 0 & \text{else} \end{cases}$$

defines a density of an **exponential distribution (with parameter λ)**. A ran-

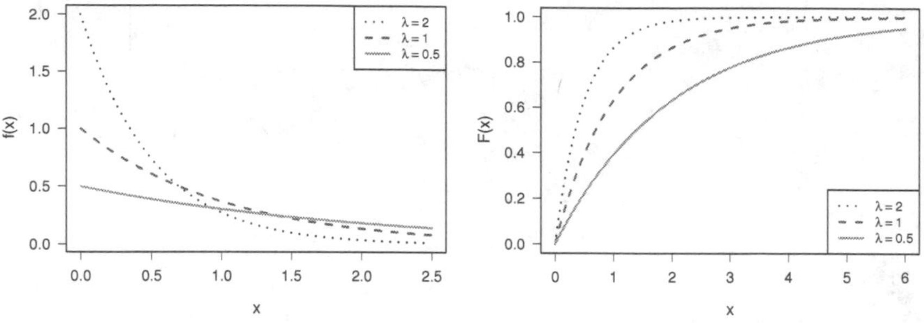

Figure 6.15: Densities and distribution functions of exponential distributions.

dom variable with such a density is called **exponentially distributed** and is abbreviated $Exp(\lambda)$.

The **distribution function of an exponential distribution** has the form:

$$F(x) = \int_0^x \lambda e^{-\lambda y} dy = \left[-e^{-\lambda y}\right]_0^x = 1 - e^{-\lambda x}, \qquad x \geq 0$$

See Figure 6.15 for densities and distribution functions of the exponential distribution.

Random Realizations of an Exponential Distribution
By the inversion method (Lemma 6.1)

$$x = -\frac{\log(1-u)}{\lambda}$$

is a realization of the exponential distribution with parameter λ when u is a realization of the uniform distribution in $[0, 1)$.

Obviously, this is true since

$$u = 1 - e^{-\lambda x} = F(x).$$

6.2.5.6 Lognormal Distribution

A lognormal distribution is an example of a distribution derived from the normal distribution.

If a continuous right-skewed distribution is expected, i.e. with a skewness coefficient $\frac{\mu_3}{\sigma^3} > 0$, by taking the logarithm often a distribution can be achieved that is nearly symmetric and even similar to a normal distribution. If $\log X$ is normally distributed, then X is called **lognormally distributed**.

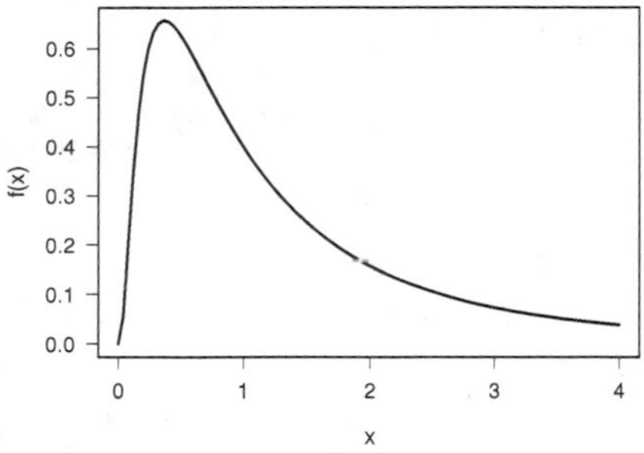

Figure 6.16: Density of a lognormal distribution.

Definition 6.31: Lognormal Distribution
Each continuous density function of the type

$$f(x) = f(x;\mu,\sigma^2) = \begin{cases} \frac{1}{\sqrt{2\pi}\sigma}\frac{1}{x}e^{-\frac{1}{2}\left(\frac{\log x-\mu}{\sigma}\right)^2} & x > 0, \text{ where } \sigma > 0 \text{ and } \mu \in \mathbb{R} \\ 0 & \text{else} \end{cases}$$

defines a density of a **lognormal distribution with parameters μ, σ^2**. A random variable with such a density is called **lognormally distributed** (cp. Figure 6.16).

The distribution function of the lognormal distribution has the form:

$$F(x) := \int_{-\infty}^{x} f(z)dz = \int_{0}^{x} \frac{1}{\sqrt{2\pi}\sigma z} e^{-\frac{1}{2}\left(\frac{\log z-\mu}{\sigma}\right)^2} dz = \Phi\left(\frac{\log x - \mu}{\sigma}\right)$$

Random Realizations of a Lognormal Distribution
Let z be a realization of a standard normal distribution $\mathcal{N}(0,1)$. Then

$$x := e^{\mu+z\sigma}$$

is a realization of a lognormal distribution with parameters μ, σ^2.

6.2.6 *Summary*

We began our survey of univariate random number generators by looking at different discrete uniform RNGs. These formed the basis on which we built all further generators for other discrete and continuous distributions. That is why we emphasized verification of these discrete uniform RNGs. For this we used a battery of statistical test procedures.

6.3 Theory: Multivariate Randomization[2]

6.3.1 Motivation and Overview

The methods for the generation of random numbers introduced until now directly generate only points of the desired distribution and fail especially in the multidimensional case. Since the 1980s, however, quite another type of **method** is more and more in use **that tries try to generate tentative points of a desired density that are either accepted or rejected by means of simplified approximating densities.**[3]

As an example for a multivariate stochastic application let us consider integration by means of a stochastic method. For example, let us assume that we want to calculate

$$\int_B g(\boldsymbol{x})d\boldsymbol{x} \quad (B \subseteq \mathbb{R}^d). \tag{6.14}$$

To do this, we write the integral in the form

$$\int_B g(\boldsymbol{x})d\boldsymbol{x} = \int_B \underbrace{\frac{g(\boldsymbol{x})}{p(\boldsymbol{x})}}_{f(\boldsymbol{x})} p(\boldsymbol{x})d\boldsymbol{x} \tag{6.15}$$

with the density function $p(\boldsymbol{x})$ (the zeros of $p(\boldsymbol{x})$ being a Lebesgue null set). Then, the ratio $f(\boldsymbol{x}) = g(\boldsymbol{x})/p(\boldsymbol{x})$ has a distribution induced by the random variable $\boldsymbol{X}$ with density $p(\boldsymbol{x})$.

The integral can now be determined (estimated) by generating random points from the distribution corresponding to $p(\boldsymbol{x})$ and by calculating the mean of the corresponding values $f(\boldsymbol{x})$.

$p(\boldsymbol{x})$ should be chosen so that $f(\boldsymbol{x})$ (and thus the value of the integral) has the smallest variance possible, i.e. so that the value of the integral is calculated as exact as possible. Obviously, $p(\boldsymbol{x}) \approx g(\boldsymbol{x})$ would be a good choice.

Note that in statistical applications such an approximation of $g(\boldsymbol{x})$ is hardly ever necessary since the density is prescribed. Examples for such integrations are the calculation of expected values, variances, and normalization constants in Bayes' statistics or marginal distributions.

What remains is the generation of random points according to $p(\boldsymbol{x})$. Two of the most flexible and promising types of methods are the **rejection method** and the **Markov Chain Monte Carlo (MCMC) methods** which will be discussed in the following up to implementation details. We will introduce two MCMC methods, the Gibbs and the Metropolis-Hastings algorithms. Note

[2]This is partly based on Röhl (1998).

[3]For a history of MCMC methods see, e.g., Robert and Casella (2011).

that both the rejection and the MCMC methods do not need normalization constants, i.e. can work with a normalizable function, meaning that the density function does not have to be normalized to integral 1.

The pros and cons of these two types of methods can be summarized in advance as follows. The **rejection method** delivers independent random points. However, their generation is generally very inefficient in high dimensions, i.e. the acceptance rate of the generated points is very low.

This problem is avoided by the **MCMC methods** by generating random points by so-called Markov chains converging against the desired distribution, i.e. ultimately generating (almost) only points from the desired distribution. However, the construction principle of the **MCMC method** also induces the two substantial disadvantages of this method:

1. the generated points are not stochastically independent and the convergence has to be assessed,
2. in the so-called "burn-in" phase, until convergence the generated points have to be rejected altogether.

These disadvantages are, however, compensated by the big flexibility of the construction of the Markov chain and its simple programming. Moreover, in high dimensions MCMC methods are often the only possibility to generate random points from complicated multivariate distributions.

In statistics **MCMC methods** have been applied since the 1980s. Only since the beginning of the 1990s, however, their **convergence** and convergence speed have been studied intensively. The results are theoretically satisfying implementations which guarantee convergence toward the stationary distribution and which allow for the calculation of upper limits for convergence speed. Unfortunately, these limits are often very conservative, resulting in unnecessarily long burn-in times. Therefore, statistical "online" methods were developed (Geyer, 1992; Gelman and Rubin, 1992) that try to clarify convergence by means of the generated points themselves.

The theoretical **convergence rate** of MCMC methods will be discussed by introducing an adequate distance measure between two distributions or densities, respectively. We will study the convergence rate for both discrete and continuous Markov chains. Moreover, the relation of the acceptance rate of the rejection method and the convergence rate of a special MCMC algorithm, the Independence Chain MCMC (ICMCMC), will be studied.

Let us stress once more that in MCMC algorithms, similar to the rejection method, tentative points of the desired distribution are generated by means of simplified, approximating densities. The tentative points are either accepted or rejected. In contrast to the rejection method, however, for MCMC methods convergence to the desired distribution can be shown under weak conditions so that at the end all points can be accepted.

Decisive for the statistical simulation with MCMC methods are the **three C questions**:

1. How do we **C**onstruct effective chains that quickly cover the whole feasible region?
2. Does the constructed chain **C**onverge?
3. How high is **C**onvergence speed, i.e. the convergence rate? When can we stop burn-in? The answer to this question requires the introduction of adequate distance measures.

The construction of effective chains (1) and the analysis of their **convergence rate** (3) are complementary in difficulty:

1. Local chains (a transition $\boldsymbol{x} \to \boldsymbol{y}$ is only possible if $\boldsymbol{y}$ is near $\boldsymbol{x}$) are easier to analyze but often ineffective (because of high autocorrelation).
2. For global chains (large jumps $\boldsymbol{x} \to \boldsymbol{y}$ are permitted) with not so low acceptance rates it is vice versa.

This makes the task to be solved somewhat complicated.

6.3.2 *Rejection Method*

Let $\pi(\boldsymbol{x})$ be a **continuous, non-negative normalizable function** on $B \subseteq \mathbb{R}^d$, i.e. $\int_B \pi(\boldsymbol{x})d\boldsymbol{x} < \infty$. In order to generate points from the density

$$\frac{\pi(\boldsymbol{x})}{\int_B \pi(\boldsymbol{x})d\boldsymbol{x}}, \tag{6.16}$$

one considers a density $q(\boldsymbol{x})$ ($\boldsymbol{x} \in B$) from which generation of random points is easy. The so-called **envelope**

$$kq(\boldsymbol{x}) \quad \text{with} \quad k \geq \sup_{\boldsymbol{x} \in B} \frac{\pi(\boldsymbol{x})}{q(\boldsymbol{x})} \tag{6.17}$$

always dominates $\pi(\boldsymbol{x})$. Such a k exists if, e.g., $q(\boldsymbol{x})$ has stronger tails than $\pi(\boldsymbol{x})$. To guarantee that $\pi(\boldsymbol{x})/q(\boldsymbol{x}) \leq \text{const} < \infty \; \forall \boldsymbol{x} \in B$, one can, e.g., take a rectangular distribution as $q(\boldsymbol{x})$ on the interval where $\pi(\boldsymbol{x}) > q(\boldsymbol{x})$. On the remaining domain the condition is automatically fulfilled for arbitrary $k \geq 1$.

Theorem 6.4: Generation of Random Points from the Density (6.16)
N random points from density (6.16) can be generated by the rejection algorithm (6.5).

Algorithm 6.5 Rejection Method: Generation of N Random Points from Density $\frac{\pi(\boldsymbol{x})}{\int_B \pi(\boldsymbol{x})d\boldsymbol{x}}$

Require: N, continuous, non-negative normalizable function $\pi(\boldsymbol{x})$, $k \geq 1$, density $q(x)$

Initialize $i = 0$
while $i < N$ **do**
  Generate independently $\boldsymbol{x}$ from q.
  Generate independently u from the uniform distribution $R[0,1)$.
  if $u \leq \frac{\pi(\boldsymbol{x})}{kq(\boldsymbol{x})}$ **then**
    Accept $\boldsymbol{x}$ as a random number from (6.16) and set $i = i+1$
  else
    Reject $\boldsymbol{x}$
  end if
end while

Proof. Let $B' \subseteq B$ and B' be Borel measurable. Then

$$\begin{aligned} P(\boldsymbol{x} \in B', \boldsymbol{x} \text{ accepted}) &= P\left(\boldsymbol{x} \in B', 0 \leq u \leq \frac{\pi(\boldsymbol{x})}{kq(\boldsymbol{x})}\right) \\ &= \int_{B'} \int_0^{(\pi(\boldsymbol{x})/kq(\boldsymbol{x}))} q(\boldsymbol{x}) 1\, du\, d\boldsymbol{x} \\ &= \frac{1}{k} \int_{B'} \pi(\boldsymbol{x}) d\boldsymbol{x} \end{aligned} \tag{6.18}$$

and

$$P(\boldsymbol{x} \text{ accepted}) = \int_B \int_0^{(\pi(\boldsymbol{x})/kq(\boldsymbol{x}))} q(\boldsymbol{x}) 1\, du\, d\boldsymbol{x} = \frac{1}{k} \int_B \pi(\boldsymbol{x})\, d\boldsymbol{x}. \tag{6.19}$$

Then, the interesting probability is

$$P(\boldsymbol{x} \in B' | \boldsymbol{x} \text{ accepted}) = \frac{P(\boldsymbol{x} \in B', \boldsymbol{x} \text{ accepted})}{P(\boldsymbol{x} \text{ accepted})} = \frac{\int_{B'} \pi(\boldsymbol{x}) d\boldsymbol{x}}{\int_B \pi(\boldsymbol{x}) d\boldsymbol{x}}, \tag{6.20}$$

i.e. the above algorithm delivers random points from (6.16). □

In the special case when $Vol(B) < \infty$ and $q(\boldsymbol{x})$ is the uniform distribution on B the density is enclosed in a $(d+1)$-dimensional frame with base B and constant height (see Figure 6.17).

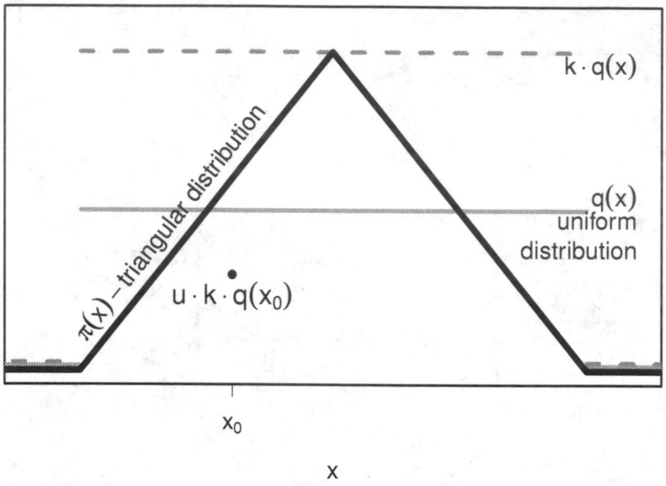

Figure 6.17: Random number generation by the rejection method.

Discrete Case

For discrete distributions $\pi(x)$, the approximating distribution $q(x)$ has to represent a known discrete distribution, from which generation of random points is easy, and the integrals have to be replaced by sums.

Example 6.8: Rejection Method

Let $\pi(x)$ be the uniform distribution on $\{1, 2, \ldots, 2^{31} - 1\}$, $q(x)$ representing a basic uniform random number generator (RNG) on $\{0, 1, 2, \ldots, 2^{31} - 1\}$, and $k = 2^{31}/(2^{31} - 1)$. Then, a random number $\in \{1, 2, \ldots, 2^{31} - 1\}$ is accepted if $u \leq \frac{\pi(x)}{kq(x)} = 1$, i.e. in any case. In contrast, the random number 0 is only accepted with probability 0 because of the condition $u \leq 0$.

Remarks

- The rejection method delivers points exactly distributed as (6.16), which are independent by construction.
- The share of accepted points is

$$P(x \text{ accepted}) = \frac{1}{k} \int_B \pi(x) dx. \tag{6.21}$$

- The larger k is, the more trial points are rejected. Therefore, $q(x)$ should have a similar form as $\pi(x)$.

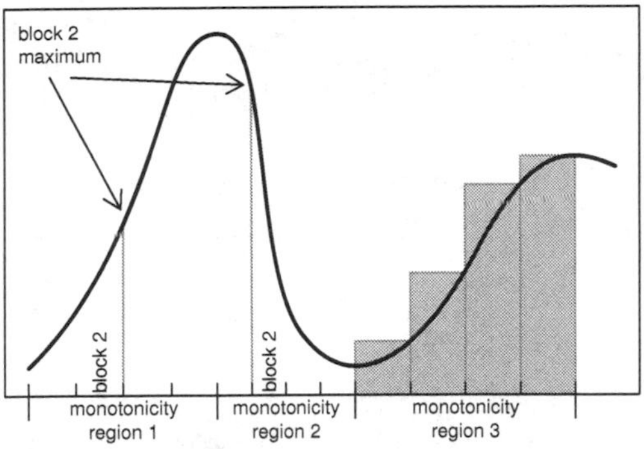

Figure 6.18: One-dimensional rejection method.

– The above special case using only one rectangular approximating distribution is thus most of the time very inefficient. Especially in the distribution tails hardly any trial point would be accepted. The higher the dimension, the more difficult it is to find an effective envelope.

One-Dimensional Case

In one dimension today's computer capacities are completely sufficient for the rejection method (see Figure 6.18). The method is applied as follows:

1. For one dimension there are effective methods for the determination of minima and maxima. Therefore, the density (6.16) on B can be partitioned into m monotonicity regions.
2. In each monotonicity region the density can be equidistantly divided into m_i ($i = 1, \ldots, m$) blocks. Then, there are points $x_1, \ldots, x_{m_i+1}$ defining intervals $[x_i, x_{i+1}]$ as blocks with minima $f(x_i)$ and maxima $f(x_{i+1})$ for monotonically increasing regions or maxima $f(x_i)$ and minima $f(x_{i+1})$ for monotonically decreasing regions.
3. In the individual blocks, the envelope q is constant and proportional to the corresponding maximum (cp. rectangle-wedge-tail method). The proportionality constant k is determined by normalization of q overall on B, since q should be a density on the whole B.
4. The points are then stepwise generated by the choice of
   - a monotonicity region with probability $1/m$,
   - a block inside the monotonicity region with probability $1/m_i$, and
   - a point inside the block by the rejection method.

6.3.3 Gibbs Algorithm

Let $\pi(\boldsymbol{x})$ be a **discrete or continuous, non-negative normalizable function** on $B \subseteq \mathbb{R}^d$, i.e. $\sum_B \pi(\boldsymbol{x}) < \infty$ or $\int_B \pi(\boldsymbol{x})d\boldsymbol{x} < \infty$. In order to generate points from the corresponding density we will construct a Markov chain $\boldsymbol{X}^0, \boldsymbol{X}^1, \ldots, \boldsymbol{X}^t, \ldots \in E \subseteq B$ with $E = \{\boldsymbol{x} | \boldsymbol{x} \in B \text{ with } \pi(\boldsymbol{x}) > 0\}$. The restriction on E only serves to exclude exceptions in the following theorems.

Definition 6.32: Markov Chain
A sequence of random variables $\boldsymbol{X}^0, \boldsymbol{X}^1, \ldots, \boldsymbol{X}^t, \ldots$ is called a (homogeneous) **Markov chain** if the conditional distribution of $\boldsymbol{X}^k$ with the density $f(\boldsymbol{X}^k | \ldots)$, given the whole history, only depends on $\boldsymbol{X}^{k-1}$, i.e. if

$$f(\boldsymbol{X}^k | \boldsymbol{X}^{k-1}, \boldsymbol{X}^{k-2}, \ldots, \boldsymbol{X}^0) = f(\boldsymbol{X}^k | \boldsymbol{X}^{k-1}).$$

(Since the density f does not depend on k, the chain is called homogeneous).

In the case of distributions with finitely many possible values with probability > 0, the probability of the transition from one state $\boldsymbol{x}_i^{k-1}$ to another state $\boldsymbol{x}_j^k$ only depends on the value $\boldsymbol{x}_i^{k-1}$ and not on the previous values of the sequence, $i, j = 1, \ldots, n$. The transition probabilities p_{ij} for such transitions can be collected in a matrix, the so-called **transition matrix** $\boldsymbol{P}$.
In the infinite case the term **transition function** $q(\boldsymbol{x}, \boldsymbol{y})$ is used.

In what follows the two most important Markov chain based methods will be introduced, which only differ by the choice of the transition probabilities. First, we will discuss the Gibbs algorithm, and afterwards the Metropolis-Hastings algorithm.

The notation for the Gibbs algorithm follows Roberts and Smith (1994). The coordinates of a vector $\boldsymbol{x} \in \mathbb{R}^d$ can be arbitrarily partitioned into k blocks. Often, the Gibbs algorithm is restricted to the special case that every coordinate of $\boldsymbol{x}$ builds a (one-dimensional) block of its own. In this case the random points are generated coordinate-wise (one-dimensional). The art is to use convenient conditional distributions.

Now, let $\pi(\boldsymbol{x}) = \pi([\boldsymbol{x}_1 \; \boldsymbol{x}_2 \; \ldots \; \boldsymbol{x}_k]^T)$ with $1 < k \leq d$ be the probability function or density of the considered distribution from which points should be generated. Let the k blocks be defined as

$$\begin{aligned} &\boldsymbol{x}_i = [x_{i1} \; \ldots \; x_{i,n(i)}]^T, \quad i = 1, \ldots, k, \\ &n(1) + n(2) + \ldots + n(k) = d, \end{aligned} \tag{6.22}$$

where x_{ij} are the components of $\boldsymbol{x}_i$.

The conditional densities $\pi(\boldsymbol{x}_i | \boldsymbol{x}_{-i})$ with the notation $\boldsymbol{x}_{-i} = [\boldsymbol{x}_j | j \neq i]$ serve as a means for the generation of random points.

Starting from an arbitrary starting value $\boldsymbol{x}^0 = [x_1^0 \ \dots \ x_k^0]^T \in E$ in one iteration $\boldsymbol{x}^1 = [x_1^1 \ \dots \ x_k^1]^T$ is generated successively by Algorithm 6.6.

Algorithm 6.6 Gibbs Algorithm: One Iteration

Require: An arbitrary starting value $\boldsymbol{x}^0 = [x_1^0 \ \dots \ x_k^0] \in E$
1: Generate x_1^1 from $\pi(x_1|\boldsymbol{x}_{-1}^0)$
2: **for** i = 2 to k-1 **do**
3: Generate x_i^1 from $\pi(x_i^1|x_1^1, \dots x_{i-1}^1, x_{i+1}^0, \dots, x_k^0)$
4: **end for**
5: Generate x_k^1 from $\pi(x_k|\boldsymbol{x}_{-k}^1)$

The random variables $\boldsymbol{X}^0, \boldsymbol{X}^1, \dots, \boldsymbol{X}^t, \dots$ with the realizations $\boldsymbol{x}^0 \to \boldsymbol{x}^1 \to \dots$ build a Markov chain with the transition matrix (discrete case) or the transition function (continuous case)

$$P_G(\boldsymbol{x}, \boldsymbol{y}) = \prod_{l=1}^{k} \pi(y_l|y_j, \ j < l, \ x_j, \ j > l) \tag{6.23}$$

for a transition $\boldsymbol{x} \to \boldsymbol{y}$. In what follows the finite and the continuous cases will be discussed separately.

6.3.3.1 *Finite Case*

Let $B \subseteq \mathbb{N}^d$ and $\pi(\boldsymbol{x})$ be a probability function. After t steps the transition matrix has the form:

$$P_G^{(t)}(\boldsymbol{x}, \boldsymbol{y}) = \sum_{\boldsymbol{z} \in E} P_G^{(t-1)}(\boldsymbol{x}, \boldsymbol{z}) P_G(\boldsymbol{z}, \boldsymbol{y}) \tag{6.24}$$

with the initialization $P_G^{(1)}(\boldsymbol{x}, \boldsymbol{y}) := P_G(\boldsymbol{x}, \boldsymbol{y})$.

In order to establish the relationship to the conventional representation of Markov chains on a finite number of states, all $\boldsymbol{x} \in E$ are numbered by the bijection $\boldsymbol{x} \leftrightarrow i \in \{1, \dots, n\}$ (n is the number of points in E) with $\pi(\boldsymbol{x}) \leftrightarrow \pi_i$.

If a Markov chain is converging, the limiting distribution has the following property:

Definition 6.33: Stationary Distribution
Let the transition matrix $P = [P_{ij}] := [P_G(i, j)]$ have the property:

$$\pi_j = \sum_{i=1}^{n} \pi_i P_{ij}. \tag{6.25}$$

Then, π is called a **stationary distribution**.

Obviously, the limiting distribution of $P^{(t)}$ has this property if the Markov chain is converging.

The two central terms in the theory of Markov chains, needed here, are introduced in the next definition.

Definition 6.34: Irreducible and Aperiodic Markov Chains: Finite Case
In the finite case a Markov chain is called **irreducible** if every state $i \in E$ is reachable from any other state $j \in E$, i.e. if $P_{ji}^{(t)} > 0$ for some $t \in \mathbb{N}_0$. A state $i \in E$ is called **periodical** with period $c > 1$ if $P_{ii}^{(t)} > 0$ for at most $t \in \{0, c, 2c, \ldots\}$. Otherwise, the state is called **aperiodic**. The Markov chain is called **aperiodic** if every state is aperiodic.

The two terms irreducible and aperiodic together mean that one can get any time from each state to an arbitrary other state in an unpredictable number of steps.

Based on the $L1$ norm quantifying the distance between two probability vectors of the n elements in E, one can prove convergence for such Markov chains. Theorem 6.5 delivers a sufficient condition.

Theorem 6.5: Convergence of Gibbs Sampling: Finite Case (Roberts and Smith, 1994)
Let f_0 be the row vector of the probabilities of the starting distribution and π the row vector of the desired probabilities. If the Markov chain defined by the Gibbs algorithm is irreducible and aperiodic, then

$$\|(f^0 P^{(t)} - \pi)^T\|_1 \to 0 \quad \text{for} \quad t \to \infty \quad \forall \boldsymbol{x}^0 \in E, \tag{6.26}$$

i.e. for every point $\boldsymbol{x}^0 \in E$ from the starting distribution the probability function $f^{(t)} := f^0 P^{(t)}$ converges with the t-step transition matrix $P^{(t)}$ toward the desired probability function π.

The probability function of $\boldsymbol{X}_i^t$ converges toward the corresponding marginal probability function

$$\pi_{\boldsymbol{x}_i} = \sum_{\boldsymbol{x}_{-i}} \pi(\boldsymbol{x}_1, \boldsymbol{x}_2, \ldots, \boldsymbol{x}_k), \quad i = 1, \ldots, k. \tag{6.27}$$

The expected value of any π absolutely summable function g can be estimated by the arithmetical mean of the realizations over time:

$$\frac{1}{t}\{g(\boldsymbol{X}^1) + \ldots + g(\boldsymbol{X}^t)\} \to \sum_{\boldsymbol{x} \in E} g(\boldsymbol{x})\pi(\boldsymbol{x}) \quad \text{for} \quad t \to \infty \quad \text{almost surely.} \tag{6.28}$$

□

Consider the case of a Markov chain generated by the transition matrix (6.23). Since $\pi(\boldsymbol{x}) > 0$, every element of $\boldsymbol{x}$ has a positive probability given the other elements. Therefore, $P_G(\boldsymbol{x}, \boldsymbol{x}) > 0$, and hence the chain is aperiodic. Then it suffices to prove irreducibility if $\pi(\boldsymbol{x}^0) > 0$.

Example 6.9: Discrete Gibbs sampling
In order to better understand the mechanism of the Gibbs algorithm, we will give a concrete example. Suppose we have $d = 2$ dimensions and $n = 4$ states. Let the above bijection be

$$1 = [0 \;\; 0]^T, \quad 2 = [1 \;\; 0]^T, \quad 3 = [0 \;\; 1]^T, \quad 4 = [1 \;\; 1]^T, \tag{6.29}$$

where $[x \;\; y]^T$ represents the common column vector notation of a two-dimensional Cartesian coordinate system.

With the abbreviations

$$p_1 := \pi([0 \;\; 0]^T), \quad p_2 := \pi([1 \;\; 0]^T), \quad p_3 := \pi([0 \;\; 1]^T), \quad p_4 := \pi([1 \;\; 1]^T) \tag{6.30}$$

with $p_1 + p_2 + p_3 + p_4 = 1$ for the probabilities, the conditional probabilities can be written in matrix form:

$$\begin{aligned}
\pi_{y|x} &= \begin{bmatrix} \pi_{y|x}(0|0) & \pi_{y|x}(1|0) \\ \pi_{y|x}(0|1) & \pi_{y|x}(1|1) \end{bmatrix} = \begin{bmatrix} \frac{p_1}{p_1+p_3} & \frac{p_3}{p_1+p_3} \\ \frac{p_2}{p_2+p_4} & \frac{p_4}{p_2+p_4} \end{bmatrix} \\
\pi_{x|y} &= \begin{bmatrix} \pi_{x|y}(0|0) & \pi_{x|y}(1|0) \\ \pi_{x|y}(0|1) & \pi_{x|y}(1|1) \end{bmatrix} = \begin{bmatrix} \frac{p_1}{p_1+p_2} & \frac{p_2}{p_1+p_2} \\ \frac{p_3}{p_3+p_4} & \frac{p_4}{p_3+p_4} \end{bmatrix}
\end{aligned} \tag{6.31}$$

The elements of the Markov chain are generated in the following order:

$$x^0, y^0, x^1, y^1, \ldots. \tag{6.32}$$

If only the chain $x^0, x^1, \ldots,$ is considered, the transition matrix has the form:

$$\pi(X^1 = x^1 | X^0 = x^0) = \sum_y \pi(Y^0 = y | X^0 = x^0) \pi(X^1 = x^1 | Y^0 = y). \tag{6.33}$$

In matrix form we have

$$\pi_{x|x} = \pi_{y|x} \pi_{x|y} \tag{6.34}$$

with the matrices (6.31). Thus, the transition matrix has the form:

$$\pi_{x|x} = \begin{bmatrix} \frac{p_1^2}{(p_1+p_3)(p_1+p_2)} + \frac{p_3^2}{(p_1+p_3)(p_3+p_4)} & \frac{p_1 p_2}{(p_1+p_2)(p_1+p_3)} + \frac{p_3 p_4}{(p_1+p_3)(p_3+p_4)} \\ \frac{p_1 p_2}{(p_2+p_4)(p_1+p_2)} + \frac{p_3 p_4}{(p_2+p_4)(p_3+p_4)} & \frac{p_2^2}{(p_2+p_4)(p_1+p_2)} + \frac{p_4^2}{(p_2+p_4)(p_3+p_4)} \end{bmatrix}. \tag{6.35}$$

One can show (see Exercise 6.3.1) that for large t, independent of the starting marginal distribution $[f_x^0(0)\ \ f_x^0(1)]$ with $f_x^0(0)+f_x^0(1)=1$,

$$[\ f_x^0(0)\quad f_x^0(1)\ ]\pi_{x|x}^t \approx [\ p_1+p_3 \quad p_2+p_4\ ] \tag{6.36}$$

Thus, in the rows of the limit of $\pi_{x|x}^t$ the desired marginal distribution corresponding to x is given. Analogously, one can show that $\pi_{y|y}$ has the desired form.

The elements of the transition matrix for the transitions $i=[x^0\ \ y^0]^T \to j=[x^1\ \ y^1]^T$ are by (6.23) given by

$$P_G([x^0\ \ y^0]^T \to [x^1\ \ y^1]^T) = \pi_{x^1|y^0}\pi_{y^1|x^1}. \tag{6.37}$$

Thus, the transition matrix has the form:

$$P_G = \begin{bmatrix} \frac{p_1^2}{(p_1+p_2)(p_1+p_3)} & \frac{p_2^2}{(p_1+p_2)(p_2+p_4)} & \frac{p_1p_3}{(p_1+p_2)(p_1+p_3)} & \frac{p_2p_4}{(p_1+p_2)(p_2+p_4)} \\ \frac{p_1^2}{(p_1+p_2)(p_1+p_3)} & \frac{p_2^2}{(p_1+p_2)(p_2+p_4)} & \frac{p_1p_3}{(p_1+p_2)(p_1+p_3)} & \frac{p_2p_4}{(p_1+p_2)(p_2+p_4)} \\ \frac{p_1p_3}{(p_3+p_4)(p_1+p_3)} & \frac{p_2p_4}{(p_3+p_4)(p_2+p_4)} & \frac{p_3^2}{(p_3+p_4)(p_1+p_3)} & \frac{p_4^2}{(p_3+p_4)(p_2+p_4)} \\ \frac{p_1p_3}{(p_3+p_4)(p_1+p_3)} & \frac{p_2p_4}{(p_3+p_4)(p_2+p_4)} & \frac{p_3^2}{(p_3+p_4)(p_1+p_3)} & \frac{p_4^2}{(p_3+p_4)(p_2+p_4)} \end{bmatrix} \tag{6.38}$$

and in the limit (see Exercise 6.3.2):

$$\lim_{t\to\infty} P_G^t = \begin{bmatrix} p_1 & p_2 & p_3 & p_4 \\ p_1 & p_2 & p_3 & p_4 \\ p_1 & p_2 & p_3 & p_4 \\ p_1 & p_2 & p_3 & p_4 \end{bmatrix}.$$

Therefore, for large t:

$$p_j \approx \sum_{i=1}^{n} p_i P_{G_{ij}}^t \qquad \forall j,$$

i.e. the distribution is stationary as desired.

6.3.3.2 *Continuous Case*

Now let $\pi(\boldsymbol{x})$ be a density function and $E \subseteq \mathbb{R}^d$. Analogous to the finite case (6.24) after t steps the transition function $P_G(\boldsymbol{x},\boldsymbol{y})$ induced by the Gibbs algorithm has the form

$$P_G^{(t)}(\boldsymbol{x},\boldsymbol{y}) = \int_E P_G^{(t-1)}(\boldsymbol{x},\boldsymbol{z})P_G(\boldsymbol{z},\boldsymbol{y})d\boldsymbol{z}. \tag{6.39}$$

Thus, the corresponding sequence of densities $f^{(t)}$ has the form:

$$f^{(t)}(\boldsymbol{x}) = \int_E P_G(\boldsymbol{y}, \boldsymbol{x}) f^{(t-1)}(\boldsymbol{y}) d\boldsymbol{y}. \tag{6.40}$$

When starting from an arbitrary starting point $\boldsymbol{x}^0$, the density is called $f^{(t)}_{\boldsymbol{x}^0}(\boldsymbol{x})$.

Using this time the $L1$ function norm (see Section 6.1.1) one can give sufficient conditions for the convergence of the densities generated by the Gibbs algorithm, i.e. for

$$\|f^{(t)}_{\boldsymbol{x}^0} - \pi\|_1 \to 0 \quad \text{for} \quad t \to \infty. \tag{6.41}$$

The two central terms in the theory of Markov chains are now generalized for the continuous case.

Definition 6.35: Irreducible and Aperiodic Markov Chains: Continuous Case
In the continuous case a Markov chain is called **π-irreducible** if for all $\boldsymbol{x} \in E$ and measurable sets A with $\pi(A) > 0$, we have $P(\boldsymbol{X}^t \in A | \boldsymbol{X}^0 = \boldsymbol{x}) > 0$ for some $t \in \mathbb{N}_0$. The chain is **aperiodic** if there does not exist a measurable partition $E = B^0, \ldots, B^{c-1}$ for some $c \geq 2$ so that $P(\boldsymbol{X}^t \in B^{mod_c(t)} | \boldsymbol{X}^0 = \boldsymbol{x}_0 \in B^0) = 1$ for all t.

Obviously, this generalizes the above definition in the finite case. The above special case corresponds to taking the single states as the sets A, c as a possible period, and the B^i again as some single states in E.

The following conditions for (6.41) can be very well checked in practice.

Theorem 6.6: Convergence of Gibbs Sampling: Continuous Case (Roberts and Smith, 1994)
Let $\pi(\boldsymbol{x})$ be lower semicontinuous at 0 (i.e. $\forall \boldsymbol{x}$ with $\pi(\boldsymbol{x}) > 0$ $\exists$ an open environment $U_{\boldsymbol{x}}$ of $\boldsymbol{x}$ and an $\varepsilon > 0$ so that $\forall \boldsymbol{y} \in U_{\boldsymbol{x}}$ $\pi(\boldsymbol{y}) \geq \varepsilon$). Let P_G be defined as in (6.39) and the corresponding Markov chain aperiodic and π-irreducible. Then, (6.41) is true and

$$\frac{1}{t}\{g(\boldsymbol{X}^1) + \ldots + g(\boldsymbol{X}^t)\} \to \int_E g(\boldsymbol{x}) \pi(\boldsymbol{x}) d\boldsymbol{x}, \quad t \to \infty \text{ (almost surely)}, \tag{6.42}$$

for every real π-integrable function g. □

The next theorem gives a first idea about convergence speed.

Theorem 6.7: Convergence Speed of Gibbs Sampling: Continuous Case (Roberts and Polson, 1994)
If there exists a non-negative function $P^*_G(\boldsymbol{y})$ so that $P^*_G(\boldsymbol{y}) > 0$ on a set of

positive Lebesgue measures and $P_G^{(t^*)}(\boldsymbol{x},\boldsymbol{y}) \geq P_G^*(\boldsymbol{y})$ for some $t^* \in \mathbb{N}$, then (6.41) is uniformly valid with convergence rate $\rho \leq (1 - \int_{\mathbb{R}^d} P_G^*(\boldsymbol{y})d\boldsymbol{y})^{1/t^*}$. □

Thus, if the joint distribution of $[\boldsymbol{x}\ \boldsymbol{y}]^T$ is limited from below by a function that only depends on the new realization $\boldsymbol{y}$, and not on the old realization $\boldsymbol{x}$, then the densities converge linearly. This property can be achieved via a condition on the conditional densities of π.

Theorem 6.8: Geometrical Convergence of Gibbs Sampling: Continuous Case (Roberts and Polson, 1994)
If there exist non-negative and lower semicontinuous functions $g_i : \mathbb{R}^i \to \mathbb{R}^+$, $1 \leq i \leq d$, such that

$$\pi(\boldsymbol{y}_i|\boldsymbol{y}_j,\ 1 \leq j < i,\ \boldsymbol{x}_j,\ i < j \leq d) \geq g_i(\boldsymbol{y}_1,\ldots,\boldsymbol{y}_i) \tag{6.43}$$

and if

$$\{\boldsymbol{y}|g_i(\boldsymbol{y}_1,\ldots,\boldsymbol{y}_i) > 0,\ 1 \leq i \leq d\} \neq \oslash \tag{6.44}$$

then $\exists M < \infty,\ 0 < \rho < 1$ with

$$\|f_{\boldsymbol{x}^0}^{(t)} - \pi\|_1 \leq M\rho^t \quad \text{(linear (geometrical) convergence).} \tag{6.45}$$

□

Thus, the conditional distributions of π should be limited from below by functions that only depend on components of the new realization and not any more on components of the previous realization. Roberts and Polson (1994) prove these conditions, e.g., for a class of hierarchical models with densities of an exponential family. So, there exist examples for linear convergence.

6.3.3.3 When Does Gibbs Sampling Converge?

In the previous two sections we have gotten to know conditions for the convergence of the Gibbs algorithm, namely, irreducibility and aperiodicity. Now, we will see that these conditions relate to the so-called **connectedness** of the region E (Arnold, 1993, pp. 602 – 605). Let us start with examples where Gibbs sampling is not functioning.

Example 6.10: Gibbs Sampling Fails: Finite Case
Let $X \sim Bin(2, 0.5)$ and $P(Y = X) = 0.5$, $P(Y = -X) = 0.5$, where $Bin(n, p)$ is the binomial distribution with n repetitions and success probability p. Then, the conditional probabilities in the Gibbs algorithm are of the form:

$$\begin{aligned} P(x|y) &= 1 \quad \text{for } x = |y|;\ \text{since } P(X=Y) = P(X=-Y) = 0.5,\ x \geq 0, \\ P(y|x) &= 0.5 \text{ for } y = \pm x. \end{aligned}$$

Starting from $x_0 = 1$, $y_0 = 1$, we see that $x_1 = 1$, $y_1 = \pm 1$, $x_2 = 1$, $y_2 = \pm 1$, etc. Actually, we always have $x_n = 1$, $y_n = \pm 1$. Thus, the distribution of the $[X_n \ Y_n]^T$ does not converge toward the joint distribution of $[X \ Y]^T$. In particular, X_n never reaches the values $0, 2$.
Moreover, for an arbitrary starting distribution for $[X_0 \ Y_0]^T$, we always have $X_n = X_0$ for all n so that the distribution of X_n always degenerates to the starting value. (Notice that X_0 can only take the values 0,1,2 and Y_0 can only be equal to $\pm X_0$.) Therefore, the distribution of X_n generally does not converge toward the distribution of X. Notice that in this example the Markov chain is not irreducible, i.e. does not reach the whole sample space!

Example 6.11: Gibbs Sampling Fails: Continuous Case
Let $Q \sim \mathcal{N}(0,1)$ and $R \sim \mathcal{N}(0,1)$ be independent. Let $X = Q$ and $Y = sign(X)|R|$, sign = sign function. Then, $P(sign(X) = sign(Y)) = 1$, and thus $P(x|y) = 1$ for $sign(x) = sign(y)$. Therefore, X always has the same sign as Y. Using the Gibbs algorithm for the joint distribution of X and Y, when X_0, Y_0 are positive, also $X_n > 0$ and $Y_n > 0$ for all n.
Moreover, $P(X > 0, Y > 0) = 0.5$, so that the distribution of $[X_n \ Y_n]^T$ cannot converge against the distribution of $[X \ Y]^T$.

Both examples have the same problem: the sample space consists of unconnected sets. If the Gibbs algorithm samples once in one of these sets, then it stays there forever. This prompts the following definition.

Definition 6.36: Connection of the Sample Space
Let $E \subset \mathbb{R}^d$ be the sample space of the random vector $\boldsymbol{X}$, i.e. the set of points on which the density function of $\boldsymbol{X}$ is positive. Consider the same partition of $\mathbb{R}^d$ into k subspaces as for the Gibbs algorithm. Such a sample space is called **connected** if for two arbitrary points $\boldsymbol{x}_0 \in E$, $\boldsymbol{x}_r \in E$ a finite sequence $\boldsymbol{x}_1, \ldots, \boldsymbol{x}_{r-1}$ of points exists so that for $i = 1, \ldots, r$:

$$\begin{aligned}
\boldsymbol{x}_i &= [\boldsymbol{x}_{i1} \ \ldots \ \boldsymbol{x}_{ik}]^T, \\
f_1(\boldsymbol{x}_{i1}|\boldsymbol{x}_{i-1,2}, \ldots, \boldsymbol{x}_{i-1,k}) &> 0, \\
f_2(\boldsymbol{x}_{i2}|\boldsymbol{x}_{i1}, \boldsymbol{x}_{i-1,3}, \ldots, \boldsymbol{x}_{i-1,k}) &> 0, \\
&\ldots, \\
f_k(\boldsymbol{x}_{ik}|\boldsymbol{x}_{i1}, \ldots, \boldsymbol{x}_{i,k-1}) &> 0,
\end{aligned}$$

where $f_i(\boldsymbol{x}_i|\boldsymbol{x}_1, \ldots, \boldsymbol{x}_{i-1}, \boldsymbol{x}_{i+1}, \ldots, \boldsymbol{x}_k)$ is the conditional density of X_i given $\boldsymbol{X}_1 = \boldsymbol{x}_1, \ldots, \boldsymbol{X}_{i-1} = \boldsymbol{x}_{i-1}$, $\boldsymbol{X}_{i+1} = \boldsymbol{x}_{i+1}, \ldots, \boldsymbol{X}_k = \boldsymbol{x}_k$.

Obviously, we have defined the connection of a sample space by the condition that from any arbitrary state the Gibbs algorithm can reach any other state so that it is not possible to be "trapped" in a subregion of the sample space.

Comparing this property with the interpretation of irreducibility in Section 6.3.3.1, the following theorem is prompted:

Theorem 6.9: Connection of Sample Space and Convergence of Gibbs Sampler

Let $\boldsymbol{X}$ be a random vector with a finite connected sample space. Let $\boldsymbol{X}^t$ be the random vector in the tth iteration step of the Gibbs algorithm. Then $\boldsymbol{X}^t \rightarrow \boldsymbol{X}$ in distribution.

Proof. By Theorem 6.5 we only have to prove irreducibility and aperiodicity. Since the sample space is connected, the Markov chain is irreducible. Moreover, the chain is also aperiodic, since for every state $P(\boldsymbol{X}_i^t = \boldsymbol{X}_i^{t-1}) > 0$, implying $P(\boldsymbol{X}^t = \boldsymbol{X}^{t-1}) > 0$. □

If $\boldsymbol{X}$ is a random vector with a finite unconnected sample space, then the Markov chain is not irreducible and the Gibbs algorithm does not converge to the distribution of $\boldsymbol{X}$, as seen in the above examples.

In the continuous case sufficient conditions for π-irreducibility are **connectedness** of E and local boundedness of each $(d-1)$-dimensional marginal distribution $\pi(\boldsymbol{x}_{-i}) = \int \pi(\boldsymbol{x}) d\boldsymbol{x}_i$ (Roberts and Smith, 1994).

6.3.4 Metropolis-Hastings Method

The Gibbs algorithm is hardly adequate for jagged probability distributions, and is therefore often only used for local distributions. The special MCMC algorithm described in the following, the **Metropolis-Hastings algorithm** (Hastings, 1970; Metropolis et al., 1953), is characterized by big flexibility and simple programmability. Again, the density to be drawn from only has to be known except for a normalizing constant.

6.3.4.1 Finite Case

Consider a Markov chain on a finite state space $\{1, 2, \ldots, n\}$ with the transition matrix $\boldsymbol{P} = [p_{ij}]_{1 \le i,j \le n}$. For the elements of the matrix let $p_{ij} > 0$. Therefore, the Markov chain is irreducible.

Additionally, a so-called **reversibility condition** is postulated:

$$\pi_i p_{ij} = \pi_j p_{ji}. \tag{6.46}$$

This immediately leads to the equation

$$\sum_{i=1}^{n} \pi_i p_{ij} = \pi_j \quad \forall j, \tag{6.47}$$

i.e. $[\pi_i]_{1 \leq i \leq n}$ represents the stationary distribution of the Markov chain, against which it may converge.

The reversibility condition describes a balance between the rate of reciprocating motion and is a sufficient, but not necessary, condition for stationarity. Thus, there might be Markov chains not fulfilling this condition but nevertheless converging toward the desired distribution $[\pi_i]$. However, then the proof of stationarity is much more difficult. Note that the above Gibbs algorithm is not reversible, but stationary in the limit under certain conditions shown in the previous chapter, whereas the following Metropolis-Hastings algorithm is reversible (and therefore automatically stationary), as will be shown below.

The Metropolis-Hastings algorithm uses the following **transition matrix**:

$$p_{ij} = q_{ij}\alpha_{ij} \quad \text{for } i \neq j, \tag{6.48}$$

$$p_{ii} = 1 - \sum_{j \neq i} p_{ij} \tag{6.49}$$

Here, the arbitrary matrix $\boldsymbol{Q} = [q_{ij}]$ with $q_{ij} \geq 0$ induces with the marginal density $q(\boldsymbol{x}, \cdot)$, an approximation for the density π given $\boldsymbol{x}$. This approximation is chosen such that random realizations can be drawn easily. p_{ii} is the probability for the rejection of the new point. The matrix $[\alpha_{ij}]_{1 \leq i,j \leq n}$ is defined by

$$\alpha_{ij} = \begin{cases} 1 & \text{if } \pi_i q_{ij} = 0, \\ \min(\pi_j q_{ji}/\pi_i q_{ij}, 1) & \text{if } \pi_i q_{ij} > 0. \end{cases} \tag{6.50}$$

In detail, the transition from $i \to j$ is realized as indicated in Algorithm 6.7.

Proposition 6.1: Reversibility Condition of the Metropolis-Hastings Algorithm
The Metropolis-Hastings algorithm fulfills the reversibility condition (6.46), i.e. $[\pi_i]_{1 \leq i \leq n}$ is the desired stationary distribution independent of the approximation $[q_{ij}]$.

Proof.

1. If $\alpha_{ij} = \min\left(\frac{\pi_j q_{ji}}{\pi_i q_{ij}}, 1\right) = \frac{\pi_j q_{ji}}{\pi_i q_{ij}}$, and therefore $\alpha_{ji} = \min\left(\frac{\pi_i q_{ij}}{\pi_j q_{ji}}, 1\right) = 1$, then $\pi_i p_{ij} = \pi_i q_{ij} \alpha_{ij} = \frac{\pi_i q_{ij} \pi_j q_{ji}}{\pi_i q_{ij}} = \pi_j q_{ji} = \pi_j p_{ji}$. $i \leftrightarrow j$ analogously.

Algorithm 6.7 Metropolis-Hastings Algorithm: One Iteration

Require: State i

1: Choose state j according to the probability distribution of the ith row of Q
2: Generate a uniformly distributed number $u \in [0,1)$
3: **if** $u \leq \alpha_{ij}$ **then**
4: Accept j {with probability α_{ij}, i.e. reject with probability $1-\alpha_{ij}$}
5: **else**
6: Reject j {If j is rejected, the next point in the chain is again i, and the choice of a new point j is repeated}
7: **end if**

2. If $\pi_i q_{ij} = 0$ and $\pi_j q_{ji} \neq 0$, then $\alpha_{ji} = \min\left(\frac{\pi_i q_{ij}}{\pi_j q_{ji}}, 1\right) = 0$, and thus $\pi_i p_{ij} = \pi_i q_{ij} = 0 = \pi_j q_{ji} \alpha_{ji} = \pi_j p_{ji}$. $i \leftrightarrow j$ analogously.
3. If $\pi_i q_{ij} = 0 = \pi_j q_{ji}$, then $\pi_i p_{ij} = \pi_i q_{ij} = 0 = \pi_j q_{ji} = \pi_j p_{ji}$.

□

6.3.4.2 *Continuous Case*

The continuous analogue to the previous section is:

$$\alpha(\boldsymbol{x},\boldsymbol{y}) = \begin{cases} 1 & \text{if} \quad \pi(\boldsymbol{x})q(\boldsymbol{x},\boldsymbol{y}) = 0, \\ \min\left(\frac{\pi(\boldsymbol{y})q(\boldsymbol{y},\boldsymbol{x})}{\pi(\boldsymbol{x})q(\boldsymbol{x},\boldsymbol{y})}, 1\right) & \text{if} \quad \pi(\boldsymbol{x})q(\boldsymbol{x},\boldsymbol{y}) > 0 \end{cases} \tag{6.51}$$

with transition probabilities

$$p(\boldsymbol{x},\boldsymbol{y}) = q(\boldsymbol{x},\boldsymbol{y})\alpha(\boldsymbol{x},\boldsymbol{y}) \quad \text{if } \boldsymbol{x} \neq \boldsymbol{y}, \tag{6.52}$$

$$p(\boldsymbol{x},\boldsymbol{x}) = 1 - \int_E p(\boldsymbol{x},\boldsymbol{y})d\boldsymbol{y}. \tag{6.53}$$

The transition function thus has the form

$$P_H(\boldsymbol{x},\boldsymbol{y}) = q(\boldsymbol{x},\boldsymbol{y})\alpha(\boldsymbol{x},\boldsymbol{y}). \tag{6.54}$$

In the finite, as well as in the continuous, case the following sufficient conditions for the **convergence of the Markov chain** are valid:

Theorem 6.10: Sufficient Conditions for Convergence of Metropolis Hastings Sampler: Continuous Case (Roberts and Smith, 1994)
If q is aperiodic, then also the Markov chain of the Metropolis-Hastings algorithm is aperiodic.
If q is (π-)irreducible and $q(\boldsymbol{x},\boldsymbol{y}) = 0$ iff $q(\boldsymbol{y},\boldsymbol{x}) = 0$, then the Markov chain of the Metropolis-Hastings algorithm is also (π-)irreducible. □

Convergence can thus be forced by an adequate choice of q. The central convergence theorem again assumes (π-)irreducibility and aperiodicity:

Theorem 6.11: Convergence of Metropolis Hastings Sampler: Continuous Case (Roberts and Smith, 1994)
Let the Markov chain corresponding to $P_H(\boldsymbol{x}, \boldsymbol{y})$ be (π-)irreducible and aperiodic. Then, for all starting points $\boldsymbol{x}^0 \in E$:

- $\|P^{(t)}_{H,\boldsymbol{x}^0} - \pi\|_1 \to 0$ for $t \to \infty$, where $P^{(t)}_{H,\boldsymbol{x}^0}$ is the density function of the Markov chain with transition function P_H after the tth iteration.
- For each real π-integrable function g

$$\frac{1}{t}\{g(\boldsymbol{X}^1) + \ldots + g(\boldsymbol{X}^t)\} \to \int_E g(\boldsymbol{x})\pi(\boldsymbol{x})d\boldsymbol{x}, \quad t \to \infty \text{ almost surely} \tag{6.55}$$

(in the finite case use sums correspondingly).

This completes the central terms and theorems on the convergence of Markov chains. Later on we will also discuss the convergence rate.

6.3.4.3 *Possible Transition Functions*

The choice of the transition matrix $\boldsymbol{Q}$ or the transition function $q(\boldsymbol{x}, \boldsymbol{y})$ is decisive for the efficiency of the algorithm. We will concentrate here on the continuous case $q(\boldsymbol{x}, \boldsymbol{y})$, where starting from the point $\boldsymbol{x}$ a new point $\boldsymbol{y}$ is generated.
If this is rejected by $\alpha(\boldsymbol{x}, \boldsymbol{y})$ too often, on the one hand, the burn-in phase takes a lot of time, and on the other hand, the Markov chain moves very slowly from one point to the other. Therefore, it takes a very long time to cover the whole space E by random points.
In contrast, if $\boldsymbol{y}$ is very often accepted, this can have two reasons: either $\boldsymbol{y}$ is chosen to be very near to $\boldsymbol{x}$ (which would lead to high autocorrelation) or $q(\boldsymbol{x}, \boldsymbol{y})$ is already very well adapted to $\pi(\boldsymbol{y})$ such that wider jumps are also accepted (which would lead to low autocorrelation).

Besides the autocorrelation, often the

$$\textbf{acceptance rate} = \frac{\text{accepted transitions } i \to j \text{ or } \boldsymbol{x} \to \boldsymbol{y}}{\text{total number of trials}}$$

is used as a measure for effectiveness of the algorithm. Obviously, too low acceptance rates and high autocorrelation slow down convergence. In the literature, under certain circumstances optimal acceptance rates are developed, meaning acceptance rates that maximize convergence speed. Unfortunately, optimal acceptance rates vary from problem to problem. Roberts et al. (1997)

showed that if the proposal densities $q(\boldsymbol{x},\boldsymbol{y})$ are normal, the optimal acceptance rate for the Markov chain can be approximately 0.234 in high dimensions. Bédard and Rosenthal (2008) gave examples where the optimal acceptance rate is even lower than 0.234. To achieve optimal acceptance rates, the variance of the proposal density family is tuned. On the other hand, if the target density π is uniform or if an independence chain (see below) is used, the acceptance rate is best 100%, but that is rarely achieved.

The following choices for $q(\boldsymbol{x},\boldsymbol{y})$ are most common:

1. **Random walk:** Let $q(\boldsymbol{x},\boldsymbol{y}) = q_R(\boldsymbol{y}-\boldsymbol{x})$ with a density q_R. Then $\boldsymbol{y} = \boldsymbol{x}+\boldsymbol{z}$, where $\boldsymbol{z}$ is q_R distributed. Possible distributions for q_R are, e.g., normal or t-distributions. Also, uniform distributions are used on a box with edge length r or a ball with this radius around $\boldsymbol{x}$. The parameter r can be used to control acceptance rate and autocorrelation in order to optimize convergence speed. For example, if r would be very small, autocorrelation would be very high.
2. **Vector autoregression:** Let $q(\boldsymbol{x},\boldsymbol{y}) = q_A(\boldsymbol{y}-\boldsymbol{\mu}-\boldsymbol{A}(\boldsymbol{x}-\boldsymbol{\mu}))$ with a density q_A as q_R above, $\boldsymbol{\mu} \in \mathbb{R}^d$, and a matrix $\boldsymbol{A} \in \mathbb{R}^{d\times d}$. Then, $\boldsymbol{y} = \boldsymbol{\mu}+\boldsymbol{A}(\boldsymbol{x}-\boldsymbol{\mu})+\boldsymbol{z}$, where $\boldsymbol{z}$ is q_A distributed. The choice $\boldsymbol{A} = -\boldsymbol{I}$ generates approximate reflections $\boldsymbol{y}-\boldsymbol{\mu} = -(\boldsymbol{x}-\boldsymbol{\mu})$ and may serve to lower autocorrelation.
3. **Independence Chain (ICMCMC):** Let $q(\boldsymbol{x},\boldsymbol{y}) = q_I(\boldsymbol{y})$ with a density q_I as above, where a location parameter is specified. The new point $\boldsymbol{y}$ is drawn independently of $\boldsymbol{x}$. The more q_I resembles the desired density π, the more effective this method is. Notice the similarity of this method with the rejection method (see Section 6.3.4.6).

More proposals for q can be found in the overview paper of Tierney (1994).

6.3.4.4 *Implementation of the Metropolis-Hastings Algorithm*

The following advice relates to the implementation of the Metropolis-Hastings algorithm if the proof of convergence is not central. The experimental convergence verification by graphical representations will be discussed, as well as the difference between uni- and multimodal distributions.

1. Let the distribution be **unimodal**:
   For $q(\boldsymbol{x},\boldsymbol{y})$ a local mechanism appears to be adequate (Random Walk), i.e. $\boldsymbol{y}$ is chosen uniformly distributed on a small box around $\boldsymbol{x}$. Because of the unimodality the Markov chain cannot miss any region of high probability.

2. Let the distribution be **multimodal**:
   In order to guarantee that $q(\boldsymbol{x}, \boldsymbol{y})$ does not miss any region of high probability of π, all modes should have been found by analysis of the formula of π or by an optimization algorithm. Then, $q(\boldsymbol{x}, \boldsymbol{y})$ should move locally as in the unimodal case and should move globally between the modes corresponding to the relative weights (heights) of the modes. In order to control acceptance rate and autocorrelation, box length and frequency of jumps should be chosen correspondingly.
3. Graphical verification:
   Every integral $I = \int_B g(\boldsymbol{x})\pi(\boldsymbol{x})d\boldsymbol{x}$ can be calculated (estimated) by the unbiased estimator

$$\hat{I}_N = \frac{1}{N} \sum_{\boldsymbol{x}_k \sim \pi}^{N} g(\boldsymbol{x}_k). \tag{6.56}$$

   The convergence can the assessed by the **graphical representation** $N \to \hat{I}_N$. Unfortunately, there is the risk that convergence is only faked since the Markov chain gets stuck in a local mode. If after a while the chain disengages from this mode, the graphical representation changes drastically. Only then does it become clear that convergence was not achieved. This is not a problem in (1) since there is only one mode, and impossible in (2) because of the jumps. The problem of fake convergence is a big disadvantage of the local Gibbs algorithm. For the calculation of integrals the burn-in phase is not a big problem since its importance is diminishing for large N in (6.56). High autocorrelation leads to higher sample sizes, because the variance increases. To avoid this, often only each, say, fifth value of the Markov chain is used. This is called **thinning**.
4. Unfortunately, graphical representation is restricted to single integrals. Different integrals need more or less points, dependent on which regions are important for the integral. The calculation of many integrals may help to assess the convergence of a chain. Security, however, gives only a distance measure, as introduced in the next section.

Notice that thinning often diminishes autocorrelation, but also deteriorates the exactness of the point estimation and its uncertainty (see, e.g., Link and Eaton (2012)).

6.3.4.5 Convergence Rate

As a formalization of a descriptive distance measure between two distributions the variation norm is used (see Section 6.1.1). First we consider the convergence rate in the finite case:

Theorem 6.12: Convergence Rate of Markov Chains: Finite Case (Diaconis and Strook, 1991)
Let $\pi(\boldsymbol{x})$ be a density on a finite set E and $P(\boldsymbol{x},\boldsymbol{y})$, the transition matrix of a reversible, irreducible, and aperiodic Markov chain with eigenvalues

$$1 = \beta_0 > \beta_1 \geq \ldots \geq \beta_{n-1} > -1, \quad n = |E|, \tag{6.57}$$

where $|E|$ = number of elements in E, then

$$\|P^{(\tilde{t})}(\boldsymbol{x},\cdot) - \pi(\cdot)\|_{Var} \leq \left[\frac{1}{2}\sqrt{\frac{1-\pi(\boldsymbol{x}^0)}{\pi(\boldsymbol{x}^0)}}\right]\beta^{\tilde{t}} =: \text{prefactor } \beta^{\tilde{t}} \tag{6.58}$$

with $\beta = \max(\beta_1, |\beta_{n-1}|)$ and an arbitrary starting point $\boldsymbol{x}^0 \in E$. □

This theorem can be interpreted as follows:

If $\tilde{t}$ is large enough, the probability to generate $\boldsymbol{y}$ having started in $\boldsymbol{x}^0$ is, up to a small error, equal to $\pi(\boldsymbol{y})$ (instead of $\boldsymbol{y}$, one can also take an arbitrary measurable set $A \subseteq E$ because of the definition of $\|\cdot\|_{Var}$). Starting then in $\boldsymbol{y}$ with a new step of the Markov chain, the probability to generate a $\boldsymbol{z} \in E$ is equal to $\pi(\boldsymbol{z})$ up to an even smaller error, since

$$4\|P^{(\tilde{t}+1)}(\boldsymbol{x}^0,\boldsymbol{z}) - \pi(\boldsymbol{z})\|^2_{Var} \leq \frac{1-\pi(\boldsymbol{x}^0)}{\pi(\boldsymbol{x}^0)}\beta^{2(\tilde{t}+1)} < \frac{1-\pi(\boldsymbol{x}^0)}{\pi(\boldsymbol{x}^0)}\beta^{2\tilde{t}}. \tag{6.59}$$

(Also, this is valid for general measurable sets $A \subseteq E$.) This means that all further points (from $\approx \tilde{t}$ on) generated by the Markov chain P are π-distributed!

The prefactor in (6.58) depends on the dimension. For an acceptable bound it should be as small as possible, i.e. $\pi(\boldsymbol{x}^0)$ as large as possible. Therefore, the Markov chain should start at the point with maximum π_i or a similar value. The larger the dimension, the larger the region the probability mass has to be distributed on, i.e. the smaller the maximum. A higher bound in (6.58) caused by a larger prefactor is, however, eventually compensated by $\beta^{\tilde{t}}$.

Example 6.12: Uniform Distribution
Let π be a uniform distribution on n points, i.e. $\pi_i = 1/n$ ($1 \leq i \leq n$). Then, using $q_{ij} = 1/n$ leads to the eigenvalues 0 and 1, i.e. the Markov chain has already converged ($\beta = 0$) and points from π are generated immediately. The acceptance rate is 100% in this special case.

Example 6.13: Nearest-Neighbor Metropolis Algorithm
Also for so-called **Nearest-Neighbor (NN) algorithms** Theorem 6.12 easily delivers the convergence rates β. Let, e.g., the transition matrix be defined for

$n = |E| = 3$ as

$$Q_3 = \begin{bmatrix} \frac{1}{2} & \frac{1}{2} & 0 \\ \frac{1}{2} & 0 & \frac{1}{2} \\ 0 & \frac{1}{2} & \frac{1}{2} \end{bmatrix} \Rightarrow \beta_3 = 0.5; \quad \text{since}$$

$$\begin{aligned} Q_3 \begin{bmatrix} 1 & 1 & 1 \end{bmatrix}^T &= \begin{bmatrix} 1 & 1 & 1 \end{bmatrix}^T, \\ Q_3 \begin{bmatrix} 1 & 0 & -1 \end{bmatrix}^T &= 0.5 \begin{bmatrix} 1 & 0 & -1 \end{bmatrix}^T, \text{ and} \\ Q_3 \begin{bmatrix} -1 & 2 & -1 \end{bmatrix}^T &= 0.5 \begin{bmatrix} -1 & 2 & -1 \end{bmatrix}^T. \end{aligned}$$

The index of Q corresponds to the number of possible points. Defining Q_5, Q_{11}, Q_{17} analogously, the corresponding β looks like: $\beta_5 = 0.81, \beta_{11} = 0.96, \beta_{17} = 0.98$.

Therefore, especially for large n, the Nearest-Neighbor (NN) Metropolis algorithm is obviously much less efficient than the algorithm with $q_{ij} = 1/n$. The convergence speed is, however, much easier to analyze for NN algorithms. Therefore, such algorithms are often analyzed in this context, e.g. in Polson (1994).

The following central theorem shows the geometrical convergence corresponding to the $L1$ norm (6.1) in the continuous case (cp. the continuous case for the Gibbs algorithm).

Theorem 6.13: Geometrical Convergence of Markov Chains: Continuous Case (Smith and Roberts, 1993)
Suppose there exists $P_H^* : E \to \mathbb{R}$ with $P_H(\boldsymbol{x}, \boldsymbol{y}) \geq P_H^*(\boldsymbol{y}) \;\; \forall (\boldsymbol{x}, \boldsymbol{y}) \in E \times E$ and

$$\rho := 1 - \int_E P_H^*(\boldsymbol{y}) d\boldsymbol{y}. \tag{6.60}$$

Then

$$\|P_{H,\boldsymbol{x}^0}^{(t)} - \pi\|_1 \leq 2\rho^t \;\; \forall t \in \mathbb{N} \text{ and } \boldsymbol{x}^0 \in E. \tag{6.61}$$

□

Theorem 6.13 applied to the Metropolis-Hastings algorithm in the special case of the **Independence Chain (ICMCMC)**, i.e. $q(\boldsymbol{x}, \boldsymbol{y}) = q(\boldsymbol{y})$, delivers, because of (6.51), the transition function

$$P_H(\boldsymbol{x}, \boldsymbol{y}) = q(\boldsymbol{y})\alpha(\boldsymbol{x}, \boldsymbol{y}) = q(\boldsymbol{y})\frac{\pi(\boldsymbol{y})q(\boldsymbol{x})}{\pi(\boldsymbol{x})q(\boldsymbol{y})} = \pi(\boldsymbol{y})\frac{q(\boldsymbol{x})}{\pi(\boldsymbol{x})} \quad \forall (\boldsymbol{x}, \boldsymbol{y}) \in E \times E, \tag{6.62}$$

which leads to a lower bound function

$$P_H^*(\boldsymbol{y}) = \min_{\boldsymbol{x}\in E} P_H(\boldsymbol{x},\boldsymbol{y}) = \pi(\boldsymbol{y}) \min_{\boldsymbol{x}\in E} \frac{q(\boldsymbol{x})}{\pi(\boldsymbol{x})}. \tag{6.63}$$

Therefore, an upper bound for (6.60) can be found:

$$\rho = 1 - \int_E P_H^*(\boldsymbol{y})d\boldsymbol{y} \le 1 - \min_{\boldsymbol{x}\in E} \frac{q(\boldsymbol{x})}{\pi(\boldsymbol{x})} \int_E \pi(\boldsymbol{y})d\boldsymbol{y}. \tag{6.64}$$

This leads to the following lemma.

Lemma 6.2: Convergence Rate for ICMCMC
For ICMCMC the convergence rate ρ can be bounded as follows:

$$\rho \le \rho_S := 1 - \frac{\int_E \pi(\boldsymbol{y})d\boldsymbol{y}}{\max_{\boldsymbol{x}\in E} \frac{\pi(\boldsymbol{x})}{q(\boldsymbol{x})}}. \tag{6.65}$$

(6.65) is interesting for the relationship to the rejection method, which is discussed in the next section.

6.3.4.6 Rejection Method and Independence Chain MCMC

If P_{accept} is the probability that for the rejection method corresponding to a density q a generated point $\boldsymbol{x}$ is accepted (q is the density used for trial point generation and $\left(\max_{\boldsymbol{x}\in E} \frac{\pi}{q}\right) q$ the envelope (cp. Section 6.3.2)), then because of Formulas (6.17) and (6.19),

$$1 - \frac{\int_E \pi(\boldsymbol{y})d\boldsymbol{y}}{\max_{\boldsymbol{x}\in E} \frac{\pi(\boldsymbol{x})}{q(\boldsymbol{x})}} = 1 - P_{accept} = P_{reject}. \tag{6.66}$$

This expression is identical with (6.65). This relationship can be motivated by the algorithms themselves. The Metropolis-Hastings algorithm generates a trial point $\boldsymbol{y}$ by $q(\boldsymbol{x},\boldsymbol{y})$. In general, q also depends on $\boldsymbol{x}$. This leads to autocorrelation. In the special case $q(\boldsymbol{x},\boldsymbol{y}) = q_I(\boldsymbol{y})$, however, the Metropolis-Hastings algorithm generates, as the rejection method, an independent point. Therefore, the two methods are similar.

Even more, the rejection method accepts $\boldsymbol{y}$ locally, i.e. in $\boldsymbol{x}$, with the probability

$$P_{RM} = \frac{\pi(\boldsymbol{y})}{kq(\boldsymbol{y})}, \quad \text{where } k \ge \sup_{\boldsymbol{x}\in B} \frac{\pi(\boldsymbol{x})}{q(\boldsymbol{x})}, \tag{6.67}$$

and the ICMCMC with probability

$$P_{IC} = \frac{\pi(\boldsymbol{y})q(\boldsymbol{x})}{\pi(\boldsymbol{x})q(\boldsymbol{y})}. \tag{6.68}$$

These expressions can be directly compared: $P_{RM} \leq P_{IC}$.

ICMCMC is thus at least as effective as the rejection method. Autocorrelation only appears with ICMCMC if y is not accepted. The accepted points build an independent sample. Note, however, that the claim that with ICMCMC the accepted points build an independent sample should not be misinterpreted that those points build an independent sample from π. Indeed, they do not, since the reject moves give weights to sample values that are essential for the stationary distribution to be π (cp. (6.49)).

A further advantage of ICMCMC is that the constant k is only needed for the bound of the convergence rate in (6.65), the algorithm itself does not need this constant. For the rejection method, however, k is an essential part of the algorithm. The worse k is estimated, the less effective is the algorithm. For both algorithms a good approximation of π by q and a high acceptance rate are important.

One consequence of this section is that for finding a bound for ρ by (6.65) the starting point of the IC Markov chain is best generated by the rejection method, since after the tth application of the rejection method

$$P(t \text{ times } \neg \text{ accepted}) = \rho_S^t. \tag{6.69}$$

While in (6.61) ρ_S^t has to be of size 0.01 (or smaller), in order to produce an acceptable error $||\cdot||_1$ (unless $||P^{(0)} - \pi||_1$ is already $\ll 1$), in (6.69) a smaller t is sufficient for the rejection method to accept at least one point. This point can then be used as the starting point for the Markov chain. This approach also has, besides the smaller t, the advantage that the chain is starting exactly in equilibrium, i.e. with $||\cdot||_{Var} = 0$.

6.3.5 *Summary*

Starting with a study of the rejection sampling method, we looked at various other means of generating multivariate random variables. These included two MCMC methods, the Gibbs sampler and the Metropolis-Hastings algorithm. Convergence properties and rates were derived for both of these methods. Finally, we showed how the rejection method compares to the independent chain MCMC method.

6.3.6 *Simple Example*

Consider an $Exp(1)$ target distribution, i.e. an exponential distribution with parameter $\lambda = 1$ (see Section 6.2.5.5), with density $\pi(x) = e^{-x} I(x > 0)$. Suppose we are interested in evaluating $\mathrm{E}_\pi X$ (where it is assumed we do

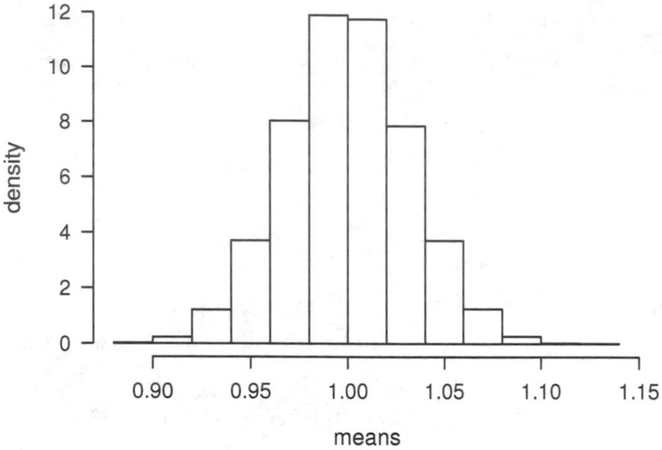

Figure 6.19: Histogram of the means of 10,000 replications from $n = 1000$ numbers of an $Exp(1)$ distribution.

not know $\mathrm{E}_\pi X = 1$). We estimate this quantity using both Monte Carlo and Markov chain Monte Carlo techniques. First, Monte Carlo estimation requires a random sample from π. To this end, we generated 10,000 iid samples of size $n = 1000$ from $Exp(1)$. Using a seed of 1, we found a point estimator $\hat{\lambda} = 1.0002$ and an interval estimator (by calculating 0.025 and 0.975 quantiles) so that $P(0.9377 < \lambda < 1.0631) = 0.95$ (see Figure 6.19).

Next, suppose it is not possible to sample directly from π. Instead, consider exploring π using the ICMCMC algorithm with an $Exp(\lambda)$ proposal distribution with a density $q(x) = \lambda e^{-\lambda x} I(x > 0)$. For each starting value $\lambda \in \{0.5, 1, 4\}$ we ran a Markov chain for 1000 iterations with a sample $\{x_1, x_2, \ldots, x_{1000}\}$ of size $n = 1000$ from an $Exp(1)$ distribution. We varied the prior distribution of λ being $Gamma(1,1)$ (see Figure 6.20, top) and $Gamma(n, \sum_{i=1}^{1000} x_i)$ (see Figure 6.20, bottom). Notice that $Gamma(\alpha, \beta)$ is the conjugate prior distribution of λ. The estimated mean is almost identical to the observed mean ($\frac{1}{\bar{x}} = \hat{\lambda} = 0.9696$) in both cases, while the estimated variance differs between the cases (see Table 6.2). We use the R package BRugs (based on OpenBUGS) to implement the ICMCMC algorithm. The BRugs code is given below:

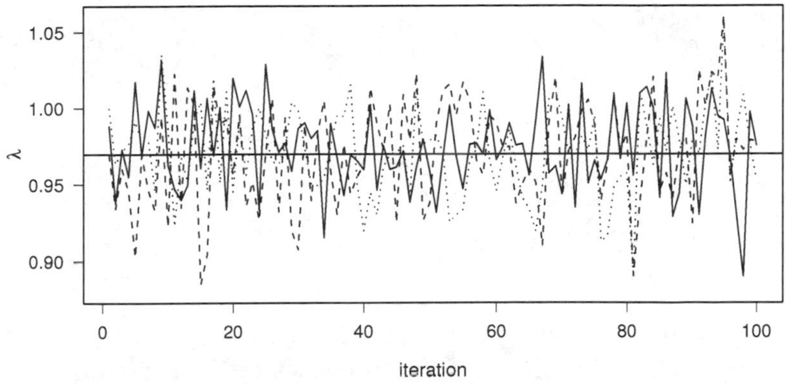

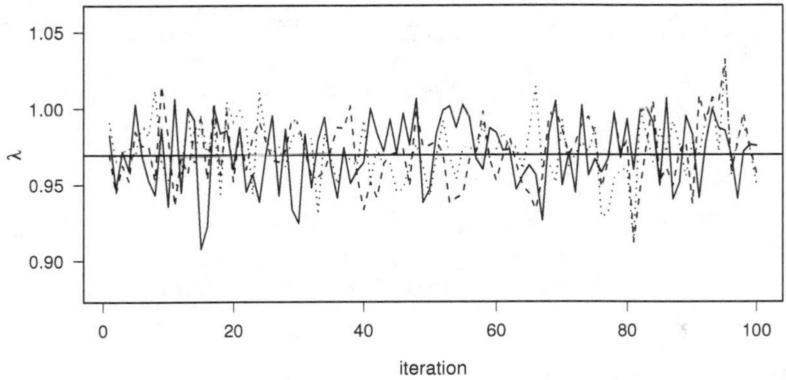

Figure 6.20: Three Markov chains for λ with different starting values, assuming $\lambda \sim Gamma(1,1)$ (top) and $\lambda \sim Gamma(N, \sum data)$ (bottom).

Table 6.2: Results of ICMCMC for $Exp(1)$

Prior Distribution	Mean	Stand. Dev.
$Gamma(1,1)$	0.9692	0.0304
$Gamma(n, \sum_{i=1}^{1000} x_i)$	0.9693	0.0215

```
library("BRugs")

## the model file:
model <- function(){
  for(i in 1:N) {
    x[i] ~ dexp(lambda)
  }
```

```
  lambda ~ dgamma(alpha, beta)
  alpha <- N        # alternatively: 1
  beta <- sum(x[]) # alternatively: 1
}

## attention: we create files in the current working directory
filename <- "model.txt"
writeModel(model, filename)
modelCheck(filename)          # check model file

set.seed(1)                   # set random seed
x <- rexp(1000)               # generate some data from Exp(1)

dat <- bugsData(list(x=x, N=length(x)), fileName = "data.txt")
modelData(dat)                # read data file
modelCompile(numChains=3)     # compile model with 3 chains

## initialize the model with 3 chains, lambda = 0.5, 1, and 4:
int <- bugsInits(list(list(lambda=0.5), list(lambda=1),
                      list(lambda=4)),
                 numChains = 3,
                 fileName = paste("inits", 1:3, ".txt", sep=""))
modelInits(int)               # read init data file

samplesSet("lambda")          # lambda is monitored
modelUpdate(1000)             # 1000 iterations (no burn-in)
samplesStats("*")             # the summarized results

## plotting the chains:
samplesHistory("lambda", mfrow = c(1, 1), end = 100, ask=FALSE)
abline(h = 1/mean(x))         # lambda = 1 / mean of the x values
```

The model is read and checked for syntax errors, the data are read and the function `modelCompile()` interprets the declared model and chooses the corresponding sampling methods, here, the exponential and gamma distributions. Then, the initial values are read for all considered Markov chains and the samples to be monitored are defined. Here, we assume that the chains have converged from the beginning (`modelUpdate(1000)`). Finally, we prepare a plot (`samplesHistory()`) with the first 100 iterations for each chain. See Section 6.4.4 for details.

6.4 Practice and Simulation: Stochastic Modeling

This section discusses an example for stochastic modeling as an application of MCMC methods. The data in the example are artificial sounds. We first discuss a statistical model for polyphonic music data, describe the usage of MCMC methods and the determination of their convergence, and assess the results. Finally, we briefly discuss the implementation of this example with the software package OpenBUGS (Spiegelhalter et al., 2004, 2012) (Bayesian inference Using Gibbs Sampling).

> "The BUGS (**B**ayesian inference **U**sing **G**ibbs **S**ampling) project is concerned with flexible software for the Bayesian analysis of complex statistical models using Markov chain Monte Carlo (MCMC) methods. The project began in 1989 in the MRC Biostatistics Unit and led initially to the 'Classic' BUGS program, and then onto the WinBUGS software developed jointly with the Imperial College School of Medicine at St Mary's, London.
> Development is now focused on the OpenBUGS project."
> (`http://www.mrc-bsu.cam.ac.uk/bugs/welcome.shtml`).

6.4.1 *Polyphonic Music Data*

Today, music is often recorded in CD quality, i.e. with $44,100$ Hz in 16-bit format, i.e. with $44,100$ observations per second and observation values between $-32,767$ and $32,768$ ($= 2^{16}$ possibilities). The resulting time series represents the total sound, i.e. all played tones together. However, e.g. with automatic transcription of musical sounds into scores, individual tones or notes, respectively, are needed. The following model thus incorporates the individual tones.

Our example data only represent a single pure fundamental frequency vibration with 440 Hz. Thus, the corresponding exact model has the form:

$$y_t = \sin(2\pi \cdot \frac{440}{11025} \cdot t), \qquad t = 1, 2, \ldots, 128,$$

i.e. per second 11,025 values are observed (= 11,025 Hz).

Figure 6.21 shows the raw data.

6.4.2 *Modeling*

In this subsubsection three different components of the complete probability model for music data analysis are developed:

– The parametric model,

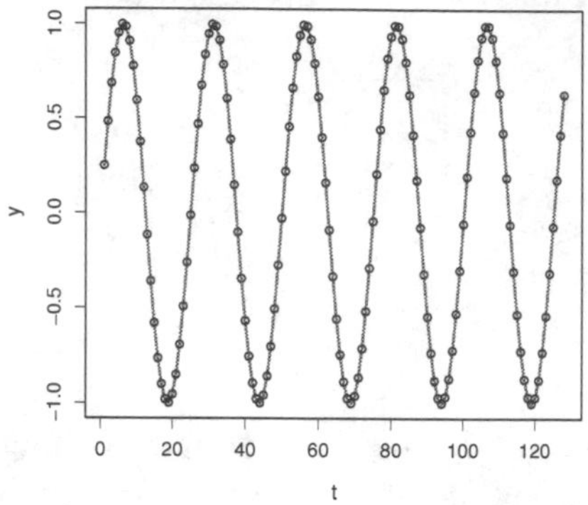

Figure 6.21: Number of 128 raw data (open circles) of a 440 Hz vibration.

– the dependency structure, and
– the a priori distributions.

Parametric Model

In general, the vibrations of a music instrument are harmonic, i.e. they are composed of a fundamental frequency and so-called overtone frequencies that are multiples of the fundamental frequency. The tones related to the fundamental frequency and to the overtone frequencies are called partial tones. In a polyphonic sound the partial tones of all involved tones are superimposed. For the identification of the individually played tones Davy and Godsill (2002) propose a general model for K simultaneously played tones with varying number M_k of partial tones. This model will be introduced here in the special case of constant volume over the whole sound length:

$$y_t = \sum_{k=1}^{K} \sum_{m=1}^{M_k} \{a_{k,m} \cos((m+\delta_{k,m})\omega_k t) + b_{k,m} \sin((m+\delta_{k,m})\omega_k t)\} + \varepsilon_t$$

where

$$\begin{aligned} K =& \text{number of simultaneously played tones,} \\ M_k =& \text{number of partial tones of the } k\text{th tone, } M = [M_1 \ \dots \ M_k]^T, \\ \Theta =& \text{amplitude vector with elements } a_{k,m} \text{ and } b_{k,m} \\ =& [a_{1,1} \ b_{1,1} \ a_{1,2} \ b_{1,2} \ \dots \ a_{1,M_1} \ b_{1,M_1} \\ & a_{2,1} \ b_{2,1} \ a_{2,2} \ b_{2,2} \ \dots \ a_{2,M_2} \ b_{2,M_2} \\ & \dots \\ & a_{K,1} \ b_{K,1} \ a_{K,2} \ b_{K,2} \ \dots \ a_{K,M_K} \ b_{K,M_K}]^T, \end{aligned}$$

$\delta_{k,m}$ =translation parameter of the mth partial tone of the kth tone,

$\delta = [\delta_1 \ \dots \ \delta_K]^T$ with $\delta_k = [\delta_{k,1} \ \dots \ \delta_{k,M_k}]^T$,

ω_k =fundamental frequency of the kth tone, $\omega = [\omega_1 \ \dots \ \omega_k]^T$, and

ε_t =error term.

We change this model insofar that we do not allow a detune in the fundamental frequency, i.e. we set $\delta_{k,1} = 0$. Therefore, the fundamental frequency of the kth tone is only determined by ω_k. The other partial tones are, however, allowed to be detuned. This is especially relevant for singers in the range of the so-called singer formant in which the singer has to be particularly loud in order to be heard when accompanied by an orchestra.

For our special data we additionally assume that only one tone is played, i.e. $K = 1$. This leads to a distinct simplification of the model:

$$\begin{aligned} y_t &= a_1 \cos(\omega t) + b_1 \sin(\omega t) + \\ &\quad \sum_{m=2}^{M} \left(a_m \cos((m+\delta_m)\omega t) + b_m \sin((m+\delta_m)\omega t) \right) + \varepsilon_t, \end{aligned} \tag{6.70}$$

where even the remaining summation term is eliminated iff $M = 1$.

Decisive for correct modeling is thus the correct determination of M. For our data, $M = 1$ would be correct. This completes the determination of the parametric model.

Structural Model

For the above parametric model the following minimal structural assumptions are made:

For each time point the expected value μ_t of y_t is a deterministic function of Θ, δ, ω, and M, where Θ and δ depend on M. M is independently drawn from a distribution with the **hyperparameter** Λ (expected number of partial tones).

Figure 6.22 shows this model in the form of a **directed acyclical graph**.

Definition 6.37: Directed Acyclical Graph (DAG)
A graph is a set of edges and vertices. A graph is **directed** if every edge has only one direction, and is **acyclical** if it has no cycle. I.e. a DAG is formed by a collection of vertices and directed edges, each edge connecting one vertex to another, such that there is no way to start at some vertex κ and follow a sequence of edges that eventually loops back to κ again. In our DAGs,

- each model variable is represented by a vertex and
- directed edges correspond to direct dependencies.

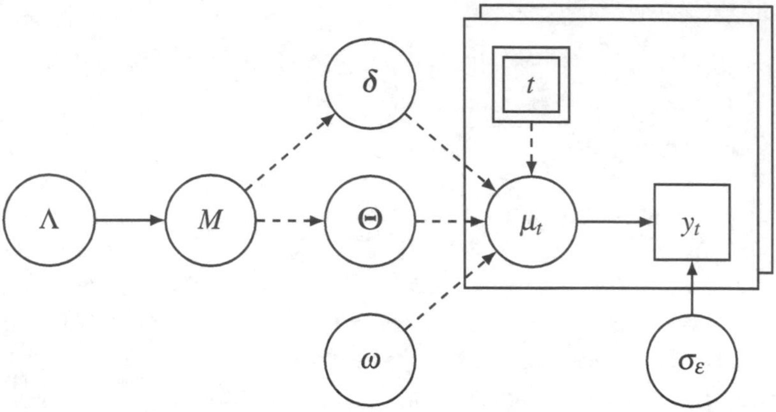

Figure 6.22: Structural model for polyphonic music data after Davy and Godsill (2002).

- Solid lines represent stochastic dependencies and
- broken lines functional (deterministic) relationships.
- Repetitive structures are indicated by stacked sheets.
- Circles represent stochastic variables, and single boxes observed data,
- constants are written in double boxes.

In order to interpret graphs, some simple notions are helpful. Let κ be a vertex in the graph and $\mathscr{K}$ the set of all vertices. We call a vertex a **parent** of κ if an edge starts in this vertex that points to κ. We call a vertex a **descendant** of κ if this vertex is reached by edges that start in κ. We call a descendant a **child** of κ if this vertex is directly reached by an edge that starts in κ. For the determination of parents and descendants deterministic connections are combined with stochastic ones.

The genetic analogy motivating the notions should be clear.

In our example, Λ, M, $\mathbf{\Theta}$, δ, ω, μ_t, and σ_ε are stochastic variables, y_t is observed, t is a constant, and, e.g., $\mathbf{\Theta}$, δ, ω, and σ_ε are parents of y_t. Moreover, all dependencies are deterministic except the dependency of M on Λ and of y_t on μ_t and σ_ε. The repetitive structure, indicated by stacked sheets, is caused by the different time points.

Probability Model

Concerning a complete probability model one can show (Lauritzen et al., 1990) that a DAG model is equivalent to the assumption of a joint distribution of the involved random variables, which is completely determined by the conditional distributions of the vertices given its parents:

$$P(\mathscr{K}) = \prod_{\kappa \in \mathscr{K}} P(\kappa | \text{parents}(\kappa)), \tag{6.71}$$

where $P(\cdot)$ stands for a probability distribution. This factorization not only allows us to combine extremely complex models from local components, but also builds the basis for the application of some MCMC methods, namely the Gibbs sampler.

In our example we thus only need the specification of the exact form of the parents-children relationships in the graph in Figure 6.22.

The likelihood terms of the model are assumed to have the following distributions:

$$y_t \sim \mathscr{N}(\mu_t, \sigma_\varepsilon^2), \tag{6.72}$$

$$\mu_t = a_1 \cos(\omega t) + b_1 \sin(\omega t) + \sum_{m=2}^{M} \left(a_m \cos((m+\delta_m)\omega t) + b_m \sin((m+\delta_m)\omega t)\right), \tag{6.73}$$

$$a_m \sim \mathscr{N}(a_0, \sigma_a^2), \tag{6.74}$$

$$b_m \sim \mathscr{N}(b_0, \sigma_b^2), \tag{6.75}$$

$$\delta_m \sim \mathscr{N}(0, \sigma_\delta^2), \tag{6.76}$$

$$M_0 \sim \text{truncated Poisson}(\Lambda, M_{\max}),\ M = M_0 + 1. \tag{6.77}$$

where $\sim$ stands for "distributed as" and $\mathscr{N}(a,b)$ is the normal distribution with expected value a and variance b. Notice that M_0, the number of overtones, is assumed to be truncated Poisson distributed. Since $M_{\max} = 12$ is chosen to be relatively big, truncation in $M_{\max}$ is not very relevant, so that the distribution is approximately Poisson.

A Priori Distributions

In order to complete the specification of the probability model we additionally need the a priori distributions in the vertices without parents in Figure 6.22: ω, a_0, b_0, σ_a^2, σ_b^2, σ_δ^2, σ_ε^2, and Λ.

In Bayes' models these distributions should not very much influence the results. However, in hierarchical models like ours it became apparent that so-called improper a priori distributions as, e.g., uniform distributions on an infinite interval should be avoided, since such distributions might lead to improper a posteriori distributions (DuMouchel and Waternaux, 1992).

Sometimes, though, such parameters are simply set a priori to a fixed value. We choose a mixture of both approaches, the specification of distributions and of constants.

Some parameters are fixed:

$$a_0 = 0.5, \quad b_0 = 0.5, \quad \sigma_a^2 = 1, \quad \sigma_b^2 = 1, \quad \sigma_\delta^2 = 0.1.$$

The other parameters vary as follows:

$$\omega \sim \text{uniform}(0, 1.7), \tag{6.78}$$

$$\sigma_\varepsilon^{-2} \sim Ga(0.5, 10), \tag{6.79}$$

$$\Lambda \sim Ga(M_0 + 0.5, 1), \tag{6.80}$$

where uniform(l, r) is a uniform distribution on the interval $[l, r]$ and $Ga(a, b)$ is the gamma distribution with expected value a/b and variance a/b^2.

The fundamental frequency ω is decided to a priori vary between 0 and 3000 Hz. Since we use the radial frequency representation, we have to transform this region by means of the relationship: $\omega = 2\pi\cdot$ Hz $/11,025$. Thus, 440 Hz corresponds to $\omega = 0.25$. The expected value Λ of the Poisson distribution of M_0 is a hyperparameter that is substantially determined by the observed M_0.

As specified, the distribution of σ_ε^2 has nearly no effect on the analysis since the inverse of the variance component, the so-called **precision**, has a variance of $0.5/100^2 = 0.00005$.

6.4.3 Model Fit by Means of MCMC Methods

We estimate our model by means of MCMC methods using the freely available OpenBUGS software (Gilks et al., 1994; Spiegelhalter et al., 2004).

In general, five steps are necessary in order to implement MCMC methods:

1. Starting values have to be provided for every unobserved vertex (parameters and missing values).
2. Complete conditional distributions have to be constructed for every unobserved vertex, and sampling methods have to be provided for all these distributions.
3. The output has to be observed in order to decide upon the length of the burn-in period and the total length of the Markov chain, or to decide that a more efficient parametrization or a more efficient MCMC algorithm is necessary.
4. Summarizing statistics have to computed in order to be able to infer the true values of the unobserved vertices.

Table 6.3: Starting Values for the Parameters in the Three Runs of the MCMC Algorithm

Parameter	Run 1	Run 2	Run 3
a_m	0.5	0.1	1.0
b_m	0.5	0.1	1.0
δ_m	0.0	0.1	−0.1
τ_ε	1.0	0.1	0.5
M_0	1.0	2.0	3.0
Λ	2.0	2.0	2.0
ω	0.1	0.4	0.2

5. Inspection of the summary statistics corresponding to goodness of fit and prediction of the model.

For a satisfying implementation of the MCMC algorithm the fifth step is crucial. In the following each of these steps will briefly be discussed.

Initialization

Ideally, the choice of the starting values is not important since the Gibbs algorithm or another sampling algorithm should run long enough to forget them. It is, however, sensible, to run some chains with different starting values to check whether the results really do not depend on them (Gelman, 1995).

On the one hand, for very extreme starting values very long burn-in-times have to be expected (Raftery, 1995). In severe cases, the algorithm might be numerically unstable in the tails of the a posteriori distribution and might even not converge.

On the other hand, starting at the mode is also no guarantee for a good mixing of the algorithm, i.e. for a fluent passing of the whole support of the a posteriori distribution.

In our example, three runs will be carried out with the starting values in Table 6.3. The first run starts in values that appear to be sensible considering Figure 6.22. The starting values of the second and third runs represent essential deviations from the first run.

Samples from Complete Conditional Distributions

When drawing samples from complete conditional distributions, the Gibbs algorithm works iteratively. The **complete conditional distribution** of a vertex is that distribution that incorporates all currently known information about the values of the other vertices in the graph. For a DAG model V we can use the structure of the joint distribution in (6.71). Then, calling the remaining vertices V_{-v} for each vertex v, it follows that the complete conditional distribution $P(v|V_{-v})$ has the form:

$$\begin{aligned} P(v|V_{-v}) &\propto P(v, V_{-v}) \\ &\propto \text{terms in } P(V), \text{ which contain } v, \\ &= P(v|\text{parents}(v)) \times \prod_{w \in \text{children}(v)} P(w|\text{parents}(w)) \end{aligned} \tag{6.81}$$

where $\propto$ means "proportional to".

Notice that the proportionality constant that guarantees that the distribution has integral 1 is generally a function of the remaining vertices V_{-v}. From (6.81) we see that the complete conditional distribution of v contains both an a priori component $P(v|\text{parents}(v))$ and likelihood components for every child of v. Therefore, the complete conditional distribution of a vertex only depends on the values of its parents, its children, and its "coparents", i.e. the other parents of the children of v.

In our model, we need complete conditional distributions for the following unobserved parameters: M, a_m, b_m, δ_m, ω, σ_ε^2, and Λ.

Consider, e.g., the cosine amplitude a_m. The general expression (6.81) tells us that the complete conditional distribution of a_m is proportional to the product of the a priori distribution of a_m in (6.74) and of the N likelihood terms given by (6.72), and (6.73), where N is the number of observations of the time series. This leads to the following expression for the complete conditional distribution of a_m:

$$P(a_m|\cdot) \propto \exp\left\{-\frac{(a_m - a_0)^2}{2\sigma_a^2}\right\} \times$$
$$\prod_{t=1}^N \exp\left\{-\frac{(y_t - a_1\cos(\omega t) - b_1\sin(\omega t) - \sum_{m=2}^M (a_m\cos((m+\delta_m)\omega t) + b_m\sin((m+\delta_m)\omega t)))^2}{2\sigma_\varepsilon^2}\right\}$$

where $\cdot$ in $P(a_m|\cdot)$ means all data and parameter vertices except a_m, i.e. V_{-a_m}. This results in a normal distribution for $P(a_m|\cdot)$ for which we will now derive the expected value and variance in the special case $M = 1$:

$$\begin{aligned} &P(a_1|\cdot) \\ &\propto \exp\left\{-\frac{(a_1 - a_0)^2}{2\sigma_a^2}\right\} \times \prod_{t=1}^N \exp\left\{-\frac{(y_t - a_1\cos(\omega t) - b_1\sin(\omega t))^2}{2\sigma_\varepsilon^2}\right\} \\ &= \exp\left\{-\frac{(a_1 - a_0)^2}{2\sigma_a^2} - \sum_{t=1}^N \frac{(y_t - a_1\cos(\omega t) - b_1\sin(\omega t))^2}{2\sigma_\varepsilon^2}\right\} \end{aligned}$$

$$= \exp\left(-\left[\frac{a_1^2 - 2a_1a_0 + a_0^2}{2\sigma_a^2} + \sum_{t=1}^{N} \frac{(y_t - b_1\sin(\omega t))^2 - 2(y_t - b_1\sin(\omega t))a_1\cos(\omega t) + a_1^2\cos(\omega t)^2}{2\sigma_\varepsilon^2}\right]\right)$$

$$= \exp\left(-\left[a_1^2\left(\frac{1}{2\sigma_a^2} + \sum_{t=1}^{N}\frac{\cos(\omega t)^2}{2\sigma_\varepsilon^2}\right) + \frac{-2a_1a_0 + a_0^2}{2\sigma_a^2} + \sum_{t=1}^{N} \frac{(y_t - b_1\sin(\omega t))^2 - 2(y_t - b_1\sin(\omega t))a_1\cos(\omega t)}{2\sigma_\varepsilon^2}\right]\right)$$

$$= \exp\left(-\frac{1}{2}\left[a_1^2\left(\frac{1}{\sigma_a^2} + \sum_{t=1}^{N}\frac{\cos(\omega t)^2}{\sigma_\varepsilon^2}\right) - \frac{2a_1a_0}{\sigma_a^2} - \sum_{t=1}^{N}\frac{2a_1(y_t - b_1\sin(\omega t))\cos(\omega t)}{\sigma_\varepsilon^2} + \frac{a_0^2}{\sigma_a^2} + \sum_{t=1}^{N}\left(\frac{y_t - b_1\sin(\omega t)}{\sigma_\varepsilon}\right)^2\right]\right)$$

$$= \exp\left(-\frac{1}{2}\left[a_1^2\left(\frac{1}{\sigma_a^2} + \sum_{t=1}^{N}\frac{\cos(\omega t)^2}{\sigma_\varepsilon^2}\right) - 2a_1\left(\frac{a_0}{\sigma_a^2} + \sum_{t=1}^{N}\frac{(y_t - b_1\sin(\omega t))\cos(\omega t)}{\sigma_\varepsilon^2}\right) + \frac{a_0^2}{\sigma_a^2} + \sum_{t=1}^{N}\left(\frac{y_t - b_1\sin(\omega t)}{\sigma_\varepsilon}\right)^2\right]\right)$$

$$= \exp\left(-\frac{1}{2}\frac{a_1^2 - \frac{2a_1\left(\frac{a_0}{\sigma_a^2} + \sum_{t=1}^{N}\frac{(y_t - b_1\sin(\omega t))\cos(\omega t)}{\sigma_\varepsilon^2}\right)}{\frac{1}{\sigma_a^2} + \sum_{t=1}^{N}\frac{\cos(\omega t)^2}{\sigma_\varepsilon^2}} + \frac{\frac{a_0^2}{\sigma_a^2} + \sum_{t=1}^{N}\left(\frac{y_t - b_1\sin(\omega t)}{\sigma_\varepsilon}\right)^2}{\frac{1}{\sigma_a^2} + \sum_{t=1}^{N}\frac{\cos(\omega t)^2}{\sigma_\varepsilon^2}}}{\frac{1}{\frac{1}{\sigma_a^2} + \sum_{t=1}^{N}\frac{\cos(\omega t)^2}{\sigma_\varepsilon^2}}}\right)$$

$$\propto \exp\left(-\frac{1}{2}\frac{\left(a_1 - \frac{\frac{a_0}{\sigma_a^2} + \sum_{t=1}^{N}\frac{(y_t - b_1\sin(\omega t))\cos(\omega t)}{\sigma_\varepsilon^2}}{\frac{1}{\sigma_a^2} + \sum_{t=1}^{N}\frac{\cos(\omega t)^2}{\sigma_\varepsilon^2}}\right)^2}{\left(\frac{1}{\sigma_a^2} + \sum_{t=1}^{N}\frac{\cos(\omega t)^2}{\sigma_\varepsilon^2}\right)^{-1}}\right).$$

The last proportionality is valid since the last term in the last but one line is independent of a_1.

Obviously, the conditional distribution is a normal distribution for which it is true that

$$\text{expected value} = \frac{\frac{a_0}{\sigma_a^2} + \frac{1}{\sigma_\varepsilon^2}\sum_{t=1}^{N}(y_t - b_1\sin(\omega t))\cos(\omega t)}{\frac{1}{\sigma_a^2} + \frac{C^2}{\sigma_\varepsilon^2}}$$

and

$$\text{variance} = \frac{1}{\frac{1}{\sigma_a^2} + \frac{C^2}{\sigma_\varepsilon^2}}, \quad C^2 = \sum_t (\cos(\omega t))^2$$

The complete conditional distributions for b_m and δ_m can be derived analogously.

The complete conditional distribution for σ_ε^{-2} can also be derived easily. Denote τ_ε for the parameter σ_ε^{-2}. The general rule (6.81) tells us that the complete conditional distribution for τ_ε is proportional to the product of the a priori distribution of τ_ε in (6.79) and the likelihood terms for τ_ε given in (6.72) for each t. These are the likelihood terms for τ_ε since the y_t are the only children of τ_ε. This leads to:

$$\begin{aligned}
P(\tau_\varepsilon|\cdot) &\propto \tau_\varepsilon^{0.5-1} e^{-10\tau_\varepsilon} \prod_{t=1}^{N} \tau_\varepsilon^{1/2} \exp\left\{-\frac{1}{2}\tau_\varepsilon (y_t - \mu_t)^2\right\} \\
&= \tau_\varepsilon^{0.5-1} \tau_\varepsilon^{N/2} \exp\{-10\tau_\varepsilon\} \prod_{t=1}^{N} \exp\left\{-\frac{1}{2}\tau_\varepsilon (y_t - \mu_t)^2\right\} \\
&= \tau_\varepsilon^{0.5-1+N/2} \exp\left\{-10\tau_\varepsilon + \sum_{t=1}^{N} -\frac{1}{2}\tau_\varepsilon (y_t - \mu_t)^2\right\} \\
&= \tau_\varepsilon^{0.5-1+N/2} \exp\left\{-\left(10\tau_\varepsilon + \frac{1}{2}\tau_\varepsilon \sum_{t=1}^{N} (y_t - \mu_t)^2\right)\right\} \\
&= \tau_\varepsilon^{0.5-1+N/2} \exp\left\{-\tau_\varepsilon \left(10 + \frac{1}{2}\sum_{t=1}^{N} (y_t - \mu_t)^2\right)\right\} \\
&\propto Ga\left(0.5 + \frac{N}{2}, 10 + \frac{1}{2}\sum_{t=1}^{N} (y_t - \mu_t)^2\right).
\end{aligned}$$

Therefore, the complete conditional distribution of τ_ε is again a gamma distribution.

In the above examples the complete conditional distributions are normal and gamma distributions, from which it is easy to draw samples (see, e.g. Ripley, 1987). For other parameters the complete conditional distributions are more complicated. However, then there are also efficient sampling procedures (see, e.g. Gilks, 1995).

Observation of the Output

The values of the unknown variables generated by the MCMC algorithm should be summarized graphically as well as statistically in order to assess mixture and convergence. As an example, we illustrate the usage of the

Gelman-Rubin statistic (Gelman and Rubin, 1992) for three runs of 5000 iterations each with the starting values given in Table 6.3. Details of the method of Gelman and Rubin are also given in Gelman (1995). We will only give a summary.

The idea of the Gelman-Rubin statistic is the construction of a scalar that indicates when parallel realizations of a distribution of a statistic Ψ deviate too much. For this we first estimate the variance of the parallel realizations distinguishing between-run variance B and within-run variance W. Assuming I parallel realizations of length n we define:

$$B = \frac{n}{I-1}\sum_{i=1}^{I}(\bar{\Psi}_{i\cdot} - \bar{\Psi}_{\cdot\cdot})^2, \text{ where } \bar{\Psi}_{i\cdot} = \frac{1}{n}\sum_{j=1}^{n}\Psi_{ij}, \; \bar{\Psi}_{\cdot\cdot} = \frac{1}{I}\sum_{i=1}^{I}\bar{\Psi}_{i\cdot} \text{ and}$$

$$W = \frac{1}{I}\sum_{i=1}^{I}s_i^2, \text{ where } s_i^2 = \frac{1}{n-1}\sum_{j=1}^{n}(\Psi_{ij} - \bar{\Psi}_{i\cdot})^2.$$

From these two variance components we construct a pooled estimator of the variance:

$$\hat{\mathrm{var}}(\Psi) = (1 - \frac{1}{n})W + \frac{1}{n}B$$

is an unbiased estimator of the variance in the case of stationarity, i.e. if the starting values of the simulations are all drawn from the target distribution. However, the variance is overestimated under the realistic assumption that the starting values scatter more than in the target distribution.

In contrast W is a variance estimator that will generally underestimate variance since individual samples most of the time cannot illuminate the whole distribution.

For $n \to \infty$ both estimates converge to the true variance, but from different sides. Therefore, we can check the convergence of the MCMC algorithm by observing the **Gelman-Rubin statistic** of the successive parallel realizations of the unknown variables:

$$\sqrt{\hat{R}} = \sqrt{\frac{\hat{\mathrm{var}}(\Psi)}{W}}$$

If the simulation converges, then this statistic should converge to 1, i.e. the realized samples should more and more overlap.

Figures 6.23 – 6.26 show the sampled values of M_0, ω, a_1, and b_1 for the three runs: All parameters converge quickly after less than 150 iterations. Notice that there is remarkable variation even in the last 500 iterations.

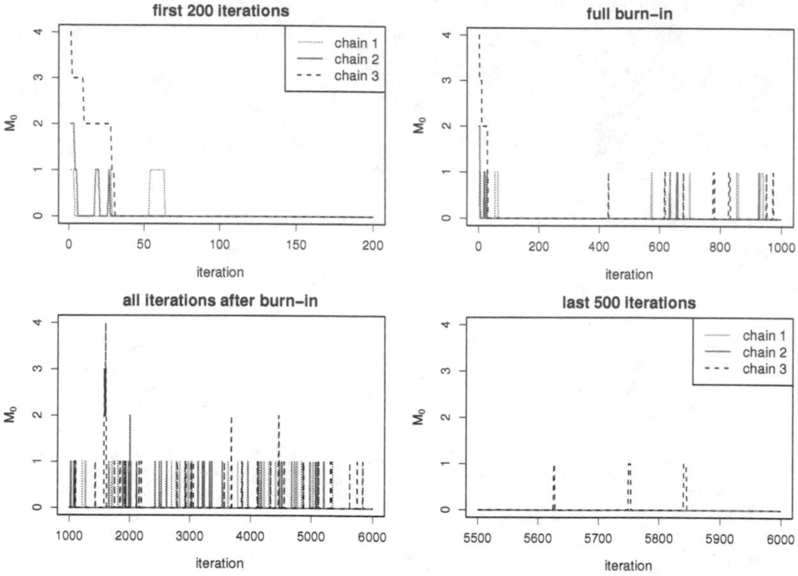

Figure 6.23: Realizations of M_0 in three runs of the MCMC algorithm applied to the model in Section 6.4.2. Starting values from Table 6.3: first 200 iterations, full burn-in phase, all iterations after burn-in, last 500 iterations.

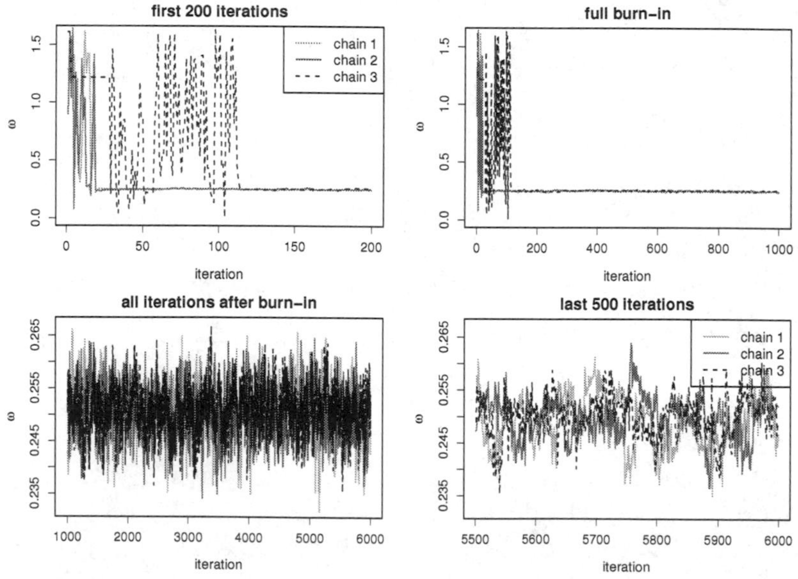

Figure 6.24: Realizations of ω in three runs of the MCMC algorithm applied to the model in Section 6.4.2. Starting values from Table 6.3: first 200 iterations, full burn-in phase, all iterations after burn-in, last 500 iterations.

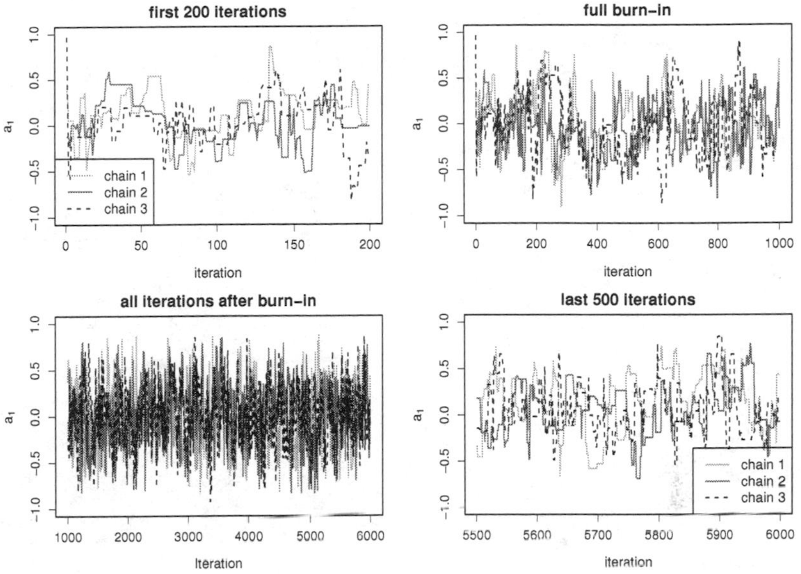

Figure 6.25: Realizations of a_1 in three runs of the MCMC algorithm applied to the model in Section 6.4.2. Starting values from Table 6.3: first 200 iterations, full burn-in phase, all iterations after burn-in, last 500 iterations.

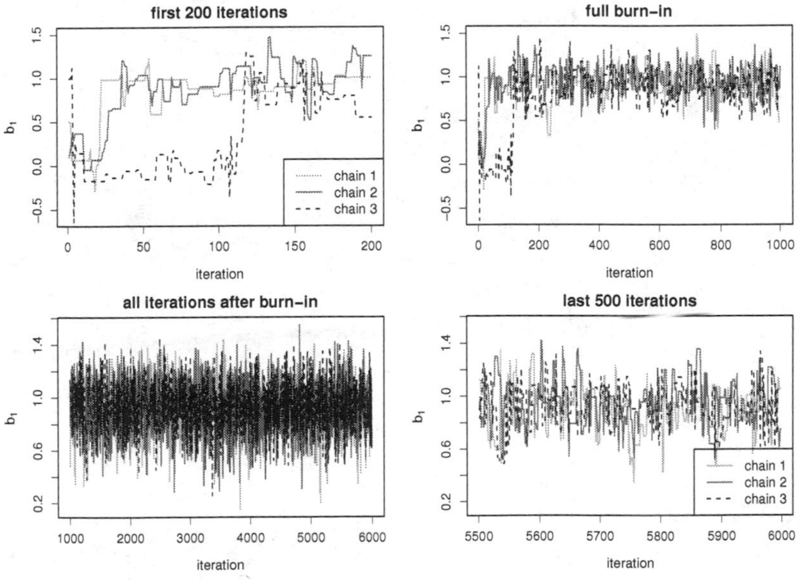

Figure 6.26: Realizations of b_1 in three runs of the MCMC algorithm applied to the model in Section 6.4.2. Starting values from Table 6.3: first 200 iterations, full burn-in phase, all iterations after burn-in, last 500 iterations.

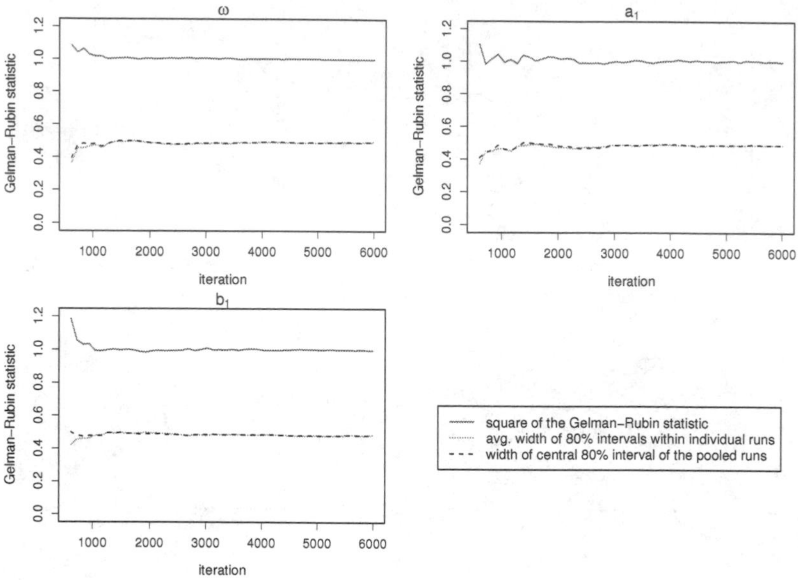

Figure 6.27: Gelman-Rubin statistic for three parameters from three parallel chains.

Figure 6.27 shows the Gelman-Rubin statistic (modified by Brooks and Gelman, 1998) for the three parameters ω, a_1, and b_1 in the three chains. Since the parameter M_0 stops varying early the Gelman-Rubin statistic is undefined. The normalized width of the central 80% interval of the pooled runs (corresponding to $\hat{\text{var}}(\Psi)$) is drawn in solid dark grey, the normalized mean width of the 80% interval of the individual runs (corresponding to W) in dashed black, and the square of the Gelman-Rubin statistic (corresponding to $\hat{R}$) in solid light grey. We aim at the convergence of $\hat{R}$ to 1 and at stable 80% intervals. Obviously, this is true in our example, and in this respect an acceptable convergence is realized.

Figure 6.28 shows plots of kernel density estimations of M_0, ω, a_1, and b_1. Notice the closeness to normal densities for ω, a_1, and b_1, and that the true values $M_0 = 0$, $\omega = 0.2507, a_1 = (0.0), b_1 = 1.0$ are all central in the densities.

Assessment of Goodness of Fit

Finally, we compare the original data with the estimated model based on the **means** of the estimated parameters over all chains:

$$\mu_t = 0.004758\cos(0.2559 \cdot t) + 0.9091\sin(0.2559 \cdot t).$$

This model and the original data nearly ideally coincide (see Figure 6.29).

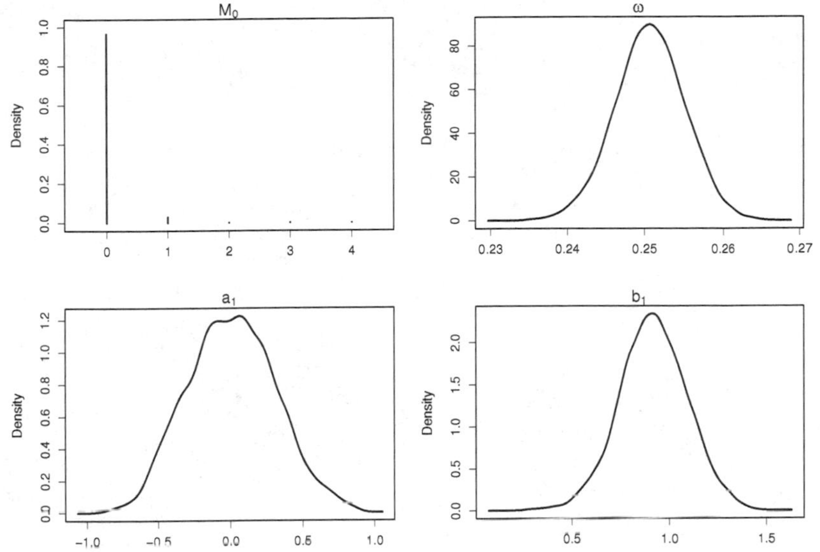

Figure 6.28: Kernel density estimations for the parameters of the model in Section 6.4.2 based on three chains.

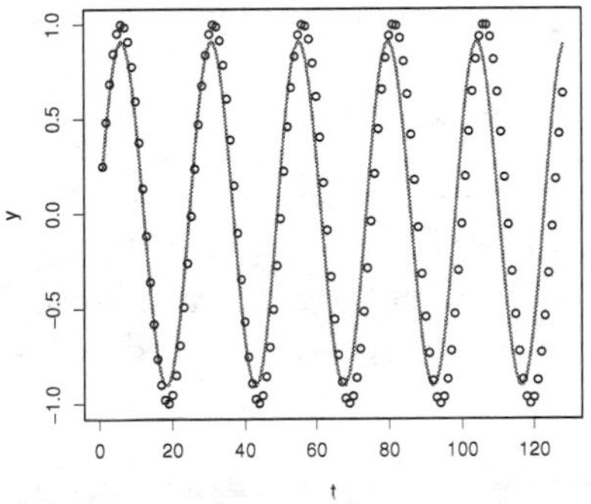

Figure 6.29: Original data (points) and fitted data (line).

Figure 6.30 shows the histogram of the model residuals. Notice that the Kolmogorov-Smirnov test (lillie.test in R) for a normal distribution with empirical mean and variance is not rejected at the 5% level ($p = 0.42$). Thus, the distribution of the residuals is similar to a normal distribution. Therefore, overall the result of the estimation is satisfactory.

Unfortunately, autocorrelations are high for a_1, b_1, ω (see Figure 6.31). Therefore, we also studied what happens to the model if we only take every

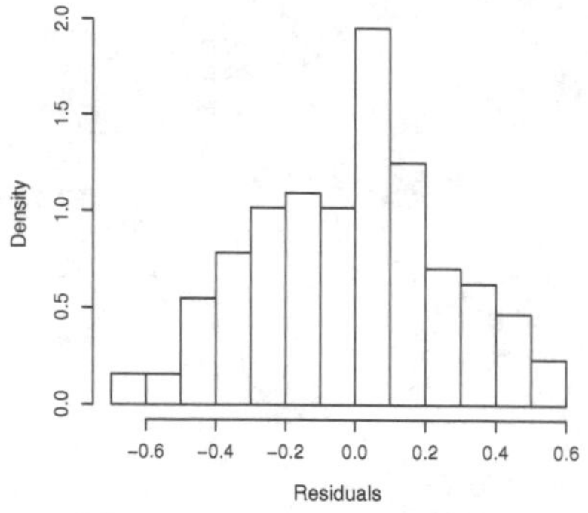

Figure 6.30: Histogram of the residuals.

30th value of the iterations. First, this eliminates the autocorrelations (see Figure 6.32). The model is changed to

$$\mu_t = 0.00425\cos(0.2505 \cdot t) + 0.9202\sin(0.2505 \cdot t).$$

This model and the original data appear to coincide even better at first sight (see Figure 6.33). Indeed, the residuals are much smaller (see Figure 6.34). However, the Lilliefors test rejects normality ($p = 0.0008$) because of too heavy tails caused by too low model peaks. Nevertheless, one can be very satisfied by the stochastic estimation.

Model Criticism

The model (6.70) of Davy and Godsill (2002) was already changed in the term for the fundamental frequency in order to get a unambiguous expression for this frequency. On closer inspection more points of criticism arise that appear to be important for more complex data:

1. The model is not identifiable, i.e. the deviations of the overtones from multiples of the fundamental frequency are not unambiguously determinable, meaning that the model is overparameterized. For example:

$$\cos\left((m+\delta_{m,1})\omega_1 t\right) = \cos\left((m+\delta_{m,2})\omega_2 t\right)$$

for special choices of $\delta_{m,1}$, $\delta_{m,2}$, ω_1, ω_2. Thus, the parameter δ_m should possibly be eliminated from the model.

2. Setting some model parameter constant makes the model inflexible, e.g. one should not choose:
$a_m \sim \mathcal{N}(a_0, \sigma_a^2)$ and $a_0 = 0.5$, $\sigma_a^2 = 1$, but, e.g.,
$a_m \sim \mathcal{N}(a_{m0}, \sigma_a^2)$, $a_{m0} \sim \mathcal{N}(0, 10000)$, $\sigma_a^{-2} \sim Ga(0.01, 0.01)$.

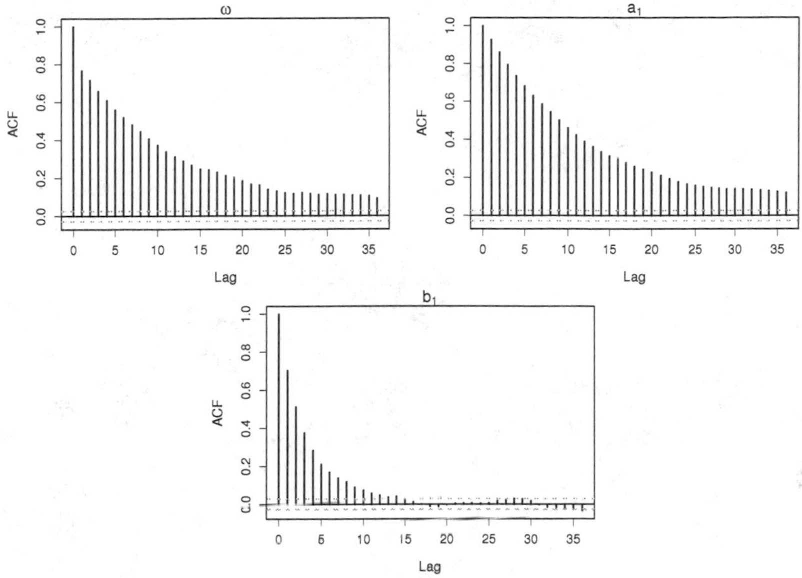

Figure 6.31: Autocorrelation functions (ACFs) for three parameters from first parallel chain after burn-in.

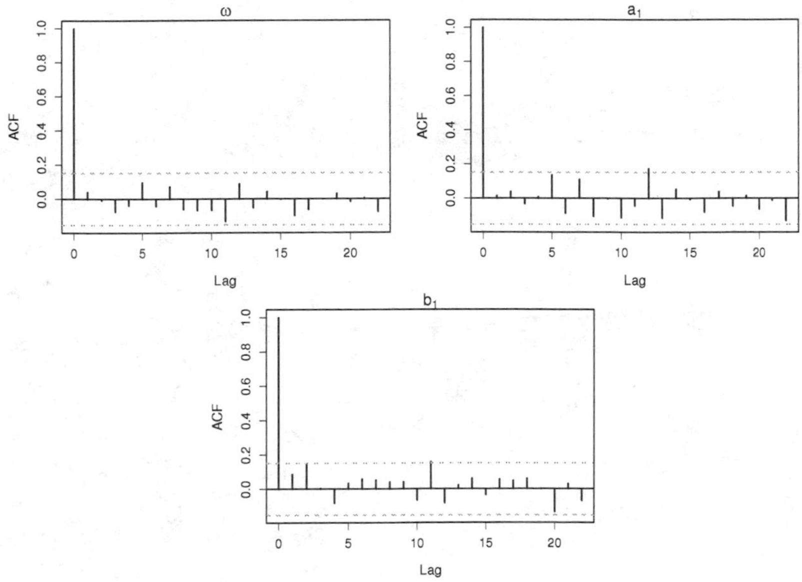

Figure 6.32: Autocorrelation functions (ACFs) for three parameters from first parallel chain after burn-in and thinning (only every 30th observation taken).

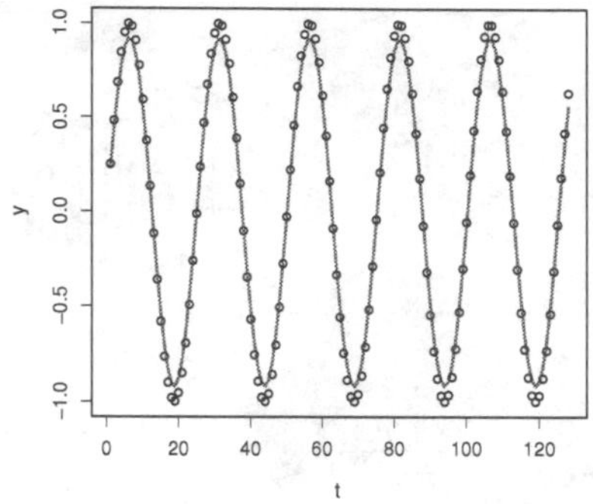

Figure 6.33: Original data (points) and fitted data (line), only every 30th observation.

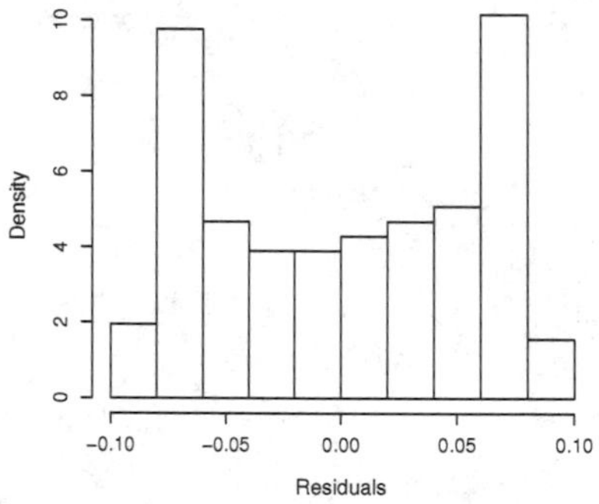

Figure 6.34: Histogram of the residuals, only every 30th observation.

6.4.4 *BUGS Implementation*

OpenBUGS is a program whose syntax allows the specification of graphical models and the generation of Markov chains. We use BRugs (Thomas et al., 2006) to control OpenBUGS from R, which allows us to automate the analysis since OpenBUGS is an interactive environment. In order to get an impression of the syntax, we will provide the definition of the data, the BRugs program for the model in Section 6.4.2 (also see Figure 6.22), the initialization of the parameters, and the control of the MCMC procedure.

```
## Define data
t <- 1:128
yvalues <- sin(2 * pi * 440 / 11025 * t)
dat <- list(T=128, mu=0, Sigma.delta=10,
            mua=0, Sigma.ampl=1, y=yvalues)
```

```
## Define Model
model <- function() {
  M0 ~ dpois(lambda)
  M <- M0 + 1
  for (m in 1:12) {
    arg[m] <- (m + (delta[m] * step(m - 2 + 0.01))) *
      step(M - m + 0.01)
    a[m] <- ampla[m] * step(M - m + 0.01)
    b[m] <- amplb[m] * step(M - m + 0.01)
    delta[m] ~ dnorm(mu, Sigma.delta)
    ampla[m] ~ dnorm(mua, Sigma.ampl)
    amplb[m] ~ dnorm(mua, Sigma.ampl)
  }
  for (t in 1:T) {
    y[t] ~ dnorm(mue[t], sigma.epsilon.quadrat)
    mue[t] <- sum(cosine[,t]) + sum(sine[,t])
    for(m in 1:12){
      cosine[m,t] <- a[m] * cos(arg[m] * omega * t)
      sine[m,t] <- b[m] * sin(arg[m] * omega * t)
    }
  }
  omega ~ dunif(0, 1.7)
  temp <- M0 + 0.5
  lambda ~ dgamma(temp, 1)
  sigma.epsilon.quadrat ~ dgamma(0.5,100)
}

## Define initial values
init1 <- list(delta=rep(0,12),
              ampla=rep(0.5,12), amplb=rep(0.5,12),
              sigma.epsilon.quadrat=1, M0=1,
              lambda=2, omega=0.1)
init2 <- list(delta=rep(0.1,12),
              ampla=rep(0.1,12), amplb=rep(0.1,12),
              sigma.epsilon.quadrat=0.1, M0=2,
              lambda=2, omega=0.4)
init3 <- list(delta=rep(-0.1,12),
              ampla=rep(1,12), amplb=rep(1,12),
              sigma.epsilon.quadrat=0.5, M0=3,
```

```
              lambda=2, omega=0.2)

## Run BRugs
library("BRugs")
writeModel(model)          # write model file
modelCheck("model.txt")    # check model file
datafile <- bugsData(dat)  # write data file
modelData(datafile)        # read data file
modelCompile(numChains=3)  # compile model (< 3 minutes!)
## write inits into some textfile:
initfiles <- bugsInits(list(init1, init2, init3), numChains=3)
modelInits(initfiles)      # read inits
# parameters monitored:
samplesSet(c("a","b","M0","lambda","omega"))
modelUpdate(1000)    # burn in  (< 10 minutes!)
modelUpdate(5000)    # more iterations ...  (< 45 minutes!)
samplesStats("*")    # summarized results

## save results
histo <- samplesHistory("*", mfrow=c(4, 2))

## plot the chains:
## plots for a[1] used in the text: parts of iteration
a1 <- histo[["a[1]"]][ , 1:3]
matplot(a1[1:200, ], type="l")        # first 200
matplot(a1[1:1000, ], type="l")       # burn-in
matplot(a1[1001:5999, ], type="l")  # all converged
matplot(a1[5501:5999, ], type="l")  # last 500
## plot of bgr statistics
samplesBgr("a[1]", mfrow=c(1,1))
## plot of densities
samplesDensity("a[1]", mfrow=c(1, 1))
## plot of autocorrelations of first chain
samplesAutoC("a[1]", 1, mfrow=c(1, 1), beg=1001, end=5999)
## again, including some thinning
samplesAutoC("a[1]", 1, mfrow=c(1, 1),
             beg=1001, end=5999, thin=30)
## densities after thinning
samplesDensity("a[1]", mfrow=c(1, 1),
               beg=1001, end=5999, thin=30)
```

Essentially, the correspondence of the BRugs syntax and model should be clear. Notice however, that the operator $\sim$ relates to "is distributed as" and `<-` to "is defined as". Notice, moreover, that in BRugs the normal distribution is parameterized by its expected value and the so-called precision (= 1 / variance). Also note that the variable M must not be used as a loop limit since it is newly drawn repeatedly. Instead, the function `step()` is used:

$$\text{step}(a) = 1 \ \text{ if } a > 0, \quad = 0 \ \text{ else.}$$

The model is read and checked for syntax errors, the data are read, and the function `modelCompile()` interprets the declared model and chooses the corresponding sampling methods, here, e.g., the normal, gamma, and Poisson distributions. Details can be found in, e.g., Gilks et al. (1994) and Spiegelhalter et al. (2004). Then, the initial values are read for all considered Markov chains and the samples to be monitored are defined. After the burn-in phase the Markov chains should have converged. This might be checked by `samplesStats()` and `samplesHistory()`. If the chains would not have converged, then more iterations have to be carried out in the burn-in phase. After convergence additional iterations are carried out for drawing samples from the target distributions. These samples are then characterized by means of time development, density, Brooks-Gelman-Rubin (BGR) statistics, and autocorrelation plots.

In our example, three runs with 6000 iterations each took approximately 1 hour on an Intel(R)CORE(TM)2 Duo CPU L7500 @ 1.6 GHz under Windows XP using BRugs.

WinBUGS 1.4.3, the successor of Classic BUGS (which is not maintained anymore), can be downloaded from the Internet from
`http://www.mrc-bsu.cam.ac.uk/bugs/welcome.shtml`,
and its successor OpenBUGS 3.2.1 (or later) from `http://www.openbugs.info`.

6.4.5 *Summary*

Using a model of polyphonic music we applied the Gibbs sampler from the previous section to estimate the fundamental frequency and the overtones of the sounds. Starting with the parametric model, we derived the dependence structure and defined a priori distributions for our parameters. Before we used WinBUGS to actually estimate the unknown parameters, we derived the complete conditional distributions of the model and the Gelman-Rubin statistic used to assess the model fit. We concluded with a presentation of the Win-

BUGS results, including a visualization of the MCMC runs, and briefly discussed the model's source code.

6.5 Implementation in R

R provides several univariate random number generators such as the default Mersenne Twister generator. The generator can be chosen by the `RNGkind()` function. Univariate random numbers for various distributions can be generated based on the chosen generator by functions whose names are constructed by the letter r followed by an abbreviation of the corresponding distribution's name, e.g. `runif()` for random numbers from a uniform distribution.

Random numbers from multivariate distributions have to be generated by functions provided by contributed packages. The package mvtnorm (Genz and Bretz, 2009; Genz et al., 2012) provides functions for the frequently used multivariate normal and t distributions.

An overview of contributed R packages related to random number generation and test suites like DieHarder (Brown et al., 2010) for RNGs is given at the CRAN Task View called "Probability Distributions" (Dutang, 2013).

Links and interfaces from R to other sampling engines are also available. Package rjags (Plummer, 2013) provides an interface to JAGS (Just Another Gibbs Sampler) (Plummer, 2003), R2WinBUGS (Sturtz et al., 2005) to WinBUGS (Spiegelhalter et al., 2004), and BRugs (Thomas et al., 2006) a more automatized interface to OpenBUGS (Lunn et al., 2009). These are tools for stochastic (Bayesian) modeling based on Gibbs sampling and MCMC methods.

6.6 Conclusion

We began our study of **randomization**, i.e. of the generation of random numbers on computers, by examining the most basic form of randomness we can generate on a computer, namely discrete uniform randomness. Several different generators for this type of randomness were presented. It became clear that their properties vary and the quality of some of the generated random number sequences is poor. This was shown both on theoretical grounds and by using empirical evidence backed by statistical testing. It cannot be overstated that the quality of the basic discrete random numbers is of utmost importance. All further generators depend on a high-quality, long period distributed RNG.

After a discussion of several generators for other common univariate distributions, we showed how MCMC methods can be used to generate samples from complex multivariate distributions. In contrast to the previously presented generators these methods usually require a burn-in period before they converge to the desired distribution. Both theoretical arguments and online monitoring are required in practice to ensure that the generator is well behaved.

Finally, we used the MCMC methods to estimate stochastic models, exemplified by a model of polyphonic musical sound using WinBUGS. This software is both a blessing and a curse. On the one hand, WinBUGS performed all the tedious and error-prone calculations required to derive all the conditional distributions, but we had to deal with choosing appropriate initial (starting) values, monitoring convergence, and specifying the model in an awkward way in a black-box software.

6.7 Exercises

Exercise 6.2.1: Devise and implement an efficient algorithm to generate a random permutation of a vector x. Make sure that each permutation is equally probable. What is the asymptotic runtime of your algorithm?

Exercise 6.2.2: A **Stirling number of the second kind** is the number of ways to partition a set of k objects into r nonempty subsets. Implement a function to calculate these numbers $S(k,r)$, $1 \leq r \leq k$, using the recursion $S(k,r) = S(k-1,r-1) + rS(k-1,r)$. By definition $S(n,n) = 1$, $S(n,j) = 0$ for $j = 0, n > 0$ or $n < j$.

We use the following notation to denote these types of numbers

$$\left\{ \begin{matrix} k \\ r \end{matrix} \right\} := S(k,r).$$

Exercise 6.2.3: Implement the **poker test**. Consider N groups of k successive numbers of the sequence $\{y_i^{(D)}\}_{i\in\mathbb{N}_0}$ and count the number of k-tuples with r different values, $1 \leq r \leq k$. Apply the chi-square test to these counts with $K = k$ categories and expected frequencies

$$\frac{D(D-1)\cdots(D-r+1)\left\{ \begin{matrix} k \\ r \end{matrix} \right\}}{D^k}$$

where $\left\{ \begin{matrix} k \\ r \end{matrix} \right\}$ are the Stirling numbers of the second kind (see Exercise 6.2.2). (The classical poker test analyzes N groups of five successive numbers of combinations that are interesting for the poker game, e.g. full house: `aaabb`, two pairs: `aabbc`, etc.) The cases $k = 4, 5, 6$ with $D = 32$ could be analyzed.

Exercise 6.2.4: Implement the **coupon collector's test**. Count the number of segments $y_{j+1}^{(D)},\ldots,y_{j+r}^{(D)}$ of length r, $D \le r < \mathscr{T}$, that are needed to gather a complete set of numbers (coupons) $0,\ldots,D-1$, and the number of such segments of length greater than or equal to $\mathscr{T}$. Apply a chi-square test to these counts with $K = \mathscr{T} - D + 1$ categories and the probabilities

$$p_r := D!\left\{\begin{matrix} r-1 \\ D-1 \end{matrix}\right\}/D^r \quad \text{and} \quad p_{\mathscr{T}} := 1 - D!\left\{\begin{matrix} \mathscr{T}-1 \\ D \end{matrix}\right\}/D^{\mathscr{T}-1},$$

where $\left\{\begin{matrix} r-1 \\ D-1 \end{matrix}\right\}$ and $\left\{\begin{matrix} \mathscr{T}-1 \\ D \end{matrix}\right\}$ are again Stirling numbers of the second kind (see Exercise 6.2.2). The case $D = 8$, $\mathscr{T} = 20$ could be analyzed.

Exercise 6.2.5: Implement the **runs test**. Determine the lengths of monotonically increasing subsequences (so-called runs up) of the sequence $\{u_i\}_{i \in \mathbb{N}_0}$. Let c_r be the number of such subsequences of length r, $l \le r \le 5$, and c_6 the number of such subsequences of length greater than or equal to six. By means of the vector c a statistic V is calculated that is approximately chi-square distributed with six degrees of freedom under the assumption that each ordering of elements in the considered sequence section is equally probable and the length of the sequence section is sufficiently long, e.g. greater than or equal to 4000. The assessment follows (6.11). The test can be applied to monotonically decreasing subsequences (so-called runs down) analogously.

Exercise 6.2.6: Implement two LCS random number generators with parameters $m = 2^{31}$, $a = 285,738,053$, $c = 453,816,693$ and $a = 65,539$, $c = 0$, respectively. Test these generators with the statistical tests given above (uniform distribution test, gap test, permutation test, maximum-of-T test, autocorrelation test, poker test, coupon collector's test, runs test). Compare the results with the inversion generators with $m = 2^{31} - 1$ and the above parameter combinations a, c. Order the generators on the basis of the classification (6.11) of the test results.

Exercise 6.2.7: Why would you avoid using $c_1 = a - 1$ and $x_1 = m - 1$ as seed for the MWC generator with parameters $a = 698,769,069$ and $m = 2^{32}$?

Exercise 6.3.1: Show by simulation that for

$$\pi_{x|x} = \begin{bmatrix} \frac{p_1^2}{(p_1+p_3)(p_1+p_2)} + \frac{p_3^2}{(p_1+p_3)(p_3+p_4)} & \frac{p_1 p_2}{(p_1+p_2)(p_1+p_3)} + \frac{p_3 p_4}{(p_1+p_3)(p_3+p_4)} \\ \frac{p_1 p_2}{(p_2+p_4)(p_1+p_2)} + \frac{p_3 p_4}{(p_2+p_4)(p_3+p_4)} & \frac{p_2^2}{(p_2+p_4)(p_1+p_2)} + \frac{p_4^2}{(p_2+p_4)(p_3+p_4)} \end{bmatrix}$$

independent of the marginal starting distribution $[f_x^0(0) \;\; f_x^0(1)]$ with

$f_x^0(0) + f_x^0(1) = 1$ in the limit,

$$[f_x^0(0) \;\; f_x^0(1)]\pi_{x|x}^t \approx [p_1 + p_3 \;\; p_2 + p_4].$$

Exercise 6.3.2: Show by simulation that for

$$P_G = \begin{bmatrix} \frac{p_1^2}{(p_1+p_2)(p_1+p_3)} & \frac{p_2^2}{(p_1+p_2)(p_2+p_4)} & \frac{p_1 p_3}{(p_1+p_2)(p_1+p_3)} & \frac{p_2 p_4}{(p_1+p_2)(p_2+p_4)} \\ \frac{p_1^2}{(p_1+p_2)(p_1+p_3)} & \frac{p_2^2}{(p_1+p_2)(p_2+p_4)} & \frac{p_1 p_3}{(p_1+p_2)(p_1+p_3)} & \frac{p_2 p_4}{(p_1+p_2)(p_2+p_4)} \\ \frac{p_1 p_3}{(p_3+p_4)(p_1+p_3)} & \frac{p_2 p_4}{(p_3+p_4)(p_2+p_4)} & \frac{p_3^2}{(p_3+p_4)(p_1+p_3)} & \frac{p_4^2}{(p_3+p_4)(p_2+p_4)} \\ \frac{p_1 p_3}{(p_3+p_4)(p_1+p_3)} & \frac{p_2 p_4}{(p_3+p_4)(p_2+p_4)} & \frac{p_3^2}{(p_3+p_4)(p_1+p_3)} & \frac{p_4^2}{(p_3+p_4)(p_2+p_4)} \end{bmatrix}$$

it is true that

$$\lim_{t \to \infty} P_G^t = \begin{bmatrix} p_1 & p_2 & p_3 & p_4 \\ p_1 & p_2 & p_3 & p_4 \\ p_1 & p_2 & p_3 & p_4 \\ p_1 & p_2 & p_3 & p_4 \end{bmatrix}.$$

Exercise 6.4.1: Write BUGS code that allows you to estimate μ and σ for a given data vector $\boldsymbol{x}$ assuming X is normally distributed. Test if you get plausible results from your code by applying it to a simulated $\boldsymbol{x}$ with 100 observations for different values of μ and σ. You may want to assume different prior distributions and try out what happens for uniform, normal, and other distributions with different sets of parameters. Do you see stable results for these different parameter sets and prior distributions?

Chapter 7

Repetition

7.1 Motivation and Overview

If you wish to obtain an impression of the distribution of, say, an estimator without relying on too many assumptions, you should repeat the estimation with different unique samples from the underlying distribution. Unfortunately in practice, most of the time only one sample is available. So we have to look for other solutions. New relevant data can only be generated by means of new experiments, which are often impossible to conduct in due time, or by a distribution assumption (see Chapter 6 for random number generation). If we do not have any indication what distribution is adequate, we should beware of assuming just any, e.g. the normal distribution. So what should we do? As a solution to this dilemma resampling methods have been developed since the late 1960s. The idea is to sample repeatedly from the only original sample we have available. These repetitions are then used to estimate the distribution of the considered estimator. This way, we can at least be sure that the values in the sample can be realized by the data generating process. In this chapter we will study how to optimally select repetitions from the original sample. After discussing various such methods, the ideas are applied to three kinds of applications: model selection, feature selection, and hyperparameter tuning.

Model Selection
In many cases several models or model classes are candidates for fitting the data. Resampling methods and the related accuracy assessment efficiently support the selection process of the most appropriate and reliable model.

Feature Selection
Often an important decision for the selection of the best model of a given model type is the decision about the features to be included in the model (e.g. in a linear model). Resampling methods can be used to assess the quality of such selections.

Hyperparameter Tuning
Most modeling strategies require the setting of so-called **hyperparameters** (e.g. starting values for learning rates in neural nets). Thus, tuning of these hyperparameters is desirable to determine settings that lead to a model of high quality. This can be evaluated using the resampling techniques introduced below.

In the following, model quality is solely reflected by model accuracy, which in our view is the most relevant aspect, although other aspects of model quality are also discussed in some settings. For example, a smooth model might be considered more appropriate than a rough model since it provides sufficient information to drive the optimization process and simultaneously diminishes the probability of being stuck in local optima. Also, interpretable models are often aimed at, which is obviously related to feature selection. It should also be noted that it is usually advisable to choose a less complex model achieving good results with the desired small sample sizes since more complex models usually require larger data sets to be sufficiently accurate.

As applications we will deal here with the two most important statistical modeling cases, i.e. the classification (see Section 7.3.1) and the regression case (see Section 7.4):

- In **classification problems** an integer-valued response $y \in \mathbb{Z}$ with finitely many possible values $y_1, \ldots, y_G$ has to be predicted by a so-called **classification rule** based on n observations of $\boldsymbol{z}_i = \begin{bmatrix} \boldsymbol{x}_i & y_i \end{bmatrix}^T$, $i = 1, \ldots, n$, where the vector $\boldsymbol{x}$ summarizes the influential factors.
- In **regression problems** a typically real-valued response $y \in \mathbb{R}$ has to be predicted by a so-called **regression model** based on n observations of $\boldsymbol{z}_i = \begin{bmatrix} \boldsymbol{x}_i & y_i \end{bmatrix}^T$, $i = 1, \ldots, n$, where the vector $\boldsymbol{x}$ summarizes the influential factors.

In both cases, the influential factors are assumed to be real-valued.
Before expanding the theory, let us discuss a simple motivating example.

Example 7.1: Classification of Two Groups of Data
Consider the problem of selecting a classification model for two labeled classes of data. We consider here the simplest case, where the data are one-dimensional. In this case we look for an optimal separating point on the data axis for the two classes. One solution is delivered by the so-called **Linear Discriminant Analysis (LDA)** which, under certain conditions, locates this separation point at the mean of the two class means (see Section 7.3.1). When applying this method to class predictions for observations not used for separation point estimation, in particular for estimating class means, the question

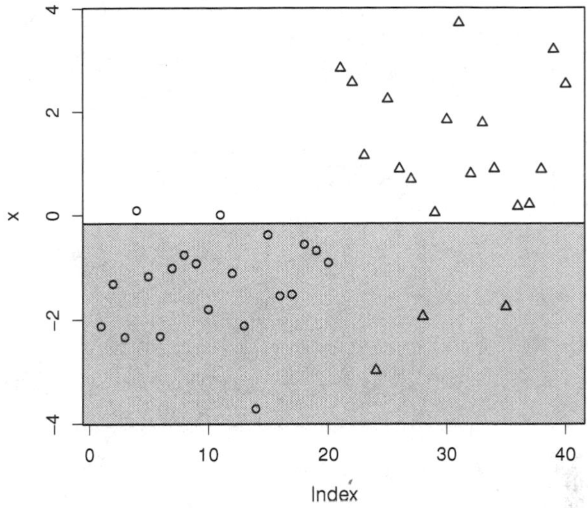

Figure 7.1: Classification by linear discriminant analysis; line = estimated class border; white or grey backgrounds refer to estimated classes; symbols refer to real classes.

is how often do we have to expect a classification error? The problem solved by resampling methods is the estimation of this error.
In order to demonstrate the ideas, we generate 20 one-dimensional observations of the two classes from $\mathcal{N}(-1.5, 1)$ and $\mathcal{N}(1, 4)$ distributions. These data are analyzed by LDA and the classification error is estimated by 10-fold cross-validation (see Section 7.2.3.1). Consider Figure 7.1 for the data situation and the estimated separation point, which is approximately $x = -0.16$ for our data. The estimated error rate is 0.125. Notice that also in Figure 7.1, 5 out of 40 observations are misclassified. However, there are separating points not producing any error in the "grey background/circles" class and only 4 errors in the "white background/triangles" class. On the other hand, though, 10-fold cross-validation indicates that concerning prediction quality 5 errors out of 40 are more realistic.

7.1.1 *Preliminaries*

In Section 7.4 we need some convergence properties of random variables together with their transformation rules. These will be introduced without proof here.

Definition 7.1: Convergence in Probability
We say that random vectors $\boldsymbol{X}_n$ indexed by sample size n **converge in probability** to a constant vector $\boldsymbol{c}$ iff

$$\lim_{n\to\infty} P(\|\boldsymbol{X}_n - \boldsymbol{c}\| > \varepsilon) = 0$$

for ε arbitrarily small. As an abbreviation for this, the term **probability limit** or **plim** is used: $\text{plim}(\boldsymbol{X}_n) = \boldsymbol{c}$.

This means that if the sample size n is increasing, it becomes more and more unlikely that $\boldsymbol{X}_n$ differs from the constant vector $\boldsymbol{c}$. The plim has the advantage that calculations are easy. This follows from the next theorem.

Theorem 7.1: Slutzki Theorem
$\text{plim}(g(\boldsymbol{X}_n)) = g(\text{plim}(\boldsymbol{X}_n))$ for any function g that does not depend on n.

From this, the calculation rules follow which are needed in Section 7.4.

Corollary 7.1: plim Calculation Rules
If X_n, Y_n are univariate, then

$$\begin{aligned}
\text{plim}(X_n \pm Y_n) &= \text{plim}(X_n) \pm \text{plim}(Y_n),\\
\text{plim}(X_n \cdot Y_n) &= \text{plim}(X_n) \cdot \text{plim}(Y_n),\\
\text{plim}(X_n / Y_n) &= \text{plim}(X_n) / \text{plim}(Y_n) \text{ if } \text{plim}(Y_n) \neq 0.
\end{aligned}$$

If $\boldsymbol{A}_n, \boldsymbol{B}_n$ are adequate matrices of random variables, then

$$\begin{aligned}
\text{plim}(\boldsymbol{A}_n^{-1}) &= (\text{plim}(\boldsymbol{A}_n))^{-1} \text{ if the inverses exist,}\\
\text{plim}(\boldsymbol{A}_n \pm \boldsymbol{B}_n) &= \text{plim}(\boldsymbol{A}_n) \pm \text{plim}(\boldsymbol{B}_n),\\
\text{plim}(\boldsymbol{A}_n \cdot \boldsymbol{B}_n) &= \text{plim}(\boldsymbol{A}_n) \cdot \text{plim}(\boldsymbol{B}_n).
\end{aligned}$$

Another convergence term is **convergence in distribution**.

Definition 7.2: Convergence in Distribution
We say that random vectors $\boldsymbol{X}_n$ with distribution functions F_n **converge in distribution** to a random vector $\boldsymbol{X}$ with a continuous distribution F iff

$$\lim_{n\to\infty} |F_n(\boldsymbol{x}) - F(\boldsymbol{x})| = 0.$$

for all $\boldsymbol{x}$ in which F is continuous. The abbreviation for this is $\boldsymbol{X}_n \xrightarrow{d} \boldsymbol{X}$.

A sort of extension of the Slutzki theorem is the following:

Theorem 7.2: Continuous Mapping Theorem
If $\boldsymbol{X}_n \xrightarrow{d} \boldsymbol{X}$, then $g(\boldsymbol{X}_n) \xrightarrow{d} g(\boldsymbol{X})$ if g is a continuous function.

This leads to the following corollary also needed in Section 7.4:

Corollary 7.2: Transformation of a Normal Distribution
If $\boldsymbol{X}_n \xrightarrow{d} \mathcal{N}(0, \boldsymbol{V})$, then $\boldsymbol{A}^T \boldsymbol{X}_n \xrightarrow{d} \mathcal{N}(0, \boldsymbol{A}^T \boldsymbol{V} \boldsymbol{A})$.

7.2 Model Selection[1]

7.2.1 General Procedure

In this section, a general procedure for model comparison and corresponding model selection will be discussed. We assume that we are interested in problems related to so-called **learning samples** $L = \{z_1, \ldots, z_n\}$ of n observations, and that there is a set of competing candidate models available. Each of these model candidates is fitted to the learning sample L leading to a function $a(\cdot|L)$. These models are then compared with respect to certain interesting properties. For example, models could be compared with respect to their ability to predict unknown response values y.

In order to identify the best model, the model candidates have to be compared by means of a problem-specific quality measure. Such a measure should depend on the model as well as on the learning data. Thus, there has to be a function $p(a, L)$ that assesses the quality of the function $a(\cdot|L)$. Since L is a random sample, $p(a, L)$ is a random variable, whose variability is induced by the variability of the possible learning samples L generated from an underlying data distribution F.

Therefore, in order to identify the best model, it is natural to compare the distributions of the quality measures. For this, it would be best to draw random samples from the distribution of the quality measure for a model a by evaluating $p(a, L)$ for different learning samples L. This will be realized by **resampling**.

Then, e.g., the statistical hypothesis of equal quality of model candidates can be tested by means of an adequate standard test (e.g. the chi-square test, see Section 6.2.2.3) if independent samples can be drawn from the distributions of the interesting quality measures. This way, we can also control the error probability of declaring a model wrongly to be best (**statistical guarantee**).

Let us define this somewhat more formally in a series of definitions.

[1]Partly based on Hothorn et al. (2003) and Bischl et al. (2012).

Definition 7.3: Learning Samples
Let $L^i = \{z_1^i, \ldots, z_n^i\} \sim F$, $i = 1, \ldots, B$, be B independent identically distributed **learning samples**, containing n observations each, generated from the underlying data distribution F.

Based on these learning samples the unknown coefficients of candidate models are learned and corresponding prediction functions are derived.

Definition 7.4: Candidate Models
Let a_k, $k = 1, \ldots, K > 1$, be **candidate models** for the solution of our problem. Based on the estimated coefficients, for each of these models a_k let $a_k(\cdot|L^i)$ be a **prediction function** of the model representing the model prediction, i.e. the predicted class for classifications problems and the predicted response value in the case of regression problems.

Since $a_k(\cdot \mid L^i)$ is based on the observations of the learning sample L^i, it is a random variable with a probability distribution A_k dependent on the data distribution F from which L^i was generated.

Definition 7.5: Distribution of Candidate Models
Let A_k be the **distribution function of** a_k, i.e.

$$a_k(\cdot \mid L^i) \sim A_k(F), \quad k = 1, \ldots, K.$$

For models a_k determined deterministically (e.g. linear models) the function $a_k(\cdot \mid L^i)$ is fixed for a fixed learning sample L^i. For models that were determined non-deterministically or for models that are based on starting values or hyperparameters (e.g. neural nets), the function $a_k(\cdot|L^i)$ is a random variable even for fixed L^i.

In order to assess the prediction quality of the candidate models we use a quality measure P.

Definition 7.6: Quality Measure
Let the scalar function P be a **quality measure** for the candidate models. Let $p_{ki} = p(a_k, L^i)$ be the values of P for a model a_k and a fixed learning sample L^i. Let the random variable p_{ki} have the distribution function P_k depending on the data distribution F:

$$p_{ki} = p(a_k, L^i) \sim P_k(F).$$

For models that are fit using a non-deterministic algorithm, it might be sensible to assess the mean quality after integration.

For each of the K models, a random sample $\{p_{k1}, \ldots, p_{kB}\}$ of B independent, identically distributed observations is drawn from the distribution $P_k(F)$. This can be used for testing hypotheses about the quality of the candidate models.

Definition 7.7: Quality Measures Hypothesis
A typical **null hypothesis** for the quality measures is

$$H_0: P_1 = \ldots = P_K,$$

postulating the equality of the considered models with respect to the distribution of their quality measures.

Obviously, this null hypothesis implies the equality of the corresponding location and variation measures.

Typically, for the definition of the corresponding alternative hypothesis a scalar-valued quality criterion Ψ (i.e. some test statistics) is assumed summarizing the distributions P_k. This leads to a definition of relative quality of the models:

Definition 7.8: Relative Quality of Models
Let the functional $\Psi(P_k)$ be a **quality criterion** of models a_k. Then, a model a_k is **better** than a model $a_{k'}$ with respect to the quality measure P if the functional $\Psi(P_k)$ is smaller for a_k, i.e. $\Psi(P_k) < \Psi(P_{k'})$.

The most common optimality criteria are based on location parameters like the expected value, $\Psi(P_k) = \mathrm{E}(P_k)$, or the median of the considered distributions. In such cases, the main interest lies in detecting the mean quality differences between the considered models. In practice, $\Psi(P_k)$ is estimated on samples $\{p_{k1}, \ldots, p_{kB}\}$. The corresponding test problem can be formalized as follows.

Definition 7.9: Quality Criterion Hypothesis
The null hypothesis for the quality criterion looks as follows:

$$H_0\colon \mathrm{E}(P_1) = \ldots = \mathrm{E}(P_K) \quad \text{vs.} \quad H_1: \exists k, k': \mathrm{E}(P_k) \neq \mathrm{E}(P_{k'})$$

This test problem is considered throughout this section.

In the next subsection we will elaborate on the case of supervised learning with the subcases classification and regression.

7.2.2 Model Selection in Supervised Learning

In supervised learning the observations $\boldsymbol{z}$ have the form $\boldsymbol{z} = \left[\boldsymbol{x}^T y\right]^T$, where y is the response and $\boldsymbol{x}$ the vector of influential factors. Learning is aimed at the determination of predictions that deliver information about the unknown response exclusively on the basis of the influential factors. Therefore, for each of the K considered models the constructed prediction function has the form $\hat{y} = a_k(\boldsymbol{x}|L^i)$. The deviation of the predicted response value $\hat{y}$ from the true response value y is typically represented by a scalar loss function $V(y,\hat{y})$.

Definition 7.10: Loss in Classification and Regression
In **classification problems**, $\hat{y}$ typically is the predicted class of observations (or the vector of estimated conditional probabilities of class memberships for each class), and the **loss** is typically $V(y,\hat{y}) = 1$ if $y \neq \hat{y}$, and $V(y,\hat{y}) = 0$ otherwise. In **regression problems**, the **loss** of the predicted response value $\hat{y}$ relative to the true response value y is typically assumed quadratic, i.e. $V(y,\hat{y}) = (y-\hat{y})^2$.

Let us now define the corresponding quality measure P.

Definition 7.11: Quality in Classification and Regression
The **quality measure** P for the predictions of classification or regression models is defined as a functional μ of the loss function, which exclusively depends on the underlying data distribution F:

$$p_{ki} = p(a_k, L^i) = \mu(V(y, a_k(\boldsymbol{x}|L^i))) \sim P_k(F).$$

In the case of a quadratic loss function $V(y,\hat{y}) = (y-\hat{y})^2$ the functional μ is generally chosen as the expected value. This is also used in the classification case, i.e. for the 0-1 loss, leading to so-called **error rates**, i.e. (number of errors)/(number of observation). The quality measure is then called conditional risk.

Definition 7.12: Conditional Risk
The **conditional risk** of the model a_k is defined as

$$p_{ki} = \mathrm{E}_{a_k} \mathrm{E}_{\boldsymbol{z}} V(y, a_k(\boldsymbol{x}|L^i)) = \mathrm{E}_{a_k} \mathrm{E}_{\boldsymbol{z}} (y - a_k(\boldsymbol{x}|L^i))^2 \sim P_k(F), \tag{7.1}$$

where $\boldsymbol{z} = \begin{bmatrix}\boldsymbol{x}^T & y\end{bmatrix}^T$ is drawn from the data distribution from which the observations of L are drawn, and the expected value E_{a_k} is only necessary in the case of non-deterministic modeling.

Other possible choices for μ are the median together with the absolute loss, and the supremum or other quantiles.

The distribution of the quality measures $P_k(F)$ of the models a_k, $k = 1,\ldots,K$, depends on the underlying data distribution F. Therefore, the drawing of the random sample from $P_k(F)$ is determined by the knowledge we have about F. In supervised learning we distinguish two situations:

1. F is known: This is true for simulation studies, where data are artificially generated, and in situations in which we can draw an infinite number of observations.
2. The finite learning sample L represents the only information about F available. In this case the empirical distribution of L represents the whole knowledge about F.

This has an impact on the concrete drawing of a random sample of the quality measure $P_k(F)$ for the model a_k. We will distinguish three different problem types:

Simulation: F is known.

Competition: A learning sample and a test sample are available, though the test sample is not usable during the development of the model.

Practice: Only a learning sample is available.

7.2.2.1 Resampling in Simulation

By means of a distribution function F artificial data are generated. The learning sample L consists of n independent observations $\boldsymbol{z}_j$, $j = 1,\ldots,n$, distributed according to F. This is denoted by $L \sim F_n$. **Resampling** is sampling of B independent learning samples from F_n:

$$L^1,\ldots,L^B \sim F_n.$$

The quality of a model a_k is assessed on the basis of the learning samples L^i $(b = 1,\ldots,B)$. Therefore, by calculating

$$p_{ki} = p(a_k, L^i) = \mu(V(y, a_k(x|L^i))), \quad b = 1,\ldots,B,$$

we get a random sample of B observations from the distribution of quality measures $P_k(F_n)$ for each a_k.

Let us now assume that μ is not calculated on the learning samples but on an extra sample drawn from F, called **test sample**.

Definition 7.13: Test Sample Quality
Let a **test sample** $T \sim F_m$ with m independent observations be drawn from F with m typically large. Then, the **test sample quality** is defined by

$$\hat{p}_{ki} = \hat{p}(a_k, L^i) = \mu_T(V(y, a_k(x|L^i))),$$

where μ_T denotes the empirical analogue of μ with respect to the test observations $\boldsymbol{z} = \begin{bmatrix}\boldsymbol{x}^T & y\end{bmatrix}^T \in T$. If μ_T is defined as in (7.1) as the expected value of the observations in the test sample (assuming, for the sake of simplicity, that a_k is determined deterministically), then in the above formula the expected value is replaced by the arithmetical mean of the loss function with respect to the observations of the test sample T:

$$\hat{p}_{ki} = \hat{p}(a_k, L^i) = \frac{1}{m} \sum_{z \in T} V(y, a_k(x|L^i)).$$

Analogously, a supremum would be replaced by a maximum and theoretical quantiles by their empirical pendants.

Also, when the models a_k are to be assessed by means of functionals Ψ of the quality measure P, expected values generally have to be approximated by empirical means over the learning samples. In the case of **classification problems** this leads to mean error rates

$$\frac{\text{number of errors}}{\text{number of observations}},$$

averaged over the learning samples or the corresponding test samples. In the case of **regression problems** the mean of the mean quadratic errors over the learning samples or test samples is used.

7.2.2.2 Resampling for Competition Data

In practice, F is most often not exactly known. Instead, only one **learning sample** $L \sim F_n$ with n observations is available, all from the same distribution F. The corresponding empirical distribution function $\hat{F}_n$ thus contains the whole knowledge about F. Therefore, F has to be imitated by means of $\hat{F}_n$. Resampling methods can be used to draw an (hopefully) independent, identically distributed sample from $\hat{F}_n$. Such methods will be introduced in the next subsection.

In statistics or machine learning competitions it is customary to have a **test sample** $T \sim F_m$ with m observations in addition to the learning sample L. However, this test sample is only used for the subsequent assessment of model quality, and is not available for model building. In such situations, model quality is finally exclusively assessed with respect to T though it is unknown to the model builder. Thus, the model builder has to rely exclusively on the learning sample to decide between the K different candidate models for the relationship between $\boldsymbol{x}$ and y (see problem type **Practice** above). Later, the quality of the chosen model is assessed with respect to the test sample.

Since quality assessment is only based on one test sample T, a model might be favored that is randomly performing very well on this data set but not on other possible test samples. This is especially true for small test samples.

7.2.2.3 *Resampling in Practice*

The most frequent situation in practice is that only one learning sample $L \sim F_n$ is available and no test sample. Also in this situation F is imitated by the empirical distribution function on the learning sample $\hat{F}_n$. From this empirical distribution function B independent learning samples are drawn by means of a resampling method:

$$L^1, \ldots, L^B \sim \hat{F}_n.$$

In order to distinguish the resampled new learning samples from the original learning sample, the new learning samples will be called **training samples** in the following.

The corresponding quality measure can be calculated as

$$\hat{p}_{ki} = \hat{p}(a_k, L^i) = \hat{\mu}(V(y, a_k(x|L^i))),$$

where $\hat{\mu}$ is an adequate empirical version of μ.

Let us discuss the choice of $\hat{\mu}$. First we will define two more "classical" choices.

Definition 7.14: Train-and-Test Method
For large n the so-called **train-and-test method** could be used, where the learning sample is divided into one learning sample L' of smaller size and one test sample T: $L = \{L', T\}$. Subsequently, the value μ_T is calculated as with problem types 1 and 2.

If n is so small that such an approach in infeasible, at first sight the following method appears to be most natural:

Definition 7.15: Resubstitution Method
The **resubstitution method** uses for each model the original learning sample also as the test sample.

Unfortunately, such an approach often leads to so-called **overfitting**, since the model was optimally fitted to the learning sample, and thus the error rate on this same sample will likely be better than on other, unseen, samples. With the help of the resampling methods described in the next subsection, such overfitting can be avoided.

7.2.3 *Resampling Methods*

The term **resampling** indicates that new samples are drawn from an existing original sample. We will discuss variants of the two most well-known resampling methods **cross-validation** and **bootstrapping**.

As a generic resampling method we will consider Algorithm 7.1, in which each of the B learning samples is split into a training sample for model fitting and a test sample for model assessment. Note that the instruction **FITMODEL** represents the fitting of the model a dependent on the model type. Also note that the elements of the set P of quality statistics are only the basis for equality tests or rankings.

Algorithm 7.1 Generic Resampling

Require: A learning sample L of n observations z_1 to z_n, the number of subsets B to generate and a loss function V.

1: Generate B subsets of L named L^1 to L^B.
2: $P \leftarrow \emptyset$
3: **for** $i \leftarrow 1$ **to** B **do**
4: $\quad \bar{L}^i \leftarrow L \setminus L^i$
5: $\quad a \leftarrow \text{FITMODEL}(L^i)$
6: $\quad p_i \leftarrow \frac{1}{|\bar{L}^i|} \sum_{z=[\boldsymbol{x}^T\ y]^T \in \bar{L}^i} V(y, a(\boldsymbol{x}))$
7: $\quad P \leftarrow P \cup \{p_i\}$
8: **end for**
9: **return** P

7.2.3.1 *Cross-Validation*

Cross-validation (CV) (Stone, 1974; Lachenbruch and Mickey, 1968) is probably one of the oldest resampling techniques. Like all other methods presented in this subsection, it uses the generic resampling strategy as described in Algorithm 7.1. The B subsets (line 1 of Algorithm 7.1) are generated according to Algorithm 7.2. Note that the instruction **SHUFFLE(L)** stands for a random permutation of the sample L. The idea is to divide the data set into B equally sized blocks and then use $B-1$ blocks to fit the model and validate it on the remaining block. This is done for all possible combinations of $B-1$ of the B blocks. The B blocks are usually called **folds** in the cross-validation literature. So a cross-validation with $B = 10$ would be called a **10-fold cross-validation**. Usual choices for B are 5, 10, and n.

The case $B = n$ is also referred to as **leave-one-out cross-validation** (LOOCV) because the model is fitted on the subsets of L, which arise if we leave out exactly one observation. With LOOCV, for a learning sample of size n a modeling method is applied to each subset of $n-1$ observations and tested on the nth observation. This leads to n different models, tested on one observation each. This way, each observation of the learning sample is used

Algorithm 7.2 Subsets for B-Fold CV

Require: A data set L of n observations z_1 to z_n and the number of subsets B to generate.

$L \leftarrow \text{SHUFFLE}(L)$
for $i \leftarrow 1$ **to** B **do**
    $L^i \leftarrow L$
end for
for $j \leftarrow 1$ **to** n **do**
    $i \leftarrow (j \bmod B) + 1$
    $L^i \leftarrow L^i \setminus \{z_j\}$
end for
return $\{L^1, \ldots, L^B\}$

exactly once as a test case for a model based on nearly the whole learning sample, neglecting nearly no information.

In classification, the error rate with respect to one resampled learning sample L^i is 1 or 0 for an incorrect or correct class prediction on the test case, respectively. As an overall quality criterion Ψ the error rate is calculated as "number of errors in test cases divided by n". For regression, the test case error is calculated as the individual quadratic loss, and the overall criterion as the mean quadratic loss.

Obviously, for large learning samples LOOCV is computer time intensive. In such cases, though, variants of the already mentioned train-and-test method, utilizing just one split of the original learning sample into a smaller new learning sample and a test sample, often produce a satisfying accuracy of the quality criterion (see also Section 7.3.2).

Also in **B-fold cross-validation** with $B < n$, the cases are randomly partitioned in B mutually exclusive groups of (at least nearly) the same size. Each group is exactly once used as the test sample and the remaining groups as the new learning sample, i.e. as the training sample. In classification, the mean of the error rates in the B test samples is called **cross-validated error rate**. Table 7.1 gives an overview of two important variants of cross-validation.

7.2.3.2 Bootstrap

The most important alternative resampling method to cross-validation is the **bootstrap**. We will discuss three variants here, which are known to have good properties: the $e0$, the .632, and the .632+ bootstrap.

The development of the bootstrap resampling strategy (Efron, 1979) is ten

Table 7.1: Variants of Cross-Validation

	Leave-One-Out	10-Fold CV
training cases	$n-1$	90%
test cases	1	10%
repetitions	n	10

years younger than the idea of cross-validation. Again, Algorithm 7.1 is the basis of the method, but the B subsets are generated using Algorithm 7.3. Note that the instruction **RANDOMELEMENT(L)** stands for drawing a random element from the sample L by means of uniformly distributed random number $\in \{1,\dots,n\}$ (see Section 6.2.2).

Algorithm 7.3 Subsets for the Bootstrap

Require: A data set L of n observations z_1 to z_n and the number of subsets B to generate.

for $i \leftarrow 1$ **to** B **do**
  $L^i \leftarrow \emptyset$
  for $j \leftarrow 1$ **to** n **do**
    $z \leftarrow \text{RANDOMELEMENT}(L)$
    $L^i \leftarrow L^i \cup \{z\}$
  end for
end for
return $\{L^1,\dots,L^B\}$

The subset generation is based on the idea that instead of sampling from L without replacement, as in the CV case, we sample with replacement. This basic form of the bootstrap is often called the **e0 bootstrap**. One of the advantages of this approach is that the size of the training sample, in the bootstrap literature often also called the in-bag observations, is equal to the actual data set size. On the other hand, this entails that some observations can and likely will be present multiple times in the training sample L^i. In fact, asymptotically only about 63.2% of the data points in the original learning sample L will be present in the training sample, since $1-(1-1/n)^n \approx 1-e^{-1} \approx 0.632$. The remaining 36.8% of observations are called out-of-bag and form the test sample as in CV.

Here the number of repetitions B is usually chosen much larger than in the CV case. Values of $B = 100$ up to $B = 1000$ are not uncommon. Do note, however, that there are n^n different bootstrap samples. So for very small n

Table 7.2: Bootstrap Method

	Bootstrap
training cases	N (j different)
test cases	$N-j$
repetitions	≥ 200

there are limits to the number of bootstrap samples you can generate. In general, $B \geq 200$ is considered to be necessary for good bootstrap estimation (cp. Table 7.2). This number of repetitions may be motivated by the fact that in many applications not only the bootstrap quality criterion is of interest, but the whole distribution, especially the 95% confidence interval for the true value of the criterion. For this, first the empirical distribution of the B quality measure values on the test samples is determined, and then the empirical 2.5% and 97.5% quantiles. With 200 repetitions, the $5th$ and the $195th$ element of the ordered list of the quality measures can be taken as limits for the 95% confidence interval, i.e. there are enough repetitions for an easy determination of even extreme quantiles. Note, however, that the bootstrap is much more expensive than LOOCV, at least for small learning samples.

The fact that with the bootstrap some observations are present multiple times in the training sample can be problematic for some modeling techniques. Several approaches have been proposed for how to deal with this. Most add a small amount of random noise to the observations (Efron, 1979).

Another problem with adding some observations multiple times to the training sample is that we overemphasize their importance. This is called **oversampling**. This leads to an estimation bias for our quality measure. A first attempt to counter this was the so-called **.632 bootstrap** procedure by Efron (1983). Here, the estimated error of the model is a weighted average of the error on the training sample and the test sample, namely $0.368 \cdot e_{\text{app}} + 0.632 \cdot e0$, where e_{app} is the so-called apparent or resubstitution error on the used training sample and $e0$ is the $e0$ bootstrap error estimator. As motivation for the .632 bootstrap it is often argued that the e0 bootstrap generally overestimates the quality criterion, and that a convex combination with e_{app} compensates the underestimation by e_{app}. Please note that e0 can be approximated by **repeated 2-fold cross-validation**, i.e. by repeated 50/50 partition of the learning sample, or by repeated **2:1 train-and-test splitting**, because the $e0$ generates roughly 63.2% in-bag (train) observations and 36.8% out-of-bag (test) observations.

The fallacy in this approach is that some modeling techniques might have an error of 0 on the training sample. An example of such a method would be an interpolating spline. To counter this, Efron proposed a further variant of the bootstrap named the **.632+ bootstrap** (see Efron and Tibshirani (1997)). This strategy is a bit more involved and deviates somewhat from the framework proposed in Algorithm 7.1. The details are given in Algorithm 7.4. The main difference here is that instead of fixed weights, as in the .632 bootstrap, the weights are individually calculated for each model to reflect how well the model can reproduce the training sample.

Algorithm 7.4 .632+ Bootstrap

Require: A data set L of n observations $\boldsymbol{z}_1$ to $\boldsymbol{z}_n$, the number of subsets B to generate and a loss function V.

1: Generate B subsets of L named L^1 to L^B

2: $P \leftarrow \emptyset$

3: **for** $i \leftarrow 1$ **to** B **do**

4: $\quad \bar{L}^i \leftarrow L \setminus L^i$

5: $\quad a \leftarrow \text{FITMODEL}(L^i)$

6: $\quad \hat{\gamma} \leftarrow \frac{1}{n^2} \sum_{p,q=1}^{n} V(y_p, a(\boldsymbol{x}_q))$

7: $\quad p^{\text{in}} \leftarrow \frac{1}{|L^i|} \sum_{\boldsymbol{z} \in L^i} V(y, a(\boldsymbol{x}))$

8: $\quad p^{\text{out}} \leftarrow \frac{1}{|\bar{L}^i|} \sum_{\boldsymbol{z} \in \bar{L}^i} V(y, a(\boldsymbol{x}))$

9: $\quad \hat{R} \leftarrow \frac{p^{\text{out}} - p^{\text{in}}}{\hat{\gamma} - p^{\text{in}}}$

10: $\quad \hat{w} \leftarrow \frac{0.632}{1 - 0.368\hat{R}}$

11: $\quad p_i \leftarrow (1 - \hat{w})p_i^{\text{in}} + \hat{w}p_i^{\text{out}}$

12: $\quad P \leftarrow P \cup \{p_i\}$

13: **end for**

The B subsets of L are again generated using Algorithm 7.3. Then, as in the general framework, the model a is calculated in line 5. Lines 6, 7, 9, and 10 are new. In line 6 the loss is estimated for the hypothetical case that our model has no predictive power. This is done by calculating the loss for each possible combination of $\boldsymbol{x}$ and y from L. The resulting quantity $\hat{\gamma}$ is called the no-information error rate, because now there is no direct dependence between $\boldsymbol{x}$ and y, and the resulting loss is the error rate the model would achieve even if there was no link between $\boldsymbol{x}$ and y, i.e. our function was pure noise. Using this and the in-bag error rate from line 7, as well as the usual out-of-bag error rate from line 8, the relative overfitting rate $\hat{R}$ is calculated in line 9. $\hat{R}$ lies between 0 and 1. If $\hat{R} = 1$, then the model is completely overfitted, i.e. it only has predictive power on the training sample. If $\hat{R}$ is almost 0, on the other hand, then the model has great predictive power on the test sample and we can use the error on the training sample to increase the accuracy of our error estimate. The final quality measure is the weighted average of the in-bag and out-of-bag error rates as calculated in line 11 using the weight derived in line 10 from the relative overfitting rate. Note that this weight is near 0.632 for a value of $\hat{R}$ almost 0.

7.2.3.3 Subsampling

Subsampling (SS) is very similar to the classical bootstrap. The only difference is that observations are drawn from L without replacement (see Algorithm 7.5). Therefore, the training sample has to be smaller than L or no observations would remain for the test sample. Usual choices for the subsampling rate $|L^i|/|L|$ are $4/5$ or $9/10$. This corresponds to the usual number of folds in cross-validation (5-fold or 10-fold). Like in bootstrapping, B has to be selected a priori by the user. Choices for B are also similar to bootstrapping, e.g. in the range of 200 to 1000.

Algorithm 7.5 Subsets for Subsampling

Require: A data set L of n observations z_1 to z_n, the number of subsets B to generate and the subsampling rate r.

$m \leftarrow \lfloor r|L| \rfloor$
for $i \leftarrow 1$ **to** B **do**
    $L' \leftarrow L$
    $L^i \leftarrow \emptyset$
    for $j \leftarrow 1$ **to** m **do**
        $d \leftarrow \text{RANDOMELEMENT}(L')$
        $L^i \leftarrow L^i \cup \{d\}$
        $L' \leftarrow L' \setminus \{d\}$
    end for
end for

7.2.3.4 Further Resampling Methods

Many variants, extensions, and combinations of the above algorithms exist, which we do not want to present here. We only want to mention three such methods.

Stratified cross-validation (SCV) ensures that all folds of cross-validation include a share of observations roughly equal to the corresponding share in the learning sample per region of the input space, e.g. per class in classification. Several methods to achieve this have been proposed, for an example see Diamantidis et al. (2000). Stratified sampling is often combined with other sampling types like the following two.

Repeated cross-validation (RCV) performs a usual B-fold CV multiple times and aggregates the results (normally by the mean) to reduce the variance of randomly splitting the data. An R times repeated B-fold cross-validation leads to the following estimate of the quality criterion:

$$\hat{\Psi}^{RCV}(R,B) = \frac{1}{R}\sum_{i=1}^{R}\hat{\Psi}_i^{CV}(B),$$

where $\Psi_i^{CV}(B)$ are the individual cross-validated quality criteria.

The partitions are random but stratified if we set $\Psi_i^{CV}(B) := \Psi_i^{SCV}(B)$. Hence, we gain comparability and a guaranteed number of observations within each class.

Bootstrap cross-validation (BCV) generates B bootstrap samples of L, performs cross-validation on each bootstrap sample, and then aggregates the resulting quality criteria by averaging. This has been shown to be advantageous for small sample sizes. Again, stratification is applied.

For B bootstrap samples $L^1,\dots,L^B$ of size n and R-fold cross-validation on these samples, bootstrap cross-validation finally leads to the following mean of quality criteria:

$$\hat{\Psi}^{BCV}(B,R) = \frac{1}{B}\sum_{i=1}^{B}\hat{\Psi}_{L^i}^{CV}(R),$$

where $\Psi_{L^i}^{CV}(R)$ are the individual quality criteria generated by R-fold cross-validation on L^i.

7.2.3.5 *Properties and Recommendations*

Properties of Leave-One-Out and Cross-Validation

Leave-one-out (LOO) cross-validation has better properties for the squared loss in regression than for its 0-1 counterpart in classification and is an almost unbiased estimator for the mean loss (Kohavi, 1995). Its near unbiasedness makes LOO an attractive candidate among the presented algorithms, especially when only few samples are available. But one should be aware of the following facts: LOOCV has a high variance (Kohavi, 1995; Weiss and Kulikowski, 1991) as estimator of the mean loss, meaning quite different values may be produced if the data used for cross-validation slightly change. It also tends to select too complex models. In Shao (1993) theoretical reasons for this effect are presented, and subsampling and balanced leave-k-out CV are shown to be superior estimators in a simulation study. Kohavi (1995) arrives at similar results regarding LOO and demonstrates empirically that 10-fold CV is often superior. He suggests a stratified version.

For these reasons we recommend LOO mainly for efficient model selection, keeping in mind that this might lead to somewhat suboptimal choices. Repeated and stratified CV will usually produce more reliable results in practice.

Properties of the Bootstrap

While the $e0$ bootstrap is pessimistically biased in the sense that it bases its performance values on models that use only about 63.2% of the data, the .632 bootstrap can be optimistic in a much worse way, as complex models can easily achieve $p^{in} = 0$. Both estimators are known to have a low variance, and $e0$ is especially good when the sample size is small and the error or noise in the data is high (Weiss and Kulikowski, 1991). The **.632+** bootstrap combines the best properties of both estimators and can generally be trusted to achieve very good results with small sample sizes. Its main drawback is that it might result in an optimistic bias, when more complex models are considered (Kim, 2009; Molinaro et al., 2005).

Bootstrapping or Subsampling?

When combining model or hyperparameter selection with bootstrapped data sets in the outer loop of nested resampling (see Section 7.2.5), repeated observations can lead to a substantial bias toward more complex models. This stems from the fact that in the inner tuning loop measurements will occur both in the training set and in the test set with a high probability because some observations appear multiple times in the bootstrap sample so that more complex models "memorizing" the training data will seem preferable. In Binder and Schumacher (2008) subsampling was proposed and evaluated as a remedy.

Independence, Confidence Intervals and Testing

In general, the generated training and test samples, and therefore the obtained performance statistics, will not be independent when sampling from a finite data set. This has negative consequences if confidence intervals for the performance measure should be calculated. The dependence structure is especially complicated for the commonly used cross-validation, where the split-up of the data in one iteration completely depends on all other split-ups. It can be shown that in this setting no unbiased estimator of the variance exists (Bengio and Grandvalet, 2004) and pathological examples can be constructed, where the variance estimator performs arbitrarily bad. Nadeau and Bengio (2003) propose a new variance estimator for CV that takes the dependence between sampled data sets into account and provides a much better foundation for interval estimators and subsequent statistical tests regarding location parameters.

7.2.4 Feature Selection

Let us now discuss other applications of resampling besides general model selection, namely feature selection and hyperparameter tuning. Let us start with feature selection. Feature selection (variable selection) is obviously a special case of model selection, and is assumed to be especially important because of at least two reasons:

Feature selection improves generalization performance: It is well-known that in the setting of many noisy, highly correlated, and possibly irrelevant covariates the predictive power of a model fitted on the whole feature set will be suboptimal.

Smaller models improve model interpretation: If one is mainly interested in understanding the data, one might even accept a substantial loss in predictive performance to achieve a smaller model.

We will concentrate here on the most popular class of feature selection algorithms, the so-called **wrappers**, because of their general applicability and strength to build very predictive models.

Definition 7.16: Wrapper Algorithms for Feature Selection
Wrapper algorithms for feature selection use a model type as a black-box and search for an optimal set of input features w.r.t. a quality criterion by repeatedly adjusting the variable set, fitting a model, and evaluating it.

In a basic (naive) algorithm we would go through each subset of features, solve the problem, and select the best subset. Unfortunately, this is often impossible because the set of subsets of a set Q of q features, called the **power set** of Q, has 2^q elements (including both the set itself and the empty set). This follows from the fact that the total number of distinct k-element subsets on a set of q elements is given by the binomial sum $\sum_{k=0}^{q} \binom{q}{k} = 2^q$. Therefore, computation time would be much too high, even for a medium number of features. The alternative we will discuss here is the so-called **greedy forward selection** (see Algorithm 7.6).

In each iteration, a feature is selected once it improves the quality criterion even more than all other features that are not yet in the set S of selected features. This feature selection algorithm is called **greedy** since a feature once

Algorithm 7.6 Greedy Forward Selection

Require: Maximum number of features to select $q_{\max}$, full set of features Q.

$S = \emptyset$ {S is the current set of selected features, $|S|$ the current number of selected features}
while $|S| < q_{\max}$ **do**
  for $v \in Q$ **do**
    $S_t \leftarrow v \cup S$
    Train the model with S_t and keep the test sample performance
  end for
  Select v_o, the feature with the best test sample performance
  $S \leftarrow v_o \cup S$
  $Q \leftarrow Q \setminus v_o$
  Keep the test performance obtained with current S
end while
Return the best set S

selected, is never removed in a subsequent step, i.e. will stay in the list of selected features until the end of the procedure.

7.2.5 Hyperparameter Tuning

The definition of hyperparameters is, unfortunately, unclear in the literature. Only in Bayesian statistics are they clearly defined as the parameters of prior distributions. This prompts our definition:

Definition 7.17: Hyperparameters
Let **hyperparameters** be model or model estimator parameters that might influence model selection but have to be chosen prior to it.

An example for such a hyperparameter is, e.g., the starting value of the learning rate in the backpropagation algorithm for parameter estimation in neural nets. This parameter is typically chosen before model estimation and model selection, and does not restrict the model class of neural nets, i.e. the model selection task is unchanged. On the other hand, the number of nodes in the hidden layer of a neural net is not a hyperparameter in our definition since this restricts model choice and is a part of it.

Since hyperparameters are to be fixed prior to model selection, they have to be varied in an extra sampling process, and one ends up with a **nested sampling** process. As an example, consider using subsampling with $B = 100$ in an outer loop for model evaluation and 5-fold cross-validation in an inner loop

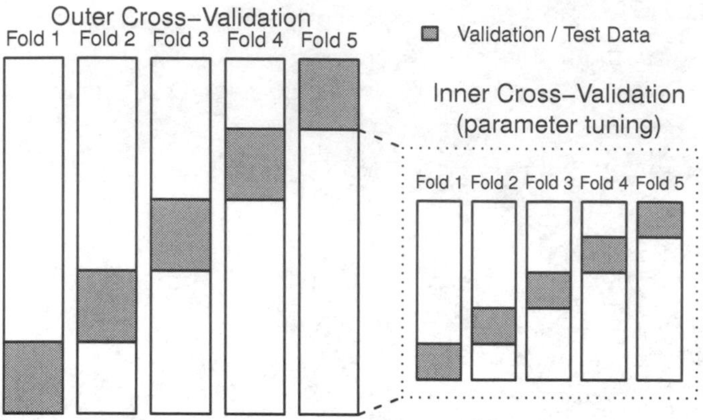

Figure 7.2: Nested resampling with two nested cross-validations.

for hyperparameter selection. For each of the 100 training samples L^i from subsampling, a 5-fold cross-validation on the training sample is employed as internal fitness evaluation to select the best setting for the hyperparameters of the model. The best obtained hyperparameters are used to fit the model on the complete training sample and calculate the quality measure on the test sample of the outer resampling strategy. Figure 7.2 shows this process schematically for two nested cross-validations, namely 5-fold CV in both the inner and the outer resampling.

7.2.6 Summary

In this section we introduced a general procedure for model selection with resampling methods for the realization of repetitions in sampling. We discussed different variations of such methods, namely cross-validation, bootstrapping, and subsampling. We showed how these methods can be used for model selection, feature selection, and hyperparameter tuning. In the next sections we will apply these general ideas to model selection in classification and regression.

7.3 Model Selection in Classification

7.3.1 The Classification Problem[2]

We start the discussion of the application of resampling methods with classification problems. The term **classification** is used in two ways in the literature:

1. For a given data set the aim is to identify classes of homogeneous objects. This is also called **unsupervised learning** or **cluster analysis**.
2. Based on a known classification in a learning sample the aim is to identify a so-called **classification rule** that is used to assign a new observation to one of the possible classes. This is also called **supervised learning** or **discrimination**.

In this section we will exclusively consider supervised learning. For an example of cluster analysis see, e.g., Section 5.2.4.

Let us specify the problem again more formally:

Definition 7.18: Classification Problem
In **classification problems** an integer-valued response $y \in \mathbb{Z}$ with finitely many possible values $y_1, \ldots, y_G$ has to be predicted by a so-called **classification rule** based on n observations of $\boldsymbol{z}_i = \left[\boldsymbol{x}_i^T y_i\right]^T$, $i = 1, \ldots, n$, where the vector $\boldsymbol{x}_i$ summarizes the influential factors.

Let us stress that finding a classification rule is a **prediction problem**. The aim is to predict classes of future observations on the basis of known classes in a learning sample. Therefore, we do not have to assess the quality of fit on the learning sample, but the ability of the classification rule to generalize to future observations (**predictive power**).

This is not the place to discuss the diversity of classification methods, since we want to concentrate on the discussion of resampling methods in supervised classification. In order to have a basis for this, we just introduce two very simple types of classification methods and the **linear discriminant analysis (LDA)** in the simplest case, i.e. for two classes.

1. **Data-independent rules**: Such a rule ignores all information in the new observation for class assignment. For example, it assigns the most frequent class of the learning sample or the classes according to their frequency in the learning sample.
   Since the application of such methods is very simple and fast, they are often

---

[2]Partly based on Michie et al. (1994, pp. 6 – 16, 107 – 124), Weiss and Kulikowski (1991, pp. 17 – 49).

used for comparison, as something like a yardstick, with more elaborate methods.

2. **Nearest-neighbor rules**: The learning sample is searched for the observation with the greatest similarity (in a predefined sense) to the factor vector of a new observation. The class of this nearest observation is then assigned to the new observation. Note that a distance measure has to be specified a priori.
   Although this rule is easy to understand, its application can be very time-consuming, especially if a large learning sample has to be searched for the nearest-neighbor. Variants of this rule use that class for prediction that occurs most often in k nearest-neighbors (**kNN method**).

3. **Linear discriminant analysis (LDA)** is based on the following assumptions:

   L1: The distributions of influential factors inside the classes are normal distributions with different expected values $\boldsymbol{\mu}_i$ but an identical covariance matrix $\boldsymbol{\Sigma}$ for all classes $y_1, \ldots, y_G$. This leads to different densities f_i for the classes.

   L2: The misclassification costs are equal for all classes.

   L3: The a priori probabilities of the classes may be different.

   For 2 classes this leads to the following **Bayes decision rule**:
   Choose class 1 iff $\frac{f_1(\boldsymbol{x})}{f_2(\boldsymbol{x})} > \frac{\pi_2}{\pi_1}$ for a priori class probabilities $\pi_i, i = 1, 2$.
   Utilizing the densities of the normal distribution, this is equivalent to

$$\frac{\exp(-0.5(\boldsymbol{x}-\boldsymbol{\mu}_1)^T\boldsymbol{\Sigma}^{-1}(\boldsymbol{x}-\boldsymbol{\mu}_1))}{\exp(-0.5(\boldsymbol{x}-\boldsymbol{\mu}_2)^T\boldsymbol{\Sigma}^{-1}(\boldsymbol{x}-\boldsymbol{\mu}_2))} > \frac{\pi_2}{\pi_1} \tag{7.2a}$$

$$-0.5(\boldsymbol{x}-\boldsymbol{\mu}_1)^T\boldsymbol{\Sigma}^{-1}(\boldsymbol{x}-\boldsymbol{\mu}_1)+0.5(\boldsymbol{x}-\boldsymbol{\mu}_2)^T\boldsymbol{\Sigma}^{-1}(\boldsymbol{x}-\boldsymbol{\mu}_2) > \log\left(\frac{\pi_2}{\pi_1}\right) \tag{7.2b}$$

$$\boldsymbol{x}^T\boldsymbol{\Sigma}^{-1}(\boldsymbol{\mu}_2-\boldsymbol{\mu}_1) < \log\left(\frac{\pi_1}{\pi_2}\right)+0.5\boldsymbol{\mu}_2^T\boldsymbol{\Sigma}^{-1}\boldsymbol{\mu}_2-0.5\boldsymbol{\mu}_1^T\boldsymbol{\Sigma}^{-1}\boldsymbol{\mu}_1. \tag{7.2c}$$

   A further simplification is achieved if the a priori probabilities of the classes are equal: Choose class 1 iff $f_1(\boldsymbol{x}) > f_2(\boldsymbol{x})$. Then the above inequality (7.2c) can also be written in the following way:
   Let $a = \boldsymbol{\Sigma}^{-1}(\boldsymbol{\mu}_2-\boldsymbol{\mu}_1)$, then

$$\boldsymbol{a}^T\boldsymbol{x} < \frac{\boldsymbol{a}^T(\boldsymbol{\mu}_1+\boldsymbol{\mu}_2)}{2}.$$

Unknown parameters are empirically estimated. This leads to:

$$\hat{\boldsymbol{a}}^T \boldsymbol{x} < \frac{\hat{\boldsymbol{a}}^T(\bar{\boldsymbol{x}}_1 + \bar{\boldsymbol{x}}_2)}{2},$$

meaning that the separation is linear, and a projection of the mean of the empirical group means is the estimated border between the classes.
LDA separates the data linearly, more exactly by a projection on a vector orthogonal to a separating hyperplane. On the projections $\hat{\boldsymbol{a}}^T \boldsymbol{x}$, a separating point for the two classes is looked for. In the simplest case we discussed above, LDA locates this separation point at the projection $\frac{\hat{\boldsymbol{a}}^T(\bar{\boldsymbol{x}}_1+\bar{\boldsymbol{x}}_2)}{2}$ of the mean of the two class means. Note that hyperplanes of one-dimensional data are points, and that the projection line is then the original data axis.

7.3.2 Classification Evaluation

For the evaluation of classification rules, generally so-called **misclassification rates** or just error rates are used. In this subsection we discuss reliable estimation of such rates. For this, we have to specify the term **classification error** more precisely.

Classification rules aim at the successful classification of new observations that were not used for the construction of the rule. This leads to the following definition:

Definition 7.19: Error Rates
The **true error rate** of a classification rule is defined as the limit of the relative error of the rule for a steadily growing number of new observations. This limit should be equal to the error rate on the whole population.

The relative classification error for a finite number of observations is called **empirical error rate**:

$$\text{(empirical) error rate} := \frac{\text{number of errors}}{\text{number of observations}}.$$

The most important question is whether one can infer from empirical error rates in small samples the true error rates. Unfortunately, this question cannot be answered universally. Decisive is the choice of the sample. Let us discuss different kinds of samples used for error estimation.

1. Frequently, the so-called **apparent or resubstitution error rate** is used as the quality criterion, i.e. the error in the learning sample. For a steadily growing learning data set, this would approximate the true error rate. In practice, however, the apparent error rate is most often a bad estimator for

the future quality of the rule. In general, the apparent error rate underestimates the true error rate considerably, since the rule was constructed just to reproduce the learning sample, and might not be comparably good for other samples. Such a classification rule is called to **overfit** the learning sample or to be **overspecialized** on this.

2. Better suited for the estimation of the true error rate are the so-called **train-and-test methods**, which choose one **training sample** as well as one **test sample** randomly from the original learning sample. Since the test sample is ignored in the construction of the classification rule, the determination of the empirical error rate is based on real new data. The result is called **test sample error rate**.
If training and test samples are **independent random samples**, the test sample error rate has relevant properties. In particular, reasonably exact error estimation can even be based on a not too big number of observations n. This can be seen by considering the confidence interval for the true error rate, which does not depend on the distribution of the observations. For this, one only has to realize that the **test sample error rate** as an **estimator** for the **true error rate** is in any case **binomially distributed**, more precisely $Bin(n, p)/n$ distributed, where p is the true error rate.
In order to get an idea about the size of the true error rate, we consider the **upper limit of the one-sided 95% confidence interval** for the true error rate.
Figure 7.3 shows such limits for different test sample error rates. For example, for a test sample error rate of 0% and 30 test cases there is a reasonable chance that the true error rate is 10%, whereas the true error rate is very probably $< 1\%$ for 1000 test cases. Moreover, for a test sample error rate of 20% and 30 or 50 test cases there is still a large probability that the true error rate is about 30%, whereas for 1000 test cases the true error rate is most probably less than 23%. Obviously, 1000 test cases always lead to rather exact estimates. 1000 test cases are, thus, typically sufficient. In practice, however, the available test sample is most often much smaller.
As important as a sufficiently large **test sample** is for the determination of the error rate, a sufficient size of the **training sample** is at least as important for the determination of a reliable classification rule. Often, a random **partition of the learning sample** at a ratio of 2 to 1 is chosen, i.e. 2/3 training sample and 1/3 test sample. Note the relationship to bootstrap samples where on average 36.8% of the learning observations are left over for testing.
However, especially in the case of a small learning samples the training

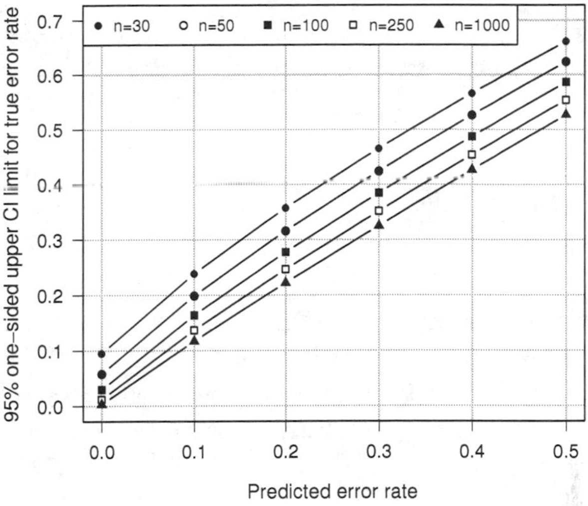

Figure 7.3: Upper 95% confidence limits for the true error rate and test sample sizes *n*.

sample should be chosen not much smaller, in order to support the classification rule as broadly as possible.

A disadvantage of such train-and-test methods is that relatively many observations of the learning sample are held off for the estimation of the error rate, and that only one partition into training and test samples is used. This leads to loss of valuable information for the determination of the classification rule. In practice, the error rates estimated by the **2:1 rule** turned out to be **relatively pessimistic estimates of the true error rate**.

3. Much better error estimates can be obtained by means of the **resampling methods** discussed in the previous section. Let us recall the recommendations from the last section:
   - We recommend LOOCV (Leave-One-Out Cross-Validation) mainly for efficient model selection, keeping in mind that this might lead to somewhat suboptimal choices. Repeated and stratified CV will usually produce more reliable results in practice.
   - .632+ bootstrap combines the best properties of the discussed bootstrap estimators and can generally be trusted to achieve very good results with small sample sizes. Its main drawback is that it might result in an optimistic bias when more complex models are considered.
   - In order to avoid repeated observations in training samples, subsampling may be preferred to bootstrapping.

Let us conclude this section with some simulations and examples in order to illustrate the above evaluation methods in practice.

7.3.3 *Simulation: Comparison of Error Rate Estimators*[3]

In this subsection we will compare different resampling-based error estimators concerning their ability to correctly estimate the true error rate.

Methods

We restrict ourselves to repeated cross-validation (RCV) with $R = B = 10$ and bootstrap cross-validation (BCV) with $B = 100, R = 5$ (both stratified), as well as the two basic estimators 10-fold cross-validation (CV) and .632+ bootstrap (.632+). Also, we restrict ourselves to two classes, and in order to study one local and one global classification method, to the 3NN estimator and LDA.

Experimental Design

The aim of the study is to control the data sets by systematically varying certain characteristics of the two classes in order to cover a reasonable part of the space of typical data situations. Compare Luebke and Weihs (2011) for basic considerations about such experimental design. Here, we restrict ourselves to the variation of the following characteristics:

1. dimension p of data,
2. minimum distance of the two class means in one data dimension,
3. maximum distance of the two class means in one data dimension,
4. skewness of the distribution in each individual data dimension,
5. number n of observations.

As the **simulation design** a **Latin Hypercube Design (LHD)** is used with 100 different combinations of these characteristics. In each design point 10 different data sets are generated. A **Latin hypercube design** is structured so that each characteristic has the same number of levels. For each level of each characteristic there is exactly one design point. The idea is, to divide the range of p characteristics in M equally probable intervals. Then, M design points are placed to satisfy the Latin hypercube requirements. Note that the number of divisions, M, is assumed equal for each characteristic. Also note that this design does not require more design points for more characteristics. This independence is one of the main advantages of this design. Further note that by serially numbering the levels, the same LHD can be used for arbitrary characteristics and levels, as long as p and M are equal.

[3]Partly based on Zentgraf (2008).

Example 7.2: Latin Hypercube Design
As an example, let us consider an LHD for 3 characteristics with 4 levels each (see Table 7.3). This design is graphically represented in Figure 7.4.

Table 7.3: Latin Hypercube

Point	x_1	x_2	x_3
1	1	1	2
2	2	4	3
3	4	2	4
4	3	3	1

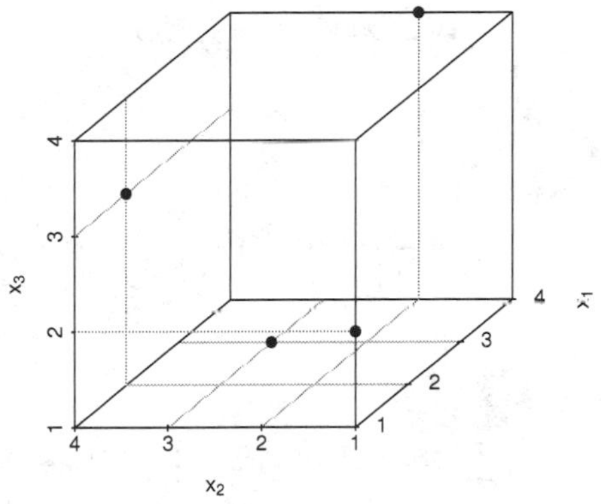

Figure 7.4: Latin hypercube design.

Let each of our five data characteristics first be projected to the interval $[0,1]$, where 0 represents the minimal possible level and 1 the maximal. For each characteristic this interval is divided into $M = 100$ equally sized subintervals. The individual design points are drawn from these subintervals. In order to cover the spectrum of possible constellations as well as possible, the design points should be as different as possible. To be more precise, the minimal distance between two different design points in our LHD is constructed to be maximum among all LHDs with five characteristics and $M = 100$ design points. This property characterizes a so-called **maximin design**.

Assume now that we have generated values $f_1, f_2, \ldots, f_5 \in [0,1]$. These values are then back-transformed into real levels of the five characteristics by the following scheme:

- $f_1 \rightarrow p := \lceil 10 \cdot f_1 \rceil \cdot 2$ (leading to realizations $\in \{2,4,6,8,\ldots,20\}$),

- $f_2 \to \min := f_2$ (the smallest minimum distance can be 0, the greatest 1),
- $f_3 \to max := 1 + 2 \cdot f_3$ (the smallest maximum distance can be 1, the greatest 3),
- $f_4 \to skewness := f_4 \cdot 1.6$ (skewness lies between 0 and 1.6),
- $f_5 \to n := \lfloor 150 - 120\sqrt{f_5} \rfloor$ (n takes values in $\{30, 31, 32, \ldots, 150\}$; since $\sqrt{x}$ is greater than x in $[0,1)$, greater sample sizes will probably appear less often than smaller ones.

Random Data

For each design point, we need a procedure to draw random realizations from distributions with the specified skewness. Such a procedure is proposed by Fleishman (1978). Let w.l.o.g. $X \sim \mathcal{N}(0,1)$ (the expected value can be shifted afterwards). In order to generate an X^* from a distribution with skewness γ_1 and kurtosis γ_2, we apply a transformation

$$X^* = a + b \cdot X + c \cdot X^2 + d \cdot X^3. \tag{7.3}$$

In our case we use $\gamma_2 = 3$, the kurtosis of the standard normal distribution. In order to determine the unknowns b, c, d dependent on γ_1 and γ_2, one can show that we have to solve the following equation system:

$$c = \frac{\gamma_1}{2(b^2 + 24bd + 105d^2 + 2)} \tag{7.4}$$

$$\gamma_2 = 24\left(bd + c^2(1 + b^2 + 28bd) + d^2(12 + 48bd + 141c^2 + 225d^2)\right) \tag{7.5}$$

$$2 = 2b^2 + 12bd + \frac{\gamma_1^2}{(b^2 + 24bd + 105d^2 + 2)^2} + 30d^2 \tag{7.6}$$

Since this is a complicated nonlinear problem, numerical methods have to be used for solving (7.4)–(7.6). Here, we use the Nelder-Mead search method (see Nelder and Mead (1965)) with 1000 different starting value combinations. The parameter a is finally determined by $a = -c$ so that the expected value remains zero.

For the p dimensions $2n$ random data are generated independently, n observations for each of the two classes. For the first dimension the minimum distance f_2 between the class means is chosen, and for the last dimension the maximum distance f_3. All other $p - 2$ dimensions are chosen such that the distances of class means are equidistantly spaced between f_2 and f_3. These distances of class means are added to the observations of the second class.

Quality Criteria

For the comparison of error rates generated by the different resampling methods, the same data sets are used for all methods. This way, the results become more comparable, and the addition of one more method would be unproblematic. Comparison is based on the following criteria:

– **Mean Relative Deviation**:

$$MRD := \frac{1}{1000} \sum_{P=1}^{100} \sum_{i=1}^{10} \frac{|\hat{e}_i^P - e^P|}{e^P} \rightarrow \text{min!}$$

– **Excess Rate**:

$$ER := \frac{1}{1000} \sum_{P=1}^{100} \sum_{i=1}^{10} I_{\hat{e}_i^P > e^P}, \quad |ER - 0.5| \rightarrow \text{min!}$$

Note that the index P represents the design points, and the index i the replicates. Obviously, the mean relative deviation MRD of estimated error rates $\hat{e}_i^P$ from the true error rates e^P should be minimal. Then, the estimator gives the most reliable error estimates. Moreover, if the excess rate ER is very different from 0.5, then the considered error rate estimator is tending to mainly over- or underestimate the true error rate.

In order to determine the true error rates e^P for each $P = 1, \ldots, 100$, a number of 10^6 new observations for each of the two classes are generated according to the feature values for P. Then, on these new observations the error rate is determined by discretization of the p independent variables. For the discretized variables histograms are estimated. A joint density is estimated by multiplication of the one-dimensional densities given by the histograms for each of the two classes. The true error rate e^P is found by individual decisions for all observations, i.e. we calculate the mean number from all the $2 \cdot 10^6$ new observations for which the corresponding density value of the incorrect class is larger than the one of the correct class. A direct comparison of the two discretized complete joint densities is not easily possible, since we had to calculate values for the number of grid points to the power of the dimension p.

In our simulation the true error rates range from 0% to 26.7%. We ignore 30 out of 1000 starting value combinations with true error rates of exactly 0% for calculation of the MRD, because the latter is undefined in such a case. Then, the smallest true error rate is 0.00005%.

Results

For LDA, simulation results are given in Table 7.4, and for 3NN in Table 7.5. BCV appears to be most reliable with respect to mean relative deviation of error rates (MRD smallest), CV is most symmetric with respect to over- and underestimation (ER nearest to 0.5). This makes clear that different quality criteria might be optimized by different resampling methods. BCV and CV

Table 7.4: Quality of Classification for LDA

LDA	*MRD*	$\|ER-0.5\|$
CV	185.6	**0.187**
.632+	178.4	0.333
BCV	**152.7**	0.315
RCV	190.8	0.220

Table 7.5: Quality of Classification for 3NN

3NN	*MRD*	$\|ER-0.5\|$
CV	244.6	**0.214**
.632+	232.8	0.346
BCV	**213.6**	0.282
RCV	256.0	0.228

yield better estimates than .632+ and RCV in this simulation study in that the latter are never best. In this study, LDA delivers smaller MRDs than 3NN.

7.3.4 *Example: Discrimination of Piano and Guitar*[4]

In **music information retrieval** a typical task is to identify the instruments involved in a recorded piece of music. Here, our task will be the classification of audio signals of single tones into the classes "piano" and "guitar". For this, we will utilize the audio features described in the next subsubsection. In Section 7.3.4.2, we will discuss the data material and the classification procedure based on Linear Discriminant Analysis (LDA), in particular the feature selection from the long list of possibly influential high-level audio features, as well as the classification results. In Section 7.3.4.3 we will discuss hyperparameter tuning for Support Vector Machines (SVM) applied to the same data.

7.3.4.1 *Groups of Features*

Each single analyzed tone has a length of 1.2 seconds and is given as a waveform (wav-) signal with sampling rate (ρ) 44100 Hz and samples x_n, $n \in \{1,\ldots,52920\}$. The signal is framed by half overlapping windows containing 4096 samples each. This results in 25 different windows, except for the absolute amplitude envelope that uses a different window width of 400. Note that piano and guitar tones are assumed to be harmonic in that they consist only of a fundamental frequency and so-called overtones, i.e. multiples of the fundamental frequency (cp. Section 6.4). Other frequencies included in the tone are ignored. Moreover, note that in time series analysis typically two kinds of features are in use: features in time space and features in fre-

[4]Thanks to Markus Eichhoff for providing the data and the basic feature description, and to Bernd Bischl for producing the classification results.

quency space. One possible transformation between these two representations is the Fourier transformation (see below). Here, we will introduce the Absolute Amplitude Envelope (AAE) as an envelope of the observations in time space, whereas all the other features are in frequency space. All four types of features included in the study, and described below for better understanding, are so-called high-level audio features in that they represent musically relevant properties of sound. The two envelopes represent the main structure of the tone in time and frequency space, and PiP and MFCC characterize the timbre of the tone. Note that we have not included the more technical so-called low-level features, like, e.g., the number of zero-crossings in time space representation, in the analysis since the number of possible features of this kind is very big (more than 1000 are in use) and music-related interpretation is often unclear. The following paragraphs defining the features in more detail can be skipped by readers not interested in a deeper understanding of automatic music analysis. However, such readers should also be aware of the deepness of knowledge of signal analysis needed to define such features. Indeed, complicated tasks most of the time need advanced features for good solutions.

Absolute Amplitude Envelope (AAE)
The **absolute amplitude envelope (AAE)** $e_{AAE} \in \mathbb{R}^{132}$ represents the upper and lower shapes of the energy (i.e. amplitude) envelope of a tone using the absolute values $|x_n|$ of the digital wav-signal X and is defined as follows using non-overlapping frames of size 400:

$$e_{AAE} = \left[\max_{1 \leq i \leq 400} \{|x_i|\} \quad \max_{401 \leq i \leq 800} \{|x_i|\} \quad \dots \quad \max_{52800-399 \leq i \leq 52800} \{|x_i|\} \right]^T .$$

A visualization of the absolute amplitude envelope is given in Figure 7.5 for a piano tone.

Pitchless Periodogram (PiP)
The **pitchless periodogram** describes the distribution of overtones of a tone. It is based on the **discrete Fourier transformation** (DFT)

$$DFT_X(k) = \sum_{j=0}^{N-1} x_j e^{-2\pi i \frac{k}{N} j}, \; i := \sqrt{-1},$$

where $X = \{x_1, \dots, x_N\}, N =$ number of time samples, is a given sequence of samples and k/N, $k = 1, \dots, N/2$, the so-called **Fourier frequencies**.

The DFT maps the time signal into the so-called frequency space, i.e. represents it as intensities of involved frequencies. The value $DFT_X(k)$ is obviously a complex number. In order to obtain a real number summary, the corresponding **periodogram** P_X of X is used, i.e. the square of the absolute value of $DFT_X(k)$ divided by N:

$$P_X(k) = \frac{1}{N}\left|\sum_{j=0}^{N-1} x_j e^{-2\pi i \frac{k}{N} j}\right|^2 .$$

Based on estimates of the fundamental frequencies $\hat{f}_0^{w_s}$ per window w_s, $s \in \{1,\dots,25\}$, overtones can be calculated as $\hat{f}_i = (i+1)\cdot\hat{f}_0$, $i \in \{0,\dots,12\}$. Note that we restrict analysis to 13 overtones only.

In order to concentrate on mean behavior in larger blocks, we decided to aggregate the windows in blocks of five. Building medians, the estimated block fundamental frequencies $\hat{f}_0^{b_{\lceil \frac{r}{5} \rceil}}$ and block overtones $\hat{f}_i^{b_{\lceil \frac{r}{5} \rceil}}$ are calculated as

$$\hat{f}_0^{b_{\lceil \frac{r}{5} \rceil}} = \text{median}\left(\hat{f}_0^{w_r}, \hat{f}_0^{w_{r+1}}, \dots, \hat{f}_0^{w_{r+4}}\right) \text{ and } \hat{f}_i^{b_{\lceil \frac{r}{5} \rceil}} = (i+1)\hat{f}_0^{b_{\lceil \frac{r}{5} \rceil}},$$

where $r \in \{1,6,11,16,21\}$ and $i \in \{0,1,\dots,12\}$.

After calculating the block fundamental frequency and the 13 block overtones the **pitchless periodogram (PiP)** $\boldsymbol{p} \in \mathbb{R}^{70}$ is calculated. This periodogram is called **pitchless** because the value of the pitch of the tone, i.e. of its fundamental frequency, is ignored in the representation, only the periodogram heights are considered on an equidistant scale $i \in \{0,\dots,13\}$. This way, the overtone structure is represented on the same scale for all fundamental frequencies. The PiP is defined as follows:

$$p = \left[p_1^{k_0}\ p_1^{k_1}\ \cdots\ p_1^{k_{13}}\ p_2^{k_0}\ \cdots\ p_2^{k_{13}}\ \cdots\ p_5^{k_0}\ \cdots\ p_5^{k_{13}}\right]^T,$$

$$\boldsymbol{p}_{\lceil \frac{r}{5} \rceil}^{k_i} := \text{median}\left(P_{x_{w_r}}(k_i), P_{x_{w_{r+1}}}(k_i), \dots, P_{x_{w_{r+4}}}(k_i)\right),$$

with k_i defined by

$$\left|\hat{f}_i^{b_{\lceil \frac{r}{5} \rceil}} - k_i/4096\cdot\rho\right| = \min_{1\le j\le 2048}\left|\hat{f}_i^{b_{\lceil \frac{r}{5} \rceil}} - j/4096\cdot\rho\right|,$$

where $i \in \{0,1\dots,13\}$, $r \in \{1,6,11,16,21\}$, and ρ denotes the sampling rate.

Thus, a pitchless periodogram block feature is the median over five consecutive windows w_r of the periodogram values $P_{x_{w_r}}(k_i)$ of the Fourier frequencies k_i nearest to the block frequencies $\hat{f}_i^{b_{\lceil \frac{r}{5} \rceil}}$. In the study, overall 70 pitchless periodogram block features are used, i.e. 1 fundamental frequency and 13 overtones for each of the 5 blocks. In Figure 7.6 the pitchless periodogram of one piano and one guitar tone (1st block) can be seen for 10 overtones. A log-transformation of the original pitchless periodogram feature vector is carried out to improve visualization.

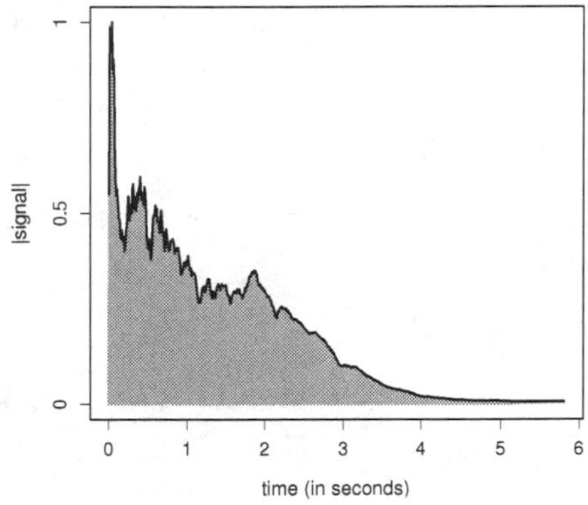

Figure 7.5: Musical signal and its envelope.

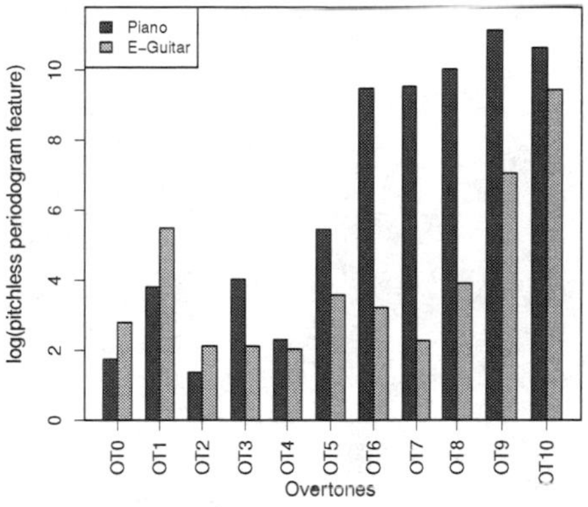

Figure 7.6: Pitchless periodogram, OT = overtone, OT0 = fundamental frequency.

Mel Frequency Cepstral Coefficients (MFCC)

The **Mel frequency cepstrum (MFC)** is a representation of a sound that has proved to be very useful in speech recognition. The MFC is based on a linear cosine transform of a log power spectrum on a nonlinear Mel scale of frequency. **Mel frequency cepstral coefficients (MFCCs)** are the coefficients of an MFC. Let us look in more detail at the definition of the MFCC.

The Mel scale of the frequencies is a perceptual scale of pitches judged by listeners to be equal in distance from one another. The power spectrum is the square of a windowed DFT where the outer parts of the windows are downgraded by Hamming weights (cp. Figure 7.7). The power spectrum is taken after transformation to Mel scale in order to analyze frequencies on a speech-relevant scale. This spectrum is summarized in (overlapping) windows, using triangular weights (MFC filter bank, cp. Figure 7.8). These triangular Mel filters are placed on the frequency axis so that each filter's center frequency follows the Mel scale. The filter bank mimics the critical bands representing the different perceptual effects at different frequency bands. To achieve this, the edges are placed so that they coincide with the center frequencies in adjacent filters. The output of the filter bank is called Mel spectrum. Logarithms are taken in order to avoid extreme peaks. A linear discrete cosine transform is a real-valued analogue of a DFT, used here for mapping the Mel spectrum to Mel frequency cepstrum coefficients. A cepstrum is, thus, in a way a spectrum of a logarithmized Fourier spectrum.

We use the Matlab implementation (Lartillot et al., 2008) of MFCC with 16 MFC coefficients. As in the case of the pitchless periodogram the MFCCs have been evaluated for each window, and the median of each MFCC over each five consecutive time frames has been calculated. Thus, in this study, overall 80 MFCC block features are used, i.e. 16 features for each of the 5 blocks. Figure 7.9 shows an example of the MFCC.

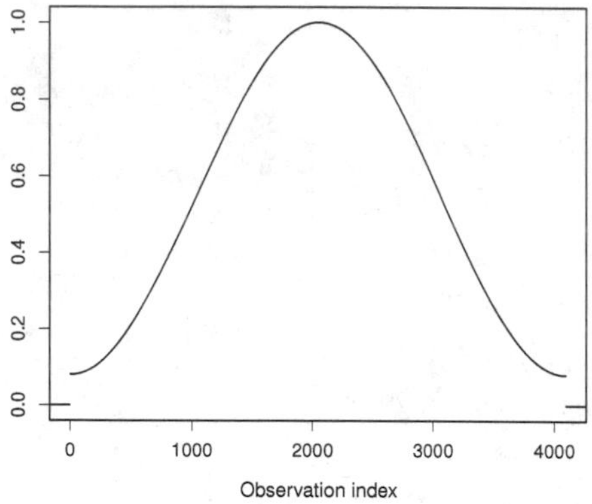

Figure 7.7: Hamming Filter.

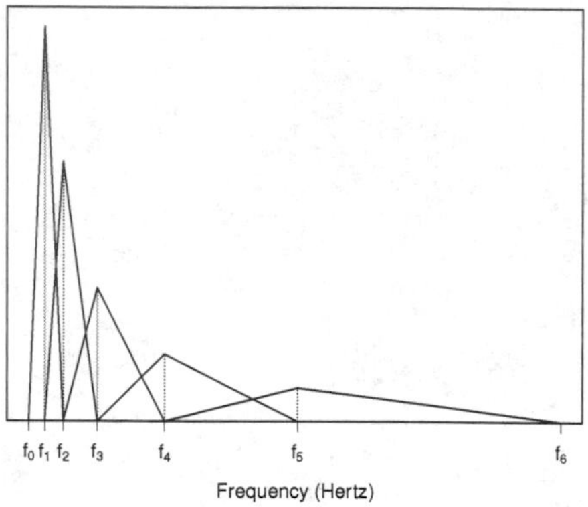

Figure 7.8: MFC triangular filter bank; f_i are the center frequencies on the Mel scale.

LPC Simplified Spectral Envelope

The **Linear Prediction Coding (LPC)** simplified spectral envelope is a smoother of the spectral envelope.

The current value of the signal x_n^s in segment s is estimated by p past samples:

$$\hat{x}_n^s = -a_2^s x_{n-1}^s - a_3^s x_{n-2}^s - \ldots - a_{p+1}^s x_{n-p}^s \text{ with } p = \lfloor 2 + \rho/1000 \rfloor = 46$$

where ρ denotes the sampling rate.

The resulting vector $\boldsymbol{a}^s = [a_2^s \; \ldots \; a_{p+1}^s]^T$ is transformed into frequency space resulting in the following complex 512-point so-called frequency response vector $\boldsymbol{H}^s$, which can be interpreted as a transfer function of the harmonic wave

$z_k = e^{i\omega_k} = \cos(\omega_k) + i \cdot \sin(\omega_k)$ for 512 different frequencies ω_k:

$$H^s(\omega_k) = \left(\sum_{j=1}^{p+1} a_l^s e^{-i\omega_k j} \right)^{-1}, \; s \in \{1, \ldots, 25\}, \; k \in \{1, \ldots, 512\}, \; a_1^s = 1,$$

where the ω_k are 512 equidistant frequencies covering $[0, N/2)$, and $a_l^s, l > 1$, are the above linear prediction coefficients $\boldsymbol{a}^s$. Taking element-wise logarithms of the absolute values yields a matrix $\boldsymbol{K} \in \mathbb{R}^{512\times25}$ with $\boldsymbol{K}_{.,s} = 20\log_{10}|\boldsymbol{H}^s|$, $s \in \{1,\ldots,25\}$. Taking medians in the blocks leads to

$$\boldsymbol{v}^r := \text{median}\left(\boldsymbol{K}_{.,r}, \boldsymbol{K}_{.,r+1}, \boldsymbol{K}_{.,r+2}, \boldsymbol{K}_{.,r+3}, \boldsymbol{K}_{.,r+4}\right)$$

with $r \in \{1,6,11,16,21\}$. This yields the matrix $\boldsymbol{V} = \begin{bmatrix}\boldsymbol{v}^1 & \boldsymbol{v}^6 & \boldsymbol{v}^{11} & \boldsymbol{v}^{16} & \boldsymbol{v}^{21}\end{bmatrix} \in \mathbb{R}^{512\times5}$. The **LPC simplified spectral envelope** $\boldsymbol{s} \in \mathbb{R}^{125}$ is then a reduced version of $\boldsymbol{V}$ taking the maximum of each subsequent 20 rows for all 5 columns of $\boldsymbol{V}$:

$$\boldsymbol{s} = \begin{bmatrix} \max\limits_{1\le j\le 20}\{V_{j,1}\} & \max\limits_{21\le j\le 40}\{V_{j,1}\} & \cdots & \max\limits_{481\le j\le 500}\{V_{j,1}\} \\ \vdots & & & \\ \max\limits_{1\le j\le 20}\{V_{j,21}\} & \max\limits_{21\le j\le 40}\{V_{j,21}\} & \cdots & \max\limits_{481\le j\le 500}\{V_{j,21}\} \end{bmatrix}^T.$$

This leads to the overall 125 LPC block features used in the study, i.e. to 25 LPC block features in 5 blocks each. Figure 7.10 shows an example of the LPC simplified spectral envelope of a piano tone.

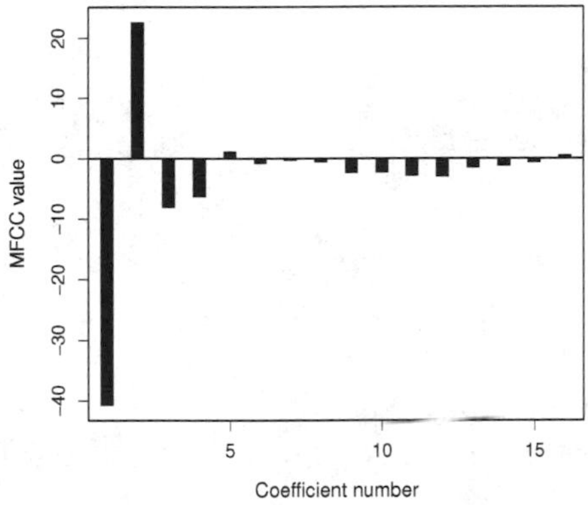

Figure 7.9: MFCC example.

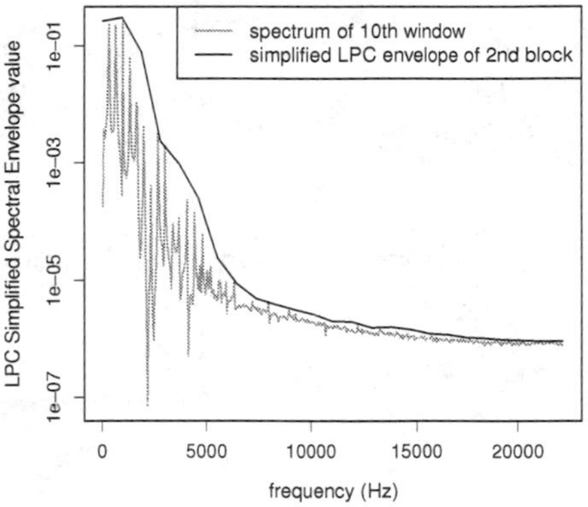

Figure 7.10: LPC simplified spectral envelope.

7.3.4.2 *Feature Selection Results*

Overall, the previous subsubsection leads to 407 numeric non-constant features not including any missing values. The classification problem is binary (piano (1) vs. guitar (0)) with 5654 observations, namely 4309 guitar and 1345 piano tones. The tones were taken from the following three data bases: the McGill (2010) master samples collection on DVD, the RWC music data base (Goto and Nishimura, 2003), and the Iowa musical instrument samples (University of Iowa, 2011).

The goal is to find a simple classifier with as few features as possible for ease of interpretation. To achieve this, we will restrict the training data set to 600 observations only, as in reality only this few labeled data might be available. We will use the MisClassification Error rate (MCE) on the remaining data (test set) as a measure for classification quality.

Experiment for Feature Selection in Piano-Guitar Classification

1. Select Linear Discriminant Analysis (LDA) as a classifier.
2. Subsampling: Randomly select 600 observations from the full data set as a training set for model selection. Retain the rest as a test set.
3. Apply sequential forward search on these 600 observations with LDA. Performance is measured by 5-fold CV and MCE. Search stops when MCE cannot be improved by 0.005. Note: The data partitioning of the CV is held fixed, i.e. is the same for all visited feature sets to reduce variance in comparisons.
4. Store the selected features.
5. Train classifier with selected features on all 600 instances of the training set.
6. Predict classes in the test set and store the test error.
7. Repeat steps (2)-(6) 50 times.

This wrapper procedure thus comprises nested resampling for feature selection with 5-fold CV in inner resampling and 50 iterations subsampling of 600 observations in outer resampling. This way, we get 50 sets of selected features and 50 unbiased MCE values. Nested resampling was chosen in order to be able to show variation in estimates and error rates caused by different (small) training samples.

The results on our data set are summarized in Figures 7.11–7.13. Figure 7.11 indicates, on the one hand, that most often an error around 5.5% is realized. On the other hand, also much larger error rates appear. Figure 7.13 gives an importance ordering of the involved features. Note that only six features were selected in more than 10% of the cases, i.e. more than 5 times. Also note the structure of the feature names starting with the type of feature (AAE, PiP, MFCC, or LPC), followed by the block number and a counter inside the block. Note that for AAE blocks are not used. From this, we see that the eight most important features are MFCC features on different blocks of the time series. A plot of the three most often chosen features is given in Figure 7.12 for a rotation angle best illustrating the linear separation of the two classes in these three features.

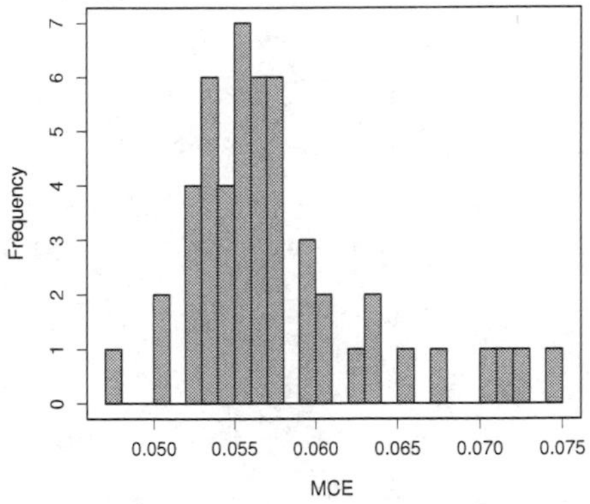

Figure 7.11: Histogram of misclassification rates from outer resampling for feature selection.

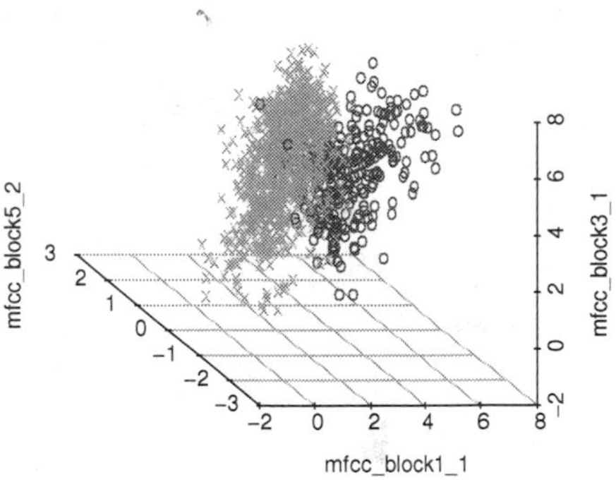

Figure 7.12: 3D plot of best three features (just 1000 samples to improve visualization).

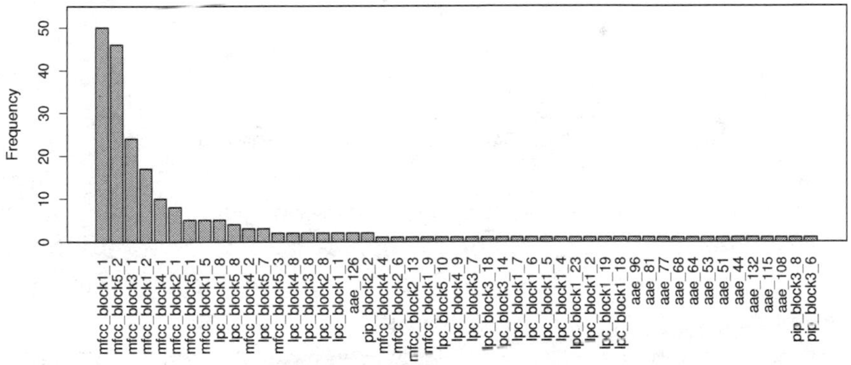

Figure 7.13: Barplot of importance of variables.

7.3.4.3 *Hyperparameter Tuning Results*

In order to demonstrate nested resampling for hyperparameter tuning, we show results for the Support Vector Machine (SVM) with radial basis kernel (see Section 4.6.4). For this SVM we vary the kernel width w and the error weight C.

Experiment for Hyperparameter Tuning in Piano-Guitar Classification

1. Select the Support Vector Machine (SVM) with radial basis kernel as a classifier.
2. Subsampling: Randomly select 600 observations from the full data set as a training set for model selection. Retain the rest as a test set.
3. Apply grid search on all powers of 2 in $[2^{-20}, 2^{20}]^2$ on these 600 observations with SVM. Performance is measured by 5-fold CV and MCE. Note: the data partitioning of the CV is held fixed, i.e. is the same for all grid points to reduce variance in comparisons.
4. Store the hyperparameters w and C with optimum MCE.
5. Train classifier with selected hyperparameters on all 600 instances of the training set.
6. Predict classes in the test set and store the test error.
7. Repeat steps (2)–(6) 50 times.

The results show that the chosen SVM without feature selection is better than LDA with feature selection. In particular, the realized empirical distribution of the MCEs covers almost exclusively lower MCE values than for LDA in the previous subsubsection (see Figure 7.14).

7.3.5 Summary

In this section we applied model selection to **classification problems**. Particularly, we discussed the train-and-test method, doing completely without repetitions for such problems. In a simulation, we compared different error rate estimators. As criteria we utilized not only the correctness of estimated error rates, but also the excess rate assessing whether the error rate estimator is tending to mainly over- or underestimate the true error rate. Finally, in an example on the discrimination of piano and guitar in audio data sets we demonstrated how deep knowledge of the analyzed topic can be utilized to generate excellent solutions of the problem at hand.

7.4 Model Selection in Continuous Models

Now, we switch to model selection in continuous models, i.e. in regression models.

In **regression problems** a typically real-valued response $y \in \mathbb{R}$ has to be predicted by a so-called **regression model** based on n observations of $z_i = [\boldsymbol{x}_i \ y_i]^T$, $i = 1, \ldots, n$, where the vector $\boldsymbol{x}$ summarizes the influential factors.

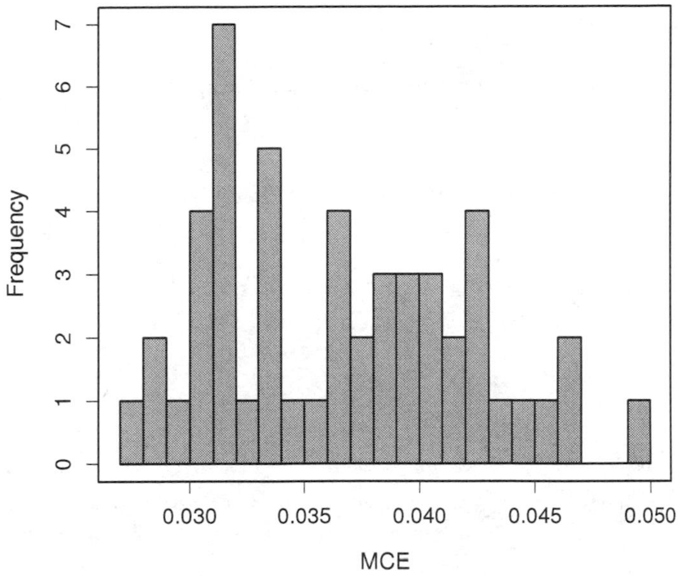

Figure 7.14: Histogram of misclassification rates from outer resampling for SVM hyperparameter tuning.

As with classification models we are interested in selecting models with best predictive power. In order to be fully prepared for the assessment of predictive power in regression models we will start with an introduction into the theory of (nonlinear) predictions in regression models (see Section 7.4.1).

Based on that, we will further discuss two kinds of continuous models: neural nets and PLS models. In Section 4.5.3 we already indicated that an interpretation of coefficients in neural nets should be avoided. For prediction, however, neural nets are very well suited if the activation function satisfies condition 2 (see Section 4.5.3). This will be clarified in Section 7.4.2. In particular, the choice of the number of so-called "hidden nodes" in the neural net is discussed.

The PLS method frequently aims to explain the responses by as few as possible so-called latent variables. The determination of relevant PLS components is explained in Section 7.4.3. Both tasks, the identification of the relevant number of hidden nodes in neural nets and of the relevant number of PLS components, are feature selection problems.

7.4.1 Theory: Nonlinear Prediction[5]

In prediction we distinguish between point prediction and interval prediction of the response for a new observation $x_0 := [x_{01} \; \dots \; x_{0K}]^T$ of influential factors. Let us first introduce the corresponding terms.

If one is interested in an **optimal prediction** of a **response** by means of several **influential factors**, we refer to a **prediction** with **multiple models**.

Definition 7.20: Nonlinear Multiple Statistical Model
A **nonlinear multiple statistical model** is defined by

$$Y = f(X_1, \dots, X_K; \beta_1, \dots, \beta_L; \varepsilon)$$

for a **response** Y dependent on K **influential factors** $X_1, \dots, X_K$ and the unknown coefficients $\beta_1, \dots, \beta_L$ as well as an **error term** ε. The function f is assumed to be at least twice continuously differentiable in all arguments.

After having estimated the unknown coefficients of the model by $\hat{\beta}_1, \dots, \hat{\beta}_L$, the point prediction is obtained by zeroing the error term, assuming that $\mathrm{E}(\varepsilon) = 0$.

Definition 7.21: Point Prediction and Prediction Intervals
The **point prediction** of a response Y for values $x_{01}, \dots, x_{0K}$ of the influential factors $X_1, \dots, X_K$ is defined as

$$\hat{y} := f(x_{01}, \dots, x_{0K}; \hat{\beta}_1, \dots, \hat{\beta}_L; 0).$$

Prediction intervals are intervals around point predictions, which cover the "true" value of the response to be predicted with a certain probability.

The point prediction is thus the value of the deterministic model component for certain values of the influential factors and the estimated values of the unknown coefficients.

For the determination of prediction intervals the distribution of the **prediction error**

$$Y(\boldsymbol{x}_0) - \hat{Y}(\boldsymbol{x}_0) := Y(\boldsymbol{x}_0) - f(x_{01}, \dots, x_{0K}; \hat{\beta}_1, \dots, \hat{\beta}_L; 0)$$

is needed for fixed values $\boldsymbol{x}_0 := [x_{01} \; \dots \; x_{0K}]^T$ of the influential factors $X_1, \dots, X_K$ for the assessment of the variation of the response around the point prediction.

[5]Partly based on Weihs (1987, pp. 49 – 55).

In order to analyze the distribution of the prediction error, it proves to be useful to introduce an auxiliary value

$$\tilde{Y}(\boldsymbol{x}_0) := f(x_{01}, \ldots, x_{0K}; \beta_1, \ldots, \beta_L; 0),$$

which represents the point prediction for known coefficients. With this, the prediction error can be split into two components:

$$\begin{aligned}
Y(\boldsymbol{x}_0) - \hat{Y}(\boldsymbol{x}_0) &= (Y(\boldsymbol{x}_0) - \tilde{Y}(\boldsymbol{x}_0)) + (\tilde{Y}(\boldsymbol{x}_0) - \hat{Y}(\boldsymbol{x}_0)) \\
&= (f(x_{01}, \ldots, x_{0K}; \beta_1, \ldots, \beta_L; \varepsilon_0) - (f(x_{01}, \ldots, x_{0K}; \beta_1, \ldots, \beta_L; 0)) \\
&\quad + (f(x_{01}, \ldots, x_{0K}; \beta_1, \ldots, \beta_L; 0) - (f(x_{01}, \ldots, x_{0K}; \hat{\beta}_1, \ldots, \hat{\beta}_L; 0) \\
&=: \delta_{0\varepsilon} + \delta_{0\beta}.
\end{aligned}$$

Obviously, the first component $\delta_{0\varepsilon}$ represents the prediction error caused by neglecting the model error (ε-component), whereas the second component, $\delta_{0\beta}$, represents the prediction error caused by using the estimated coefficients $\hat{\beta}$ instead of the true coefficients β (β-component).

Assuming independence of the model errors in prediction and the model errors in estimation leads to independency of the two components of the prediction error, since estimation of coefficients only depends on the model errors ε_i, $i = 1, \ldots, n$, and not on the model error ε_0 in the prediction situation, and ε_0 is independent of the ε_i. Therefore, the two components can be independently analyzed, and the variance of the prediction error is the sum of the variances of $\delta_{0\varepsilon}$ and $\delta_{0\beta}$.

In what follows, as often in statistics, we additionally use a **normality assumption:** $\varepsilon_i \sim i.i.\mathcal{N}(0, \sigma^2)$, $i = 0, 1, \ldots, n =$ number of observations.

Linear models
For linear models $y = \boldsymbol{X}\boldsymbol{\beta} + \boldsymbol{\varepsilon}$, $\boldsymbol{X}$ having maximum column rank, we get:

$$\begin{aligned}
\mathrm{var}(\delta_{0\varepsilon}) &= \mathrm{var}(\varepsilon) = \mathrm{var}(Y) = \sigma^2 \qquad \text{and} \\
\mathrm{var}(\delta_{0\beta}) &= \mathrm{var}(\boldsymbol{x}_0^T\boldsymbol{\beta} - \boldsymbol{x}_0^T\hat{\boldsymbol{\beta}}) = \boldsymbol{x}_0^T\,\mathrm{Cov}(\hat{\boldsymbol{\beta}})\boldsymbol{x}_0 = \sigma^2\boldsymbol{x}_0^T(\boldsymbol{X}^T\boldsymbol{X})^{-1}\boldsymbol{x}_0 \\
&= \mathrm{var}(\hat{Y}).
\end{aligned}$$

By summing up the two variances we get
$Y(\boldsymbol{x}_0) - \hat{Y}(\boldsymbol{x}_0) \sim \mathcal{N}(0, \sigma^2[1 + \boldsymbol{x}_0^T(\boldsymbol{X}^T\boldsymbol{X})^{-1}\boldsymbol{x}_0])$,
i.e. we obtain the following well-known representation of the $\alpha \cdot$ *100%* **pre-**

diction interval for linear models:

$$\Big[\hat{Y}(\boldsymbol{x}_0) - t_{n-K;(1+\alpha)/2}\hat{\sigma}\sqrt{1+\boldsymbol{x}_0^T(\boldsymbol{X}^T\boldsymbol{X})^{-1}\boldsymbol{x}_0},$$

$$\hat{Y}(\boldsymbol{x}_0) + t_{n-K;(1+\alpha)/2}\hat{\sigma}\sqrt{1+\boldsymbol{x}_0^T(\boldsymbol{X}^T\boldsymbol{X})^{-1}\boldsymbol{x}_0}\Big],$$

where $t_{n-K;(1+\alpha)/2}$ is the $(1+\alpha)/2$ quantile of the t-distribution with $n-K$ degrees of freedom and

$$\hat{\sigma}^2 := \frac{1}{n-K}\sum_{i=1}^{n}\hat{\varepsilon}_i^2 \quad ,$$

is the estimator for σ^2.
Note that in our notation the constant $X_1 = 1$ would be one of the influencing factors so that, e.g., the number of degrees of freedom is not, as often stated, $n-K-1$, but $n-K$. Also note that in our notation α typically takes values $0.95, 0.90$ so that the quantiles we are interested in can be calculated by $(1+\alpha)/2$, typically leading to corresponding values $0.975, 0.95$.

In what follows we simplify the nonlinear model by only considering **additive model errors**: $Y = f(X_1,\ldots,X_K;\beta_1,\ldots,\beta_L)+\varepsilon$.
For this model we will now derive the variances of the β-component and the ε-component of the prediction error.

Let us start with the **variance of the β-component**. In the linear case, we utilized the unbiasedness and the normality of the least squares estimator $\hat{\beta}$ of β for the determination of $\text{var}(\delta_{0\beta})$ and of the distribution of the prediction error. However, already when the matrix $\boldsymbol{X}^T\boldsymbol{X}$ is not invertible, the least squares estimator is not unbiased anymore. In case of bias, we typically switch to asymptotic statements. Also for nonlinear models we can only assume consistency and asymptotic normality of the estimator.

Definition 7.22: Consistency and Asymptotic Normality
Let $\boldsymbol{\theta}$ be a parameter vector of a statistical model. An estimator $\boldsymbol{t}_n$ of $g(\boldsymbol{\theta}) \in \mathbb{R}^q$ based on n observations is called **consistent** iff for all $\eta > 0$: $P(\|\boldsymbol{t}_n - g(\boldsymbol{\theta})\| > \eta) \to 0$ for $n \to \infty$, i.e. $\text{plim}(\boldsymbol{t}_n) = g(\boldsymbol{\theta})$, where plim stands for the probability limit (see Definition 7.1).

An estimator $\boldsymbol{t}_n$ of $g(\boldsymbol{\theta}) \in \mathbb{R}^q$ based on n observations is called **asymptotically normal** iff there is a sequence of nonsingular matrices $\boldsymbol{A}_n$ and vectors $\boldsymbol{a}_n$ so that $(\boldsymbol{A}_n\boldsymbol{t}_n - \boldsymbol{a}_n)$ converges in distribution to a multivariate normal distribution $\mathcal{N}(\boldsymbol{0},\boldsymbol{\Sigma})$, $\boldsymbol{\Sigma}$ nonsingular, for $n \to \infty$ (see Definition 7.2). In the simplest case $\boldsymbol{a}_n = A_n \cdot \text{E}(\boldsymbol{t}_n)$, where A_n is a scalar $\in \mathbb{R}$.

Let $\hat{\boldsymbol{\beta}}$ be a consistent and asymptotically normal estimator of $\boldsymbol{\beta}$, more precisely let $\sqrt{n}(\boldsymbol{\beta}-\hat{\boldsymbol{\beta}})$ be asymptotically $\mathcal{N}(\mathbf{0},\sigma^2\boldsymbol{B}^{-1})$ distributed, where σ^2 is the variance of the model error ε. Since $\tilde{Y}$ is the same (continuously differentiable) function of $\boldsymbol{\beta}$ as $\hat{Y}$ of $\hat{\boldsymbol{\beta}}$, $\sqrt{n}(\tilde{Y}-\hat{Y})$ is asymptotically $\mathcal{N}(\mathbf{0},\sigma^2\nabla f_0^T\boldsymbol{B}^{-1}\nabla f_0)$ distributed, where $\nabla f_0 := \frac{\partial f}{\partial \boldsymbol{\beta}}(\boldsymbol{x}_0;\boldsymbol{\beta})^T$ is the gradient vector of f with respect to $\boldsymbol{\beta}$ in $\boldsymbol{x}_0$ (see Corollary 7.2). It can be consistently estimated by $\widehat{\nabla f_0} := \frac{\partial f}{\partial \boldsymbol{\beta}}(\boldsymbol{x}_0;\hat{\boldsymbol{\beta}})^T$. Therefore, $\widehat{\mathrm{var}}(\delta_{0\beta}) = \hat{\sigma}^2\widehat{\nabla f_0}^T(n\hat{\boldsymbol{B}})^{-1}\widehat{\nabla f_0}$ is a consistent estimator of $\mathrm{var}(\delta_{0\beta})$ if $\hat{\boldsymbol{B}}$ is a consistent estimator of $\boldsymbol{B}$ and $\hat{\sigma}^2$ is a consistent estimator of the variance σ^2 of the model error (see Corollary 7.1).

What is left for the computation of $\widehat{\mathrm{var}}(\delta_{0\beta})$ is the determination of a consistent and asymptotically normal estimator $\hat{\boldsymbol{\beta}}$ of $\boldsymbol{\beta}$ and the determination of the asymptotic covariance matrix $\boldsymbol{B}$ and of a consistent estimator $\hat{\boldsymbol{B}}$ of $\boldsymbol{B}$.

Fortunately, the **nonlinear least squares estimator** is asymptotically normal under weak regularity conditions with asymptotic covariance matrix $\sigma^2\boldsymbol{B}^{-1}$,

$$\boldsymbol{B} := \lim_{n\to\infty}\frac{1}{n}\sum_{i=1}^{n}\frac{\partial f}{\partial \boldsymbol{\beta}}(\boldsymbol{x}_i;\boldsymbol{\beta})^T\frac{\partial f}{\partial \boldsymbol{\beta}}(\boldsymbol{x}_i;\boldsymbol{\beta}),$$

for the learning sample $\boldsymbol{x}_i, i=1,\ldots,n$ if the limit exists and is invertible. Then

$$\hat{\boldsymbol{B}} := \frac{1}{n}\sum_{i=1}^{n}\frac{\partial f}{\partial \boldsymbol{\beta}}(\boldsymbol{x}_i;\hat{\boldsymbol{\beta}})^T\frac{\partial f}{\partial \boldsymbol{\beta}}(\boldsymbol{x}_i;\hat{\boldsymbol{\beta}})$$

is a consistent estimator for $\boldsymbol{B}$ (see, e.g., Jennrich (1969)) and

$$\hat{\sigma}^2 := \frac{1}{n-L}\sum_{i=1}^{n}\hat{\varepsilon}_i^2 \quad .$$

is a consistent estimator of the variance σ^2 of the model errors.

Altogether, this leads to the following estimator for the $\boldsymbol{\beta}$-component of the prediction error:

$$\begin{aligned}\widehat{\mathrm{var}}(\delta_{0\beta}) &= \hat{\sigma}^2\widehat{\nabla f_0}^T(n\hat{\boldsymbol{B}})^{-1}\widehat{\nabla f_0}\\ &= \hat{\sigma}^2\frac{\partial f}{\partial \boldsymbol{\beta}}(\boldsymbol{x}_0;\hat{\boldsymbol{\beta}})[\sum_{i=1}^{n}\frac{\partial f}{\partial \boldsymbol{\beta}}(\boldsymbol{x}_i;\hat{\boldsymbol{\beta}})^T\frac{\partial f}{\partial \boldsymbol{\beta}}(\boldsymbol{x}_i;\hat{\boldsymbol{\beta}})]^{-1}\frac{\partial f}{\partial \boldsymbol{\beta}}(\boldsymbol{x}_0;\hat{\boldsymbol{\beta}})^T.\end{aligned}$$

Let us now consider the determination of the **variance of the ε-component** of the prediction error. Since $Y = f(X_1,\ldots,X_K;\beta_1,\ldots,\beta_L)+\varepsilon$, we have $\delta_0 = \varepsilon_0$, and analogous to the linear model:
$\mathrm{var}(\delta_0) = \mathrm{var}(\varepsilon) = \mathrm{var}(Y) = \sigma^2$.
Therefore, $\hat{\mathrm{var}}(\delta_0) = \hat{\sigma}^2$ is a consistent estimator of $\mathrm{var}(\delta_0)$.

Linear Model (cont.)
In the case of a **linear model** we get
$Y = f(X_1,\dots,X_K;\beta_1,\dots,\beta_L)+\varepsilon = [X_1 \ \dots \ X_K]\boldsymbol{\beta}+\varepsilon$. Therefore:
$\frac{\partial f}{\partial \boldsymbol{\beta}}(\boldsymbol{x}_i;\boldsymbol{\beta}) = \frac{\partial f}{\partial \boldsymbol{\beta}}(\boldsymbol{x}_i;\hat{\boldsymbol{\beta}}) = (x_{i1} \ \dots \ x_{iK}) = \boldsymbol{x}_i^T$, $i = 0,1,\dots,n$.
In this notation, we then get

$$\boldsymbol{X} = \begin{bmatrix} \boldsymbol{x}_1^T \\ \vdots \\ \boldsymbol{x}_n^T \end{bmatrix}.$$

Therefore, we have:

$$\boldsymbol{X}^T\boldsymbol{X} = [\boldsymbol{x}_1 \ \cdots \ \boldsymbol{x}_n]\begin{bmatrix} \boldsymbol{x}_1^T \\ \vdots \\ \boldsymbol{x}_n^T \end{bmatrix} = \sum_{i=1}^{n} \boldsymbol{x}_i\boldsymbol{x}_i^T = n\hat{\boldsymbol{B}} \quad .$$

Obviously, this leads to the above representation of the variance of the prediction error.

Overall, we have proven the following form of the **prediction interval for the nonlinear model with additive model error**:

Theorem 7.3: Prediction Interval for Nonlinear Model
For the nonlinear model with additive model error
$Y = f(X_1,\dots,X_K;\beta_1,\dots,\beta_L)+\varepsilon$, the $\alpha \cdot$ *100%* **prediction interval** of a response Y for fixed values $x_0 := [x_{01}\dots x_{0K}]^T$ of the influential factors $X_1,\dots,X_K$ is given by

$$\Bigg[\hat{Y}(\boldsymbol{x}_0) - t_{n-L;(1+\alpha)/2}\hat{\sigma}\sqrt{1+\widehat{\nabla f}_0^T(n\hat{\boldsymbol{B}})^{-1}\widehat{\nabla f}_0},$$
$$\hat{Y}(\boldsymbol{x}_0) + t_{n-L;(1+\alpha)/2}\hat{\sigma}\sqrt{1+\widehat{\nabla f}_0^T(n\hat{\boldsymbol{B}})^{-1}\widehat{\nabla f}_0}\Bigg],$$

with the estimated gradient vector $\widehat{\nabla f}_0 := \frac{\partial f}{\partial \boldsymbol{\beta}}(\boldsymbol{x}_0;\hat{\boldsymbol{\beta}})^T$, the nonlinear least squares estimator $\hat{\boldsymbol{\beta}}$ of $\boldsymbol{\beta}$, the estimated asymptotical covariance matrix

$$\hat{\boldsymbol{B}} := \frac{1}{n}\sum_{i=1}^{n}\frac{\partial f}{\partial \boldsymbol{\beta}}(\boldsymbol{x}_i;\hat{\boldsymbol{\beta}})^T\frac{\partial f}{\partial \boldsymbol{\beta}}(\boldsymbol{x}_i;\hat{\boldsymbol{\beta}})$$

assumed invertible, the $(1+\alpha)/2$ quantile $t_{n-L;(1+\alpha)/2}$ of the t-distribution with $n-L$ degrees of freedom, and the estimated error variance

$$\hat{\sigma}^2 := \frac{1}{n-L}\sum_{i=1}^{n}\hat{\varepsilon}_i^2 .$$

The $\alpha \cdot 100\%$ **prediction limits** are the upper and lower interval limits of the $\alpha \cdot 100\%$ prediction interval.

7.4.2 *Model Selection in Neural Nets: Size of Hidden Layer*[6]

Let us now apply this to nonlinear prediction with neural nets. Please recall the form of the corresponding nonlinear model:

$$\begin{aligned}
Y &= \alpha_0 + \sum_{i=1}^{d} \alpha_i g(\beta_i^T X + \beta_{i0}) + \varepsilon =: f(X;\Theta) + \varepsilon, \text{ where} \\
\boldsymbol{X} &= [X_1 \dots X_K]^T \text{ is the vector of input signals,} \\
\beta_i &= [\beta_{il} \dots \beta_{iK}]^T \text{ is the vector of the weights of the input signals for the ith node of the hidden layer,} \\
\alpha &= [\alpha_1 \dots \alpha_d]^T \text{ is the vector of the weights of the nodes of the hidden layer for the output,}
\end{aligned}$$

α_0 is an overall constant, and

ε is a random variable with expected value 0.

Obviously, the vector of $L = 1 + 2d + dK = 1 + (2+K)d$ unknown **model coefficients** of this model has the form:

$$\Theta = [\alpha_0 \dots \alpha_d \ \beta_{10} \dots \beta_{d0} \ \beta_1^T \dots \beta_d^T]^T.$$

In Section 4.5.3 we have derived conditions for which

$$\boldsymbol{B}_E := E\left[\frac{\partial f}{\partial \Theta}(\boldsymbol{X};\Theta)^T \frac{\partial f}{\partial \Theta}(\boldsymbol{X};\Theta)\right]$$

is invertible.

Since the nonlinear least squares estimator is asymptotically normal with covariance matrix

$$\boldsymbol{B} := \lim_{n\to\infty} \frac{1}{n} \sum_{i=1}^{n} \frac{\partial f}{\partial \Theta}(\boldsymbol{x}_i;\Theta)^T \frac{\partial f}{\partial \Theta}(\boldsymbol{x}_i;\Theta)$$

and the mean converges to the expected value (strong law of large numbers), $\boldsymbol{B}$ is invertible as well as, asymptotically,

$$\hat{\boldsymbol{B}} := \frac{1}{n} \sum_{i=1}^{n} \frac{\partial f}{\partial \Theta}(\boldsymbol{x}_i;\hat{\Theta})^T \frac{\partial f}{\partial \Theta}(\boldsymbol{x}_i;\hat{\Theta}).$$

---

[6]Partly based on Cross et al. (1995) and Hwang and Ding (1997).

Therefore, assuming the conditions of Section 4.5.3 hold, we can use the above form of the prediction interval for neural networks as well.

For the case that the **structure of the neural network** is fixed, i.e. in our case the number d of nodes in the hidden layer, the $\alpha \cdot 100\%$ prediction interval was given in the previous Section 7.4.1. However, how do we choose the number of nodes in the hidden layer? We will give two criteria that might be used. Both criteria maximize aspects of **prediction quality**.

Definition 7.23: Prediction Quality Criteria
The **coverage** of the prediction interval should be as close to $\alpha \cdot 100\%$ as possible, i.e. the prediction interval should, if possible, cover exactly $\alpha \cdot 100\%$ of the distribution of $Y(\boldsymbol{x}_0)$.

The **length of the prediction interval** should be as small as possible, i.e. the uncertainty about the location of the true value of $Y(\boldsymbol{x}_0)$ should be as low as possible.

We can use **(leave-one-out) cross-validation** (cp. Section 7.2.3) for prediction quality evaluation. Thus, each observation of the learning sample is held out individually as a test sample so that n neural nets with fixed structure (= number of nodes in the hidden layer) are estimated based on $n-1$ observations each, where for all response variables **point predictions** for the hold-out observation are built. In this way, for each realized value combination of the influential factors we have both true and predicted values of responses available and a corresponding **prediction interval**. These are used to determine the **prediction quality measures** coverage and mean interval length.

The **optimal neural net** for a response Y can be constructed by means of cross-validation as described in Algorithm 7.7.

Example 7.3: Neural Net for Electric Load Prediction
In the paper of Hwang and Ding (1997) an example is given for predictions with a neural net for the electrical load at noon on Tuesdays using the system loads and the temperatures available at 8:00 a.m. that day. Seven influential factors were used, namely 3 loads and 4 temperatures. The sample size was $n = 341$. Table 7.6 shows covering percentages and mean prediction interval lengths for 1–4 nodes in the hidden layer using the logistic activation function. The nets with 2 or 3 nodes are good candidates for practical application. The choice is dependent on which criterion is preferred.

Algorithm 7.7 Construction of Optimal Neural Net

$q_0 \leftarrow 0$
$l_0 \leftarrow \infty$
$d \leftarrow 1$
loop
  Evaluate neural net with d hidden units by means of cross-validation.
  $q_d \leftarrow$ coverage for model, i.e. the percentage of true response values in the $\alpha \cdot 100\%$-prediction interval.
  $l_d \leftarrow$ mean length of prediction interval
  if $|q_{d-1} - \alpha \cdot 100\%| \leq |q_d - \alpha \cdot 100\%|$ **and** $l_{d-1} \leq l_d$ **then**
    return the neural net with the best criteria.
  end if
  $d \leftarrow d + 1$
end loop

Table 7.6: Prediction Quality Criteria for Different Structures

Number of Nodes (d)	1	2	3	4
coverage for $\alpha = 0.9$	90.6%	**89.7%**	89.1%	86.5%
coverage for $\alpha = 0.95$	**93.5%**	**93.5%**	92.1%	90.9%
mean length for $\alpha = 0.9$	269.1	213.1	**207.4**	368.8
mean length for $\alpha = 0.95$	320.6	254.0	**247.1**	439.4

7.4.3 Model selection for PLS: Selection of Latent Variables[7]

Let us, finally, apply feature selection (see Section 7.2.4) to dimension reduction in the space of PLS components, i.e. to the selection of relevant PLS components. The aim is to explain the response variables with as few as possible latent variables, i.e. PLS components. As a quality criterion, we propose another measure of prediction quality, namely the so-called cross-validated coefficient of determination R^2_{cv}.

Definition 7.24: Predictive Power based on Cross-Validation
For a multivariate linear model with response variables Y_j, $j = 1, \ldots, M =$ number of response variables, the **predictive power** for response Y_j is defined by the **cross-validated coefficient of determination** R^2_{jcv} (R_j^2 **cross-**

[7] Partly based on Weihs and Jessenberger (1998, pp. 170 – 172).

validated), based on the **prediction errors**

$$\hat{v}_{ij} = y_{ij} - \hat{y}_{ij} \quad i = 1, \ldots, m,$$

m being the number of observations, where

$$R^2_{jcv} := 1 - \frac{RSS_{jcv}}{\sum_{i=1}^m (y_{ij} - \bar{y}_{ij})^2}, \quad \text{with } RSS_{jcv} := \sum_{i=1}^m \hat{v}_{ij}^2,$$

y_{ij} = ith observation of the jth response variable,

$\hat{y}_{ij} = \boldsymbol{z}^i \hat{\boldsymbol{a}}_j(i)$ = point prediction of the ith observation of the jth response variable,

$\boldsymbol{z}^i$ = ith row of matrix $\boldsymbol{Z}$ of the scores of the PLS-components,

$\hat{\boldsymbol{a}}_j(i)$ = jth column of the matrix of coefficients $\boldsymbol{A}$ for the PLS-components Z based on all observations except the ith.

Note that we have defined the cross-validated coefficient of determination individually for each response variable. This way, we allow for individual dimension reduction for each response. For each response variable, based on greedy forward selection, we can include PLS components into the **prediction model** as long as a suitable **functional of all the cross-validated coefficients of determination** $R^2_{jcv}, j = 1, \ldots, M$, increases. For example, such a functional can be the mean, the median, or the minimum or the maximum of the individual coefficients.

Note that the criterion for the construction of PLS components is different from the cross-validated coefficient of determination. In particular, for the multivariate case, PLS components are not chosen individually for the different responses but jointly for all responses together. Therefore, it might happen that for prediction the components are not chosen in the ordering given by the numbering of the components.

Example 7.4: Dyestuff Production

We now continue Example 5.2 where we calculated the PLS-components on the basis of correlations and considered the following models for the two response variables:

$$\begin{aligned} HUEREM(AL) - \overline{HUEREM(AL)} &= \beta_1 PLS1 + \varepsilon \text{ and} \\ HUEREM(AL) - \overline{HUEREM(AL)} &= \beta_1 PLS1 + \beta_2 PLS2 + \varepsilon, \end{aligned}$$

where PLS1 and PLS2 are the first and second PLS components, respectively.

Table 7.7: Goodness of Fit and Predictive Power for PLS Components

	HUEREM		HUEREMAL	
	PLS1	PLS1+2	PLS1	PLS1+2
R^2_{cv}	**0.46**	**0.59**	**0.73**	**0.79**
R^2	0.47	0.62	0.74	0.81

Table 7.7 extends Table 5.1 and shows a comparison of the goodness of fit and the predictive power of the two models for the two response variables. Obviously, the improvement of predictive power from only including the 1st to also including the 2nd PLS component is higher for HUEREM.

Let us now decide on the number of PLS components to be used for optimal prediction of the hues. We applied the variable selection method to optimize the predictive power to the PLS components. As a result, for HUEREMAL the first three components were chosen in the order PLS1, PLS2, PLS3. With these three PLS components the predictive power is 0.805 whereas with one more PLS component the predictive power slightly shrinks to 0.804. For HUEREM, six components were chosen in the order PLS1, PLS2, PLS3, PLS4, PLS17, PLS15. Altogether, this leads to a predictive power of 0.6217, whereas with one more PLS component the predictive power only slightly shrinks to 0.6215.

7.4.4 *Summary*

In this section we applied model selection to **regression problems**. In order to derive quality criteria for continuous models, we introduced the theory of nonlinear prediction for such models. To determine the prediction intervals, the distribution of the **prediction error** was studied. Coverage and length of such intervals are used as quality criteria in an application to neural nets aiming at the determination of the size of hidden layers. As another quality criterion, predictive power is defined by cross-validation in an application to the PLS method aiming at the determination of the number of latent variables. Please note that prediction quality should not be measured by goodness of fit, since this would be the resubstitution quality, known to be too optimistic. Only for many observations is goodness of fit a good approximation to predictive power.

7.5 Implementation in R

As a general starting point, CRAN offers the task view "Machine Learning & Statistical Learning" by Hothorn (2013). It lists a multitude of packages for statistical learning models and associated methods. Among these are tree-based methods (packages rpart (Therneau et al., 2013), C50 (Kuhn et al., 2013a), and randomForest (Liaw and Wiener, 2002)), neural networks (package nnet (Venables and Ripley, 2002)), support vector machines (packages e1071 (Meyer et al., 2012) and kernlab (Karatzoglou et al., 2004)), boosting (packages adabag (Alfaro-Cortes et al., 2012) and gbm (Ridgeway, 2013)), an interface to all algorithms in the popular WEKA Java (Hall et al., 2009) software for machine learning, and many more.

Many of these packages already include procedures to cross-validate their implemented models, but most of the time only this specific resampling procedure is provided and the concrete splits cannot be predefined by the user. Exceptions to this rule are the packages caret (Kuhn et al., 2013b) and DMwR (Torgo, 2010), which use a more generic approach and allow more convenient model comparisons. DMwR, for example, offers the usual hold-out, subsampling, and bootstrapping procedures, in addition to cross-validation, and furthermore allows statistical model comparisons through integrated hypothesis tests. Basic generic hyperparameter tuning (mainly by grid search) is included in the package caret and by the function `tune()` in package `e1071`.

For feature selection, a larger number of feature filtering algorithms is provided in the FSelector (Romanski, 2013) package, while Boruta (Kursa and Rudnicki, 2010) offers a specific wrapper algorithm for the selection of all relevant features. For many models, feature importance measures are available (which could also be used for feature filtering), e.g., see the function `importance` in the package randomForest for a popular example. The package caret again tries to generalize this principle to a greater number of classification and regression algorithms.

7.6 Conclusion

Overall, in this chapter, we demonstrated how repetitions can be utilized to simulate different samples from one learning data set only. In particular, we introduced how such repetitions can be utilized in a general framework for model selection being a crucial task in applied statistics. Also, in this chapter we introduced resampling-based prediction quality criteria for the two most important supervised learning tasks, discrimination and regression. Naturally, we discussed only a small excerpt from the tremendous amount of literature on this topic. Let us stress again that we do not want to give an overview of all important statistical methods, but that we want to give an introduction to basic ideas of statistical algorithms, in this section of repetition in order to simulate repeated sampling from a population whose distribution is not known.

7.7 Exercises

Exercise 7.2.1: We consider the Leave-One-Out (LOO) estimator of the error rate for a data-independent rule. Consider i.i.Bernoulli(0.5) distributed labels $Y \in \{0,1\}$ as training data (ignore the influential factors). Let $Y_1, \ldots, Y_n$ be the training data. Let our data-independent rule be of the following form: If the number of $1s$ in the training data is odd, then constantly predict 1 for new data, else 0.

a. What is the true expected error rate of this rule?
b. Let us estimate this expected error rate by means of LOO. What is the expectcd value and the variance of this estimator?
c. IIow do you interpret these results?

Exercise 7.2.2: Consider the two-sided one-sample t-test has not been invented, we have lots of data, and we go for the test statistic $b := |\bar{x} - \mu|$. Implement a test procedure based on the bootstrap that evaluates the test statistic 200 times, reports the test decision about μ being the correct expected value, and calculates a bootstrapped 95% confidence interval for μ.

Exercise 7.3.1.1: Consider a classification problem with two classes where influential data stem from two independent normal distributions with $\mu_1 = 0, \mu_2 = 1$, and $\sigma_1 = \sigma_2 = 1$, and the a priori probabilities π_1 and π_2 of the two classes are equal.

a. Plot the two density functions $\Phi(x|\mu_1;\sigma_1^2)$ and $\Phi(x|\mu_2;\sigma_2^2)$ in a joint diagram.
b. Compute the a posteriori probabilities of the two classes and plot them in a joint diagram.
c. Formulate the Bayes rule under equal misclassification costs for the two classes. Plot the corresponding decision limit into the above diagrams.
d. Calculate the error rate of the Bayes rule.
e. Now assume that $\pi_2 = 4/5$, i.e. that the a priori probability of class 2 is higher. Plot the two functions $\pi_1\Phi(x|\mu_1;\sigma_1^2)$ and $\pi_2\Phi(x|\mu_2;\sigma_2^2)$ as well as the two a posteriori probabilities in a joint diagram, correspondingly.

f. How does the optimal classification rule change? Draw the optimal decision limit into the above diagrams. How does the error rate change?

Exercise 7.3.1.2: Let us analyze the well-known Iris data set (data available in R, just type `iris`, or from the exercises section under `http://www.statistik.tu-dortmund.de/fostal.html`) with 50 examples each for three different types of Iris plants (Iris setosa, Iris versicolour and Iris virginica). Four features are collected describing length and width of sepals and petals.

a. Plot a scatterplot matrix. Assess the data concerning their linear separability. Which combination of features appears to be most suited for linear separation?
b. Determine the error rate of 1NN by means of 5-fold cross-validation. Interpret the result.
c. Also determine the error rate for kNN with $k = 3, 5, \ldots, 79$ by means of 5-fold cross-validation. Compare all the results in a table and in a scatterplot of k vs. error rate. Interpret the result.

Exercise 7.3.1.3: Consider the artificial 2-class classification problem with the training set "orange.train" and the test set "orange.test" (available from the exercises section under `http://www.statistik.tu-dortmund.de/fostal.html`) containing the class variable ("class") and 10 influential features $F1 - F10$ (see the "Skin of an orange" example in Hastie et al. (2001, pp. 384 – 385)). In R the data can be generated by

```
library("ElemStatLearn")
orange.train <- orange10.test[[1]]
orange.test <- orange10.test[[2]]
```

a. Train an LDA on the training set and calculate the test data error rate.
b. Determine which features are useful for class separation (feature selection). Use both the wrapper strategy in Section 7.2.4 and graphics as, e.g., a scatterplot matrix or histograms of single features distinguishing the two classes.
c. Is it sensible to apply LDA? What kind of separation do you expect from the graphics?

Exercise 7.4.1: Based on the PLS implementation in Exercise 5.1.1 apply the greedy feature selection in Section 7.2.4 to all 18 PLS-components of Example 5.2 for the two response variables HUEREM and HUEREMAL together.

Compare the results to the results of Examples 5.2 and 7.4. For the calculation of the PLS components use the data available from the exercises section under `http://www.statistik.tu-dortmund.de/fostal.html`.

Chapter 8

Scalability and Parallelization

8.1 Introduction

In Chapter 2 we studied what is computable and how much effort it takes to compute certain things. For this we used the Turing machine, an abstract model of a computer that **sequentially** executes a sequence of instructions stored on a tape. Real-world central processing units (CPUs) in computers are conceptually similar. They sequentially read instructions and data from memory, process them, and write them back to memory. Increasing the execution speed of such a machine amounts to increasing the number of instructions or data words read and processed per second.

In this chapter we will study the empirical scalability of statistical algorithms and especially how the availability of parallel computing resources has changed and will continue to change the way we develop and deploy statistical methods. In the previous chapters of this book, we considered the scalability of algorithms only on a theoretical level by characterizing the number of basic operations performed by the algorithm as a function of the size of the input. Some algorithms, such as those in Chapter 2, have similar theoretical runtime characteristics, but scale quite differently in practice.

It is important to stress that when we study the scalability of an algorithm in this chapter, we will usually need to restrict ourselves to a concrete implementation of that algorithm because even different implementations of the same algorithm can have widely different runtime characteristics. This is partially due to the fact that most algorithms do not specify every detail needed for the implementation, so that programmers implementing the algorithm are free to choose different strategies to realize these details. Another factor that can have an influence on the runtime of an algorithm is the language it is written in and the compiler or interpreter used to run the program. When studying the empirical scalability of an algorithm, this should always be kept in mind. But before we dive into detailed discussion of the programming challenges posed by modern computing systems, we should step back and ask ourselves why a statistician should have a basic working knowledge of scalability, parallel programming, and high-performance computing; after all, there are specialized computer scientists that work in this field.

There are two main reasons why statisticians should know more about parallel computing and scalability. First, we are now living in the so-called **information age** which has brought upon us large and unstructured data sets from which we derive our information. In an ideal world, a statistician would be involved in all parts of the data gathering process. This would allow him or her to apply methods from the field of design of experiments to minimize the amount of data gathered and to ensure that the data gathered contains all the required information to answer the question at hand. This is not always the case, even if the questions asked nowadays are much more complex to answer than those asked 10 or 20 years ago. Tackling these jobs has only recently become possible with the widespread availability of computers and the associated development of more complex and powerful statistical algorithms. So statisticians should know which algorithms they can apply to what types of data set and how long a statistical analysis will approximately take given a limited amount of computing resources.

Second, the development of new statistical methods has shifted away from studying designs that are purely grounded in theory to methods that incorporate and possibly even actively build upon the vast computing resources available today. A good example of this is the new breed of ensemble learning methods being developed that actively exploit the availability of parallel computing resources by training many classifiers or regression models in parallel on, possibly smaller, data sets and then combining these models into a more powerful predictor. A nontrivial example of this is the cascading SVM classifier(see Graf et al., 2004).

Both of these reasons should of course not hinder a statistician from consulting a specialist in the field. Often times a different approach using a different algorithm to solve the same problem can lead to even greater speedups. Always remember that using the techniques described in this chapter we cannot change the underlying complexity of the algorithm but only hope to change the constants of the implementation of the algorithm. It is therefore always advisable to research into more efficient algorithms.

The remainder of this chapter is structured as follows. We start by giving a history of the development of high-performance and parallel computing systems, and then we introduce the concept of high-throughput computing, which is a simplified model of parallel computing for homogeneous or heterogeneous clusters of computers that requires no changes to the underlying statistical algorithm. When modifying the algorithm to run efficiently on a parallel machine is an option, high-performance computing comes into play. We discuss several different paradigms of parallel computing and show how these can be applied to some of the algorithms we have studied so far. We conclude the chapter with a short outlook on what the future might hold for statistical programmers. This includes the use of General Purpose Graphics Processing Units (GPGPU) or other special purpose processors for certain analytical tasks.

8.2 Motivation and History

Much of the technology underlying parallel computing today has existed for many years as part of the infrastructure of high-performance computing (HPC), but only the last couple of years, with the widespread availability of multicore CPUs, has brought these technologies into the mainstream. This in turn has made them available to statisticians, who are not necessarily experts in computer science or HPC.

The benefits of this trickling down of parallel computing technology are twofold. First, it has allowed statisticians to work on larger data sets or more complex analysis than previously possible, and second, statisticians can now explore ideas and algorithms that had previously been considered infeasible due to their, at that time, overwhelming computational costs. One area where this applies in particular is in Bayes statistics. Here one often needs to sample from or integrate over complex distributions, which was not possible in the past due to the limited availability of computing resources.

8.2.1 Motivation

The scalability of a purely sequentially operating computer is limited by, among other things, the speed of light. Why is this so? Because we cannot transmit information faster than the speed of light, and neither can we shrink computers down in size infinitely. One nice way to look at this is the approximate distance light travels during 1 nanosecond, which is equivalent to 1 GHz.[1] This is approximately 1 foot. So no matter what we do, a processor running at 1 GHz cannot move information around by more than 1 foot during one instruction cycle. This is one of the reasons why processor clock speeds have not continued to increase like they did in the 1980s and 1990s. In fact, clock speeds have not increased much since 2000, and from 2005 to

[1] A cycle time of 1 nanosecond is equivalent to a clock frequency of 1 GHz.

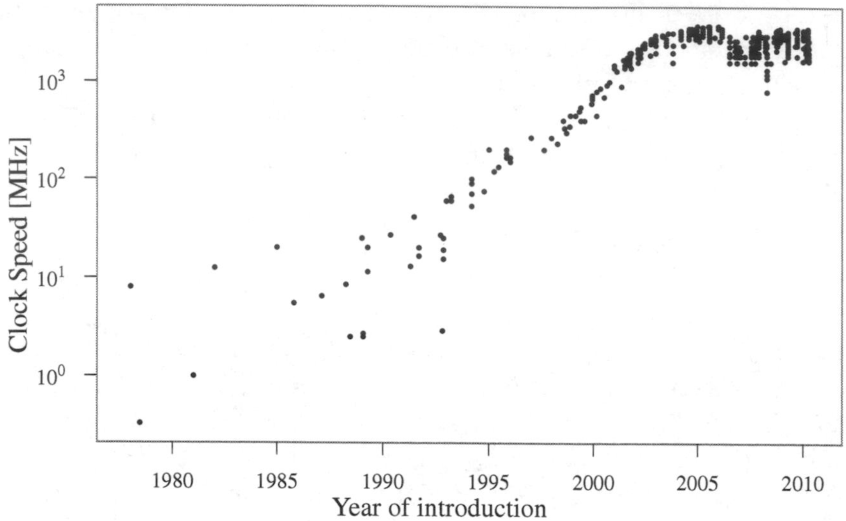

Figure 8.1: Year of introduction and clock frequency of a selected subset of Intel CPUs.

2007 even decreased on average (see Figure 8.1). But processors, with a few exceptions,[2] have not been getting slower.

Why is this? For one thing, most manufacturers of processors started to introduce extensions to their instruction sets. In the case of the x86 architecture those were the MMX instruction set introduced by Intel in 1996, the SSE1 to SSE4 instruction sets introduced in 1999, 2001, 2004, and 2007, and the now obsolete 3Dnow! instruction set introduced by AMD in 1998. So how do these instructions differ from regular instructions, and why do they improve performance? All these instruction sets have one thing in common: their instructions do more work by applying the same operation to multiple pieces of data in parallel. They are called **single instruction multiple data** (SIMD) type instructions. As the name implies, instead of performing one operation on a single datum, they perform the same operation on several data points **in parallel**. This is something fundamentally new compared to the model of computation we have studied so far. Previously we had assumed that the machine on which we run our algorithms behaves just like the Turing machine we studied in Chapter 2.

[2]One notable exception is Intel's Atom processor, which is not optimized for speed but for power consumption.

For modern processors this is certainly not the case, but something else happened in the last years. Computer manufacturers started placing multiple processing cores onto a single processor die. These so-called multicore systems have become ubiquitous. They can run multiple instruction streams in parallel. Therefore, both data parallelism and task parallelism have become available to almost all branches of computing. Both of these forms of parallelism have been studied in the field of HPC. We will therefore give a brief overview of the history of supercomputing and high-performance computing.

8.2.2 Early Years

8.2.2.1 Hardware

The early history of supercomputing is tightly interwoven with the early history of computing. Many early computers were also the supercomputers of their time because there were not that many computers and the computers that were in use were usually used to tackle computational problems that could not be solved by hand. A good example of this is the Colossus computer built during World War II to break the FISH cipher. But it was not a Turing complete machine (cp. Section 2.2.1); it was purpose-built to solve one problem and one problem only. The first fully electronic computer built that is Turing complete was the ENIAC. It was designed in 1943 to calculate artillery tables for the U.S. Army. Building the machine took until 1946 and it was not until 1947 that it was actually in use by the U.S. government. A good account of the early history of computing is given in Rojas and Hashagen (2002).

We jump ahead now into the early 1960s, when the introduction of the transistor had revolutionized the way computers are built. At this time the Control Data Corporation (CDC) was the market leader for supercomputers, building the fastest machines in the world. It had been set up by a few former employees of the Sperry Corporation. The most notable of these was Seymour Cray, a name that is now synonymous with high-performance computing. His ideas and designs would shape the way supercomputers were built for the next 30 years. Arguably the first supercomputer introduced by CDC was its model 6600. It had a central processor running at 10 MHz combined with 10 auxiliary processors that dealt with I/O. The successor to the 6600 was the model 7600, which ran at 36 MHz and was able to perform about 36 million floating-point operations per second (MFLOPS).

After the design of the 7600, trade began work on a model 8600, but this was never completed. A competing design team at CDC had started work on what would become the STAR-100 and funding for the 8600 project was cut.

At this time Cray left CDC to form his own company—Cray Research. The STAR-100 was introduced in 1974 and was again the fastest computer of its time, with an aggregated processing speed of 150 MFLOPS. Two years after the introduction of the STAR-100, the first computer from Cray Research, the Cray-1 was introduced. It provided an aggregated processing speed of 250 MFLOPS. Within a time span of 10 years the speed of the fastest computer had therefore increased 10-fold. Both the Cray-1 and the STAR-100 are examples of vector processing machines. These processors could be told, with a single instruction, to fetch several long vectors of numbers from memory, add or multiply these and write the result back to memory. This style of computing can be considered a precursor to the SIMD instructions found in almost all modern microprocessors.

We have left out one other supercomputer of that era; this is the ILLIAC IV, built by the University of Illinois for the Burroughs Corporation. It used a radically different design than the STAR-100 or Cray-1. By combining many small and simple processing engines using a sophisticated mesh network, the ILLIAC IV was able to achieve a sustained processing speed of 150 MFLOPS. However, it was quite difficult to program at the time, since the problem had to be decomposed in order to profit from the parallel processing capabilities of the machine.

The 1980s saw a boom in the supercomputing business. Many new companies entered into the market, each using a slightly different approach by combining different levels of parallelization and vector processing in their machine. Some of these companies were Thinking Machines, NEC, Hitachi, and Convex.

There was also a new breed of computers that evolved during this time frame, the workstation and the personal computer. Especially later model workstations from manufacturers like Sun, SGI, HP or DEC are conceptually not very different from scaled-down supercomputers. Since they were sold in much higher volume, it became increasingly difficult for companies such as Cray to justify the cost of developing their own microprocessors. Instead, many manufacturers began to use common off-the-shelf components to build high-performance systems. The thing setting them apart from a cluster of workstations was the high-speed interconnect between the different nodes of the system. There were two possible configurations for this type of machine. It could be a shared memory system, where each processor had access to all of the system memory, or it could be a distributed memory system, where each processor or small cluster of processors had a certain amount of local memory available and in order to access nonlocal memory, the processor would need to query the other nodes so that the data could be sent to the querying node over the interconnect.

8.2.2.2 Software

At this point, it might be a good idea to go back and look at the development of the supporting software ecosystem that happened during this time frame. Just as the hardware was highly specialized in the early years of high-performance computing, so was the software usually tailored to the problem at hand. This was in part necessitated by the different architectures and floating-point formats of the machines and partially caused by the lack of libraries or collections of common subroutines. Adding to this problem, no single programming language had gained enough traction in the numerical computing realm and oftentimes users were forced to write critical sections of their code in non portable assembly language to squeeze every last drop of performance out of the available hardware. By the end of the 1960s, however, FORTRAN had established itself as the dominant numerical programming language. In turn, this allowed the development of larger bodies of portable FORTRAN code that encapsulated standard numerical procedures so that in the mid to late 1970s, in short succession EISPACK (Smith et al., 1976), a collection of routines for dense eigenvalue problems, LINPACK (Dongarra et al., 1979), a library of linear algebra routines and MINPACK (Moré et al., 1980), a suite of nonlinear optimization routines were published free of charge by the Argonne National Laboratory. These libraries were designed to be portable across a wide range of machines, from smaller DEC PDPs up to and including the large Cray and CDC machines.

The LINPACK routines were also the first larger software package to use the Basic Linear Algebra Subroutines (BLAS, see Lawson et al., 1979), which had been developed at NASAs Jet Propulsion Laboratory and released to the public in 1979. The initial BLAS version (later termed BLAS Level 1) contained many kernels for vector-vector type operations. The idea behind BLAS was that each manufacturer or computing site could provide their own optimized copy of the routines for a higher-level library, such as the LINPACK package, to use. They would therefore all profit from any new optimizations done to these low-level routines. This idea was so powerful that to this day, the BLAS are still a de facto standard application programming interface (API) for basic vector and matrix computations, and all the major companies that build microprocessors or complete supercomputers provide their own optimized copy of the BLAS. Over time the initial BLAS release of vector-vector operations was extended with the so-called Level 2 routines of matrix-vector

Table 8.1: List of All First Placed Computers on the TOP500 List

Date	Machine	# cores	R_{max}	R_{peak}
1993-06 – 1993-06	CM-5/1024	1024	59	131
1993-11 – 1993-11	Num. Wind Tunnel	140	124	235
1994-06 – 1994-06	XP/S140	3680	143	184
1994-11 – 1995-11	Num. Wind Tunnel	140	170	235
1996-06 – 1996-06	SR2201/1024	1024	220	307
1996-11 – 1996-11	CP-PACS/2048	2048	368	614
1997-06 – 1998-11	ASCI Red	7264	1068	1453
1999-06 – 1999-06	ASCI Red	9472	2121	3154
1999-11 – 2000-06	ASCI Red	9632	2379	3207
2000-11 – 2000-11	ASCI White	8192	4938	12288
2001-06 – 2001-11	ASCI White	8192	7226	12288
2002-06 – 2004-06	Earth-Simulator	5120	35860	40960
2004-11 – 2004-11	BlueGene/L	32768	70720	91750
2005-06 – 2005-06	BlueGene/L	65536	136800	183500
2005-11 – 2007-06	BlueGene/L	131072	280600	367000
2007-11 – 2007-11	BlueGene/L	212992	478200	596378
2008-06 – 2009-06	Roadrunner	122400	1026000	1375780
2009-11 – 2010-06	Jaguar	224162	1759000	2331000
2010-11 – 2010-11	Tianhe-1A	186368	2566000	4701000
2011-06 – 2011-06	K computer	548352	8162000	8773600
2011-11 – 2011-11	K computer	702024	10510000	11280400
2012-06 – 2012-06	Sequoia	1572864	16324800	20132700
2012-11 – 2012-11	Titan	560640	17590000	27112500
2013-06 – ...	Tianhe-2	3120000	33826000	54902400

Note: R_{max} is the maximum attained speed on the LINPACK benchmark in GFLOPS and R_{peak} the theoretical maximum, again in GFLOPS. All values are rounded to the nearest GFLOP. The table is current as of November 2013.

operations in 1984 and support for matrix-matrix operations was added to the BLAS in 1988 as the Level 3 functionality. Much of the history of these collections of routines is given in Dongarra (2005).

The LINPACK routines are also the basis of another de facto standard in the supercomputing world. They are used as a benchmark to rank machines for the TOP500 (Meurer and Dongarra, 2010) list of fastest supercomputers of the world. This list has been released every June and November since 1993, and it tracks the speed of the fastest computers on earth by measuring the time it takes them to solve a system of dense linear equations using the LINPACK routines. From that the number of floating-point operations per second performed is derived, and that number is used to rank the systems. Additionally, a theoretical maximum for the number of floating-point operations that can be performed in one second is determined for each machine by counting the number of floating-point operations a single processor can complete in one instruction cycle multiplying that by the clock frequency of the processor and the number of processors in the system. Combining these two values, allows one to judge the efficiency of the machine.

Table 8.1 list all machines that at one time were ranked first in the TOP500 list. Looking at the maximum attained speed on the LINPACK benchmark, we see that the number of floating-point operations performed per second has grown exponentially over time. This can also be seen in Figure 8.2, where apart from the best system, the 10th ranked, 125th ranked, 250th ranked, and 500th ranked systems are also shown. From that figure, we can also see that it takes approximately 10 years for a system to drop out of the TOP500 list or put another way, a systems rank doubles approximately every year. A somewhat dated, but still accurate, more in-depth analysis of the TOP500 data is performed by Feitelson (1999).

One final software package that needs to be mentioned is the LAPACK (Anderson et al., 1999) set of highly-portable routines developed as a replacement for LINPACK and EISPACK and introduced in 1990. By incorporating new design ideas to better utilize parallel machines the newer LAPACK routines tend to achieve a higher efficiency than their older counterparts from both the EISPACK and the LINPACK collection of routines. This was achieved by using not just the Level 1 BLAS routines but also the newer Level 2 and Level 3 routines, thereby expressing matrix-vector and matrix-matrix type operations more naturally and offering both the compiler and the BLAS implementer more freedom to optimize these routines. This collection of high-level linear algebra routines is to this day the de-facto standard in high-performance computing and widely used in all types of numerical software. It has undergone two major revisions in its lifetime. Version 2 was initially released in 1994, and version 3 was released in 1999. Apart from the software routines, the LAPACK team also provides an extensive manual (Anderson et al., 1999) for the software which includes a whole chapter on the numerical accuracy and stability of the included routines.

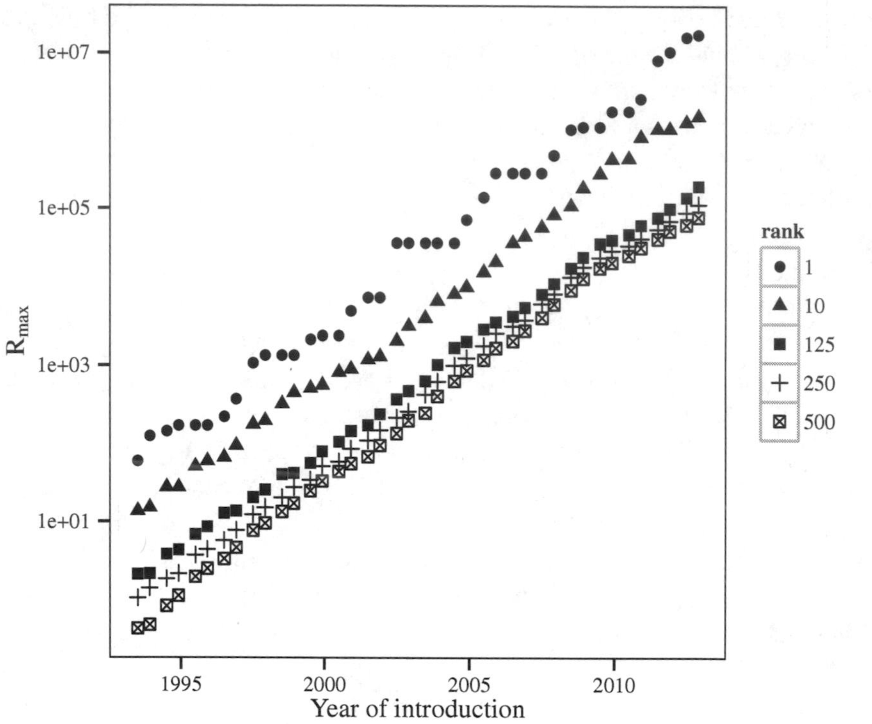

Figure 8.2: Development of R_{max} of the TOP500 supercomputers over time. Maximum achieved processing speed of each system on the LINPACK benchmark (in GFLOPS) is plotted against the year and month the list was published.

8.2.3 Recent History

Looking at the TOP500 list for the last 10 years shows a clear trend in supercomputing. Instead of building highly specialized vector processors, as was the case in the 1970s, 1980s, and early 1990s, nowadays, off-the-shelf processors or derivatives of common processors are used to build very large clusters on which the computations take place. IBM took this idea to the extreme with their BlueGene series of supercomputers introduced in 2004. These were built from low power PowerPC 440 processors that had been modified so that several processing cores would fit on a single die with the associated networking peripherals. Not only did these chips run at a, for their time, moderately low clock frequency of initially 700 MHz, but they were also very energy efficient compared to other designs of that era. The current, as of 2012, leader of the

TOP500 list is a U.S. system named Titan, which combines general purpose CPUs with specialized GPUs, in this case Nvidia Tesla cards, to achieve its speed of more than 17.5 trillion floating-point operations per second. The idea of using specialized coprocessors to speedup calculations is by no means new, but the availability of low-cost and high-performance graphics processors that can be repurposed for scientific calculations is dramatically changing scientific computing. Because of their low cost and availability, GPUs are used not only in high-end supercomputers, but also in ordinary desktop PCs to, for some calculations, drastically cut computation times.

In the previous section we described the push for standardized libraries of numerical routines in the early years of supercomputing. This trend has continued to this day, but the APIs in use have not changed much. In fact, the BLAS and LAPACK routines are still in common use. Instead of designing new libraries, software development has focused on optimizing these routines for each architecture. With the advent of massively parallel machines with many thousands, if not tens or hundreds of thousands, of processors another area had to be standardized. Since most of these systems are distributed memory parallel computers, a standardized API to transfer data from one node in the cluster to another had to be specified. Initially there were two main contenders in this area, on the one hand there was the Message Passing Interface (MPI, see Message Passing Interface Forum, 2009) API and on the other hand we had the Parallel Virtual Machine (PVM, see Geist et al., 1994) API. Development of the latter has largely ceased and it seems that MPI has won the "race". Almost all clusters and supercomputers provide an implementation of the MPI for programmers to use.

This concludes our brief tour of the history of scientific computing and supercomputing in particular. We have focused our attention on the U.S. development of the supercomputing scene, but similar developments can be traced back in Japan and Central Europe. Especially Japan is a traditionally strong market for vector processing. Both NEC and Hitachi have a long history of building vector processors and supercomputers using these vector processors. Not much is known about the developments of supercomputer technology in Soviet Russia. What is known today is that the focus of the Russian industry was somewhat different than in the West. Instead of creating ever faster microprocessors, the Russian scientists spent more time researching compilation strategies, which would allow them to better utilize their existing computing infrastructure. Building custom-tailored processors for certain tasks that encoded the algorithm in hardware instead of software was also more popular in the USSR, especially among the Russian military. Finally we point out that in our discussion of the available numerical software we have focused on freely available libraries and subroutine collections. There are quite a few established commercial numerical libraries that provide similar functionality. The two most prominent examples are the NAG and IMSL libraries.

8.2.4 Summary

This section gave a brief review of the historical development of scientific computing. It highlighted some of the major milestones along the way: the first electronic computing machines, the development of vector processors, the introduction of the PC, and the advent of cheap and ubiquitous microprocessors, leading to massively parallel computers. In parallel, the software used matured and standardized around a core set of APIs. Of most interest to statisticians are the BLAS and LAPACK libraries for basic matrix and linear algebra computations. Finally, we hinted at the future, which will see a rise in the use of GPUs and other special purpose hardware to tackle the computing challenges of tomorrow.

8.3 Optimization

After our overview of the high-performance computing history, and before we dive into the different paradigms of high-performance computing we should briefly touch on the topic of optimization. Optimization in this context is not related to the subject covered in Chapter 4. Instead, we want to improve the runtime of our code not by switching to a faster algorithm, but rather by making the algorithm's implementation more efficient. Given that the exercises of this book are meant to be solved using the programming environment R, we will focus on tips and techniques that are applicable to this environment.

8.3.1 Examples in R

Because R is an interpreted language, it is important that each statement that is run by the interpreter do as much work as possible. R provides many facilities to express loops and other repetitive types of statements as so-called vector operations. The clever use of the `sapply`, `lapply`, and `mapply` functions and the vectorized versions of addition, multiplication, subtraction, and division can often lead to dramatic performance increases. However, prematurely optimizing code blocks can lead to a code base that is difficult to understand and maintain. It is therefore important to thoroughly test each optimization, document it, and make sure that it has no detrimental side effects. One area in particular that needs to be looked after in R is the increase in memory usage that the vectorization of a given code block incur.

Example 8.1: Mean Calculation
Assume for a minute, that the `mean` function did not exist in R. Here are four different possible implementations.

One based on a simple `for` loop:

```
for_mean <- function(x) {
  sum <- 0
  for (xx in x) {
    sum <- sum + xx
  }
  sum / length(x)
}
```

One using vector-vector multiplication (`%*%`):

```
cp_mean <- function(x)
  drop(rep(1, length(x)) %*% x / length(x))
```

One functional approach using `Reduce`:

```
reduce_mean <- function(x)
  Reduce(`+`, x) / length(x)
```

And finally, one using `sum`:

```
sum_mean <- function(x)
  sum(x) / length(x)
```

Which one would you consider to be the fastest? Are any faster than `mean`? How is the runtime influenced by the length of `x`? See Exercises 8.3.1 and 8.3.2.

Using more specialized functions is another area of possible code optimizations in R. Let us again look at a very simple example of this.

Example 8.2: Vector of Integers of `length(x)`
Given an R vector `x`, generate the vector of integers from 1 to `length(x)`. Here are a few possibilities:

```
1:length(x)
seq(1, length(x))
seq.int(1, length(x))
seq_len(length(x))
seq_along(x)
```

On a 3 GHz Intel Core2Duo machine running Windows XP, R 2.13.0

and with `length(x)` = 10,000, the following median of 100 runtimes [3] is measured:

Expression	Time [ns]	Rank
`1:length(x)`	16996.5	3
`seq(1, length(x))`	37001.0	5
`seq.int(1, length(x))`	18804.5	4
`seq_along(x)`	16122.0	2
`seq_len(length(x))`	13067.5	1

Three things are surprising. First, the intuitive expression `1:length(x)` is not the fastest approach. It is beat by both `seq_along(x)` and `seq_len(length(x))` expressions. Secondly, `seq_along(x)` is slower than `seq_len(length(x))`. Since `seq_along` is the most specialized function, we might have assumed that it would be the fastest approach. And finally `seq.int(1, length(x))` is noticeably slower than `1:length(x)`, although both functions[4] provide similar functionality.

Both of the previous examples have been somewhat contrived to simplify the presentation. It was fairly clear what we wanted to optimize. This is generally not the case. We might have a code base with many hundreds of lines of code distributed over tens of files and functions. Which ones are worth optimizing? There are two ways to tackle this problem:

1. Look at the structure of the source code and identify basic blocks or functions that are called in tight loops. Look for code that is vectorized and operates on large data sets. These building blocks are worthwhile targets for optimizations.
2. Profile the code. Use `Rprof` and `Rprofmem` functions of R to identify hotspots in the code base. However, sometimes the output of the profiler might be misleading. If we do not run the program with realistic input, it might behave completely different. Remember that we are trying to fine-tune an **implementation** for a certain workload.

After manually or automatically identifying the hotspots of the code, we can go about optimizing them. The first step should always be a review of the algorithm used. Is there a better-fitting algorithm for our use-case? Remember that Bubble Sort might be faster than Quick Sort if the input is small for example. On the other hand, some algorithms might break down if the input size becomes too large. If we think a different algorithm could significantly

[3]Runtimes measured using the `microbenchmark` package.

[4]Yes, : is a regular R function; try `‘:‘(1, 10)` as an example.

benefit us, we need to implement it and then empirically determine if and to what degree it speeds up the calculation. This step is crucial. We should not believe theoretical promises about an algorithm's superiority. Instead, we must verify that it leads to the same numerical results even for worst case input (cf. Chapter 3), and that the speed gains are substantial enough to warrant the change.

If we still need to improve the execution speed, we can profile our new code once more and now attempt some of the micro-optimizations detailed above by comparing the speed of different implementation strategies. Do not expect to gain much in this step. Improvements of 10% to 20% might be possible, but more is rare. We should only invest time into scaling out an algorithm at this stage if we truly see a benefit in such a comparatively small reduction of the execution time. The trade-off here is one of productivity versus performance. If we spend more time optimizing our code than we save later on when running our experiment, we gain nothing.

Should more speed be required, we now have two options. We can either reimplement our algorithm in a compiled language that is generally orders of magnitude faster than R for some operations or we can try to run our algorithm in parallel on multiple CPUs or even computers. Parallelization will be dealt with in depth in the next section so we will not discuss it here, but rewriting core functions in C or FORTRAN might buy us the speedup we are looking for. But this is a two-edged sword; on the one hand, we potentially decrease runtimes and on the other we increase the complexity of our software system. If we are already using vectorized R functions the expected improvement is not that large because internally the vector operations are already implemented in efficient C code. Larger gains can be expected for operations that are not easily vectorized, such as `for` or `while` loops whose loop body is not independent.

8.3.2 Guidelines

We conclude this section with a list of practical advice to increase the scalability of your code. While it is again focused on R projects, much of it applies to Matlab, Python or any other interpreted language as well.

1. Have tests in place to ensure that optimizations do not decrease numerical accuracy or stability.
2. Have code in place to accurately time your code with realistic input. Realistic input is important because you will want to tune your implementation to a certain type of input. Do not make the mistake to always use the same

input. You will be overly optimistic about the results you achieved. This is akin to the generalization error we wanted to estimate in Chapter 7.

3. Set your goal regarding the desired decrease in runtime and the amount of time you are willing to invest to achieve such a decrease.
4. Make sure you know the hotspots of your code. Use profilers or manual source code review to find them and verify that indeed these are the code parts where most of the execution time is spent.
5. Investigate if a change of algorithms might be all you need. Often this will lead to the most drastic improvements in performance.
6. Vectorize your hotspots if possible. Vectorization lets the interpreter do more work per evaluated expression. This is for the most part a good thing because the vectorized code is usually written in a compiled language and heavily optimized. Also, any new optimization of a vectorized primitive immediately benefits your code. Make sure you weight in the possible increase in memory consumption when vectorizing code.
7. If all else fails, consider implementing core functions in C or FORTRAN. This may look daunting at first, but remember that you do not need to write a full-blown C program, and most of the mundane tasks, such as reading data in from a file, checking arguments etc. can be done in R. You only need to implement the CPU-intensive parts in C. See also Exercise 8.3.6.

By following the above advice, you should be able to solve most scalability problems. If, on the other hand, you are in the situation that your data sets are so large or your computations so complex that even after optimizing your code you feel the expected runtime is still too high, you may want to explore parallelization as another way to improve runtime. In the next section we will explore some of the many ways that we can take a sequential program and turn it into a parallel one.

8.3.3 Summary

We started our discussion of optimization of program code with a motivating example in R. From the observations made, we looked at other examples where we can increase the efficiency of our software at little or no cost. It turned out that some of the small changes required were entirely non-intuitive. This led to a set of guidelines for program optimization. Even though code can often be sped up using many of these (micro)optimizations, we should heed the following warning:

> We should forget about small efficiencies, say about 97% of the time: premature optimization is the root of all evil.
>
> **Donald Knuth**

8.4 Parallel Computing

If after optimizing our implementation according to the guidelines of the previous section our code is still not fast enough, we can literally throw hardware at the problem by using not just one CPU or computer, but many, to solve the problem. This is called **parallel computing** and has a long history in the field of HPC. As we saw in the section on the history of supercomputing, the idea of using multiple CPUs to tackle a problem is almost as old as the field of supercomputing itself. However, it was not until modern microprocessors started to contain more than one CPU core that the general public took notice of this development. Parallel computing used to be a specialized field for a select few algorithm designers and engineers who worked on some of the toughest computing problems around. Nowadays almost anyone has access to a few CPUs, and it is not uncommon for a department at a university or a medium-sized firm to have a compute cluster containing tens, maybe even hundreds, of CPUs that is available to anyone wishing to use it. So what types of statistical problems might benefit from parallel computing, and what kind of reductions in runtime can we expect?

8.4.1 Potential

At first, parallel computing might look like a silver bullet to solve all our resource and runtime problems, but we must remember that if we throw n CPUs at a problem, the runtime will, at best, be reduced by a factor of n:

$$t_{\text{parallel}} = \frac{t_{\text{serial}}}{n}.$$

For this to be the case, we have to assume that **everything** in our algorithm can be done in parallel. This is obviously never the case, although there are certainly algorithms that come very close, like bootstrapping or subsampling. So the first thing to remember is that using more compute nodes will not necessarily decrease computation times. If we assume a more realistic model for our algorithm, namely that a proportion π_{parallel} can be parallelized and the remainder $\pi_{\text{serial}} = 1 - \pi_{\text{parallel}}$ has to be executed sequentially, then we arrive at Amdahl's Law (Amdahl, 1967).

Theorem 8.1: Amdahl's Law
Given an algorithm of which a portion π_{parallel} can be executed in parallel

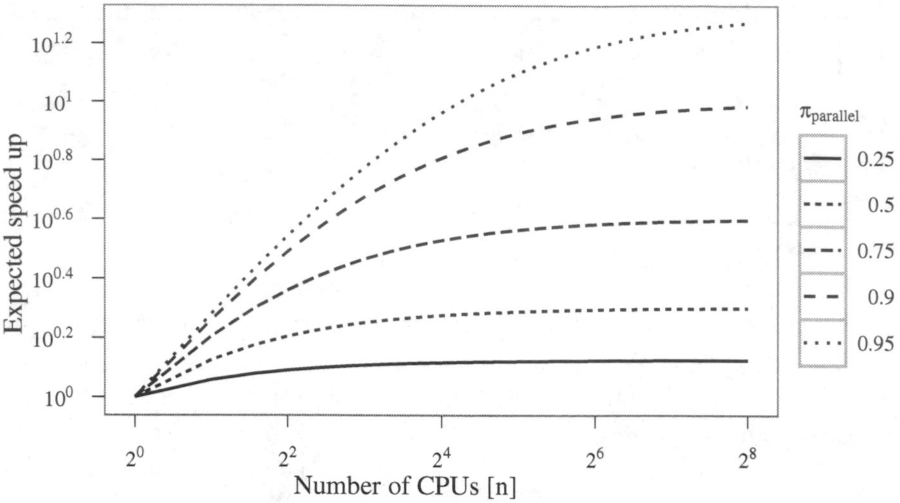

Figure 8.3: Amdahl's law for different values of π_{parallel}. Note the logarithmic axes.

and the remainder $\pi_{\text{serial}} = 1 - \pi_{\text{parallel}}$ has to be executed sequentially, then the maximum speedup a computer with n CPUs can achieve relative to a sequential computer is given by

$$S_{\text{Amdahl}}(\pi_{\text{parallel}}) = \frac{1}{(1 - \pi_{\text{parallel}}) + \frac{\pi_{\text{parallel}}}{n}} = \frac{1}{\pi_{\text{serial}} + \frac{\pi_{\text{parallel}}}{n}}.$$

The expected speedup according to Amdahl's law is shown in Figure 8.3 for different values of π_{parallel}. It is clear that even for moderately large values of π_{parallel}, say 0.75, even throwing hundreds of CPUs at the problem will not yield a 10-fold increase in processing speed. This is one of the reasons why optimization should be the first step when scaling an algorithm. It is especially crucial to optimize serial parts of a program or reduce the amount of work done in the sequential sections.

There are arguments to be made why more CPUs might still be better and one of them goes like this. Amdahl's law only applies if the input to the algorithm is fixed; i.e., we do not increase the data set we wish to handle. So what happens if we consider π_{parallel} and π_{serial} to be functions of the input size k. The expected speedup is then given by

$$S_{\text{Gustafson}}(k) = \frac{1}{(1 - \pi_{\text{parallel}}(k)) + \frac{\pi_{\text{parallel}}(k)}{n}} = \frac{1}{\pi_{\text{serial}}(k) + \frac{\pi_{\text{parallel}}(k)}{n}}.$$

If we assume, as is reasonable, that $\pi_{\text{parallel}}(k)$ is monotonically increasing in

k, then as our problem size increases, the time we spend in the parallel part of the algorithm increases, and we therefore have potentially more use for further processors. This model was first proposed by Gustafson (1988). It is therefore often referred to as **Gustafson's law**.

Let us look at how Gustafson's law applies to a very simple algorithm, namely, calculating the distance matrix that is used in a hierarchical clustering.

Example 8.3: Distance Matrix
Given a data set `data` with k observations, the Algorithm 8.1 calculates the $k \times k$ distance matrix D that contains all pairwise distances between two observations:

Algorithm 8.1 Distance Matrix Calculation

Require: Data set `data` with k observations $\mathtt{data}_1$ to $\mathtt{data}_k$.

for $i = 1$ **to** $k-1$ **do**
  for $j = i+1$ **to** k **do**
    $D_{ij} \leftarrow \text{distance}(\mathtt{data}_i, \mathtt{data}_j)$
  end for
end for
return D

The body of the outer `for` loop of this algorithm is independent in the sense that it may be executed in parallel for all i. Even the inner `for` loop could be parallelized once the value of i is known. So on a machine with $(k-1)(k-2)/2$ processors the execution time of this algorithm could be reduced to the time it takes to execute one distance calculation.

In Example 8.3 we glossed over one detail while studying the scalability of the algorithm. If we were to indeed run the algorithm on multiple CPUs, each CPU would need access to the data set and would have to be able to write into the distance matrix variable. In practice, when running on multiple CPUs that might not even be part of the same machine, there is an overhead associated with this read and write access. For some types of algorithms, this one included, this **communication overhead** can become overwhelming. It is an additional factor we need to consider when talking about practical scalability.

That concludes our introductory thoughts on the theoretical limits of parallelization. We have seen that there are limits to what can be achieved by throwing hardware at a problem. We have also seen that optimizing our code before we think about parallel computing is an important step since the sequentially executing parts of our algorithm will ultimately limit our scalability. Next we will describe the setup we will assume for the rest of this chapter so that we can look at some practical examples of parallel computing.

8.4.2 Setting

For the next sections we will need to fix some context in which we want to run parallel computations. In the context of this book, we will not deal with extreme scalability to many hundreds or thousands of processors. Instead, we will assume that we have access to a moderately sized cluster with tens, maybe even hundreds, of nodes. Each node will contain several processors each with several cores. One node might be a two processor machine with four cores per processor for a total of eight cores per node. The nodes should be networked with regular gigabit Ethernet or some other form of inexpensive interconnect. We do not assume a low latency interconnect, such as Infiniband or Myrinet, for the following, although it might help for some problems. We do not make any assertions about the amount of memory per node; we only assume that all cores on a node have access to all memory of that node and that separate nodes have separate address spaces, meaning they cannot directly access each other's memory. All nodes should have access to a shared file system although we do not explicitly need this. It makes it easier since we do not have to worry about the distribution of our programs and data to each node. The easiest way to provide such a file system would be NFS, but Lustre and other parallel file systems might yield better performance.

On the software side we will restrict ourselves to nodes running Linux or some other Unix-like operating system. This includes Mac OS X, but not Windows with Cygwin. Windows is a world of its own when it comes to HPC and while almost all the software mentioned later does exist in one form or another for Windows, it is much more common to see clusters running under Linux. The cluster should have a working MPI implementation and may have some sort of batch system, such as Torque, Slurm, or Platform LSF running on it. We will not cover the use of such systems, although their use is common and makes job scheduling and resource allocation much easier and more transparent. Finally, we assume R is available on all cluster nodes.

8.4.3 Data Parallel Computing

In the previous sections we have looked at some rather trivial algorithms and how these might be optimized. In this section we will look at an algorithm of moderate complexity and how we can reduce its runtime using techniques from parallel computing. The algorithm in question is the k-means algorithm, one possible iterative method to find a solution to the k-means clustering problem.

8.4.3.1 *Example: k-means Clustering*

Given n observations $\boldsymbol{x}_1, \boldsymbol{x}_2, \ldots, \boldsymbol{x}_n$, the goal of the k-means algorithm is to find cluster centers or **means** $\boldsymbol{m}_1$ to $\boldsymbol{m}_k$ so that the sum of the squared distances between each observation and its closest mean becomes minimal. We can then cluster the data by viewing the means as our cluster centers and forming clusters by collecting all observations that are closest to the cluster center in one cluster. This problem is NP-hard[5] if we use the L_2 metric no matter what value of k we choose. There are, however, heuristics to find "good" means. One of them is the k-means algorithm. It uses the EM framework of Chapter 5 by first choosing k means at random, then assigning each observation to its **expected** cluster if the means were the true cluster centers. Next, we recalculate the cluster centers as the mean of all members of the cluster. This is the **maximization** step. A sequential version of the algorithm is given formally given in Algorithm 8.2.

Algorithm 8.2 Sequential k-means Algorithm

Require: Number of clusters k
Require: Observations $\boldsymbol{x}_1$ to $\boldsymbol{x}_n$
Require: $k \leq n$
Choose k initial means $\boldsymbol{m}_1$ to $\boldsymbol{m}_k$
Initialize cluster sets c_1 to c_k as empty
while not converged **do**
  for $i = 1$ to n **do**
    $j \leftarrow \arg\min_j \|\boldsymbol{x}_i - \boldsymbol{m}_j\|$
    $c_j \leftarrow c_j \cup \{i\}$
  end for{Expectation step}
  for $j = 1$ to k **do**
    $\boldsymbol{m}_j \leftarrow |c_j|^{-1} \sum_{i \in c_j} \boldsymbol{x}_i$
    $c_j \leftarrow \emptyset$
  end for{Maximization step}
end while
return $\boldsymbol{m}_1$ to $\boldsymbol{m}_k$

Before we begin to parallelize this algorithm, we need to step back and

[5]It is conjectured that an NP-hard problem cannot be solved in polynomial time (see Definition 2.6).

look at its complexity. The algorithm is composed of two `for` loops, the expectation and the maximization step respectively, that are repeated until some convergence criterion is met. Just by looking at the `for` loops, we see that the expectation step is more costly than the maximization step[6] because there are more loop iterations to perform and there is even more work to be done per iteration. Finding the minimum d_{min} of $\|\boldsymbol{x}_i - \boldsymbol{m}_j\|$ amounts to calculating every norm unless we use specialized data structures, which we won't for this example. Our goal will be to scale Algorithm 8.2 to large values of n by using techniques from parallel computing. We will start off with some very low-level optimizations and work our way up to more high-level optimizations as we go.

Our first concern is that we should fully exploit the instruction set of our CPU. It is safe to assume that the processors in our cluster have some form of SIMD instruction set and that we could benefit from this. The question is, what benefits and how do we use it? There are two operations that likely benefit the most. The calculation of the norm could be vectorized as well as the averaging of the cluster members to calculate the cluster center. Since we are presumably working with an interpreted language (R), there is no way for us to control what exact instructions the interpreter will issue to calculate these quantities. What we can do however is make sure that we are using an optimized BLAS routine that exploits these instructions for vector-vector and vector-matrix type operations. This is easy and gives us our first form of **data parallel computing** almost for free. The hard part of managing what values to put into which register at what time has already been done for us by experts who are likely more knowledgeable in this area. This type of parallelization is called **implicit** parallelization because we do not explicitly state what is executed in parallel on which processor, or in this case, processing element in the processor.

Before we continue, we should define what we mean by data parallel computing, the term we used to describe the parallelization technique we just applied. Data parallel computing executes the same program, subprogram or operation for different input values in parallel. This model is sometimes also referred to as the **single program, multiple data** (**SPMD**) model of computing. As we scale out our k-means algorithm, it will become clearer why it is called this.

The next target for parallelization is the calculation of the minimum of the norm. Instead of implicitly writing this optimization step, we could also transform it into an explicit `for` loop as shown in Algorithm 8.3. Initially it

[6]For this we assume that $n \gg k$.

might seem like this cannot be parallelized, because the loop body depends on the value of d_{min} and it might change in each iteration. There are two ways to deal with such a data dependence. The first is to use synchronization primitives to ensure only one `if` block can be executed at any time. For this to be efficient we need two requirements. First, we need a shared memory system so that each processor can access the memory location holding d_{min} and j_{min} and second, we need some sort of fast locking scheme. Both of these requirements are met if we restrict ourselves for the moment to executing on one node of our cluster. Still, getting the locking right is tricky and can lead to all kinds of problems if done incorrectly. Errors in locking can lead to so-called race conditions, where the result of an operation is dependent on the sequence or timing of other uncontrollable events, a class of errors that is hard to detect and even harder to debug. It is therefore advisable to look for a different approach to parallelize the algorithm.

Algorithm 8.3 Assign Cluster to an Observation in the Sequential k-means Algorithm

---

Require: k cluster centers $\boldsymbol{m}_1$ to $\boldsymbol{m}_k$
Require: Observation $\boldsymbol{x}$

$d_{min} \leftarrow \infty$
$j_{min} \leftarrow 0$
for $j = 1$ to k **do**
  $d \leftarrow \|\boldsymbol{x} - \boldsymbol{m}_j\|$
  if $d < d_{min}$ **then**
    $d_{min} \leftarrow d$
    $j_{min} \leftarrow j$
  end if
end for
return j_{min}

How this might be done will become clearer once we look at a slightly rewritten version of Algorithm 8.3. In Algorithm 8.4 we have split the `for` loop into two consecutive `for` loops. This may look odd at first, because not only have we introduced another `for` loop, but we have also increased the memory consumption of our algorithm from $O(1)$ to $O(k)$. What we have effectively done is separate out the hard part, computing the norm, and the easy part, determining the minimum. The increase in memory consumption is negligible compared to the memory required to store our data set. The beauty of this reformulation is that the body of the first loop now only depends on the inputs. We can compute any d_j without knowing anything about the other

$d_{j'}$. This gives us the option to execute the `for` loop in parallel. Say we have p processors available on our node; then we only need to calculate $\lceil k/p \rceil$ norms per processor and our execution time is decreased proportionally. One other reason why this parallelization fits nicely is that both k and p are usually small.

Algorithm 8.4 Assign Cluster to an Observation in the Sequential k-means Algorithm (Version 2)

Require: k cluster centers $\boldsymbol{m}_1$ to $\boldsymbol{m}_k$
Require: Observation $\boldsymbol{x}$

$d_{min} \leftarrow \infty$
$j_{min} \leftarrow 0$
for $j = 1$ to k **do**
    $d_j \leftarrow \|\boldsymbol{x} - \boldsymbol{m}_j\|$
end for
for $j = 1$ to k **do**
    if $d_j < d_{min}$ **then**
        $d_{min} \leftarrow d$
        $j_{min} \leftarrow j$
    end if
end for
return j_{min}

Where it gets tricky is when we want to actually implement this. No standard way of expressing this type of parallelism has yet matured to a point were it is ubiquitous. There is one approach that is applicable to compiled languages such as C and FORTRAN. It is called OpenMP. The beauty of OpenMP is that it only relies on annotations that are added to the source code to signal the compiler that a loop body should be run in parallel using multiple threads on the same node. The algorithm is still written in a sequential style and can still be run sequentially if the compiler does not support OpenMP or the machine only has one processor. While the annotations explicitly state what should be executed in parallel, the programmer does not have to deal with workload distribution among the available processing cores, and does not have to spawn the workers or explicitly collect the results.

The closest we can get to this style of programming in R is using the foreach package (Revolution Analytics, 2011c). Using it, we can write explicit parallel loops that execute **for each** element of some series. It has several back ends to run these loops in parallel, such as doMC (Revolution Analytics, 2011a), doSNOW (Revolution Analytics, 2011b), doMPI (Weston, 2010), which use different approaches to distribute the workload across multiple processors. However, the package is not quite in the spirit of OpenMP because all of these methods do not exploit the fact that we are assuming that the program is running on a shared memory system. That is, instead of implicitly communicating through shared memory between the workers, there is explicit communication between distinct processes using some form of interconnect. Again, as when exploiting the SIMD instructions of our processor, as a statistician it is best to look for compiled code that has been explicitly parallelized with OpenMP or some other framework in this situation. The overhead incurred with a framework such as foreach will likely trump any gains in performance we might achieve due to the parallel execution of the loop body. Since there are only a few loop iterations, any initialization time, which is part of the sequentially executing portion of the program, will likely further hinder scaling (see Theorem 8.1).

Since we just saw how the foreach package can be used to turn a regular `for` loop, given each iteration does not depend on any results of the previous iteration, into a parallel loop that is executed on several CPUs of the same node or even across many nodes of a cluster, we might be tempted to go ahead and turn every `for` loop of Algorithm 8.2 into a parallel loop. While it is technically possible to do this if we rewrite the loops slightly to eliminate any data dependencies, this is not wise. It does not increase the scalability of our algorithm much because we still have some **master** node that needs to load all the data into memory and coordinate the execution of the parallel sections. So this solution will quickly see its limits when the number of observations (n) increases. If there is enough available memory, the communication overhead of distributing the observations to the nodes in each `while` loop iteration of the algorithm will ultimately overwhelm all other computational costs.

To overcome this, we will reformulate Algorithm 8.2 yet again as a collection of workers that execute independently on a subset of the data and only communicate when necessary to synchronize the current cluster centers. This is deemed the **message passing style** of parallel programming and is the classical example of the SPMD programming paradigm mentioned earlier.

Before we examine a parallel version of the k-means algorithm, we need to discuss how **message passing** works. Our model will be based on MPI, although we cannot and will not cover this API in its full breath[7]. Conceptually MPI assumes that each node in the cluster is running the same program. Each instance of the program is initially part of the global communicator

[7]MPI contains over 200 functions for message passing and parallel IO, of which only a small subset is required to start writing efficient message passing style parallel code.

and has an associated unique rank. Communicators are an MPI concept that groups a number of processes that can pass messages to each other. For our purposes we will not need more than the global communicator, but highly parallel code that has to scale to many hundreds of nodes and thousands of processors would use different communicators to minimize communication overhead by exploiting the physical topology of the cluster when grouping nodes into communicators. Ranks are assigned consecutively within a communicator, starting with rank 0. Each instance of the program is free to use additional levels of parallelization, namely, SIMD instructions or OpenMP, but must take care that the required additional resources are available.

There are many different ways that nodes in an MPI communicator can pass messages to each other, but we will only be concerned with three types here. First, we can send a message from one instance to another. This is called **point-to-point** communication and is facilitated by the `MPI_Send` and `MPI_Recv` functions. We will use the more readable names `send` and `receive` in our pseudocode. If we want to send a message to every process in a communicator, we can use the broadcast functionality of MPI (`MPI_bcast`, in pseudocode `broadcast`). This is useful to distribute common data to all processes in an efficient manner. Finally, we will use a reduce operation named `MPI_allreduce` (pseudocode `all_reduce`). It takes a value and an aggregation function and returns the aggregate of all the values in the communicator to all nodes. It is an efficient way to calculate the sum or product of the result of a calculation performed by each node and distribute the result back to all nodes.

Using these three forms of message passing, we can now write a parallel version of Algorithm 8.2 that scales to large n by adding additional processors. We assume that our data set has been divided into p approximately equally sized chunks that will be processed by the p nodes in our cluster. While not explicitly shown, all previous optimizations are still applicable to the parallel version of the algorithm. The idea of Algorithm 8.5 is that all we need to communicate between the nodes are the current cluster centers. Given the cluster centers, each node can do one expectation and maximization step before it needs any information from another node.

One detail we need to handle specifically is the initial choice of cluster centers. Here we have chosen to let the first node choose the cluster centers and send them to all the other workers using a broadcast. After that, all nodes execute the same EM step and then combine their respective cluster centers to form the new centers. In our code all nodes report the final solution, but in practice we would probably add another guard so that only the first node emits the final result.

Algorithm 8.5 Parallel k-means Algorithm

Require: Number of independent workers p
Require: Rank of this worker r ($r = 0$ to $p-1$)
Require: Number of clusters k
Require: Observations $\boldsymbol{x}_1$ to $\boldsymbol{x}_{n_r}$

if r is 0 **then**
  Choose k initial means $\boldsymbol{m}_1$ to $\boldsymbol{m}_k$
  `broadcast`$(\boldsymbol{m}_1, \dots, \boldsymbol{m}_k)$
else
  `receive`$(\boldsymbol{m}_1, \dots, \boldsymbol{m}_k)$
end if
Initialize cluster sets c_1 to c_k as empty
while not converged **do**
  for $i = 1$ to n_r **do**
    $j \leftarrow \arg\min_j \|\boldsymbol{x}_i - \boldsymbol{m}_j\|$
    $c_j \leftarrow c_j \cup \{i\}$
  end for{Expectation step}
  for $j = 1$ to k **do**
    $n_j \leftarrow |c_j|$
    $\boldsymbol{s}_j \leftarrow \sum_{i \in c_j} \boldsymbol{x}_i$
    $c_j \leftarrow \emptyset$
  end for{Maximization step}
  for $j = 1$ to k **do**
    $n_j \leftarrow$ `all_reduce`$(+, n_j)$
    $\boldsymbol{s}_j \leftarrow$ `all_reduce`$(+, \boldsymbol{s}_j)$
    $\boldsymbol{m}_j \leftarrow \boldsymbol{s}_j / n_j$
  end for
end while
return $\boldsymbol{m}_1$ to $\boldsymbol{m}_k$

We have ignored how convergence is detected and especially how all nodes know that we have converged. Convergence detection usually amounts to monitoring the change in location of the cluster centers. If this change is close to zero, we say the procedure has converged. Since all nodes know the current cluster centers at all times, they will also all terminate at the same time under this regime. So while convergence detection could again fall to one node that then broadcasts the result to all others, we have, in the interest of simplicity, opted to have each node check for convergence.

8.4.3.2 *Summary and Outlook*

Summary

This concludes the section on data parallel computing. Using the k-means algorithm as an example, we have looked at ways to parallelize an algorithm at several different levels. We saw how SIMD instructions can be used to improve the performance of critical primitives of an algorithm, how `for` loops can be distributed across several cores of a machine in the presence of data independence. If data independence is not given, we use a trick to achieve it by using additional memory to save temporary results. Finally, we scaled the algorithm to data set sizes so large that a single node of our hypothetical cluster would not have been able to handle it. In the process we learned about OpenMP and MPI as two programming APIs that can be used to explicitly express parallel algorithms on a machine.

Outlook

Some details have been glossed over. On a true cluster, we would have used a batch processing system to handle the orchestration of our MPI runs for us. Setup and teardown of parallel jobs are a notoriously tricky business best left to software. We did not talk about some of the finer points of measuring scalability, such as **strong scaling**, **weak scaling**, and **isoefficiency**.

8.4.4 *Task Parallel Computing*

Task parallel computing is conceptually much simpler than data parallel computing. Instead of having to worry about how to partition the data over the available processing resources, the program is divided into several independent computational tasks. Each task is then scheduled to run on one node of the cluster. Since no data are shared between the tasks, this style of computing is much simpler to work with and much easier to reason about. The challenges are not so much in writing the programs than in providing an infrastructure to efficiently execute these types of programs. The biggest problem is one of resource starvation. What happens if the number of available nodes is less than the number of tasks? Then we need some algorithm that will distribute the workload across the nodes for us and schedule them in an optimal manner.

In its simplest form, task parallel computing amounts to decomposing one large program into many smaller ones and then running these as multiple jobs using a batch scheduling system on a cluster. In this scenario, the scheduler does all the hard work of finding a free node, starting our job and monitoring its execution. If we need more fine grained control, we are of course free to use a message passing style of programming to distribute the workload to other nodes. This amounts to writing an MPI program where the master, i.e. rank 0, does all the work, and when it wants to start a task in parallel it finds another free node, sends it a message detailing the job to execute, and then later collects the answer that was sent back from the node at the end of the calculation.

There are two primary uses for task parallel computing in statistics. The first is interactive data analysis. This may not sound like a classical domain of parallel computing, but when we explore a data set interactively, we do not want to be stopped in our tracks by a long-running computation. We can use task level parallelism to run one type of analysis **in the background**, on another CPU of our machine or possibly even on a different node, while we continue with our exploration on the master node. Later, when the analysis is finished, we collect the results and visualize them or draw our conclusions from them. At first, this might sound like a trivial usage, but in using the R package multicore, this has become such a trivial job that we, the authors, use it quite often. The other main area where task level parallelism is exploited is in large studies or computational experiments.

Example 8.4: Choosing k in the k-means Algorithm
For choosing k in the k-means algorithm we need to run the algorithm for different values of k and compare the quality of the clustering in some manner. Instead of sequentially looping over the values of k, we could just as well run k parallel tasks and then collect the respective quality measures to decide on a value of k.

Example 8.5: Cross-validation
More complex examples would be choosing a classification or regression method for a particular machine learning task. Here we would need to do, e.g., a cross-validation (see Chapter 7) for each learner to assess its predictive power. Again, these can be run in parallel.

Because task level parallelism is easy to detect and exploit, it is generally better to first look for ways to apply it to a problem before looking for opportunities to apply data parallel computing. The reason is that data parallel computing always entails communication among the nodes, and this communication might easily become the bottleneck of the computation. We should also consider the effort required to implement and debug an algorithm in such a way that it can exploit data level parallelism. Task parallel computing is much simpler in these regards. We can always run a single task locally and observe its computations if we need to debug it. Adding more processors to a task parallel problem will always speedup the computation as long as there are still tasks to be worked on.

Finally, combining both task and data parallel computing is of course a viable option. If, say, we already had a data parallel algorithm and wanted to tune some of its hyperparameters using grid search, there is nothing stopping us from doing so using a task parallel computing approach to evaluate several grid points in parallel. This nesting does require some experience because it is not clear how to balance the resources between the task parallel execution of jobs and the data parallel algorithm. Deciding if an additional CPU should go toward the task or data parallel part is tricky. Devising these hierarchies of nested parallel algorithms is currently still more of an art than a science, and little practical advice is to be found that generalizes well. Often configurations are published that have been devised for one kind of cluster with a certain resource limit and communication topology. Whether this configuration can be adapted to other types of clusters or similar problems is uncertain.

Before we conclude this chapter on scalability and parallel computing, we will shine a light on some modern concepts of parallel computing that are currently the area of intense research.

8.4.5 Modern Paradigms

In this section we will look at some parallel computing concepts that have not been covered so far and which have only recently gained traction in the HPC world. There are two topics we wish to discuss. First there is the upcoming trend of using **graphics processing units (GPUs)** for general purpose computing (GPGPU computing). Secondly we will take a look at the ideas and concepts behind the **MapReduce paradigm** as popularized by Google (Dean and Ghemawat, 2008) and its open-source variant Hadoop.

The first modern paradigm really is just an application of an old idea to a new class of hardware. Instead of using a general purpose processor to do all calculations early computers often had coprocessors to take care of some specialized operations. In the history of supercomputing these were often IO processors meant to relieve the main CPU of the boring task of moving data to and from stable storage. Later on, the first PCs had coprocessors to handle floating-point math, and some even had dedicated digital signal processors or RISC CPUs to offload some computations from the main CPU. All of these attempts were ultimately doomed because they were too expensive, too slow or were outdated by the progress of technology before their full potential could be harnessed. Not so with GPUs. The market for ever-faster graphics processing units has seen fierce competition in the past 10 years, and there is a steady increase in their performance. Modern GPUs are much more powerful than their CPU counterparts when it comes to floating-point math. But their power can only be harnessed if the problem can be formulated in a massively parallel manner since GPUs are composed of many hundreds of simple processing elements that all execute the same code but using different input. Reformulating existing algorithms or inventing new algorithms that harness the power provided by these GPUs for tasks not related to 3D graphics is the field of **general purpose GPU (GPGPU)** computing. One of the earliest examples of GPGPU computing from the machine learning literature is Steinkrau et al. (2005), where they trained a neural network using a GPU.

Developing for these massively parallel hardware platforms is a major challenge. On the one hand, these systems are so fast that it is hard to transfer data to and away from the graphics processor fast enough, and on the other hand, the functions that can be implemented are quite limited. Until recently for example, conditional branching was not possible in a GPU program. Often clever pre- and post-processing is necessary on the CPU to prepare and digest the input and output of a GPU algorithm. There is also still quite a bit of movement in how these processors should be programmed. The currently most popular option is Nvidia's CUDA toolkit, but it is only available for GPUs from that vendor and requires careful tuning of each program for each new GPU introduced to the market. OpenCL, the Open Computing Language, is a cross-platform and cross-vendor API specified to rectify this situation. However, it has not caught on so far in the HPC sector.

The other modern paradigm that we have not covered so far is **MapReduce** style parallel computing. The ideas and principles behind MapReduce are not new; both the Map as well as the Reduce function are well-established ideas in functional programming. The innovation of MapReduce and its open-source incarnation Hadoop is more technical and methodological. So what makes MapReduce different? The core idea of MapReduce is to divide a computation into a series of Map and Reduce steps. During the mapping phase a function is applied to each observation in a data set. This function can return one or more key value tuples. These tuples are then passed to the reduce function, which combines or aggregates them to the final result. Typical reductions could be summing the output of the map function or calculating other aggregates but could also be complex operations such as the fitting of a classification model in cross-validation (see Chapter 7). This result could then either be the input to the next map step or be reduced to a single result.[8]

[8]For a sequential implementation of the two functions map and reduce, see the R functions by the same name.

In the cross-validation case, the map step would fit the model to each fold of the data and the reduce step would then aggregate the estimated errors.

By definition each function application to an observation in the mapping step is independent of all other function applications and can therefore be executed in parallel to them. The innovation that MapReduce brings to the table is technical. Instead of moving the data around the cluster, MapReduce evaluates the function on a node that is topologically close to the data. That is, apart from the execution framework, one of the core components of any MapReduce system is a filesystem or storage layer that distributes massive data sets across all the nodes.

This idea has allowed some of the many innovative companies to scale their businesses to tens or hundreds of terabytes of data that need to be processed. The type of processing done ranges from such mundane tasks as mining log files for anomalies to computations on the social graph of the members of a social network. When dealing with such data sets, it is not wise to translate established algorithms to this framework. Instead, new types of models are required for such large data sets, and certain types of analysis have only become possible because of the vast and rich amount of data that is collected and stored nowadays. One interesting project that is working on this front, based on the Hadoop framework, is the Apache Mahout project which plans to implement many classical and modern statistical methods as scalable Hadoop procedures. See Ingersoll (2011) for a general overview of the project.

There are many more parallel programming paradigms that are currently being researched. We expect the future to be full of new and innovative approaches because the current trends in CPU design will lead to many-core CPUs in the future so that just as we now face the problems that marked the forefront of HPC 10 to 15 years ago, the current state-of-the-art in parallel computing will become a commodity within the next years. This will likely entail drastic changes in the way we program these computers since almost all languages in use today were crafted to describe sequentially executed sets of statements. Reasoning about large programs that use message passing becomes increasingly difficult, and we hope that the future will bring us tools and languages that ease the burden of parallelization and make most parallelization implicit instead of the current explicit formulation of parallelism.

8.4.6 Summary

In this section we have looked at different forms of parallel computing. Before we looked at different parallel computing architectures, we assessed the potential for parallelization by looking at Amdahl's law, which quantifies the maximum speedup we can expect when parallelizing an operation. When looking at parallel computing architectures, we divided the possible approaches into so-called **data parallel computing** and **task parallel computing** paradigms. In the former, a single operation is applied to multiple pieces of data in parallel. Using data parallel computing is usually easy and often happens automatically when using modern compilers. Task parallel computing, on the other hand, requires that we explicitly break down our overall program into smaller tasks and schedule their execution in parallel. We concluded the section with a short review of modern parallel computing paradigms such as MapReduce where operations are mapped over the data and the outcome of these operations then reduced to the final result.

8.5 Implementation in R

There are myriad available options to implement parallel algorithms using R. These would be the Rmpi (Yu, 2010), snow (Tierney et al., 2011), snowfall (Knaus, 2010), and foreach packages for distributed computing using message passing, although the latter packages abstract away the message passing interface and usually provide `apply` style functions to express explicit parallelism. The Rdsm (Matloff, 2011) package provides an API that simulates a distributed shared memory computing environment on top of R. The RScaLAPACK (Samatova et al., 2009b) and sprint (University of Edinburgh SPRINT Team, 2011) packages are somewhat special in that they provide data parallel algorithms implemented using MPI which have been heavily optimized. RScaLAPACK is an interface to the ScaLAPACK routines that provide data parallel versions of some of the LAPACK routines. sprint currently provides only a few statistical algorithms but these have been tuned for extreme scalability by the authors. Finally, the multicore (Urbanek, 2011) and taskPR (Samatova et al., 2009a) packages both provide convenient APIs for **task parallel computing**.

8.6 Conclusion

In the final chapter of this book we have looked at the issues of scalability and parallelization.

After discussing the hardware developments, we studied the evolution of the software available to us for scalable and parallel computing. Various APIs were presented with increasing degrees of abstraction.

Next, we looked at some of the ways a sequential program can be sped up, culminating in a list of practical guidelines to optimize our code.

After optimizing our code, three levels of parallel computing were presented. SIMD instructions can be used to perform work in parallel at the processor level. Explicit and implicit loop parallelization is one way to make use of the multiple processors in a cluster node, a so-called shared memory system, and finally, we can use message passing to distribute our computation across multiple nodes in a cluster.

All of these strategies relied on some form of data parallelism. If we cannot exploit this in our algorithms, there is still task parallel computing to decrease runtimes. Here a computation is split into several, often many, separate tasks that can be executed independently of each other. These are then distributed across multiple processors in a node or even across many nodes in a cluster. This framework proved especially useful because it is, if applicable, much easier to implement from an existing code base.

Finally we gave an outlook on some new and promising paradigms in parallel computing. These include GPGPU computing and MapReduce.

Let us conclude this chapter with the reminder that while scaling an algorithm is desirable and parallel computing can be fun and rewarding, we should not overdo it. Optimization, scaling, and parallel computing know-how are tools for the statistician to use when they are required. Before starting on any scaling endeavor, we should evaluate our goals and decide if they are feasible. Once we reach these goals, we should restrict ourselves and refrain from any further optimizations.

8.7 Exercises

Exercise 8.1.1: Describe what is meant by (empirical) scalability and how parallelization plays into scalability.

Exercise 8.1.2: Compare the empirical scalability of the sort algorithms described in Chapter 2.

Exercise 8.1.3: What are advantages and disadvantages of a symmetric multiprocessing machine compared to a cluster of machines using message passing.

Exercise 8.2.1: Research what algorithm is used for the QR factorization in the LAPACK routine `DGELS` which can be used to solve a LLS problem.

Hint
It is not one of the algorithms discussed in Chapter 2.

Exercise 8.2.2: What other routines does LAPACK include to solve LLS problems?

Exercise 8.3.1: Using the `microbenchmark` package, answer the questions posed in Example 8.1. What do the difference stem from? Can you deduce general principles from the example for writing efficient code? Does this help or hinder readability?

Exercise 8.3.2: Can you think of a simple optimization to further speedup the `cp_mean` function of Example 8.1?

Exercise 8.3.3: Research why the R function `‘:‘` is faster than `seq.int` on most systems for similar inputs.

Exercise 8.3.4: Why might the R function `seq_along` be slower than the combination of `seq_len` and `length` as demonstrated in Example 8.2?

Exercise 8.3.5: Consider the following function to calculate the mean squared error:

```
mse <- function(predicted, true) {
  error <- predicted - true
  squared_error <- error^2
  mean(squared_error)
}
```

Assume that our input vectors `predicted` and `true` have about 10 000 entries. Look for ways to speedup this function without resorting to C or FORTRAN code.

Exercise 8.3.6: Use the following template to rewrite the `mse` function of Exercise 8.3.5 in C. The R function is reduced to

```
dyn.load(paste0("mse", .Platform[["dynlib.ext"]]))
mse_in_c <- function(predicted, true)
  .Call("do_mse", as.numeric(predicted),
        as.numeric(true))
```

and the C template is given by

```
#include <R.h>
#include <Rinternals.h>
#include <Rmath.h>
```

```
SEXP do_mse(SEXP s_predicted, SEXP s_true) {
  double mse;
  /* Unpack the 's_predicted' vector:
   * n_predicted - length of the 's_predicted'
   * predicted - pointer to the values
   */
  const R_len_t n_predicted = length(s_predicted);
  double *predicted = REAL(s_predicted);

  /* Unpack the 's_true' vector:
   * n_true - length of 's_true'
   * true - pointer to the values
   */
  const R_len_t n_true = length(s_true);
  double *true = REAL(s_true);

  /* Your code goes here */

  /* And finally return the value stored in 'mse' */
  return ScalarReal(mse);
}
```

The C code can be compiled using `R CMD SHLIB mse.c`. Make sure your code is well behaved in the face of `NA`, `NaN` and `Inf` values. How much faster is your code compared to the naive R implementation?

Exercise 8.3.7: In Section 8.2 we learned about the BLAS and how it has become a de facto standard for basic linear algebra operations. R uses the BLAS and LAPACK libraries extensively to accelerate certain types of operations. Apart from the obvious of `*`, `%*%`, `eigen`, and `qr`, what other common R functions use the BLAS and LAPACK routines to speedup calculations?

Exercise 8.3.8: Find out if R uses an optimized BLAS on your system. If yes, what version is used? If not, find out if any optimized BLAS is available for your system (hint: there probably is).

Exercise 8.3.9: Name some of the reasons why compiled code written in C or FORTRAN is generally faster than interpreted R code. Why don't we use C or FORTRAN to write all our software, but instead rely on tools such as R for our data analysis?

Exercise 8.4.1: Give a concise argument for Amdahls law (Theorem 8.1).

Exercise 8.4.2: Devise an experiment to estimate π_{parallel} of an unknown algorithm.

Exercise 8.4.3: Investigate the overhead the foreach package adds to a simple `for` loop. Use a loop body with a known execution time such as

```
library("foreach")
## Load and initialize some parallel back-end for
## foreach

foreach(i=1:10, .combine=c) %dopar%
  Sys.sleep(0.1)
```

How do you think the overhead added by the foreach package influences the possible performance gains when parallelizing tight inner loops?

Exercise 8.4.4: Why does the run time of Algorithm 8.5 stay the same, or possibly get worse, if we have one node check for convergence and then broadcast the result to all nodes?

Bibliography

M. al Chwarizmi. *Kita-b al-muchtasar fi hisab al-dschabr wa-l-muqabala ("The Compendious Book on Calculation by Completion and Balancing").* 830.

C. Alamprese, L. Datei, and Q. Semeraro. Optimization of processing parameters of a ball mill refiner for chocolate. *Journal of Food Engineering*, 83 (4):629 – 636, 2007. ISSN 0260-8774.

E. Alfaro-Cortes, M. Gamez-Martinez, and N. Garcia-Rubio. *adabag: Applies multiclass AdaBoost.M1, AdaBoost-SAMME and Bagging*, 2012. URL `http://cran.r-project.org/package=adabag`. R package version 3.1.

G.M. Amdahl. Validity of the single processor approach to achieving large scale computing capabilities. In *Proceedings of the April 18-20, 1967, spring joint computer conference*, AFIPS '67 (Spring), pages 483–485, New York, NY, 1967. ACM.

E. Anderson, Z. Bai, C. Bischof, S. Blackford, J. Demmel, J. Dongarra, J. Du Croz, A. Greenbaum, S. Hammarling, A. McKenney, and D. Sorensen. *LAPACK Users' Guide*. Society for Industrial and Applied Mathematics, Philadelphia, PA, third edition, 1999.

S.F. Arnold. Gibbs sampling. In C.R. Rao, editor, *Handbook of Statistics*, volume 9, pages 602–605. Elsevier, Amsterdam, 1993.

J.W. Backus, R.J. Beeber, S. Best, R. Goldberg, H.L. Herrick, R.A. Hughes, L.B. Mitchell, R.A. Nelson, R. Nutt, D. Sayre, P.B. Sheridan, H. Stern, and I. Ziller. *The FORTRAN Automatic Coding System for the IBM 704 EDPM: Programmer's Reference Manual*, 1956.

Y. Bard. On a numerical instability of Davidon-like methods. *Mathematics of Computation*, 22:665–666, 1968.

Y. Bard. *Nonlinear Parameter Estimation*. Academic Press, New York, 1974.

F.L. Bauer. Elimination with weighted row combinations for solving linear equations and least-squares problems. *Numerische Mathematik*, 7:338–352, 1965.

M. Bédard and J.S. Rosenthal. Optimal scaling of metropolis algorithms: Heading toward general target distributions. *Canadian Journal of Statistics*, 36:483–503, 2008.

A. Ben-Israel and T.N.E. Greville. *Generalized Inverses: Theory and Applications*. Springer, New York, 2 edition, 2003.

Y. Bengio and Y. Grandvalet. No unbiased estimator of the variance of k-fold cross-validation. *Journal of Machine Learning Research*, 5:1089–1105, 2004.

H.-G. Beyer and H.-P. Schwefel. Evolution strategies - A comprehensive introduction. *Natural Computing*, 1(1):3–52, May 2002.

H. Binder and M. Schumacher. Adapting prediction error estimates for biased complexity selection in high-dimensional bootstrap samples. *Statistical Applications in Genetics and Molecular Biology*, 7(1):12, 2008.

B. Bischl, O. Mersmann, H. Trautmann, and C. Weihs. Resampling methods in model validation. *Evolutionary Computation*, 20:249–275, 2012.

A Björck. Solving linear least-squares problems by Gram-Schmidt orthogonalization. *BIT*, 7:1–21, 1967.

M. Blum, R.W. Floyd, V. Pratt, R.L. Rivest, and R.E. Tarjan. Time bounds for selection. *J. Comput. Syst. Sci.*, 7(4):448–461, August 1973.

K.H. Böhling. *Grundlagen der Informatik I*. 1971. Script.

S. Boyd and L. Vandenberghe. *Convex Optimization*. Cambridge University Press, New York, NY, USA, 2004.

Richard P. Brent. *Algorithms for Minimization without Derivatives*. Prentice-Hall, Englewood Cliffs, N.J., 1973.

S.P. Brooks and A. Gelman. Alternative methods for monitoring convergence of iterative simulations. *Journal of Computational and Graphical Statistics*, 7:434–455, 1998.

K.M. Brown and J.E. Jr. Dennis. A new algorithm for nonlinear least-squares curve fitting. In J.R. Rice, editor, *Mathematical Software*, pages 391–396. Academic Press, New York, 1971.

R.G. Brown, D. Eddelbuettel, and D. Bauer. DieHarder, 2010. URL `http://www.phy.duke.edu/~rgb/General/dieharder.php`.

C.G. Broyden. Quasi-newton methods and their application to function minimization. *Mathematics of Computation*, 21:368–381, 1967.

P. Cattin. Estimation of the predictive power of a regression model. *Journal of Applied Psychology*, 65(4):407, 1980.

A. Cauchy. Méthode générale pour la résolution des systemes dÉquations simultanées. *Comp. Rend. Sci. Paris*, 25(1847):536–538, 1847.

T.F. Chan, G.H. Golub, and R.J. LeVeque. Updating formulae and a pairwise algorithm for computing sample variances. In H. Caussinus et al., editor, *Compstat 1982, Proceedings of the 5th Symposium*, pages 30–41, Toulouse, 1979.

T.F. Chan, G.H. Golub, and R.J. Leveque. Algorithms for computing the sample variance: Analysis and recommendations. *The American Statistician*, 37:242–247, 1983.

T.F.C. Chan and J.G. Lewis. Rounding error analysis of algorithms for computing means and standard deviations. Technical Report 284, The Johns Hopkins University, Department of Mathematical Sciences, 1978.

S.S. Cross, R.F. Harrison, and R.L. Kennedy. Introduction to neural networks. *The Lancet*, 346:1075–1079, 1995.

G.B. Dantzig. Programming of interdependent activities: II mathematical model. *Econometrica, Journal of the Econometric Society*, pages 200–211, 1949.

G.B. Dantzig. *Maximization of a Linear Function of Variables Subject to Linear Inequalities, in Activity Analysis of Production and Allocation*, chapter XXI, pages 339–347. Wiley, New York, 1951.

M. Davy and S.J. Godsill. Bayesian harmonic models for musical pitch estimation and analysis. Technical Report 431, Cambridge University, Engineering Department, Cambridge, 2002.

J. Dean and S. Ghemawat. MapReduce: simplified data processing on large clusters. *Commun. ACM*, 51:107–113, January 2008.

A.P. Dempster, N.M. Laird, and D.B. Rubin. Maximum-likelihood from incomplete data via the EM algorithm. *Journal of the Royal Statistical Society, Series B*, 39:1–38, 1977.

J.E. Jr. Dennis. Some computational techniques for the nonlinear least squares problem. In G.D. Bryne and C.A. Hall, editors, *Numerical Solutions of Systems of Nonlinear Algebraic Equations*, pages 157–183. Academic Press, New York, 1973.

L. Devroye and L. Devroye. *Non-Uniform Random Variate Generation*, volume 4. Springer, New York, 1986.

P. Diaconis and D. Strook. Geometric bounds for eigenvalues of markov chains. *The Annals of Applied Probability*, 1:36–61, 1991.

N.A. Diamantidis, D. Karlis, and E.A. Giakoumakis. Unsupervised stratification of cross-validation for accuracy estimation. *Artificial Intelligence*, 116(1-2):1–16, 2000.

W.J. Dixon. Further contributions to the problem of serial correlation. *Annals of Mathematical Statistics*, 15:119–144, 1944.

J. Dongarra. Oral history interview by Thomas Haigh, 2005. URL `http://history.siam.org/oralhistories/dongarra.htm`.

J.J. Dongarra, C.B. Moler, J.R. Bunch, and G.W. Stewart. *LINPACK Users' Guide*. Society for Industrial and Applied Mathematics, Philadelphia, PA, 1979.

J.A. Doornik. Conversion of high-period random numbers to floating point. *ACM Trans. Model. Comput. Simul.*, 17, January 2007.

W. DuMouchel and C. Waternaux. Discussion an hierarchical models for combining information and for meta-analyses (by C.N. Morris and S.L. Normand). In J.M. Bernardo, J.O. Berger, A.P. Dawid, and A.F.M. Smith, editors, *Bayesian Statistics*, volume l, pages 338–341. Oxford University Press, 1992.

C. Dutang. *CRAN Task View: Probability Distributions, Version 2013-06-12*, 2013. URL `http://cran.r-project.org/web/views/Distributions.html`.

B. Efron. Bootstrap methods: Another look at the jackknife. *The Annals of Statistics*, 7(1):1–26, 1979.

B. Efron. Estimating the error rate of a prediction rule: Improvement on cross-validation. *Journal of the American Statistical Association*, 78(382): 316–331, 1983.

B. Efron and R. Tibshirani. Improvements on cross-validation: The 0.632+ bootstrap method. *Journal of the American Statistical Association*, 92 (438):548–560, 1997.

J. Eichenauer and J. Lehn. A non-linear congruential pseudo random number generator. *Statistical Papers*, 27(1):315–326, 1986.

Euclides and C. Thaer. *Euklid: Die Elemente*. Akademische Verlagsgesellschaft, 1937.

D.G. Feitelson. On the interpretation of TOP500 data. *Int. J. High Perform. Comput. Appl.*, 13:146–153, May 1999.

A. I. Fleishman. A method for simulating non-normal distributions. *Psychometrika*, 43(4):521–532, 1978.

R. Fletcher. A technique for orthogonalization. *IMA Journal of Applied Mathematics*, 5(2):162–166, 1969.

R. Fletcher. A new approach to variable metric algorithms. *The Computer Journal*, 13(3):317–322, 1970.

R. Fletcher and M.J.D. Powell. A rapidly convergent descent method for minimization. *The Computer Journal*, 6(2):163–168, 1963.

C. Fraley and A.E. Raftery. Model-based clustering, discriminant analysis and density estimation. *Journal of the American Statistical Association*, 97:611–631, 2002.

M. Frank and P. Wolfe. An algorithm for quadratic programming. *Naval Research Logistics Quarterly*, 3:95, 1956.

M.M. Gaber, A. Zaslavsky, and S. Krishnaswamy. Mining data streams: a review. *ACM SIGMOD Record*, 34:18–26, 2005.

R. Gandy. Church's thesis and principles for mechanisms. In J. Barwise, H.J. Keisler, and K. Kunen, editors, *The Kleene Symposium*. North-Holland, Amsterdam, 1980.

C.F. Gauß. Theoria motus corporum coelestium in sectionibus conicis solem ambientium, 1809.

A. Geist, A. Beguelin, J. Dongorra, W. Jiang, R. Manchek, and V. Sunderman. *PVM: Parallel Virtual Machine — A User's Guide and Tutorial for Networked Parallel Computing*. MIT Press, Cambridge, MA, 1994.

P. Geladi and B.R. Kowalski. Partial least-squares regression: a tutorial. *Analytica Chimica Acta*, 185:1–17, 1986.

A. Gelman. Inference and monitoring convergence. In W.R. Gilks, S. Richardson, and D.J. Spiegelhalter, editors, *Markov Chain Monte Carlo in Practice*, pages 131–143. Chapman & Hall, London, 1995.

A. Gelman and D.B. Rubin. Inference from iterative simulation using multiple sequences. *Statistical Science*, 7:457–511, 1992.

S. Geman and D. Geman. Stochastic relaxation, Gibbs distributions and the Bayesian restoration of images. *Pattern Analysis and Machine Intelligence, IEEE Transactions on*, PAMI-6(6):721–741, 1984.

A. Genz and F. Bretz. *Computation of Multivariate Normal and t Probabilities*, volume 195 of *Lecture Notes in Statistics*. Springer-Verlag, Heidelberg, 2009.

A. Genz, F. Bretz, T. Miwa, X. Mi, F. Leisch, F. Scheipl, and T. Hothorn. *mvtnorm: Multivariate Normal and t Distributions*, 2012. URL `http://cran.r-project.org/package=mvtnorm`. R package version 0.9-9994.

C.J. Geyer. Practical markov chain monte carlo. *Statistical Science*, 7:473–511, 1992.

W.R. Gilks. Full conditional distributions. In W.R. Gilks, S. Richardson, and D.J. Spiegelhalter, editors, *Markov Chain Monte Carlo in Practice*, pages 75–88. Chapman & Hall, London, 1995.

W.R. Gilks, A. Thomas, and D.J. Spiegelhalter. A language and program for complex bayesian modeling. *The Statistician*, 43:169–178, 1994.

D. Goldberg. What every computer scientist should know about floating-point arithmetic. *ACM Comput. Surv.*, 23:5–48, March 1991. doi: 10.1145/103162.103163.

S.M. Goldfeld and R.E. Quandt. *Nonlinear Methods in Econometrics*. North-Holland, Amsterdam, 1972.

S.M. Goldfeld, R.E. Quandt, and H.F. Trotter. Maximization by quadratic hill-climbing. *Econometrica*, 34:541–551, 1966.

G.H. Golub. Numerical methods for solving linear least squares problems. *Numerische Mathematik*, 7:206–216, 1965.

G.H. Golub and J.H. Wilkinson. Note on the iterative refinement of least squares solution. *Numerische Mathematik 9*, pages 139–148, 1966.

M. Goto and T. Nishimura. Rwc music database: Music genre database and musical instrument sound database. In *ISMIR 2003*, pages 229–230. 2003.

H.P. Graf, E. Cosatto, L. Bottou, I. Durdanovic, and V. Vapnik. Parallel support vector machines: The cascade SVM. In *NIPS*, 2004.

T.N.E. Greville. Some applications of the pseudoinverse of a matrix. *SIAM Review*, pages 15–22, 1960.

B. Grün and F. Leisch. FlexMix version 2: Finite mixtures with concomitant variables and varying and constant parameters. *Journal of Statistical Software*, 28(4):1–35, 2008.

J.L. Gustafson. Reevaluating amdahl's law. *Communications of the ACM*, 31: 532–533, 1988.

M. Hall, E. Frank, G. Holmes, B. Pfahringer, P. Reutemann, and I. H. Witten. The WEKA data mining software: An update. *SIGKDD Explorations*, 11 (1), 2009.

T. Hastie, R. Tibshirani, and J. Friedman. *The Elements of Statistical Learning*. Springer, New York, 2001.

W.K. Hastings. Monte carlo sampling methods using markov chains and their applications. *Biometrika*, 57(1):97–109, 1970.

P. Hellekalek. Inversive pseudorandom number generators: concepts, results, and links. In C. Alexopoulos, K. Kang, W.R. Lilegdon, and D. Goldsman, editors, *Proceedings of the 1995 Winter Simulation Conference*, pages 255–262. 1995.

D.M. Himmelblau. A uniform evaluation of unconstrained optimization techniques. In F.R. Lootsma, editor, *Numerical Methods for Nonlinear Optimization*, pages 69–97. Academic Press, London, 1972.

J. H. Holland. *Adaptation in Natural and Artificial Systems*. University of Michigan Press, Ann Arbor, MI, USA, 1975.

R. Hooke and T.A. Jeeves. Direct search solution of numerical and statistical problems. *Journal of the ACM (JACM)*, 8(2):212–229, 1961.

K. Hornik, D. Meyer, and S. Theussl. *ROI: R Optimization Infrastructure*, 2011. URL `http://cran.r-project.org/package=ROI`. R package version 0.0-7.

A. Höskuldsson. PLS regression methods. *Journal of Chemometrics*, 2:211–228, 1988.

T. Hothorn. *CRAN Task View: Machine Learning & Statistical Learning, Version 2013-05-27*, 2013. URL `http://cran.r-project.org/web/views/MachineLearning.html`.

T. Hothorn, F. Leisch, A. Zeileis, and K. Hornik. The design and analysis of benchmark experiments. Technical Report 82, SFB Adaptive Informations Systems and Management in Economics and Management Science, 2003.

A.S. Householder. Unitary Triangularization of a Nonsymmetric Matrix. *J. ACM*, 5(4):339–342, October 1958.

J.T.G. Hwang and A.A. Ding. Prediction intervals for aritificial neural networks. *Journal of the American Statistical Association*, 92:748–757, 1997.

G. Ingersoll. Apache Mahout: Scalable machine learning for everyone, 2011. URL `http://www.ibm.com/developerworks/java/library/j-mahout-scaling/`.

K.E. Iverson. *A Programming Language*. Wiley, New York, 1962.

R.I. Jennrich. Asymptotic properties of non-linear least squares estimators. *Annals of Mathematical Statistics*, 40:633–643, 1969.

G.G. Judge, W.E. Griffiths, R.C. Hill, H. Lütkepohl, and T.C. Lee. *The Theory and Practice of Econometrics*. Wiley, New York, 2 edition, 1981.

A. Karatzoglou, A. Smola, K. Hornik, and A. Zeileis. kernlab – an S4 package for kernel methods in R. *Journal of Statistical Software*, 11(9):1–20, 2004. URL `http://www.jstatsoft.org/v11/i09/`.

W.J. Jr. Kennedy and J.E. Gentle. *Statistical Computing*. Marcel Dekker,

New York, 1980.

J. Kiefer. Sequential minimax search for a maximum. *Proceedings of the American Mathematical Society*, 4(3):502 – 506, June 1953.

J.-H. Kim. Estimating classification error rate: Repeated cross-validation, repeated hold-out and bootstrap. *Computational Statistics and Data Analysis*, 53(11):3735–3745, 2009.

F.H. Kishi. On line computer control techniques and their application to reentry aerospace vehicle control. *Advances in Control Systems: Theory and Applications*, 1:245–354, 1964.

J. Knaus. *snowfall: Easier cluster computing (based on snow)*, 2010. URL `http://cran.r-project.org/package=snowfall`. R package version 1.84.

D.E. Knuth. *The Art of Computer Programming*, volume II. Addison-Wesley, Amsterdam, 1969.

D.E. Knuth. *The Art of Computer Programming*, volume III. Addison-Wesley, Amsterdam, 1998.

R. Kohavi. A study of cross-validation and bootstrap for accuracy estimation and model selection. In *International Joint Conference on Artificial Intelligence (IJCAI)*, pages 1137–1143, 1995.

M. Kuhn, S. Weston, N. Coulter, and R. Quinlan. *C50: C5.0 Decision Trees and Rule-Based Models*, 2013a. URL `http://cran.r-project.org/package=C50`. R package version 0.1.0-14.

M Kuhn, J. Wing, S. Weston, A. Williams, C. Keefer, A. Engelhardt, and T. Cooper. *caret: Classification and Regression Training*, 2013b. URL `http://cran.r-project.org/package=caret`. R package version 5.15-61.

M. B. Kursa and W. R. Rudnicki. Feature selection with the Boruta package. *Journal of Statistical Software*, 36(11):1–13, 2010. URL `http://www.jstatsoft.org/v36/i11/`.

P. A. Lachenbruch and M. R. Mickey. Estimation of error rates in discriminant analysis. *Technometrics*, 10(1):pp. 1–11, 1968. URL `http://www.jstor.org/stable/1266219`.

E. Landau. *Handbuch der Lehre von der Verteilung der Primzahlen*, volume 1. B.G. Teubner, Leipzig, 1909.

O. Lartillot, P. Toiviainen, and T. Eerola. A matlab toolbox for music information retrieval. In C. Preisach, H. Burkhardt, L. Schmidt-Thieme, and R. Decker, editors, *Data Analysis, Machine Learning and Applications*, pages 261–268. Springer, 2008.

S.L. Lauritzen, A.P. Dawid, B.N. Larsen, and H.-G. Leimer. Independence properties of directed markov fields. *Networks*, 20:491–505, 1990.

C.L. Lawson and R.J. Hanson. *Solving Least Squares Problems*. Prentice Hall, Englewood Cliffs, NJ, 1974.

C.L. Lawson, R.J. Hanson, D. Kincaid, and F.T. Krogh. Basic linear algebra subprograms for FORTRAN usage. *ACM Trans. Math. Soft.*, pages 308–323, 1979.

P. L'Ecuyer and R. Simard. TestU01: A C library for empirical testing of random number generators. *ACM Trans. Math. Softw.*, 33, August 2007.

D.H. Lehmer. Proceedings of a second symposium on large-scale digital calculating machinery, jointly sponsored by the navy department bureau of ordnance and harvard university at the computation laboratory, 13-16 september, 1949, 1951.

F. Leisch and S. Grün, B. *CRAN Task View: Cluster Analysis & Finite*, 2013. URL `http://cran.r-project.org/web/views/Cluster.html`.

A. Liaw and M. Wiener. Classification and regression by randomForest. *R News*, 2(3):18–22, 2002. URL `http://cran.r-project.org/doc/Rnews/`.

W.A. Link and M.J. Eaton. On thinning of chains in mcmc. *Methods in Ecology and Evolution*, 3:112–115, 2012.

K. Luebke and C. Weihs. Linear dimension reduction in classification: adaptive procedure for optimum results. *Advances in Data Analysis and Classification*, 5:201–213, 2011.

D.G. Luenberger. *Introduction to linear and nonlinear programming*. Addison-Wesley, Amsterdam, 1973.

T. Lumley. *biglm: bounded memory linear and generalized linear models*, 2011. URL `http://cran.r-project.org/package=biglm`. R package version 0.8.

D. Lunn, D. Spiegelhalter, A. Thomas, and N. Best. The BUGS project: Evolution, critique, and future directions. *Statistics in Medicine*, 28:3049–3067, 2009.

P. Mair and M. Hudec. *mixPHM: Mixtures of proportional hazard models*, 2008. R package version 0.7.0.

D.W. Marquardt. An algorithm for least squares estimation of nonlinear parameters. *Journal of the Society for Industrial and Applied Mathematics*, 11:431–441, 1963.

G. Marsaglia. Random numbers fall mainly in the planes. In *Proceedings of the National Academy of Sciences of the United States of America*, volume 61, pages 25–28, September 15 1968.

G. Marsaglia. Multiply-With-Carry generators, 1996. URL `http://www.stat.fsu.edu/pub/diehard/cdrom/pscript/mwc1.ps`.

G. Marsaglia. Random numbers. *Journal Of Modern Applied Statistical Methods*, 2(1):2–13, 2003.

G. Marsaglia, M.D. MacLaren, and T.A. Bray. A fast procedure for generating random variables. *Communications of the ACM*, 7(1):4–10, 1964.

N. Matloff. *Rdsm: Threads Environment for R*, 2011. URL `http://cran.r-project.org/package=Rdsm`. R package version 1.1.0.

M. Matsumoto and T. Nishimura. Mersenne twister: a 623-dimensionally equidistributed uniform pseudo-random number generator. *ACM Trans. Model. Comput. Simul.*, 8:3–30, January 1998.

H. Maurer. *Theoretische Grundlagen der Programmiersprachen*. Bibliographisches Institut, 1969.

McGill. *McGill University Master Samples*, 2010. URL `http://www.music.mcgill.ca/resources/mums/html/MUMS_dvd.htm`. DVD set.

A.G. McKendrick. Applications of mathematics to medical problems. *Proc. Edin. Math. Soc.*, 44:98–130, 1926.

G.J. McLachlan and T. Krishnan. *The EM Algorithm and Extensions*. Wiley, New York, 1997.

X. Meng and D. van Dyk. The EM algorithm - an old folk-song sung to a fast new tune. *Journal of the Royal Statistical Society, Series B*, 59:511–567, 1997.

O. Mersmann and B. Bischl. *soobench: Single Objective Optimization Benchmark Functions*, 2012. URL `http://cran.r-project.org/package=soobench`. R package version 1.0-73.

Message Passing Interface Forum. *MPI: A Message-Passing Interface Standard*, Version 2.2 edition, September 2009. URL `http://www.mpi-forum.org/docs/mpi-2.2/mpi22-report-book.pdf`.

N. Metropolis, A.W. Rosenbluth, M.N. Rosenbluth, A.H. Teller, and E. Teller. Equations of state calculations by fast computing machines. *J. Chem. Phys.*, 21:1087–1091, 1953.

H. Meurer and J. Dongarra. TOP500 Supercomputer Sites, 2010. URL `http://www.top500.org`.

B.-H. Mevik and R. Wehrens. The pls package: Principal component and partial least squares regression in R. *Journal of Statistical Software*, 18(2): 1–24, 2007.

D. Meyer, E. Dimitriadou, K. Hornik, A. Weingessel, and F. Leisch. *e1071: Misc Functions of the Department of Statistics (e1071), TU Wien*, 2012. URL `http://cran.r-project.org/package=e1071`. R package version 1.6-1.

D. Michie, D.J. Spiegelhalter, and C.C. Taylor. *Machine Learning, Neural and Statistical Classification*. Ellis Horwood, London, 1994.

A. J. Miller. Algorithm AS 274: Least squares routines to supplement those of Gentleman. *Appl. Statist.*, 41(2):458–478, 1992.

A.M. Molinaro, R. Simon, and R.M. Pfeiffer. Prediction error estimation: a comparison of resampling methods. *Bioinformatics*, 21(15):3301–3307, 2005.

J.J. Moré, B.S. Garbow, and K.E. Hillstrom. User guide for MINPACK-1. Technical Report ANL-80-74, Argonne National Laboratory, Argonne, IL, August 1980.

C. Nadeau and Y. Bengio. Inference for the generalization error. *Machine Learning*, 52(3):239–281, 2003.

J.C. Nash and R. Varadhan. Unifying optimization algorithms to aid software system users: optimx for R. *Journal of Statistical Software*, 43(9):1–14, 2011.

J.A. Nelder and R. Mead. A simplex method for function minimization. *Computer Journal*, 7:308–313, 1965.

M.B. Nevel'son, B. Silver, and R.Z. Hasminski. *Stochastic Approximation and Recursive Estimation*, volume 47 of *Translations of Mathematical Monographs*. American Mathematical Society, 1976.

M. Newman. Kantorovich's inequality. *Journal of Research of the National Bureau of Standards - B. Mathematics and Mathematical Physics*, 64B(1): 33–34, 1959.

S.I. Newton. *Analysis per Quantitatum Series, Fluxiones, Ac Differentias: Cum Enumeratione Linearum Tertii Ordinis*. Ex Officina Pearsoniana, 1711.

S.K. Ng, T. Krishnan, and G.J. McLachlan. The EM algorithm. In J.E. Gentle, W. Härdle, and Y. Mori, editors, *Handbook of Computational Statistics*, pages 137–168. Springer, Heidelberg, 2004.

H. Niederreiter. New developments in uniform pseudorandom number and vector generation. In H. Niederreiter and P.J.-S. Shiue, editors, *Monte Carlo and Quasi Monte Carlo Methods in Scientific Computing, Volume 106 of Lecture Notes in Statistics*. Springer, New York, 1995.

J. Nocedal and S.J. Wright. *Numerical optimization*. Springer series in operations research. Springer, 1999.

OpenMP Architecture Review Board. *OpenMP Fortran Application Program Interface, Version 1.0*, 1997. URL `http://www.openmp.org/mp-documents/fspec10.pdf`.

A. Ostermeier, A. Gawelczyk, and N. Hansen. A derandomized approach to self-adaptation of evolution strategies. *Evolutionary Computation*, 2(4): 369–380, 1994.

T. Ottman and P. Widmayer. *Algorithmen und Datenstrukturen*. Spektrum, Heidelberg, 1996.

F. Panneton, P. L'Ecuyer, and M. Matsumoto. Improved long-period generators based on linear recurrences modulo 2. *ACM Trans. Math. Softw.*, 32: 1–16, March 2006.

G. Peters and J.H. Wilkinson. The least-squares problem and pseudoinverses. *The Computer Journal 13*, pages 309–316, 1970.

M. Plummer. JAGS: A Program for Analysis of Bayesian Graphical Models Using Gibbs Sampling. In K. Hornik, F. Leisch, and A. Zeileis, editors, *Proceedings of the 3rd International Workshop on Distributed Statistical Computing, March 20–22*, Vienna, 2003. Technische Universität Wien. URL `http://www.ci.tuwien.ac.at/Conferences/DSC-2003/Proceedings/`.

M. Plummer. *rjags: Bayesian graphical models using MCMC*, 2013. URL `http://cran.r-project.org/package=rjags`. R package version 3-10.

N.G. Polson. Convergence of markov chain monte carlo algorithms. *Proceedings of the fifth International Meeting on Bayesian Statistics*, pages 483–512, 1994.

W.H. Press, S.A. Teukolsky, W.T. Vetterling, and B.P. Flannery. *Numerical Recipes in C*. Cambridge University Press, Cambridge, 1995.

R.E. Quandt. Computational problems and methods. In Z. Griliches and M. Intriligator, editors, *Handbook of Econometrics*, volume 1. North-Holland, Amsterdam, 1983.

A.E. Raftery. Hypothesis testing and model selection. In W.R. Gilks, S. Richardson, and D.J. Spiegelhalter, editors, *Markov Chain Monte Carlo in Practice*, pages 163–187. Chapman & Hall, London, 1995.

I. Rechenberg. *Evolutionsstrategie: Optimierung technischer Systeme nach Prinzipien der biologischen Evolution.* Frommann-Holzboog Verlag, Stuttgart, 1973.

Revolution Analytics. *doMC: Foreach parallel adaptor for the multicore package*, 2011a. URL `http://cran.r-project.org/package=doMC`. R package version 1.2.2.

Revolution Analytics. *doSNOW: Foreach parallel adaptor for the snow package*, 2011b. URL `http://cran.r-project.org/package=doSNOW`. R package version 1.0.5.

Revolution Analytics. *foreach: Foreach looping construct for R*, 2011c. URL `http://cran.r-project.org/package=foreach`. R package version 1.3.2.

J.R. Rice. Experiments on Gram-Schmidt orthogonalization. *Mathematics of Computation*, 20:325–328, 1966.

G. Ridgeway. *gbm: Generalized Boosted Regression Models*, 2013. URL `http://cran.r-project.org/package=gbm`. R package version 2.0-8.

B.D. Ripley. *Stochastic Simulation.* Wiley, New York, 1987.

C. Robert and G. Casella. A history of markov chain monte carlo-subjective recollections from incomplete data. *Statistical Science*, 26:102–115, 2011.

G.O. Roberts and N.G. Polson. On the geometric convergence of the Gibbs sampler. *J. R. Statist. Soc. B*, 56:377–384, 1994.

G.O. Roberts and A.F.M. Smith. Simple conditions for the convergence of the Gibbs sampler and Metropolis-Hastings algorithms. *Stochastic Processes and their Applications*, 49:207–216, 1994.

G.O. Roberts, A. Gelman, and W.R. Gilks. Weak convergence and optimal scaling of random walk metropolis algorithms. *Annals of Applied Probability*, 7:110–120, 1997.

C.M. Röhl. *Computerintensive Dimensionsreduktion in der Klassifikation.* J. Eul, Lohmar, 1998. Dissertation Universität Dortmund.

R. Rojas and U. Hashagen, editors. *The First Computers – History and Architectures.* MIT Press, Cambridge, MA, 2002.

P. Romanski. *FSelector: Selecting attributes*, 2013. URL `http://cran.r-project.org/package=FSelector`. R package version 0.19.

H.H. Rosenbrock. An automatic method for finding the greatest or least value of a function. *The Computer Journal*, 3:175–184, 1960.

N.F. Samatova, D. Bauer, and S. Yoginath. *taskPR: Task-Parallel R Package*, 2009a. URL `http://cran.r-project.org/package=taskPR`. R package version 0.34.

N.F. Samatova, S. Yoginath, D. Bauer, and G. Kora. *RScaLAPACK: A seamless interface to perform parallel computation on linear algebra problems using the ScaLAPACK library*, 2009b. URL `http://cran.r-project.org/package=RScaLAPACK`. R package version 0.6.1.

H.-P. Schwefel. *Evolutionsstrategie und numerische Optimierung*. Dr.-Ing. Dissertation, Technische Universität Berlin, Fachbereich Verfahrenstechnik, 1975.

R. Sedgewick. A new upper bound for Shellsort. *J. Algorithms*, 7(2):159–173, June 1986.

J. Shao. Linear Model Selection by Cross-Validation. *Journal of the American Statistical Association*, 88(422):486–494, 1993.

J. Sherman and W.J. Morrison. Adjustment of an inverse matrix corresponding to a change in one element of a given matrix. *The Annals of Mathematical Statistics*, 21(1):pp. 124–127, 1950. URL `http://www.jstor.org/stable/2236561`.

R.C. Singleton. Algorithm 347: an efficient algorithm for sorting with minimal storage. *Commun. ACM*, 12(3):185–186, March 1969.

A.F.M. Smith and G.O. Roberts. Bayesian computation via the Gibbs sampler and related Markov Chain Monte Carlo methods. *J. R. Statist. Soc B*, 55: 3–23, 1993.

B.J. Smith, B.T. Smith, J.M. Boyle, B.S. Garbow, Y. Ikebe, and V.C. Klema. Matrix eigensystem routines–EISPACK guide. *Mathematics of Computation*, 30, 1976.

C.A.B. Smith. Some examples of discrimination. *Journal of Eugenics*, 13(1): 272–282, 1946.

I. Sommerville. *Software Engineering*. Addison-Wesley, Harlow, England, 9. edition, 2010.

W. Spendley, G.R. Hext, and F.R. Himsworth. Sequential application of simplex designs in optimization and evolutionary operation. *Technometrics*, 4: 441–461, 1962.

D. Spiegelhalter, N. Best, A. Thomas, and D. Lunn. *Bayesian Analysis using BUGS: A Practical Introduction*. Chapman & Hall, New York, 2012.

D.J. Spiegelhalter, A. Thomas, N.G. Best, and D. Lunn. *WINBUGS: User Munual*. Cambridge: Medical Research Council Biostatistics Unit., 2.0

edition, 2004.

D. Steinkrau, P.Y. Simard, and I. Buck. Using GPUs for machine learning algorithms. In *ICDAR*, pages 1115–1119, 2005. doi: 10.1109/ICDAR.2005.251.

I. Steinwart and A. Christmann. *Support Vector Machines*. Springer, Heidelberg, 2008.

J. Stoer. *Einführung in die Numerische Mathematik I*. Springer, Heidelberg, 1972.

M. Stone. Cross-validatory choice and assessment of statistical predictions. *Journal of the Royal Statistical Society, Series B*, 36(1):111–147, 1974.

S. Sturtz, U. Ligges, and A. Gelman. R2WinBUGS: A Package for Running WinBUGS from R. *Journal of Statistical Software*, 12(3):1–16, 2005.

R. Sundberg. Maximum likelihood theory for incomplete data from an exponential family. *Scandinavian Journal of Statistics*, 1:49–58, 1974.

T. Therneau, B. Atkinson, and B. Ripley. *rpart: Recursive Partitioning*, 2013. URL `http://cran.r-project.org/package=rpart`. R package version 4.1-1.

S. Theussl. *CRAN Task View: Optimization and Mathematical Programming, Version 2013-02-14*, 2013. URL `http://cran.r-project.org/web/views/Optimization.html`.

A. Thomas, B. O'Hara, U. Ligges, and S. Sturtz. Making BUGS open. *R News*, 6(1):12 – 17, 2006.

L. Tierney. Markov chains for exploring posterior distributions. *The Annals of Statistics*, 22:1701–1762, 1994.

L. Tierney, A.J. Rossini, N. Li, and H. Sevcikova. *snow: Simple Network of Workstations*, 2011. R package version 0.3-5.

L. Torgo. *Data Mining with R, learning with case studies*. Chapman and Hall/CRC, 2010. URL `http://www.liaad.up.pt/~ltorgo/DataMiningWithR`.

J.F. Traub. *Iterative Methods for the Solution of Equations*. Prentice-Hall, Englewood Cliffs, N.Y., 1964.

H. Trautmann, O. Mersmann, and D. Arnu. *cmaes: Covariance Matrix Adapting Evolutionary Strategy*, 2011. URL `http://cran.r-project.org/package=cmaes`. R package version 1.0-11.

A. Turing. On computable numbers, with an application to the Entscheidungsproblem. *Proceedings of the London Mathematical Society*, 42, 1936.

A.M. Turing. Rounding-off errors in matrix processes. *The Quarterly Journal of Mechanics and Applied Mathematics*, 1(1):287–308, 1948.

University of Edinburgh SPRINT Team. *sprint: Simple Parallel R INTerface*, 2011. URL `http://cran.r-project.org/package=sprint`. R package version 0.3.0.

University of Iowa. *University of Iowa: Electronic Music Studios. Musical instrument samples*, 2011. URL `http://theremin.music.uiowa.edu/`.

S. Urbanek. *multicore: Parallel processing of R code on machines with multiple cores or CPUs*, 2011. URL `http://cran.r-project.org/package=multicore`. R package version 0.1-5.

A. van der Sluis. Stability of the solutions, of linear least squares problems. *Numerische Mathematik 23*, pages 241–254, 1975.

W. N. Venables and B. D. Ripley. *Modern Applied Statistics with S*. Springer, New York, fourth edition, 2002. URL `http://www.stats.ox.ac.uk/pub/MASS4`. ISBN 0-387-95457-0.

V.E. Vinzi, W.W. Chin, J. Henseler, and H. Wang, editors. *Handbook of Partial Least Squares*. Springer, Berlin, Heidelberg, 2010.

J. von Neumann. Various techniques used in connection with random digits. *Applied Mathematical Series*, 12(36-38):1, 1951.

I. Wegener. *Effiziente Algorithmen*. 1991. Skript WS 1990/91.

I. Wegener. *Datenstrukturen*. 1992. Skript WS 1991/92.

C. Weihs. Kondition des linearen Ausgleichsverfahrens, Testmatrizen, Vergleich von Lösungsverfahren, 1977. Diploma thesis, Department of Mathematics, Universität Bonn.

C. Weihs. *Auswirkungen von Fehlern in den Daten auf Parameterschätzungen und Prognosen*. Physica, Heidelberg, 1987.

C. Weihs. Testing numerical methods solving the linear least squares problem. In B. Schipp and Krämer. W., editors, *Statistical Inference, Econometric Analysis, and Matrix Algebra*, pages 333–347. Physica, Heidelberg, 2009.

C. Weihs and J. Jessenberger. *Statistische Methoden zur Qualitätssicherung und -optimierung in der Industrie*. Wiley VCH, Weinheim, 1998.

S.M. Weiss and C.A. Kulikowski. *Computer Systems that learn*. Morgan Kaufmann, San Francisco, 1991.

S. Weston. *doMPI: Foreach parallel adaptor for the Rmpi package*, 2010. URL `http://cran.r-project.org/package=doMPI`. R package version 0.1-5.

M.J. Wichura. Algorithm AS 241: The percentage points of the normal distribution. *Applied Statistics*, 37:477–484, 1988.

O. Wiener, M. Bonik, and R. Hödicke. *Eine elementare Einführung in die Theorie der Turing-Maschinen.* Springer, New York, 1998.

J.H. Wilkinson. *Rounding errors in algebraic processes.* Prentice-Hall, Englewood Cliffs, 1963.

J.H. Wilkinson. *The Algebraic Eigenvalue Problem.* Oxford University Press, 1965.

H. Wold. Model construction and evaluation when theoretical knowledge is scarce. In *Evaluation of Econometric Models*, pages 47–74. Academic Press, 1980.

C.F.J. Wu. On the convergence of the EM algorithm. *The Annals of Statistics*, 11:95–103, 1983.

Y. Xiang, S. Gubian, B. Suomela, and J. Hoeng. Generalized Simulated Annealing for Efficient Global Optimization: the GenSA package for R. *The R Journal*, 2012. URL `http://journal.r-project.org/`. Forthcoming.

E.A. Youngs and E.M. Cramer. Some results relevant to choice of sum and sum-of-product algorithms. *Technometrics*, 13:657–665, 1971.

H. Yu. *Rmpi: Interface (Wrapper) to MPI (Message-Passing Interface)*, 2010. URL `http://cran.r-project.org/package=Rmpi`. R package version 0.5-9.

C.K. Yuen. Testing random number generators by Walsh transform. *Computers, IEEE Transactions on*, 100(4):329–333, 1977.

A. Zell. *Simulation Neuronaler Netze.* Oldenbourg, München, 1995.

M. Zentgraf. Optimale Schätzung der Klassifikationsfehlerrate. Diploma Thesis, Department of Statistics, TU Dortmund, Germany, 2008.

G. Zielke. A new test matrix for inverting matrices. *ACM SIGNUM Newsletter*, 8(2):22–23, 1973.

G. Zielke. Testmatrizen mit maximaler Konditionszahl. *Computing*, 13:33–54, 1974.

G. Zielke. Report on test matrices for generalized inverses. *Computing*, 36: 105–162, 1986.

K. Zuse. Die Rechenmaschine des Ingenieurs. *Interner Bericht*, 30, 1936.

Index